D1165940

THE HARD W

THE HARD WORK OF HOPE

Climate Change in the Age of Trump

Robert William Sandford
Jon O'Riordan

First Edition

RMB | Rocky Mountain Books Ltd.
rmbooks.com
@rmbooks
facebook.com/rmbooks

Cataloguing data available from Library and Archives Canada
ISBN 9781771602228 (hardcover)
ISBN 9781771602235 (electronic)

Printed and bound in Canada by Friesens

Distributed in Canada by Heritage Group Distribution and in the U.S. by Publishers Group West

For information on purchasing bulk quantities of this book, or to obtain media excerpts or invite the author to speak at an event, please visit rmbooks.com and select the "Contact Us" tab.

We acknowledge the financial support of the Government of Canada through the Canada Book Fund and the Canada Council for the Arts, and of the province of British Columbia through the British Columbia Arts Council and the Book Publishing Tax Credit.

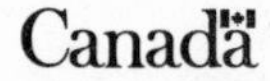

Disclaimer

For Her Honour Judith Gichon, Lieutenant-Governor of British Columbia, for her commitment to youth and for her tireless efforts in bringing the crisis at the climate nexus to the attention of British Columbians and to a wider Canadian audience through her contacts with government and community leaders across the country.

For Deborah Harford, fearless champion of societal resilience in the face of Earth system change, the global water crisis and climate instability

CONTENTS

ONE

Global Water Security & Climate Stability in the 21st Century

Changes in the composition of the Earth's atmosphere are causing water to move more energetically through the global hydrological cycle, making the world's water crises even more urgent to address. Until we lost the relative stability of the planetary water cycle, we had no idea how much we relied on that stability. Water is at the very centre of human existence, part of an intimately interwoven nexus that links the amount of water we need to sustain human life, to how much of it we take from nature to grow our food, to the amount we need for generating energy, to the increasing impact our burgeoning population is having on biodiversity worldwide.

What we are discovering is the extent to which the fundamental function of our political structures and global economy are predicated on relative hydrologic predictability, especially as it relates to precipitation patterns that define water security. As a result of the loss of relative hydrologic stability, it is not just food production, energy use and biodiversity-based Earth system function that are disrupted. Political and economic stability is also at risk in a number of regions in the world. We are only now beginning to understand how complex this issue has become. Hydro-climatic change has the potential to literally and fundamentally redraw the map of the world.

The global water crisis is more widespread than we think. First Nations and other remote communities even in Canada have suffered needless water shortages or contamination. Places here and abroad have been impacted disastrously by flooding of magnitudes never before witnessed. Droughts are occurring that are so deep and so prolonged that no one remembers such suffering ever happening before. It is questionable whether some of these places will ever be habitable again. Ours is a world of distress and danger we can help alleviate and prevent. That is why it is important to consider how a disruption in the global order right now could affect our capacity to advance and export our expertise in the management of water to where it is needed. That is why we should pay close attention to what is happening in the United States. So why be concerned? We should be concerned because we may be at a turning point in human history that will complicate our efforts to address the growing global crisis with respect to water security and climate stability.

The new American president has stated that the US will withdraw from the Paris Climate Agreement. The immediate risk is that what is happening in the US will slow the global momentum on climate action. If this happens, President Trump may be condemning future generations of Americans and the rest of the world to hell on Earth. This profound threat to the future of humanity, however, has not been mentioned in mainstream coverage of the new political road the president began immediately to plow through the White House. America's potential failure to meet its obligations with respect to the Paris Agreement ought to be of concern not just to those who worry about climate or about water resources but to anyone who cares about the world order or even about order itself. If there is to be hope, then an improved order must somehow emerge out of what will hopefully be a temporary setback.

THE DEEPLY DIVIDED PERSONALITY OF THE US REVEALED OVERNIGHT

If you were in the United States on the evening of November 8, 2016, you might have mistakenly imagined that somehow the personality of the United States changed overnight. It might have seemed to you that it was as if millions of people had suddenly abandoned the moral principles and standards to which they had previously adhered, revealing a darkness within society we knew existed but the need to live peaceably together demanded we repress. While this perception may be partly true, what really happened that night was that a different side of the American personality – a side that has existed since the beginning – suddenly assumed full political power. The potential consequences of this change in national political voice became immediately apparent.

In advance of the inauguration, rumours were purposely circulated that President Trump would not abandon US responsibility to the climate. Gossips even had it that Trump's daughter, Ivanka, a 36-year-old former reality TV star and fashion model, now a businesswoman and executive vice-president of the Trump Organization, would be charged with publicly championing the climate issue domestically. Some actually believed the rumours.

Realistically, however, any false hope that pundits might have generated, and whatever misinformation the media and others had communicated about whether or not Trump would go through with his threat to immediately walk away from the Paris Agreement, should have been tempered by Trump's nomination of Rex Tillerson, the CEO of Exxon-Mobil, as secretary of state.

Tillerson was an interesting choice, not because oil executives had never played key roles in American cabinets historically – they most certainly have – but that the CEO of the largest oil company in the United States and the fifth largest one in the world happened to be appointed secretary of state shortly after the potential cost to the global oil industry of implementing the Paris Climate Agreement became public knowledge. In a report to the Task Force on Climate-related Financial Disclosures, chaired by Bank of England governor, Mark Carney, adviser Mark Lewis of the Barclays Group estimated that "if measures to stop global warming are fully implemented, oil company revenues could fall by more than $22-trillion over the next 25 years."

It is hardly likely that one of the largest industries in the world is going to happily take losses of this magnitude sitting down. Nor is it likely that nations like Russia – whose economies are heavily dependent on oil revenues and with whom Tillerson and Exxon-Mobil have had close ties in the past – will in any meaningful way support full implementation of the conditions of the Paris Agreement. This should have been expected. It is not as if a sleeping giant was suddenly awakened; these giants never sleep. That is how they become giants.

The world just got a clear look at a hitherto largely hidden face of global politics: the face of the internationalism no longer masking as the interests of sovereign nations, the remorseless face of independent transnational economic power.

So, would the president-elect reverse his campaign stance and suddenly announce that climate change is not in fact a hoax? On June 1, 2017, he dashed any such hopes with a clear statement that the US will withdraw from the Paris Agreement, which in his view reeks of UN interference and massive economic disadvantage to the United States. Climate change will not be on this administration's agenda during this president's tenure, however long that may be.

President Trump has a number of formal avenues for withdrawing. Under the agreement, countries volunteer to take steps to reduce their impacts on climate, beginning in 2020. The US had pledged to reduce its greenhouse emissions to 26 per cent below 2005 levels by 2025, principally by moving away from coal-fired electricity generation.

The easiest option for the Trump administration was to simply abandon all the rules, incentives and programs designed to reduce emissions in the United States, which would prevent the nation from living up to its commitments under the agreement. The simplest way for the president to do that was to greatly reduce the capacity of the federal Environmental Protection Agency, by reducing its budget by nearly a third.

KILLING THE CLIMATE ACCORD

Scott Pruitt, the person Donald Trump put in charge of the EPA, is one of the country's most vociferous climate-change deniers. Pruitt has described himself publicly as a "leading advocate against the EPA's activist agenda," and, as if to prove it, he brought forward no fewer than 14 legal challenges opposing the policies of the agency while he was Oklahoma's attorney general.

Any hope that the Trump administration was not going to carry out its threat of walking away from the Paris Agreement was banished even before the incoming president finished his 16-minute inaugural speech. Eight minutes, or 732 words, into the 1,433-word speech, just as Trump was proclaiming that US policy would now be strictly "America first," a page titled "An America First Energy Plan" went up on the White House website, joining pages identifying the main environmental policies of the Obama era that the new administration had targeted as "detrimental to society."

"For too long," the plan reads, "we've been held back by burdensome regulations on our energy industry. President Trump is committed to eliminating harmful and unnecessary policies such as the Climate Action Plan and the Waters of the US rule." The document went on to announce that "the Trump administration will embrace the shale oil and gas revolution to bring jobs and prosperity to millions of Americans." The text then noted that "the Trump administration is also committed to clean coal technology, and to reviving America's coal industry, which has been hurting for too long." Four days after the inauguration, on January 24, 2017, the EPA and other government agencies were ordered to stop communicating with the public through social media or the press and to freeze all grants and contracts. Then, on March 28, President Trump signed an executive order that removed the Obama regulations on reducing carbon emissions from coal-burning electricity plants.

If you trace the recent political and legal evolution of the highly contested Clean Water Rule, you find Scott Pruitt had been working hard for some time to have the legislation gutted, with the clear aim of undermining regulation not just of the disastrous effects on mountaintop coal mining on watersheds but also the impacts of fracking and shale gas extraction and agricultural runoff on downstream water quality. While the legal and jurisdictional fight is not over by a long shot, at the time of this writing, Pruitt appeared ever closer to achieving his ends.

IMPACTS ON CANADA

President Trump's blandishments notwithstanding, there is no such thing as clean coal. Coal's time has passed. In addition to exacerbating still-persistent acid rain and other pollution-related public health problems we can't afford even now to address, renewed investment in coal will be expensive and its future uncertain. New assets could easily be stranded if, after this president has left office, the regulatory climate reverts to what it was before. If such coal investments do get made, however, there will be impacts on Canada.

The reopening of shuttered mines and expanded burning of coal in the US Midwest in an unfettered way, as Trump proposes, will produce acid rain which will fall in Ontario, Quebec and the Atlantic provinces, taking the entire region backwards to the 1970s, '80s and '90s, when bilateral agreements and strict emissions regulations were necessary in both countries to prevent the death of thousands of Canadian lakes. Water and climate are integrally linked in other ways also, as illustrated by concerns over the impacts of fracking. According to recent research by Drexel University scientists in 2017, persistent methane leaks from shale plays over time – whether from well sites, pipelines or uncontrolled gas migration from fracked rock formations – could cancel out any climate benefits from burning natural gas. The study demonstrated that methane leakage rates greater than 2 or 3 per cent of total production made the greenhouse gas footprint of shale gas extraction dirtier than coal or oil. Research in 2015 by the US National Oceanic and Atmospheric Administration had already shown that leakage rates in shale gas basins ranged from as high as 6 and 12 per cent in Utah's Uintah Basin to as low as 1 per cent in the Haynesville Shale in Louisiana. It appears the Trump administration's proposals not only risk creating more acid rain in Canada but also will result in accelerated climate warming here and globally.

What should also be of interest to Canadians is that the highly contested "Waters of the US" rule was brought into existence specifically to clarify sections of the US Clean Water Act with respect to protection of American water bodies and endangered aquatic ecosystems and species and the capacity of waters to purify themselves for human use downstream. This is no time to forget the fact that 40 per cent of the boundary between Canada and the United States either lies along shared open waters, as in the Great Lakes, or is defined by a shared watercourse, as between Alaska and Yukon. What the US does to its water supplies affects us all.

The deteriorating condition of the Great Lakes emerged as a binational issue in the late 1960s when Lake Erie had become so polluted that many scientists feared it was near ecological death. Lake Michigan and Lake Ontario were not in much better health. A Herculean international joint effort, however, proved even Lake Erie could be restored to health if the two countries worked together to identify and address contamination sources.

The resulting Great Lakes Water Quality Agreement, which the US and Canada first signed in 1972, spurred federal–provincial co-operation in Canada. The agreement worked because it went on to encourage binationalism; promoted community participation; demonstrated equality and parity in its structure and obligations; arrived at common objectives such as joint phosphorus reduction targets; facilitated joint fact-finding and research; and demanded accountability and openness in information exchange and flexibility and adaptability to changing circumstances.

Unfortunately, because of dramatic population growth in the Great Lakes Basin, increased agricultural production in the region, the introduction of invasive species, complicated problems with algal blooms, and changing hydrometeorological conditions that have resulted in more frequent flooding and greater mobilization of nutrients, Lake Erie is back to the same near-death state as it was in 1972. Renewed bilateral co-operation to restore the health of Erie and the other Great Lakes culminated in a policy called the Great Lakes Restoration Initiative, which set aside some $425-million in 2010 and $260-million in 2011 to advance coordinated research and fund local groups working on everything from toxin cleanups to the fight against invasive species. Barely two months after President Trump was inaugurated, it was Scott Pruitt again, the president's appointee as head of the Environmental Protection Agency, who proposed to curtail annual GLRI funding by 90 per cent, which, together with other cuts, would reduce the EPA's total budget by nearly a third. In the face of considerable political opposition that will likely result in legal challenges, cutbacks of a similar magnitude were also proposed for NOAA, with a special focus on reducing funding to its climate research program.

In a series of posts in March of 2017, journalist Charles Pierce commented that the Great Lakes funding cut was the largest proposed dollar reduction to any program on a list that included major cuts to climate change research, restoration funding for Puget Sound and Chesapeake Bay, research into chemicals that disrupt human reproductive and developmental systems, enforcement of pollution laws and funding for brownfield cleanups. The plan also included a $13-million cut in compliance monitoring, which the EPA uses to ensure the safety of drinking water systems. State grants for beach water quality testing would also be eliminated. "When you see these reductions, you'll be able to tie it back to a speech the president gave, or something the president had said previously," Mick Mulvaney, administration budget director, explained at a White House press briefing barely a month after Trump's inauguration, "We are taking his words and turning them into policies and dollars." Pierce made it clear in his dispatches that what Mr. Mulvaney meant was that the Trump administration was cutting other government programs in order to find the $54-billion the president wanted for expanding the Defense Department.

In addition to worrying about the Great Lakes, this may also be a time to remember that for the past two decades we have essentially ignored egregious violations of the Boundary Waters Treaty by both countries, and largely failed to use the full powers of the International Joint Commission created by the treaty as a means of identifying and preventing water disputes along our shared border. We should also be very attentive to the fact that renegotiation of NAFTA could have huge implications for water management in Canada and continentally if issues related to Canada's position on interbasin transfers and bulk water exports are not fully reasserted from the outset.

Canadians should also recognize that we are in the midst of reconsideration of the Columbia River Treaty, which many would agree is the most important negotiation we will have with the United States over water in this generation. With the election of Donald Trump as president and with libertarian sensibilities in control of both the House of Representatives and the Senate, our shared waters are being stirred. As Kathy Eichenberger, the chief negotiator for British Columbia, has maintained since both countries agreed to review the treaty in 2014, this is the time for us, all of us, to ensure we have our water house in order.

President Trump has vowed to eliminate 75 per cent of the regulations supposedly hampering businesses in the US. Depending on evolving political circumstances here in Canada, pressure for deregulation could spread northward. The threats Trump's policies pose to the quality of Canadian waters, however, are dwarfed by the impacts of these policies on action in response to the accelerating threat of climate disruption, not just in Canada but globally. Trump's abandonment of US commitment to countering the global climate threat has already begun to cow other nations into softening their approaches and delaying action on carbon taxes and other responses to rapidly accelerating changes in the composition of the Earth's atmosphere. Not surprisingly, we saw a hint of this reaction in Canada immediately after Trump's election when Saskatchewan Premier Brad Wall rejected a carbon tax on the grounds that our American neighbours will no longer adhere to the Paris Climate Agreement. While most other nations, including France, Germany and China, have reaffirmed their commitment to the pact, the risk now is that, as the US abandons it, other countries may want to hedge their bets. Morally reprehensible attempts will of course continue to be made within individual nations to divide the public over this issue as we saw in Canada when, on March 16, 2017, an editorial in the *National Post* urged Canadians to "Bid adieu to Paris." Under the threat of losing trade and other relations with the United States, denial of climate change and delay of action on it could once again become an epidemic globally, making things more difficult for countries, like Canada, which have reaffirmed their commitment to achieve agreed-upon emissions reductions.

EFFECTS ON THE REST OF THE WORLD

This trend could be hastened by the fact that, at the moment, the stability of the world order is even more fragile than it has been since the presidency of George W. Bush. At a special 2016 forum on water, conflict and involuntary human migration, organized in London by the Strategic Foresight Group, the topic turned to the proposed British exit from the European Union and what was characterized by leading foreign policy and global security experts as "the globalization of anger" over the manner in which democratic governments function.

Though carrying through with Brexit would require changes to some 776 pieces of legislation, it was the opinion of some of the forum participants that what happened with the referendum in Britain was just the tip of an iceberg of potential change. Speakers noted that, emboldened by the election of Donald Trump, radical populism based on complete distrust of governments and their ties with powerful economic elites is spreading widely. While disagreements over what each state holds to be its future in a more globalized world may be different than such disagreements in the United States, growing numbers of people in Poland, Hungary, Austria, Germany, France and England, and as far away as India, no longer believe in the political status quo. Fuelling this distrust of government is a growing sense that what the people want no longer matters, that governments pay only lip service to the public interest. As unemployment grows, particularly among youth, and economies stagnate, people are feeling the system is rigged in favour of powerful interests and that the entire edifice has to go. It is exactly these sentiments in part that Donald Trump presumably mobilized among his populist constituency in the United States. Whether Trump will provide a viable alternative remains unclear, but while that question is being answered an about-face in American political disposition and direction continues.

History has seen this many times before, usually with undesirable outcomes. Angry populists have not just been emboldened in their perspectives by Trump's election. If the ramping up of increasingly vicious push-back against science in letters to newspapers and blog comments even in this country is any indication, people with these attitudes appear to have very quickly come to feel entitled, if not politically and socially obligated, to attack whatever they don't like or understand. Scientists fear it may not be long until we are back to that terrible time at the close of the last century when climate scientists had to endure death threats and angry deniers interrupted public presentations of clear evidence of climate facts to hurl rancour at anyone who held that the climate threat was real.

Anger has been fuelled and deterioration in the quality of public discourse has been facilitated in part by the media. Americans have become media consumers. They choose the news they want. They shop for news they like. Media outlets have responded by picking news their audience might like. In order to satisfy consumer appetites for the news they want, fact-checking has become politicized. People have come to reject facts they don't like.

Public permission to attack what one does not understand or what one disagrees with has been given further licence by straight-faced pronouncements on national television news of the outright lies told by the president of the United States and members of his inner circle, beginning but far from ending with former press secretary Sean Spicer uttering falsehoods that adviser Kellyanne Conway then tried to rationalize as being merely "alternative facts." George Orwell would have been amused. The effects of employing obvious doublespeak are only slowly beginning to impact one of the most important assets of any government: its credibility.

It is not that we didn't see this coming. Many observed the direction the split was taking between those with a cosmopolitan world view who saw America as an engine of global democracy, and those who were afraid of losing their jobs, rights and very identity as Americans. Few, however, realized how quickly this split was growing and none offered prescriptions for avoiding the divisions that now characterize American politics. In 1995 one of the 20th century's most famous astronomers, Carl Sagan, wrote *The Demon-Haunted World: Science as a Candle in the Dark*, in which he decried the fact that 95 per cent of Americans did not understand science. Most Americans, he claimed, had no idea how science contributed to making their quality of life higher than any society had ever enjoyed before. Sagan was particularly troubled by the fact that so many Americans would fall for pseudo-science, uninformed opinion and wishful thinking in matters of critical importance to their future such as climate warming, air pollution, toxic and radioactive wastes, acid rain, topsoil erosion, tropical deforestation, exponential population growth and a dozen other issues science has identified, all of which we need science in tandem with due political process to address. Two decades before the present writing, Sagan dedicated his book to explaining the scientific method to lay readers, encouraging people everywhere to aspire to truly critical and skeptical thinking, going so far as to offer "a baloney detector" the average person could use to separate pseudo-scientific falsehoods from scientific fact. Such a device would have served all Americans well during the 2016 presidential election and in its wake. Sagan's baloney detector identified and illustrated the range of perilous fallacies of logic and rhetoric that science has attempted to eliminate as much as possible from the determination of scientific truth. These fallacies include ad hominem assaults where the character and credibility of the arguer and not the argument is attacked; arguments from positions of phony authority; arguments made solely from the point of view of adverse consequences; and arguments that rely on changing the subject and baiting opponents about unrelated or irrelevant matters. These fallacies also include blatant appeals to ignorance; special pleading; arguments that depend on selective observation, counting hits but ignoring misses; straw-man characterizations that make opposing positions easier to attack; and abuse of the statistics of small numbers. A good baloney detector would also protect us from false dichotomies; confusions of correlation and causation; non sequitur arguments, in which there is no logical link between one statement and the next; half-truths or suppressed evidence; and the employment of weasel words. All these are logical fallacies that unfortunately have come to characterize public discourse in the post-truth Trumpocene.

In his 2009 book *Idiot America: How Stupidity Became a Virtue in the Land of the Free*, the same Charles Pierce – whose commentary we noted earlier as to local Michigan news reports of Trump administration cuts to Great Lakes restoration and protection programs and the pending evisceration of the US Environmental Protection Agency – warned that talk radio and reality television were becoming the communications equivalent of professional wrestling. Pierce worried that this dangerous explosion of misinformation was turning public discourse in the United States into inaccurate, uninformed argument in which truth would soon be defined, not objectively, but by how many and how fervently people would attest to it. On the issue of climate change, Pierce feared that more and more Americans were buying what they were being sold, not what they were actually seeing. He worried that America was creating a public playing field on which idiocy could fight against science and win. Pierce's predictions have come true. With the election of Donald Trump, the war on science and evidence-based truth has been dramatically escalated.

Again, the ascendance of political populism in the United States could hardly be said to have come out of the blue. In *Ill Fares the Land*, fully six years before the 2016 presidential election, the great British–American historian Tony Judt warned that the major institutions of the United States had been degraded, above all by money. He outlined the dangers of no longer granting citizens protection from the ravages of the market economy and warned of a growing and depressing cynicism toward the responsibilities and capacities of the liberal state. Judt worried that the language of politics itself had been drained of substance and meaning. Citing polls in the UK and elsewhere, he observed that "disillusion with the politicians, the party machines and their policies has never been greater." We would be ill-advised, he argued, to ignore such sentiments.

Judt went on to say that "if we are going to build a better future, it must begin with a deeper appreciation of the ease with which even solidly grounded liberal democracies can founder." He observed that improvements on unsatisfactory circumstances are likely the best we can hope for and probably all we should seek at this moment. Donald Trump may prove Judt right, at least briefly. Resistance to what Trump represents has woken up Americans to the persistently and now increasingly divided state of their cherished democracy.

Among the issues America now faces, to which all democracies should harken, is the problem of what could be called militant ignorance on the part of many citizens. As Tom Nichols, a professor of national security affairs at the US Naval War College, pointed out in an article in *Foreign Affairs*, the loss of faith in expertise in America is a serious problem, not just for the US but for the entire world. Like Nichols, more and more Americans are waking up to the fact that as a society it is dangerous to make it a social norm to grant unearned respect to unfounded opinions. Nichols cites a 2015 Public Policy Polling survey that asked a broad sample of both Democrat and Republican primary voters whether they would support bombing a place called Agrabah. The poll found that nearly a third of Republican responders favoured the idea, with 13 per cent opposed. The response of Democrats was the reverse, with 36 per cent opposing the bombing and only 19 per cent in favour. Liberals argued the poll result showed the hawkish tendencies of conservatives, while conservatives countered it showed just how soft Democrats were in their pacifism. Actual experts in national security such as Nichols were unamused. Rather, they were shocked that 43 per cent of the Republicans polled and 55 per cent of the Democrats had clear, defined opinions about bombing a place that didn't exist except in a cartoon: Agrabah is a fictional place in the 1992 Disney film *Aladdin*.

Nichols goes on to point out that when we elect governments based on ignorance or misinformation, those governments rule not only over those who voted for them but over all of society. Too few citizens, he says, understand democracy to mean political equality in which all are equal in the eyes of the law. Many Americans, Nichols observes, have instead come to think of democracy as a state of actual equality, in which every opinion is as good as any other, regardless of the logic or evidence behind it. Americans are waking up to the fact that this is not how their democracy is meant to work. In this the rest of the world once again owes a debt, for what the US is learning is a lesson for democracies everywhere.

Meanwhile, the United Nations continues to work around the clock to reiterate to the other 192 signatories to the Paris Agreement the hard-won scientific evidence each of those countries has already verified that, contrary to President Trump's assertions, proves beyond any reasonable doubt that climate change is most assuredly not a hoax. The UN is reminding the world that if it backs off on efforts to stabilize the composition of the Earth's atmosphere now, it could cost our civilization the last precious years in which actions by us might have made a difference in managing a rapidly accelerating global water cycle and its potentially devastating effects on climate.

MEANWHILE THE CLIMATE CONTINUES TO CHANGE

So what have the climate and the global water cycle been doing while much of humanity was distracted by the geopolitical upheaval associated with the US election and the potential threat to the future of the European Union brought about by rising populism in Europe? Scientists in the US and around the world officially declared 2016 the hottest year on record, the third such precedent-setting year in a row. Arctic warming is already 3.54°C above the global average, so holding that region to an increase of less than 2° per the Paris Agreement is already out of the question and has been so for a decade.

The concern among climate scientists is that in the absence of Arctic sea ice, and with oceans warming, we appear to be approaching the point where we have warmed the planet enough that the Earth itself and its cold oceans have begun to literally sweat out greenhouse gases. What these scientists are talking about is the very real potential for runaway climate feedbacks in our time. The problem is that there are a lot of hydrocarbons in the ground in the Arctic, and most are kept trapped there by an imperfect cap of frozen ground and permafrost.

As we will see in the next chapter, what we appear to be

facing in the Arctic is a carbon-release time bomb. Only by keeping the Arctic cold can we prevent that bomb from going off, and if it does go off, Canadians will be the first to feel its effects. In this regard, what is happening politically in the United States is unfortunately not just another predictable pendulum shift to the right like we have seen before. Many argue there is no reason to worry, as undoubtedly the pendulum will swing back. We may not have time for that to happen, however.

One might liken this particular swing to the right as the political equivalent of the loss of hydrologic stationarity. What the US did in the past, even under ostensibly conservative governments, is no longer a guide to the future. What is just as new as this political movement to the right in the United States is that it is happening at a time when the global hydrological cycle is not only accelerating but experiencing unprecedented vulnerability to disruption. What is also different and dangerous about this situation is that America's current political agenda comes at the worst possible time in that it corresponds with a growing crisis in the stability of Earth system function.

THE EMERGENCE OF EARTH SYSTEM SCIENCE

Earth system science arose in the 1990s and early 2000s as the planet began to be understood as a complex, evolving, unified system that was more than the sum of its parts. Crucial to the emergence of this new way of thinking was a dawning awareness about two fundamental elements of the way integrated Earth system functions support life. The first was that the Earth itself is a single system, within which the biosphere is an active and critical component. In other words, the presence of life itself on Earth is critical to the creation of the conditions that make this life possible. More than that, the system itself is created and sustained by biodiversity: the sum total of all the immensely variegated life on the planet.

The second key realization was that human activities are now so pervasive and profound in their consequences that they affect Earth system function at a global scale "in complex, interactive and accelerating ways," in the words of the International Geosphere–Biosphere Programme. Humanity, by virtue of its numbers and its needs, now has the capacity to alter Earth system function in ways that threaten the very processes – biotic and abiotic – upon which humans depend. This is what scientists now refer to as "global change," of which climate change is only one manifestation. As the Amsterdam Declaration on Global Change put it, "In terms of some key environmental parameters, the Earth system has moved well outside the range of natural variability exhibited over the last million years at least. The nature of changes ... their magnitudes and rates of change are unprecedented. The Earth system is currently operating in a no-analogue state." In other words, we are moving into an era of conditions on Earth for which humanity has no previous experience.

This has led in turn to three linked breakthroughs in Earth system science. The first of these is the concept of the Anthropocene, the idea that we have entered a new geological epoch defined by human influence on Earth system function. The second breakthrough is the concept of the Great Acceleration, the extraordinary increase in human impacts on Earth system function since the end of the Second World War. And third, there is the notion of Planetary Boundaries, limits within which we need to stay if we are to create a safe operating place for humanity so we can continue to assure the stability of the planetary conditions we can live with now and endure in the future.

THE HARD WORK OF HOPE

So where does hope reside in all this? Today, at the outset of the Anthropocene, Canada remains a place where headway can still be made on water and water-related climate issues. After ten years of stalled progress and environmental deregulation similar to what is now occurring in the US, the policy environment has changed dramatically in Canada, both nationally and in many provinces. While it is difficult to know which way the political winds will blow in the future, there is opportunity for Canadians to mitigate and adapt to the new conditions we will face in the very dangerous epoch in which the world presently finds itself.

We should be reminded that, as outlined in the next chapter, all hopes for achieving the goals of the UN's global sustainable development agenda depend on what each member state is able to achieve at the national and subnational level. We should remember that even though it may not seem like it in the moment, great opportunity still very much exists, not just to change the world but to make it a better place, a place where human and planetary health are seen to mirror one another, where human equality and rights are respected and expanded, where treaties between peoples are honoured and where development is restorative rather than destructive.

It is of great import that these remain qualities of nationhood to which we continue to relentlessly aspire in Canada. By ceaselessly continuing our efforts to achieve true and lasting sustainability despite the troubles that surround us, we will become a beacon for others. If we just stay the course – and by our example help others to do the same – there is no question that, if we want it to be, this could be Canada's moment, its chance to shine. This is Canada's opportunity to lead the world.

It may be that now is the time to show the rest of the world that much, much more can be done when hope is no longer focused on the future but instead becomes an electrifying force for action in the present. This could be the time to resolutely support and promote factual, intelligent, thoughtful and persistent dialogue as a hallmark of what Canada will stand for in this new epoch. And this is clearly a time when, for the sake of the future, we must uphold scientific principles and defend the scientific method in all matters related to water and climate change..We need to support, report, communicate and celebrate the scientific research undertaken in Canada by government agencies and universities, and to share and showcase advances in leading-edge technology and practice in the private sector.

Most importantly, however, this is a time when we need to get our own water house in order, not just because of growing concerns over water in the United States but because our own future prosperity and place in the world will depend on how we deal with the accelerating changes of hydro-climatic regimes that will affect every part of this immense country. By acknowledging and celebrating successes in making the world better – successes that have come about by cultivating collective memory of past social injustices and environmental problems and recognizing how far we have come in addressing many of those – we demonstrate how we are in fact already actively clarifying and realizing a vision of where and how we want to live in the Anthropocene.

Climate change is not a hoax. Though it is easy to wish otherwise, climate change is real and it is a threat to us all. But if we work together to manage water better, we can deal with that threat. With this challenge before us, those who understand how food production, energy use and biodiversity-based Earth system function revolve around water and how water is related to climate are going to be very important to the future. Such is the hard work of hope in our time.

TWO

Sustaining Global Water Security and Climate Stability in a Post-Truth Trumpocene

With the smoke clearing and clearer heads prevailing, we are beginning to understand what happened in the United States in 2016 and why. Considered one of America's most influential magazines over the past 95 years, *Foreign Affairs* began with Woodrow Wilson, who was president of Princeton University before he became president of the United States. While he was US president, his leadership, ideals and eloquence struck a deep chord with many Americans. When he went to the Paris Peace Conference in January 1919, he carried with him the fervent hopes not only of the American people but of most of the world: that a just and lasting peace might emerge with an appropriate international organization created to help achieve and sustain it. Unfortunately, the establishment of such an organization would have to wait until the horror of another world war came to an end. Wilson did return from France with a strong feeling, however, that there was a new and important role for private institutions to play in international relations. After a proposal for a high-level, Anglo–American public policy institution lost traction, a group of Americans met through 1920 and 1921 to set up in New York the body that would become the Council on Foreign Relations. The members of this elite panel actively participated in meetings and group discussions designed to enlarge the understanding of the state of the world, and in 1922 they began publishing the results of their deliberations. The print magazine continues today as a bimonthly with a paid circulation of more than 200,000 around the world, plus a website that receives nearly a million unique visitors each month.

Like the Council on Foreign Relations from which it sprang, *Foreign Affairs* tolerates wide differences of opinion but prides itself in expressing the views of many of the best-informed people in the world. The March–April 2017 issue was dedicated to "Trump Time" and set out to offer the widest range of perspectives on what happened in the US that led to Trump's election; the potential impacts on the economy; implications for America's relations with Russia, China and North Korea; how Trump's election might alter American foreign policy in the Middle East; and how the US will manage its war on terror and the consequences for the international order should President Trump actualize his most consistently expressed campaign promises on any of these matters.

The editor of *Foreign Affairs* opened the issue by expressing profound concern about the deep divisions in American society. Gideon Rose urged "bubble-wrapped cosmopolitans" to broaden their perspectives and "engage in the full reality of their fellow citizens' lives." On the other hand, he also enjoined angry populists that had risen to power to understand that the election was over and that it was time to stop trying to score political points by bashing the establishment of which they were now a part. Rose urged everyone to understand America's responsibility for shaping the fate of not only millions of people at home but billions abroad. What was not mentioned in either the editorial or any of the ensuing essays were the changes in weather and precipitation patterns that are already affecting water security, and the climate disruption that will exacerbate all the political tensions presently dividing America. That said, the essays offered a very compelling picture of the current state of democracy, not just in the US but around the world.

If we are to achieve any meaningful level of sustainability in the face of Earth system breakdown, we need to understand the roots of the discord that presently exists in the US; ensure that we address those before they explode onto the political stage in Canada; and find ways to make water, food and energy security and the protection and restoration of biodiversity-based Earth system function into non-partisan political issues of the greatest urgency.

In that same issue of *Foreign Affairs* was a piece by Walter Russell Mead, one of the world's most trusted and respected observers of US national political trends and their potential influence on that nation's foreign policy. "The Jacksonian Revolt: American Populism and the Liberal Order" offers one of the best early analyses yet of what happened in the US that culminated in the election of Donald Trump. Mead's article begins by pointing out that since the Second World War, US foreign policy has been shaped by two major schools of thought, both of them focused on achieving a stable world order, with the US at the centre of course.

One of these fiercely competing schools of thought, named

for Alexander Hamilton, began as a movement that argued that it was in the American interest to, as one of Woodrow Wilson's advisers, Edward House, put it, replace the United Kingdom as "the gyroscope of world order." Following the Second World War, the Hamiltonians put in place a financial and security architecture aimed at creating a global liberal order understood primarily in economic terms.

The rival faction, on the other hand – the Wilsonians, named for the former president – conceived of this global liberal order in terms of values rather than economics. Seeing corrupt and authoritarian regimes abroad as an ongoing threat to global security, the Wilsonians sought world peace and American hegemony by example through the promotion of human rights, democracy and the rule of law. By the time the Cold War was drawing to a close, the Wilsonians had bifurcated into two camps: the liberal institutionalists, who focused on promoting international co-operative bodies as the means of continuing global integration; and the neoconservatives, who brought to the fore the opposing belief that the answer to a stable world order and American hegemony was aggressive, unilateral American support for greater global economic integration brought about through the voluntary assent of like-minded national partners.

When the Soviet Union fell, however, the unquestioned grip of the globalists on American foreign policy began to weaken. The American public, Mead points out, was becoming increasingly disenchanted with costly failures of American attempts to control the global order and began to challenge the directions the foreign policy establishment in Washington was taking the country. Suddenly, Mead explains, other early schools of thought regarding America's place in the world that had been out of favour during the heyday of the incumbent political order came back, as he says, "with a vengeance."

Mead claims Trump "sensed something" in the American mindset that his political rivals completely missed: the distinctly populist nationalism that remains rooted in the political ideals of America's first populist president, Andrew Jackson. For contemporary Jacksonians, Mead observes, "the United States is not a political entity created and defined by a set of intellectual propositions rooted in the Enlightenment and oriented toward the fulfillment of a universal mission." That is not what the Jacksonians who form the foundation of Trump's support want. They believe the chief business of the American nation state is to take care of business at home. It is their view that, notwithstanding the American national vision of transforming the world, the singular commitment of the government should be to the equality, freedoms and dignity of individual American citizens. Jacksonians believe the responsibility of the US government is, first and foremost, to attend to the physical security and economic well-being of American citizens at home. These people genuinely believe the American way of life is under siege, that its fundamental values are under attack and the future of the country is at risk. Mead observes that while Americans with a more cosmopolitan and liberal world view, who believe they are working for improvements in the human condition in general, see Trump supporters as backward and ignorant, Jacksonians fear that the cosmopolitan elite, by way of their morally contemptible failure to put their own country and its citizens first, are no longer reliably patriotic, to the point of being almost guilty of treason. Many Jacksonians are deeply frustrated that they find themselves in a nation they helped build which now purportedly offers economic advantages based on concessions to ethnic authenticity that extend to everybody but them. "Political correctness" is anathema to growing numbers of Jacksonians. As Mead explains, if people are constantly being accused of being racist for simply trying to have others recognize and appreciate their own identity, then sooner or later they may decide that racist is what they are and that they ought to just live with it. This is one of the reasons, Mead notes, why Trump supporters see immigration as part of a deliberate and conscious effort to marginalize them in their own country.

Mead also points out that Jacksonians believe the Second Amendment, the right to bear arms, to be the most important in their country's Constitution. They see the right of revolution – enshrined in the Declaration of Independence as the right of a free people to defend themselves – as a last bastion protecting them from tyranny. Trump supporters are among those who fear that Democrats and centrist Republicans are trying to disarm them, which, Mead explains, is why mass shootings and consequent calls for gun control stimulate spikes in arms sales, even though crime rates in most places in the United States are falling.

The American election of 2016 made it abundantly clear that globalization and automation have shattered the socio-economic aspirations upon which postwar prosperity, peace and order were to be founded, domestically as well as internationally.

In short, Mead offers, Trump voters deeply believed they were saving themselves and their country from catastrophe. While some may well have seen shortcomings in the candidate himself, what Trump supporters were expressing was their lack of confidence in the status quo and their deep opposition to what appears to them to be an inexorable movement of their country toward self-destruction. These and related matters, including trade and foreign policy, are concerns to which Canada and other Western countries should pay close attention, lest the unaddressed grievances of the forgotten and left behind explode into similar divides elsewhere.

In surveying the road ahead, Mead contends Western policy-makers in general have bought into some dangerously oversimplified notions about capitalism, not the least of which is that somehow the system has been perfected and we no longer have to fear social, economic or political upheavals in a globalized democracy. As Canadian author Jennifer Welsh noted in her remarkable 2016 book and CBC Massey Lecture, *The Return of History*, it was the demise of the Soviet Union and the end of the ideological confrontation between East and West in 1989 that prompted American political commentator Francis Fukuyama to write his famous essay, and later book, entitled *The End of History*. Fukuyama's principal argument was that the end of the Cold War signalled the end point in the socio-cultural and ideological evolution of humanity and that the result would be the "universalization of Western liberal democracy as the final form of human government." The triumph of liberal democracy, Fukuyama forecasted, would result in the exhaustion and ultimate transformation of power politics, the end of large-scale conflict and the realization of a path toward a more peaceful world. This, he offered, was essentially the end of history as humanity had known it.

Not so fast, said Jennifer Welsh, in the run-up to the 2016 US election. History has not ended. Don't be fooled, she wrote: history has come back, as Mead too indicated, "with a vengeance." It is not the end of history in Russia or the Middle East. Nor at the UN. Democracy has not remained inevitably the most common political system. There is widespread backsliding in Thailand, Turkey and even parts of Europe. History has also returned in the form of "total war" and targeting of civilians in interminable major conflicts. History, Welsh argues, has also come back in some twisted new forms: the wide flaunting of violations of international law; the return of barbarism and mass flight; the return of Cold War and widespread global economic uncertainty. Growing inequality generated by globalization, Welsh asserts, has become the kiss of death for the preservation of law and liberty and a source of rising populism globally. Months before the American election, Welsh explained the logic of rapidly rising populism in the US and widely around the world. In her conclusion, she warned that while democracies can withstand shocks, what must be avoided at all costs are complacency and overconfidence. We need to reread history and bring fundamental human values of fairness and equality back into our current post-truth context. The question is whether we need to do this before we address a growing global Earth system crisis – the looming climate change disaster that Bill McKibben clearly described in his take on the end of history in his landmark 1989 book *The End of Nature.*

One of the most serious of all foreign policy issues completely forgotten during the American election of 2016 was natural Earth system function: the need for humanity to pay attention to the planet's self-regulating life-support system. The failure to address water security issues and come to terms with climate disruption will not only hurt at the local level, which matters most to Trump supporters, but will also undermine the values of those with more cosmopolitan world views. The hard work of hope demands that we find a way to have both solitudes realize that we all need to work together to prevent that from happening.

We need to build a functioning bridge in Canada and the United States between those committed to outward thinking, whose principal focus is on cosmopolitan values, and those who are passionately inward looking and focused on individual rights and freedoms and who no longer trust the status quo. The bridge we need to build must take us in the immediate direction of non-partisan co-operation on water, agricultural practices, energy use and biodiversity protection. We have to do this before society tears itself apart and before climate disruption and attendant refugee and other crises widen the split between us. The first step toward this end may be to realize that the world we live in now is no longer like the one we were born into. We have entered a new epoch of human presence in the world. We cannot live in this era in the same way as we have in the past.

EARTH SYSTEM URGENCY: THE ANTHROPOCENE

The biodiversity-based planetary life support system we depend on for the stability of the conditions necessary for humanity to feed itself, create nations, mount armies, and establish political systems and local, regional and global economies doesn't follow American elections and doesn't care who the president of the United States might be or who is lying or who is telling the truth in and around the White House. The indifference of Earth system function to human interest is deceiving.

The concept of Earth system function is based on evidence that the conditions that make life possible and tolerable on this planet as we know it today have been brought about and are being maintained by biodiversity. Life, in interaction with the physical elements of the planet, has, over billions of years, created a homeostatic, co-evolutionary system that ensures optimal conditions for life even in the face of periodic shocks. We have disrupted that homeostasis. It should never be forgotten that an Earth system malfunction could abruptly wipe out all of humanity.

Though directly linked, water security and climate disruption are only two of many related threats to social, economic and political stability globally. Our collective cumulative and compound effects on every component of the Earth system, including the global hydrological cycle and the world's climate patterns, appear to have reached a tipping point. Human impacts on the function of the Earth system are now such that scientists are arguing we have entered an entirely new geological epoch in the history of the Earth, which they propose to call the Anthropocene.

This epoch, geologically at least, is marked by the laying down of an entirely new stratum in the geological record wholly distinguishable from earlier, Holocene deposits. Again, it is not just about climate. The distinguishing features of the Anthropocene include evidence in the fossil record of a rise in extinctions at a rate 100 to 1,000 times faster than in background strata. Well, you might say, doesn't every species eventually become extinct? The reality is that with extinctions in the past, species often didn't die at all but instead evolved into two or more daughter species. The accelerated extinction rates brought about by human impacts do not allow time for new species to emerge from the old. The problem here is that our planet's self-regulating nature – the Earth system – is a single system, within which the biosphere is an active and critical component. In other words, the presence of life itself is in turn critical to the maintenance of the conditions that make that life possible. More than that, the system itself is created and sustained by biodiversity: the sum total of all life on the planet. If you take out too many parts, its function will change or cease.

As sociobiologist E.O. Wilson has observed, we are on our way to being alone in the world. The most vulnerable habitats of all, with the highest extinction rates per unit area, include rivers, streams and lakes in both tropical and temperate regions. Between 1898 and 2006, 57 species of freshwater fish became extinct in North America alone. The risk here is that we will find ourselves in a much diminished world in which we will have to invent artificial means of providing the basic Earth system services that natural systems provided for free, in order to ensure we have air we can breathe, water we can drink and food to eat as our population continues to grow. Providing these services will be very expensive, if we can even learn how to provide them at all. The uncertainty is that we will not likely recognize the thresholds that could lead to the irreversible collapse of natural Earth system function until long after we have crossed them.

Given the extent of our accelerating impacts, it is not unreasonable to say that, in addition to our seemingly endless political woes and setbacks, we as a global society face some very substantial and very complex immediate threats to the sustainable presence of the global social order as it exists today. While these threats have for the past 50 years been viewed as environmental and largely shunted aside, to be dismissed as irrelevant by economists and politicians, Earth system damage is now so substantial that it threatens the very foundation of the economy of nearly every nation in the world. The damage is serious enough now that it is beginning to impact not just global prosperity but the very idea of prosperity itself. As Bill McKibben would note, if you think the end of history will demand your full attention, just wait until you see the end of nature.

ECONOMIC AND POLITICAL URGENCY

Concern related to potential Earth system change is growing globally, not just in environmental circles but at the highest levels of economic and political influence. But it is not growing fast enough. A proposed water and water-related climate agenda was put forward by the InterAction Council, a group of more than 30 former heads of state and government for the G20 conference held in September 2016 in Hangzhou, China.

The G20 is an informal but very powerful group of 19 countries plus the European Union, with representation also from the International Monetary Fund and the World Bank. One of the main themes of the conference was moving the G20 countries in the direction of an "innovative, invigorated, interconnected and inclusive world economy." Also on the G20 agenda for 2016 were discussions concerning "green finance" in support of the implementation of the United Nations 2030 *Transforming Our World* global sustainable development agenda. One of the other goals of the G20, if not all nations, is to find ways to chart a course to the future that balances considerations too often and too easily cast in opposition.

It is increasingly seen that a better route forward is to

rethink the connections that link water, food and energy security to human well-being and planetary health as expressed by biodiversity. A nexus has in fact been revealed – an interconnecting point where water, food, energy and biodiversity become one; where what we do to and with any one of these factors affects the others. This nexus has already become ground zero in effective action against climate and hydrological change.

Where is our growing understanding of the nexus taking us?

One place it is taking us is to new realizations of just how critically important a relatively predictable global water cycle is to global economic and political stability. Without stable water and climate regimes, sustainability will forever remain a moving target and so will global peace and order. We need to rethink how we value, use and manage water in order to define a safe place in terms of sustainability to which all of humanity can aspire.

In the past decade, we have learned that sustainable development as we have defined it in Canada is not enough. We cannot simply constantly accommodate development through passive mitigation; if we want continued prosperity, we have to actively put vital Earth system function back in place in everything we do. This suggests we may need a new global business model that demands that we stop causing damage to natural systems, rather than simply creating a market to deal with the uncalculated, and likely incalculable, damage we are doing to the Earth system as a matter of prescribed course.

So where do we go next? It is one thing to ignore the elephant in the room: population. It is quite another to ignore a dangerous monster like the one we have created out of the cumulative and compounding effects of human numbers and activities on Earth system function. In addition to altering our planet's global nutrient cycles, we are also causing changes in the chemistry, salinity and temperature of our oceans and the composition of our atmosphere. Alterations in the relative ratios of the various constituents of the atmosphere, in tandem with land use changes and our growing water demand, have not only altered the manner and rate at which water moves through the global hydrologic cycle but have disrupted our climate.

Sooner or later we have to stop sugar-coating this. At the time of this writing, Canadians were experiencing the 385th consecutive month in which global temperatures had been above the 20th-century average. The ocean-driven evaporation–precipitation cycle, which largely defines global climate, accelerated by 4 per cent between 1950 and 2000. Earth system change also appears to be accelerating. With respect to hydro-meteorological change, 2030 is the new 2050; and 2050 is the new 2100.

The G20 barely paid lip service to these concerns or to the recommendations of the InterAction Council of former heads of state and government, focusing instead on further breaking down barriers to global trade and investment. It appears the G20 leaders still believe we have to get rich before we can get smart.

The immediate-term priorities of the G20 leaders notwithstanding, the problem of hydro-meteorological change is not going away. We see now that the kind of policy attention these issues demand keeps assuming new shapes. The urgency is different than we initially thought and keeps changing in ways we didn't expect. Some of these new iterations of perspective are troubling.

In late July 2015, Michael Ellison, at that time one of the editors of *Eos*, the official scientific journal of the American Geophysical Union, published a blogpost on the importance of meaning the same thing when we use the same words to describe climate change. Citing recent research, Ellison noted that despite all our concern about "tipping points" in the climate system, we are currently largely ignoring how even non-tipping points in climate and other parts of the Earth system might cause truly radical transformations in our social and political systems. What is becoming increasingly apparent is that abrupt social and economic tipping points may be just as important as – or even more important than – slower moving but equally critical hydro-climatic thresholds. Ellison noted we have yet to pay enough attention to identifying how climate disruption can trigger social, economic and political shocks.

At present, much of our research on climate damage is focused on easier-to-characterize changes, such as those associated with the threat of sea level rise on coastal cities; potential impacts of extreme weather events on expensive infrastructure; possible changes in crop yield or the effects of warming mean atmospheric temperatures on precipitation and water supply. But these are not the full picture.

The more pressing questions – the ones we are largely ignoring – relate to how climate disruptions of these kinds will more immediately impact our fragile, unsuspecting political institutions, our vulnerable global economy, and already tense international relations in an ever more crowded and rapidly warming world.

Some of the harder questions we need to ask and answer relate to what adaptation is really meant to achieve and what resilience really means. What we are presently witnessing is a derangement of hydrology globally. What we now face – as a consequence of human-induced desertification; the pollution, contamination and eutrophication of fresh water; rising greenhouse gas emissions; and accelerating biodiversity loss – is nothing less than a cascade of effects that could result in global hydrotrophic collapse. We have to avert this.

We have a moral obligation, not just to fix things for ourselves but to restore the system for everyone, including our descendants. Our success in resolving the growing crisis at the nexus of water, food, energy and biodiversity, and our reputation in the world for how we did it, will therefore be defined not just by hard engineering but by how we manage water and land as living systems.

The fact that the G20 gave only passing attention to accelerating hydro-meteorological change does not by any means suggest that the urgency of doing so has gone away. The G20 may not have understood this, but the World Economic Forum did.

In January of each year, the WEF, held in Davos, Switzerland, releases its annual Global Risks Report. As part of the survey, nearly 750 experts assess 29 separate global risks for both impact and likelihood out to a 10-year time horizon. The risk with the greatest potential impact in 2016 was found to be the failure of climate change mitigation and adaptation.

A haunting image of the extent and nature of contemporary risks was put forward at the 2016 forum by the global insurance giant Munich Re. The image was a map of interconnections between various economic, environmental, geopolitical, societal and technological risks associated with the failure to effectively and meaningfully adapt to climate change. What the map illustrated was the cascading effect of the failure to adapt to hydro-climatic change. On a global scale, failure leads first to greater vulnerability to extreme weather events, food crises, water crises, large-scale involuntary migration and further man-made environmental catastrophes, which in turn lead to biodiversity loss and Earth system collapse.

The 2017 Global Risks Report came at a critical moment for the world as we know it. The report noted that unprecedented forces are reshaping society, economics, politics and our planet itself in ways some might not have predicted when the first risks report was launched a decade ago. These tumultuous times, the report acknowledges, presented humanity with groundbreaking opportunities for changing how we can create a better world and how we need to operate within it. Such change brings with it some significant, globally interconnected risks, however, the foremost of which are, once again, environmental.

The report urged that even though the risk will play out over the long term, actions have to be immediate and long-lasting to have any hope of reversing the trajectory of climate change.

The environment, the report concluded, dominates the entire 2017 global risk landscape in terms of impact and likelihood, with extreme weather events and large natural disasters, as well as failure of mitigation and adaptation to climate change, being the most prominent global risks. Unlike the threat of nuclear war or pandemic disease, however, climate change ranks among the highest in terms of likelihood as well as impact. It will be one of the top three trends shaping global developments over the next ten years, the reported noted, and remains one of the truly existential risks to our world. What was happening in the Arctic in the months leading up to the 2017 World Economic Forum no doubt had an influence on the decision to rank climate impacts as a fundamental threat to humanity's continued presence on this planet. Climate circumstances in the Arctic, which are changing faster than humans likely have ever witnessed before, attest to the seriousness of this threat.

THE RISK OF RUNAWAY CHANGE: THE ARCTIC MELTDOWN

Whether we want to admit it or not, Canada is the canary in the coal mine for climate change. The challenges are greatest in western and northern parts of the country. According to the Canadian Rockies Hydrological Observatory, we have lost two full months of annual snow cover in western Canada over the past 50 years. Average winter temperatures in Alberta have risen 5.5°C since the 1960s. Then there is the Arctic, where the changes are even more dramatic.

What was happening in the Arctic in the fall and winter of 2016 suggests we should stop worrying so much about conditions down south and start paying serious attention to the situation up north. On November 17, 2016, temperatures over the North Pole and the Arctic Ocean were 20° above normal and sea ice formation was the lowest on record. Climate scientists fear this prolonged and persistent warming of the Arctic may mean the climate problem and the further acceleration of the global water cycle could be about to get away on us.

Arctic snow cover has been declining at a rate of 22 per cent per decade, Arctic sea ice at 12 per cent per decade. But now something else is happening. Multi-year ice is disappearing and not as much new sea ice is forming. There are two linked concerns here: the impact this loss of ice will have on weather patterns at mid-latitudes; and the potential for methane presently entombed by Arctic cold to be spontaneous released into the atmosphere.

ARCTIC SEA ICE RETREAT AND THE JETSTREAM

The increasingly rapid loss of sea ice is not yet a global political issue. That will come soon enough. There is, however, urgent concern among climate scientists, who worry that the accelerating rate of melt and thaw occurring in the Arctic could further destabilize global weather patterns.

To understand the problem, some very basic climate science knowledge is necessary. A "positive feedback" in the climate system is a chain of cause and effect in which two or more circumstances begin to accelerate one another. A positive feedback loop enhances or amplifies changes, tending to move a climate system away from its equilibrium and make it unstable. It appears that a positive feedback has been created between warming in the Arctic and the behaviour of the northern hemisphere jet stream – the westerly winds that arise as the world spins inside the rumpled blanket of its own atmosphere. To put it simply, the loss of northern snow cover and diminishment of Arctic sea ice are causing the jet stream to slow and become wavier. We see now that a wavier northern hemisphere jet stream brings warmer air to the Arctic more frequently. Temperature anomalies as significant as 30°C above normal were recorded during the fall and throughout the winter of 2016 in the Arctic and parts of northern

Canada. These warmer temperatures are leading to more sea ice loss, which in turn makes the jet stream even wavier, which brings even more warm air into the Arctic. This makes the jet stream even wavier yet again, resulting in still more warm air being transported into the Arctic, which causes further sea ice loss. As this feedback loop accelerates, the Arctic Ocean absorbs more heat, which could also affect the behaviour of deep-ocean currents, which would impact climate globally.

Sea ice, scientists have discovered, is a central element in the natural thermostat that regulates the temperature of the entire hemisphere, and once this thermostat is turned up, weather down to the mid-latitudes and beyond quickly becomes more variable and erratic.

So why does this matter?

The relationship between warming air and how much moisture this air can carry is critical. One of the fundamental laws of atmospheric physics decrees that for every degree Celsius the atmosphere warms, it can carry 7 per cent more water vapour. Water vapour is a powerful greenhouse gas in its own right. Its increased presence in the atmosphere adds to the warming produced by the transport of warmer air northward. But according to climate scientist Dr. Jennifer Francis, this is not the only effect increased water vapour has had on the accelerated warming of Arctic temperatures. The more water vapour there is, the more clouds form. Cloud cover holds heat in, which increases evaporation from the ocean, further accelerating warming. The Arctic is now warming five to eight times faster than the equatorial region. By reducing the temperature gradient between the poles and the tropics, these feedbacks are disrupting the entire weather system in the northern hemisphere. A slower, wavier jet stream is causing lock-ins of weather extremes that result in horrific rainfalls in some places and deep and persistent drought in others. What scientists fear is that the climate system in this part of the world is destabilizing. If this were happening to a business of yours, you would call this degree of pending destabilization "hemorrhaging" and act immediately to stop it. Then there is the even greater threat posed by natural methane releases from the warming land into the air and into inland and coastal waters.

CANADA'S CARBON TIME BOMB

The well-recognized global risk associated with the loss of carbon storage in frozen ground is the feedback amplification of anthropogenic warming due to greenhouse gases released from thawing permafrost. This feedback begins with the thawing of organic matter containing carbon dioxide and methane, which in itself increases the surface temperature, further warming the top layer of frozen soil, which accelerates further melt and further carbon dioxide and methane release.

Until we experienced this recent and unprecedented acceleration of warming in the Arctic, we didn't think this would be a big problem until at least the end of the century, by which time, we hoped, we would have climate warming under relative control by means of a wide range of techniques. Unfortunately, the cold caps that are keeping the carbon in the ground are thawing faster than we thought. It should be further noted that feedback from Arctic carbon sinks was not included in the warming scenarios for the future put forward by the Intergovernmental Panel on Climate Change in its latest five-year Assessment Report, released in 2015.

Essentially, what we have here, as University of Alaska

researcher Dr. Katie Walter put it, is "a very leaky cryospheric cap" that is deteriorating rapidly under accelerated warming. At present, permafrost caps prevent methane and carbon dioxide releases, but the caps are melting and glaciers are disappearing, causing decompression of subsurface geological structures and the speeding up of what's called "isostatic rebound." This "bouncing back" of the Earth from being compressed under the great weight of glacial ice results in releases of thermogenic methane. Similar processes, Walter notes, are clearly active in areas where glaciers are disappearing in Greenland and Iceland.

Permafrost contains approximately 1400 to 1700 gigatonnes of carbon, or roughly twice the amount presently found in the atmosphere. It has also been surmised that there is an additional 400 gigatonnes of carbon stored in "deep terrestrial permafrost sediments" – not to mention an as yet unknown amount in subsea permafrost on shallow continental shelves such as in the East Siberian Sea.

A veritable sink of biogenic and geological sources of methane is leaking into the atmosphere, and the only thing preventing wholesale release and runaway warming feedbacks is winter cold, which slows release and keeps a leaky cap of permafrost partially in place. This cap, however, is breaking down, turning the Arctic into a source rather than a sink for carbon. But, as noted above, thawing permafrost is only one concern.

Embedded in the frozen marine sediments of the Arctic are methane hydrates and clathrates. As the Arctic Ocean warms, the sediments containing these compounds thaw, producing methane gas, which has begun to rise to the surface in great bubble plumes. In deep water these plumes oxidize and disappear before they reach the surface, but at depths of less than 50 to 100 metres the methane does not have time to dissolve. Instead it emerges almost intact into the atmosphere, where in the immediate term it appears it may have a greenhouse effect that can be as much as 100 to 200 times that of carbon dioxide.

We are also discovering in the Arctic that it is possible to warm the world without increasing emissions. All you have to do is reduce the reflectivity of ice and snow and the planet will warm by itself. Alarmingly, scientists have found that the overall combination of ice and snow loss in the northern hemisphere may contribute an additional 50 per cent to the direct global heating effect caused by the addition of CO_2 to the atmosphere.

The importance of this has yet to be generally realized. We are reaching the point at which we should no longer simply say that adding carbon dioxide to the atmosphere by way of our emissions is warming our planet. Instead we should say that the carbon dioxide we have added to the atmosphere has *already* warmed our planet to the point where the feedback mechanisms related to loss of reflectivity of ice and snow are themselves increasing the effect of those emissions by a further 50 per cent. *This means that carbon dioxide may not be the only driver of climate change.* What this suggests is that we are not far from the moment when the feedbacks will themselves be driving the change – that is, we will not need to add more carbon dioxide to the atmosphere at all but will get warming anyway.

Peter Wadhams, one of the world's leading experts on the relationship between sea ice and climate, characterized the risk associated with these kinds of feedbacks this way: "When

Jimi Hendrix played the guitar he had the ability to play passages using feedback alone – his fingers didn't pluck the strings but he manipulated electronic feedback to produce the sounds. We are fast approaching the stage when climate change will be playing the tune for us while we stand back and watch hopelessly, with our reductions in carbon dioxide having no effect."

SO WHAT DO WE DO?

The first thing we should do is clearly recognize that notwithstanding US President Trump's claim, climate change is not a hoax. Nor is it something we can wish away. As a quote often attributed to Winston Churchill puts it: "The truth is incontrovertible. Malice may attack it, ignorance may deride it, but in the end, there it is." We don't just live in an economy; we also live in an ecosystem and in a society where our real prosperity is linked to our health. We have to stop what has effectively become an uncontrolled oil spill into the Earth's atmosphere. The accelerating disruption in climate patterns in the northern hemisphere will continue until we stabilize the composition of the Earth's atmosphere and restore damaged elements of natural Earth system self-regulation. This will require a number of measures: reducing greenhouse emissions; reversing ecosystem damage, thus increasing the ecosystem's capacity to store carbon so that net emissions become zero; reversing the impacts we have on our rivers and streams; halting extinctions; and ensuring that all development is not only sustainable but restorative. All these are elements of the United Nations' new plan for global sustainability.

Transforming Our World

In responding to the urgency and the opportunity of finally getting sustainable development right, the United Nations announced its long-anticipated new framework for global action.

While it did not receive the same attention in the media as the Paris Agreement, the announcement of the UN's 2030 *Transforming Our World* global sustainable development agenda in September of 2015 was at least as important as the later climate negotiations in Paris, if only because it deals with damage we are doing to other elements of the Earth system that are exacerbating and being exacerbated by climate change. What is important about the UN's *Transforming Our World* agenda is that at last we have a universally understood and accepted definition of what sustainability really means, as well as a common timetable for implementation of clear goals aimed at achieving measurable targets both globally and nationally.

The endorsement, however tentative, of *Transforming Our World* may be the most important thing humanity has done for its future since we created the United Nations. The 2030 *Transforming Our World* agenda makes it very clear that sustainable development can no longer simply aim for environmentally neutral solutions. If we are to achieve any meaningful level of sustainable development, all development has to be not only sustainable but restorative. We can no longer simply aim to slow or stop damage to the Earth system; we have to thoughtfully restore declining Earth system function. We face so many overlapping and intersecting crises that we can no longer afford to fix them one at a time or in isolation from one another. All future development must seek double, triple, if not quadruple benefits in terms of the restoration of fundamental Earth system function as reflected in biodiversity stability, efficient water use, soil vitality, carbon storage and human and planetary health. In order to achieve these goals, it is important to further incentivize advancements in the fields of both engineering and planning.

What is unfortunate is that what is happening in the US at the time of this writing threatens the UN's already fragile efforts to ensure water security and climate stability, not just in North America but globally, in time to prevent irreversible Earth system damage and possible collapse.

From an Earth system point of view, the timing of events in the US could not be worse. The Paris Agreement broadened and deepened awareness, but as readers of this book likely already know, we are not going to get to the world we want unless scientific evidence drives public policy priorities at the national and subnational level. The risk now, from a UN perspective, is that under the threat of losing trade and other relations with the United States a lot of good people working there and elsewhere on crucial global problems like water security and related climate stability will be sidelined. That said, what is happening in the US is not the end of the world. The Paris Agreement and what it means for the management of water could eventually become "Trump-proof" if all the other countries in the world meet their promises in a timely manner and if the private sector, especially globally, steps up to the plate, which, as we just saw at COP 22 in Morocco, is not impossible. Much uncertainty remains, however.

It is against this tense and unsettled backdrop that all of us carry on the hard work of hope, the tireless task of putting scientific evidence in front of decision makers in the hope of positively advancing public policy in the direction of true sustainability.

The International Decade for Action "Water for Sustainable Development"

On December 21, 2016, the UN General Assembly unanimously adopted a resolution entitled "International Decade for Action 'Water for Sustainable Development,' 2018–2028." Sponsored by 177 UN member states, the decade aims to promote sustainable development and integrated water resources management, implementation and promotion of relevant programs and projects, and increased co-operation and partnerships to support the internationally agreed goals and targets related to water resources, including the UN's 2030 sustainable development goals.

Why another water decade? Because water security and infrastructure are a huge part of the development challenges we still face. Water relates fundamentally to natural resources and social issues; eradicating hunger and poverty will depend heavily on our attitude toward water.

The most glaring water-related challenge for the world is sanitation services: that is, the safe disposal of human waste, from toilets and sewage systems to solid waste management. Today more than 2.4 billion people do not have access to proper facilities. Poor sanitation is estimated to cost the world US$260-billion per year – more than the entire GDP of Chile. Kenya, for example, loses US$324-million annually because of sanitation: $244-million due to premature deaths resulting from diarrhea, $51-million in healthcare costs, $2.7-million in productivity costs from time absent from work and school as a result of diseases due to poor sanitation, and $26-million due to time lost looking for a place to defecate.

Water scarcity already affects millions of people around the world. Population growth, rapid urbanization, more water-intense consumption patterns, and climate change are intensifying the pressure on existing resources. Developing countries in Africa and Asia will carry the main burden of this increase in demand. Many already suffer greatly from water stress or scarcity, and lack the infrastructure and know-how to address it. Ethiopia, for example, is currently facing its worst drought in decades. More than 10 million people there need to rely on food aid. If we continue on our current path, the world may face a shortfall of 40 per cent in water availability by 2030. Poor sanitation and water scarcity will only make existing regional challenges worse and undermine our global efforts to advance sustainable development for all. In short, water is a crucial factor in all aspects of social and economic development. On the positive side, this means water is an incredibly useful medium through which many global challenges can be addressed.

Declaring 2018–2028 the "Water for Sustainable Development" decade will raise awareness of the critical state of water resources around the world and inspire more action. To that end, the current proposal has two main objectives.

First, there is a call for greater international co-operation on water-related issues. In line with the recommendations of the report from the UN secretary-general's advisory board on water and sanitation, a "new international water architecture" is needed to make financing and implementation efforts more effective.

Second, the availability and accessibility of information

is vital. Any success on this front depends on increasing our knowledge of the water situation worldwide, developing effective water management strategies, and ensuring that those who need this knowledge have access to it.

This call for action is an important initiative, of course. But it still does not adequately address the question of how to finance the sustainable development goals. Governments must invest in water infrastructure, in protecting basins and ecosystems and in treating waste water and reducing pollution. Developing countries need new water management institutions and utilities, better knowledge of integrated water resources management, and the capacity to protect water basins and ecosystems. This requires access to the best available sustainable practices, through training and partnerships. The necessary funds are estimated to be in the trillions between now and 2030. But the cost of inaction is even higher. So where does Canada stand relative to the new UN Decade for Action on the global water crisis?

Towards a national water strategy in Canada

The idea that Canada possesses 20 per cent of all the fresh water on Earth has been a source of Canadian pride since before Confederation. Unfortunately, it is just a myth. While we do have a great deal of water in some parts of the country, much of it is left over from the melting of the two-kilometre-deep ice sheets that covered almost all of Canada at the close of the last great ice age. Only 6.5 per cent of the water we claim to possess is returned to us each year through the annual precipitation cycle.

Another nagging truth eating away at the myth of limitless abundance of water in Canada is the fact that we are among the most egregious water wasters in the world. We also pollute a great deal of our water, effectively removing it from the cycle and often at great expense to natural ecosystem function. Among the biggest threats to water supply and quality in Canada are those posed by industrial agriculture.

Far outweighing these rebuttals of the myth of limitless abundance of water in Canada, however, is the fact that by way of our needs and our sheer numbers, humanity has begun to accelerate the rate and manner in which water moves through the global hydrological cycle. The statistics from the past related to how surface, subsurface and atmospheric water will act under a variety of circumstances are no longer reliable. This, we have recently discovered, is a lot more serious than we at first thought. It is now widely held that it is no longer practical or even defensible to continue to assume hydrological stability in designing our cities and infrastructure, in planning to supply water to growing populations and in growing our food. The myth of limitless abundance of water is not going to survive this assault. So what is going on out there?

The most profound changes to the hydrological cycle relate to how much more water a warmer atmosphere can hold. We have known for more than a century that for every degree Celsius of warming we can expect the atmosphere to carry 7 per cent more water vapour. This, in combination with rising sea surface temperatures, allows for extreme cloudbursts and storms with greater power that last longer and carry more punch.

What we are also discovering is that the ratio of snow to liquid water in the great seasonal redistribution of precipitation in the northern hemisphere is changing, with huge potential consequences for all of us. Nowhere is this more evident than in Canada's Rocky Mountains. Research has demonstrated that we lost some 300 glaciers in the mountain national parks of the Rockies alone in the 85 years between 1920 and 2005. This loss is expected to continue, with over 90 per cent of the ice in the interior ranges of Canada's western mountains expected to be gone by the end of this century. This loss of glacial ice is but a symptom of a much larger problem. The same warming that is causing our glaciers to disappear so quickly is reducing snowpack and the duration and extent of snow cover throughout the mountain West. Snowpack and snow cover are now declining by 17 per cent per decade. By mid-century, the Canadian West will be as changed by this as it was by European settlement.

What we are also seeing in Canada is that the loss of Arctic sea ice and the rapid contraction of the extent and duration of snow cover in the northern hemisphere are impacting the behaviour of the jet stream, which in turn translates into more erratic weather patterns. The Arctic is now well on its way to becoming a driver of, rather than just a responder to, global change. We are now beginning to understand the extent to which Arctic sea ice acts as a thermostat controlling climate right down to the mid-latitudes throughout the northern hemisphere.

What researchers are demonstrating is that none of these changes are taking place independent of one another. There is a direct systemic link between water, land and climate. Water is a good place to start in reimagining how the Earth system works. It is well known that 90 per cent of global heat dynamics are controlled by hydrological processes. The warmer the global atmosphere, the more evaporation there is and the more water the atmosphere can hold and transport. Changes in land use and cover, in combination with changes in the behaviour of the northern hemisphere jet stream brought about by the diminution of Arctic sea ice, affect where and how much water in the atmosphere falls to Earth in the form of precipitation. Continuing changes in the composition of the Earth's atmosphere brought about by fossil fuel combustion and other greenhouse emissions, in combination with the reduced capacity of human-altered landscapes to store and absorb carbon, are accelerating the rate and manner of water's motion through the global hydrological cycle. The acceleration of the global water cycle is resulting simultaneously in more catastrophic storms and deeper and more persistent droughts. The trend toward ever more disastrous weather events and worse droughts will persist until we stabilize the composition of the global atmosphere and restore the stability of the global hydrologic cycle. To rehabilitate the water cycle, we need to manage land and water as living systems.

To accomplish this, Canada urgently needs a new water ethic. And if we are to continue to prosper in rapidly changing global hydro-climatic conditions, we need it now. We know what to do; it is time to do it.

Canada's water crisis is in part due to past federal and provincial government neglect of the links between growing water quality concerns and water-related climate issues. Despite clear scientific evidence, governments have over the past two decades refused to acknowledge that we face the cascading, cumulative effects of increasingly damaging industrial, mining and agricultural water contamination; more frequent flooding brought about by inappropriate land use and development in flood plains and headwaters; and ever more destructive extremes of flood and drought due to climate disruptions to which we have contributed by altering the composition of the Earth's atmosphere.

In addition to the substantial costs of remediating water contamination, flood and drought crises, along with related storm damage and forest fire impacts, are now hammering every region of the country as climate change accelerates. Economic losses associated with drought on the Canadian prairies between 2000 and 2004 exceeded $4-billion, and the Alberta–Saskatchewan–Manitoba floods of 2011–2014, which included major inundation in Calgary, exceeded $11-billion. The cost of one single day of heavy rainfall in June 2013 in Toronto ultimately reached $1-billion. Drought extending from Mexico to Alaska and from Vancouver Island to Manitoba in 2015 resulted in massive and very costly wildfires that restricted oil production in Canada and impacted food availability widely throughout North America. How has Canada responded to this crisis?

Prior to the election of 2015, our federal government's response to this national crisis showed little foresight. Water monitoring on a national level and federal commitment to science had been gradually cut over several decades. Canada stood out then and continues to stand out in the developed world for having neither a national flood forecast system nor mandatory drinking water standards. Appalling water quality and human health conditions on First Nations and other remote downstream communities have put Canada on a notorious "third world" footing in terms of meeting global water supply and sanitation goals – unacceptable in a G7 country.

How did we get into this mess? There was a time – the 1970s and early '80s – when Canada had a world-class reputation for building our national prosperity on superbly clean and abundant water, the continuous supply of which was assured by outstanding water science and careful federal management. The decay of Canada's water management reputation began with the breakup of the Inland Waters Directorate of Environment Canada in the 1990s, followed by decimation of federal water science and gradual withdrawal from national flood damage reduction programs and drought mitigation initiatives. The near death knell of Canada's water management reputation came later with the Harper government's abrogation of responsibility for fisheries habitat, navigable waters protection and environmental impact assessment. With Ottawa missing in action, water issues in Canada now fall in a largely uncoordinated fashion to the ten provinces and three territories, resulting in atomization of water management policy. The weakness in this arrangement is that more than 75 per cent of Canadians live in boundary water basins shared either with the United States or with other provinces and territories. The failure of federal leadership ignores the reality of how water flows within our own national boundaries and leaves Canada vulnerable in an era of rapid hydroclimatic change to major water crises that are already impoverishing regional economies and now stand poised to restrict economic growth of the entire country by limiting agricultural and energy production and damaging or destroying costly infrastructure.

The Liberal government elected in 2015 has indicated a readiness to address Canada's water crisis by implementing flood and drought forecasting and management, and improving water quality and fishery protection and transboundary water management through advice based on enhanced water science and observations. The Canada First Research

Excellence Fund has invested heavily in initiatives like the Global Water Futures research program. Currently the largest university-based water research program in the world, this partnership-based, seven-year science initiative aims to transform the way communities, governments and industries in Canada and other cold regions of the world prepare for and manage an increasing number of water-related threats. But while research is necessary and valuable, what is equally important is the restructuring of water governance in Canada and abroad. One of the possible first steps would be for the federal government to appoint a secretary of state or similar position for water in Canada.

The problems the nation faces with respect to the management of its water resources cannot be addressed by simply adding additional government departments and more layers of bureaucracy, however. What is needed is stronger leadership and enhanced coordination among existing departments and jurisdictions at all levels, together with a better bridge between scientific research outcomes and the timely and orderly evolution of public policy.

The beauty of secretary of state positions, or offices like them, is that they can be constituted quickly to very specifically address both lingering and new government priorities in times of fiscal restraint without the complicated, time-consuming and expensive task of creating new bureaucracy. Secretaries of state are given mandates to use existing bureaucratic resources to achieve specific objectives in domains important to the prime minister and the government. In this context, the office of the secretary of state for water could act as a badly needed clearing house for overlapping federal, provincial and municipal water interests. The appointment of such an officer would have the further advantage of pinpointing areas of potential interjurisdictional conflict before they become full-blown disputes, thereby circumventing potential defensiveness among provinces that may feel the federal government could be trying to encroach on what they see as being clearly defined as a provincial resource. The appointment of a secretary of state for water could also help keep relations with our American neighbours on an even keel.

Many federal and provincial Cabinet ministers have responsibility for water, but in no case is that their *only* dedicated responsibility. If the federal government wants the administration of water to be commensurate with water's worth to our economy and our future, and wants to fully realize the benefits of investing in research, its objective should be to appoint a national water champion as secretary of state for water and to surround that person with a small staff, drawn from among the most knowledgeable water policy experts in the country, who would help coordinate and synergize interdepartmental efforts at the federal level on all matters related to water. But the responsibilities of such an office cannot end there. A secretary of state for water should serve as a two-way portal for everyone in the country interested in working together to improve water governance, including provincial governments, First Nations, scientists, watershed interests and other NGOs, cities, universities, the International Joint Commission and UN agencies.

It should be the goal of a secretary of state for water to determine the most pressing areas in need of interdepartmental and interjurisdictional action, and to facilitate that action as quickly as possible. Managing water in the 21st century demands such capacity. The rewards for Canada could be huge.

A tangential benefit that could accrue from taking the value of our water resources more seriously would be realization of a long awaited "national water strategy," which could emerge *ex post facto* from the work of a secretary of state for water by simply connecting the dots of "what has to be done when and by whom" to fill the gaps in water policy across all levels of government. If we are to continue to prosper in rapidly changing global hydro-climatic conditions, Canada urgently needs a new water ethic. We know what we need to do to create that ethic, and this is one way of getting started.

While water is important in itself, it is also what is connected to water that forms the nexus between the past, the present and the future of Canada. Energy is at this nexus also, as are food and, by direct association, agriculture and biodiversity, all of which are affected by climate disruption. What is emerging is the realization that you can't have a national climate strategy without a national water strategy.

THREE

Where We Are Heading: National and Provincial Low-Carbon Resilience Plans

The long-term target set by the UN Paris Agreement in December 2015 is for the globe to become carbon neutral. This means total emissions would be completely offset by increases in carbon storage by some combination of natural ecosystems such as forests and wetlands and of engineered systems associated with carbon-emitting industrial plants. In practical terms this goal will require reduction of global emissions by 50 per cent and an increase in sequestration or storage also by 50 per cent. As this strategy would have to be achieved by 2050, it has been termed in this book as the 50–50–50 approach. It represents the essence of a low-carbon resilience strategy: decarbonization of the economy, supported by healthier and more robust ecosystems. The Intergovernmental Panel on Climate Change has indicated such a target must be met if there is to be any reasonable likelihood of limiting overall global temperature increase to 2°C, and of managing the onset of the Anthropocene conditions outlined in the previous chapter. This chapter considers how Canada and its provinces are positioned to achieve this long-term goal, and examines the new challenges created by the new US administration with its commitment to leave the Paris Agreement.

Canada made little progress on reducing its carbon emissions during the decade of the Harper government, but the Trudeau government elected in 2015 has brought new life to the discussion and new policies aimed at addressing the climate threat. Trudeau convened a federal–provincial meeting in March 2016 to set the stage for developing a pan-Canadian framework on clean growth and climate change by the end of that year. This framework is based on two overarching targets consistent with the UN goals stated above, namely a reduction of carbon emissions by 30 per cent by 2030 from a base year of 2005 and by 80 per cent by 2050.

Since the election of President Trump, whose stated intention is to reverse all the climate policies of the previous administration, a number of world leaders have looked to Canada to become a bastion of good governance for tackling climate change. In March 2017, the European Union's Commissioner for Climate Action and Energy, Miguel Arias Cañete, visited Ottawa to meet with Canada's Minister for Environment and Climate Change, Catherine McKenna, to try to ensure there would not be, as Commissioner Cañete put it, "a vacuum in leadership in climate change policy" in the Trump era. The EU is looking to Canada to provide at least regional leadership in North America on climate issues. But policies supporting Canada's emissions reduction targets have been seriously undermined by recent announcements from the Trump administration.

In round figures, Canada's commitment to the Paris Agreement is to reduce our carbon emissions by approximately 270 million tonnes annually by the end of 2030. Given that Canada increased its emissions by 20 per cent from 2005 to 2016, such cuts would require unprecedented action by all levels of government. The federal government released two documents in 2016 detailing how these ambitious targets could be met. The first was the *Pan-Canadian Framework* referred to earlier, setting out its plans to meet the 2030 target. The second document, *Canada's Mid-Century Long-Term Low-Greenhouse-Gas Development Strategy*, outlines an 80 per cent reduction from 2005 levels by mid-century.

THE 2030 PLANS FOR CARBON REDUCTION

There are four cornerstone policies underpinning the Pan-Canadian framework: pricing carbon emissions; reducing emissions across the economy; increasing resilience to the effects of the changing climate; and supporting innovation and clean technology. The government initiated its pricing policy in 2018 with a rate of $10 per tonne, to increase to $50 per tonne by 2022. A key action on reducing emissions is an agreement between the US, Mexico and Canada to reduce methane emissions from the oil and gas sector by 45 per cent from 2007 levels by 2025. As noted in the previous chapter, methane is very potent, so such regulations would make a sizeable contribution to achieving the nation's greenhouse gas emission reduction targets. A further policy on carbon reduction is to replace all existing coal-fired generating plants across Canada with renewable sources – hydroelectric dams and, where practical, solar, tidal and wind facilities – by 2030. Let's consider how well these policies stand up, both under domestic politics and under the changes proposed by the Trump administration.

Four of Canada's most populous provinces have instituted some form of carbon pricing. British Columbia brought in a carbon tax in 2008 that was revenue neutral, meaning all its levies are offset by corresponding reductions in personal and corporate income tax, and stabilized the rate at $30 per tonne in 2012. Alberta introduced a carbon levy of $20 per tonne in January 2017, to be increased to $30 in 2018. Both Ontario and Quebec have introduced a cap and trade system, whereby total carbon emissions are capped by policy and emitters have the option of either staying beneath the cap or offsetting any overshoot by buying credits on the market created by other emitters who find more efficient ways to reduce their emissions. The market covers both provinces and California. For provinces that do not have a price on carbon, the federal government will impose a national levy of $10 per tonne for 2018, rising $10 per year to reach $50 by 2022. The resulting revenues are to remain in the jurisdiction of origin, to be used according to that particular province's needs, including to address impacts on vulnerable populations and sectors and to support climate-change and clean-growth goals.

The prospects of this carbon tax regime being fully implemented in Canada appear rocky. First, not all provinces are on board. Saskatchewan has refused to institute a carbon tax and is considering a court challenge to an imposition of one by the federal government. Second, there is an uneven playing field across the country. Canada's Ecofiscal Commission, an independent economic think tank, has calculated that the real price on carbon in Ontario and Quebec under their cap and trade systems will be around $19 per tonne in 2020 and

$24 per tonne in 2022. These lower prices are largely the result of a political decision in California to oversupply carbon credits to the electricity sector, which depresses the potential market for Ontario and Quebec, as all three jurisdictions are part of the same carbon market. As a result, the previous government of British Columbia balked at the federal proposal to increase prices above that province's current $30 per tonne, pending an independent review of carbon prices scheduled for 2020. However, the new provincial government elected in July 2017 is committed to resume raising carbon tax rates annually, starting in 2018.

The US administration has no intention of putting a price on carbon emissions. Consequently, there will soon be an uneven playing field for oil and gas producers in Canada, who would have to pay the national tax while their US competitors face no such levy. This means Canada will have to tread carefully when it considers whether to maintain its carbon pricing policy, as the US is essentially the sole international market for Canada's oil and gas exports.

As a result of the slump in global oil and gas prices in 2014, US producers of natural gas from fracking in the Dakotas and Texas have reduced costs dramatically compared to their Canadian counterparts and are now competitive at current prices of around US$50 per barrel of oil. As a result, oil and gas production there is booming again, and since this fits the US narrative of strengthening domestic industry, there is no chance the nation will impose any tax on its fossil fuel industry under a Trump administration. The Canadian Association of Petroleum Producers has warned the Canadian government not to get too far out of step with US policy on trade.

The future of the proposed methane reduction regulation is just as bleak as an eventual North American price on carbon. The original idea was that all three signatories to the North American Free Trade Agreement – Canada, the US and Mexico – would introduce the same regulation continent-wide. On March 28, 2017, however, President Trump signed an executive order cancelling, among other measures, the methane reduction regulation. This places the entire policy in jeopardy, as the original pact was to level the field for all producers in the three countries and thus gain traction for investing in new technologies that could then be shared across the sector. Loss of this initiative is a body blow to Canada's chances of attaining its carbon emission targets, as the methane reduction regulation was projected to contribute about 15 per cent of the total reductions. As a result, the federal government has delayed implementation of the regulation by three years. Such regulatory gaps are not trivial. In northeastern British Columbia, for example, independent research in the Montney gas field in 2015 indicated that total methane emissions could be up to 150 per cent higher than reported by the provincial government.

The policy to accelerate the phase-out of coal-fired electricity generation in Canada by 2030 has more likelihood of succeeding. About 60 per cent of Canada's electricity is already based on renewable sources, mainly hydro. Alberta has the largest coal-fired infrastructure, but it has agreed to pay the three major power producers some $1.36-billion over 14 years as compensation to meet the 2030 timetable, with the funds coming from the carbon levy. Full implementation of the national policy will require expensive new transmission lines running east and west across the country, but as yet there are no commitments by any level of government to fund such new infrastructure.

The federal government has emphasized its plan is based on balancing economic growth with environmental protection, positing that both are possible. This may be practical politics, but it does not support the tougher action required to meet prescribed emissions reduction targets. In November 2016, Ottawa approved two pipelines to transport Alberta oil to new markets: Enbridge's Line 3 to the US Midwest and Kinder Morgan's Trans Mountain pipeline to the West Coast. The latter project will triple the capacity of the existing line to 890,000 barrels per day and increase the number of tankers plying the sensitive waters of Burrard Inlet in the heart of Vancouver from five per month to 34 per month. In attempting to balance environmental interests, the government also announced a multi-billion-dollar upgrade to the Coast Guard on all three oceans to improve response to and cleanup of oil spills, and cancelled the previous government's approval of the more controversial Northern Gateway pipeline, which would have transported oil from Alberta to Kitimat on the BC coast.

The newly elected BC government has vowed to ensure that all regulatory approvals for the Trans Mountain line that are under provincial jurisdiction will be designed to protect fresh and marine waters from spills.

The federal government was also mindful of the potential for First Nations legal challenges to its pipeline decisions. There are few treaties in British Columbia to ground Aboriginal constitutional claims to land and resources, so the courts have had to resort to requiring federal and provincial governments to exercise fair and genuine procedures for consulting with First Nations and accommodating their potential claims to such land and resources as might be affected by major resource developments. Indeed, the Federal Court of Appeal ruled that the federal government had not followed such proper procedures in consulting with the First Nations affected by the Northern Gateway proposal, and ordered the government to undertake new consultations. Legal challenges against the Kinder Morgan pipeline were scheduled to be heard in October 2017, with the new British Columbia government seeking intervenor status.

Accordingly, Ottawa undertook major new initiatives to reconcile some of these matters with First Nations and specifically extended the consultation phase on the Kinder Morgan pipeline in an attempt to accommodate the concerns of First Nations along the route. Kinder Morgan had also been diligent in working with Indigenous communities along the route to secure economic development agreements and gain acceptance for expansion of the pipeline. A total of 51 Aboriginal communities – 10 in Alberta and 41 in BC – have signed mutual-benefit agreements. However, three bands have filed lawsuits against the pipeline in Federal Court on grounds of environmental and health risks and lack of adequate consultation and accommodation. It was not yet clear at the time of this writing whether these challenges would delay the scheduled September 2017 date to start construction.

In addition, the federal government gave regulatory approval for the Pacific NorthWest liquefied natural gas plant near Prince Rupert, even though the facility would add 4.3 million tonnes of carbon annually, with an additional 6.5–8.7 million tonnes being emitted from natural gas production, collection and transmission infrastructure. In July 2017, however, the developer announced it would not proceed with the project, due to continuing low international prices for natural gas. It is now clear that the federal government approved these oil and gas infrastructure projects to gain support from the governments of Alberta and British Columbia for its 2030 carbon plan and its proposed policy on a national carbon tax.

Meanwhile, the Trump administration, in one of its first decisions, approved the Keystone XL pipeline, which will move Alberta oil to the US Gulf Coast for refining and export overseas, and the Dakota Access line linking oil producers in the northern states also to the Gulf Coast. Construction of Keystone XL, however, will likely be delayed, as the State of Nebraska requires a lengthy, detailed regulatory process and a number of environmental groups in the US plan to initiate legal challenges. The Obama administration had denied approval of Keystone XL on the basis that the nation's ambitious climate targets could not be met if this oil were to be used in the US and overseas.

It is worth noting that all these large-scale infrastructure projects lock in carbon emissions for many decades. It can take up to 40 years to return initial investments of billions of dollars in these projects, and carbon emissions continue to mingle with the atmosphere for a further 40 years, so decisions taken in 2017 can have lasting consequences for almost a century.

The Angus Reid Institute, a national polling firm, tested Canadian public opinion on these pipeline decisions in March 2017. The poll found Canadians were highly critical of Trump's proposal to pull the US out of the Paris Agreement, while just less than half supported the Keystone XL decision.

Yet two-thirds of the respondents encouraged Canada to stay its course in implementing its pan-Canadian framework climate plan. There was more support in Canada for the Keystone XL pipeline than for the two Canadian pipelines – Trans Mountain and the now defunct Northern Gateway – both of which hover near the four in ten level of support.

Not surprisingly, public support for oil pipelines is strongest in the Prairie provinces, with over 75 per cent approval. Accordingly, the Alberta government has crafted a climate plan that has expansion of oil sands development built in but which also puts a price on carbon and sets a limit of 100 million tonnes of emissions a year. The government has given itself some wriggle room to continue developing the resource, as current emissions are around 75 million tonnes per year. The presently depressed price of oil has already forced greater efficiencies on carbon emissions from existing oil sands operations, and if the carbon tax continues, it can be expected to go on driving lower emissions per unit of production.

Ontario and Quebec have also drafted climate action plans to reduce emissions and assist in meeting national targets. But these plans too face challenges from the Trump administration, as they are based in part on continent-wide standards for increased fuel economy and reduced carbon emissions per vehicle, which are now under review by the US government. Both provinces are relying largely on conversion from gasoline-powered cars to electric ones and, as will be noted in the next chapter, increased energy efficiency in buildings. Initially, Ontario had set very stringent standards for the use of natural gas for home heating. However, as a general election was approached, the province had to modify its plans and use some of its cap and trade revenues to subsidize homeowners' gas and electricity prices to shore up its popularity.

British Columbia, for its part, had set ambitious targets for carbon reduction back in 2007: 33 per cent by 2020 and 80 per cent by 2050 from 2007 levels. The province was so confident it could achieve these benchmarks that it enshrined them in legislation. But, in a recent update of BC's climate plan, its independent Climate Leadership Team stated flatly it could not meet its 2020 target. Its 2050 goal depended on full implementation of the proposed federal oil and gas sector methane reduction regulation noted earlier, which is now threatened by developments in the US. Indeed, the province is more likely to increase its current carbon footprint by 2050 if any of the proposed liquefied natural gas projects actually get built, although the newly elected government indicated it would draft a new carbon action plan by the end of 2017.

Similarly, Alberta would not reduce its carbon footprint if the carbon emission cap of 100 million tonnes from oil sands production were to be reached. With Alberta and Ontario representing some 38 and 23 per cent, respectively, of the nation's carbon emissions, it will be impossible for Canada to achieve its 80 per cent reduction target by 2050 given the two provinces' current policies.

There is also political pushback to the more aggressive carbon reduction policies of the federal, Alberta and Ontario governments. All opposition parties in these jurisdictions reject both a carbon tax and increased spending on green infrastructure. As both Alberta and Ontario face provincial elections within two years, there is a high risk that even the current plans will not be fully implemented.

In the US, meanwhile, the Trump government is steadily

dismantling the carefully prepared plans of the Obama administration to meet its Paris target to reduce emissions by 26 to 28 per cent from 2005 levels by 2025. On March 28, 2017, President Trump signed an executive order overturning most of these policies. The core of the previous administration's commitment was the Clean Coal Plan, which would have reduced carbon emissions from the electricity sector by 32 per cent by 2030 by, among other measures, requiring coal-fired plants to switch to natural gas or renewables or store carbon emissions in the ground. The Trump order also removed restrictions on carbon emissions from the natural gas sector and opened up federal lands for coal, oil and gas development. While the intent of the order is supposedly to restore jobs in the US coal industry, the non-partisan US Energy Information Administration noted that coal industry employment had dropped 12 per cent in 2015, while the workforce in the US solar industry had expanded at 12 times the average rate of overall job creation in the same year.

The Trump administration has also ordered a review of proposed fuel efficiency standards for cars and trucks set for 2025 as the government panders to Americans' demand for SUVs and trucks as a result of the reduction in fuel prices. To add to this assault, the federal and some state governments are set to repeal subsidies and incentives for electric cars, even though such vehicles currently represent only 1 per cent of the market. Some states have even legislated new annual registration fees of around $200 on electrics on the basis that such vehicles are not contributing to infrastructure upgrades through gasoline taxes. Not all states have followed this trend, with both California and New York maintaining their electric vehicle incentive programs and regulatory requirements for zero emissions. The US could find itself a laggard in this industry, as China and some European countries saw electric vehicle sales surge by 50–70 per cent in 2016. In the long run, these policies may run counter to the Trump administration's policy of "buy America, hire America" as offshore car makers ramp up production of competitively priced models for burgeoning overseas markets.

At the time of writing – within three months of the inauguration – it is difficult to determine how many of these policy changes will come into effect. Changes to regulations require due process of economic assessment, technical evaluation, public comment and Congressional approvals. In addition, opposition groups can challenge some of these changes in the courts. In any event, all these procedures will eat up time, and in November 2018 there will be a fresh set of elections for the House of Representatives and a third of the Senate. Michael Bloomberg, the former mayor of New York and appointed by the UN as a special envoy for cities and climate change, has issued an "American Pledge" approved by a coalition of 18 states, 187 city mayors, business leaders and universities to retain the original carbon reduction commitment of the US and report directly to the UN in parallel with any formal submission from the president. Almost all other signatories to the Paris Agreement have reaffirmed their respective national commitments despite the withdrawal by the US.

Canada and the US are not the only developed countries facing difficulties in trying to meet the climate objectives set under the Paris Agreement. Australia established an independent Climate Change Authority in 2011 to advise it as to its target under the agreement and a strategy for achieving it.

The government set a goal of 26–28 per cent below 2005 levels of emissions by 2030 and drafted a plan called "Towards a Climate Policy Toolkit" in August 2016. Two of the leading scientists on the authority – Clive Hamilton and David Karoly – issued a minority report in which they indicated it would be impossible for Australia to meet its mid-century goal of a 50 per cent reduction in emissions and that the country was abrogating its global responsibilities. In addition, the authority capitulated to political pressure by dismissing an emissions cap and trade system and supporting an emission intensity scheme that enables increased use of coal for electricity production.

Indeed, one of the dissenters, Clive Hamilton, has expanded his concerns into a new book, *Defiant Earth: The Fate of Humans in the Anthropocene*. He feels the advent of the Anthropocene is of "monumental significance, on a par with the arrival of civilization itself." It will require us to rethink everything, as we have rendered the Earth more unpredictable and less controllable. In short, the Earth itself has become defiant – of us humans. In the final chapter of this book, we will examine ways in which humanity can constructively harness its creative powers to deal with this challenge.

The European Union is another entity that has proposed a carbon reduction plan in line with the UN's long-term goals. But it has recently taken its eyes off the ball with the turmoil created by the UK vote in June 2016 to leave the EU. In addition, pending national elections in Europe could well decide whether the EU remains a viable entity or faces a less certain future. Safe to say that despite the pleadings of the European Union's commissioner for climate action and energy to avoid a vacuum as a result of the loss of the US as a leading proponent for the Paris Agreement, it is more likely the EU will also face political challenges in achieving its stated targets until the fallout from the elections of 2017 have been fully assessed.

This leaves China and India as the other large carbon emitters to pick up the ball. China has pledged to stabilize its emissions by no later than 2030 and to reduce them thereafter. It is arguably the one major emitting nation that has held steadfast to meeting its commitment, as it has already reduced its coal-fired electricity production, due in part to serious air pollution in its major cities and in part to massive investments in solar and wind energy. In addition, China has a unitary government, which enables the country to undertake its planning commitments without facing democratic accountability tests every four years. India too is moving ahead with carbon reduction strategies through sizeable investments in renewable energy and increasing capacity of natural ecosystems to store carbon. For large-scale electricity projects, the cost of solar power is now cheaper than coal. India's latest national energy plan forecasts the share of clean energy will increase by 57 per cent by 2027.

With the withdrawal of the US from this field, China and India could, by default, become global leaders in advancing climate change policy. It is worth noting, however, that China has a record of underreporting its emissions. Indeed, there is growing concern that a number of signatories to the Paris Agreement do not have independent verification of their emissions. It will be fascinating to see how their roles play out as the world undertakes a transformation in energy production over the coming decade.

MID-CENTURY CARBON REDUCTION STRATEGIES

At the UN Climate Conference in Marrakesh in November 2016, Canada, Mexico, Germany and the US published mid-century strategies outlining how they might achieve the UN long-term goals of 50 per cent reduction in carbon emissions and 50 per cent increase in carbon sequestration. These plans are not binding, but they do offer a glimpse of the magnitude and comprehensive nature of the policy transformations and disruptive technologies outlined in the next chapter that will be required to meet these targets. Here are some of the highlights.

All electricity would be produced by renewable sources by 2035, and by 2050 such sources would also have completely replaced the fossil energy used in transportation and buildings. Urban landscapes will be redesigned for low-carbon resilience – increased densities; restoration of wetlands and flood plains; greatly increased public transit and bike infrastructure; car sharing; investment in urban agriculture and in planting forests on public lands and in urban and rural areas. Finally, there would have to be vastly increased financing in innovative clean technologies. Unfortunately, current trends and practices in carbon reduction indicate that actual performance is below stated intentions.

Unfortunately, the working group assessing Canada's capacity for innovation was harshly critical of current policies. Canada's clean-technology exports have dropped to 1.3 per cent of total exports from 2.2 per cent a decade ago. Our innovative patents have decreased by 7 per cent over the last eight years, while global patents increased by a similar percentage over the same time. China and the US have much more aggressive financing and risk-taking profiles that encourage investments in the clean technology required to power the transformation. This is an area where Canada has to smarten up, though it took an initial step in the 2017 federal budget by establishing a new agency to expedite investment in innovation.

However, there is undoubtedly the beginning of a shift in energy production toward renewables as a result of investments in China, the European Union and the United States over the past decade. The World Energy Council took a good look at these trends in a report called *World Energy Scenarios 2016: The Grand Transition*. A parallel analysis, *Perspectives for the Energy Transition: Investment Needs for a Low Carbon Energy System*, was undertaken by the International Energy Agency and its sister think tank the International Renewable Energy Agency, which advise 29 countries on strategies for energy security, economic development, environmental awareness and clean technology.

Both of these agency forecasts were based on market trends and the business models of energy production and distribution rather than on predictions of specific policies undertaken by leading energy-using nations. The reports note that energy-related carbon emissions have stabilized over the past three years, even though the global economy had grown, due to increased renewable power generation, switches from coal to natural gas, and improvements in energy efficiency. But to achieve the UN goal of keeping global temperature rise to less than 2°C, carbon emissions would have to peak before 2020 and fall by more than 70 per cent below current levels by 2050. The share of fossil fuels in total energy demand would have to fall by 50 per cent by 2050. These strategies will require an exceptional and enduring effort far beyond that currently pledged in the Paris Agreement.

The key findings of both groups echo the government report, *Canada's Mid-Century ... Strategy*, namely that by 2050 all electricity would have to be produced from low-carbon and/or renewable sources; seven out of ten cars would have to be electric compared to one in a hundred today; and the nation's entire building stock would need to be retrofitted to emit 80 per cent less carbon than it does today. Fossil fuels would still be needed to provide base load support for intermittent renewable energy sources, for transportation of heavy loads in trucks and aviation and for the petrochemical industry, and would be supported mainly by natural gas. And most importantly, renewable energy sector investments of $3.5-trillion per year – twice the current total – would be required to stimulate this transition. Interestingly, such investment represents only 0.3 per cent of the total gross domestic product forecast for 2050. The strategies and technologies required to support this transformation are explored in the next chapter.

More recently, the Bloomberg Group released its annual long-term economic forecast of the world's power sector, *New Energy Outlook 2017*, which supports the findings of the two international energy agencies. The report forecasts that China and India will lead global investment in renewables and that coal production will be slashed in Europe by 87 per cent and even in the US by 45 per cent by 2040.

It is important to note that these recommendations are being made not by environmental advocacy groups but by the energy policy establishment. Yet both agencies do not appear to believe such forecasts will actually be met, as they would require unprecedented transformation in energy policy. Despite lower costs for renewables, the proportion of total energy based on fossil fuels has been reduced by only

5 percentage points in the last 40 years, from 86 per cent in 1970 to 81 per cent in 2014. In reality, both agencies consider it unlikely that fossil fuel production will peak before 2040, due to such factors as recent global investments in new fossil fuel infrastructure; the current inability to store renewable energy from intermittent wind and solar sources, thus requiring fossil fuel sources to fill the gaps; continued government subsidies for the fossil fuel industry; and the need for oil and gas to supply the petrochemical industry and move trucks and airplanes. Indeed, natural gas consumption is forecast to increase by 50 per cent by 2040.

The key conclusion of both internationally accredited energy groups is that taking early action is critically important if the increase in global temperature is to be less than 2°C. Delaying decarbonization of the energy sector would require even greater investment in renewables over a shorter period, leading to greater likelihood of "stranding" fossil fuel assets, meaning there would no longer be a market for them. In addition, there would be a greater need for even more costly engineering technologies to remove carbon directly from the atmosphere.

Unfortunately, there is currently little indication that such a massive shift in investment is happening. Although the energy market is moving toward renewables, the pace is far too timid to make a real dent in carbon emissions in the near term. In 2015, for the first time, renewable energy accounted for more than 50 per cent of all net new electricity generation globally, led by China, Europe and the US. The cost of wind power has dropped by a third and solar power by 80 per cent over the past five years. In part, the increased proportion of renewables is a reflection of the massive retreat in the price of oil from US$100 per barrel in June 2014 to just over $30 in early 2016 and now around $50. The Canadian oil and gas sector has jettisoned over 40,000 jobs, reduced investment by over $40-billion and lost almost $9-billion before taxes. On the brighter side, the industry has cut costs and in many cases invested in innovative technology to reduce its carbon footprint.

However, oil and gas production costs in the US have declined much more dramatically than in Canada. This has given the industry there much more flexibility to react to price changes, with production increasing from 8.46 million barrels per day in September 2016 to 8.95 million barrels a day in January 2017 due to a slight uptick in world prices. Canada's challenge is that investment in new oil fields has dropped off significantly, and as it takes five to seven years to obtain regulatory approval for large-scale new developments and another five to seven years to reach full production, the more nimble US sector will be more competitive.

However, the two recent pipeline approvals have improved the potential for Alberta oil development and access to markets. The Trans Mountain and Line 3 projects are slated to add two million barrels per day capacity, while Keystone XL will contribute over 800,000 barrels per day. It is therefore no wonder that the Canadian Association of Petroleum Producers forecasts daily oil output to grow from 3.85 million barrels in 2015 to 4.93 million by 2030. Depending on whether the Alberta carbon tax measures succeed in encouraging carbon efficiencies, such increases in oil production will challenge Canada's commitment to reduce emissions by 270 Mt below 2007 levels.

The decline in global natural gas prices has also stalled liquefied natural gas development in British Columbia: only one project, at Woodfibre near Squamish, has announced a decision to start production by 2020. Other countries have already established LNG infrastructure, with the US having begun to ship gas from Houston to Brazil in 2016 and Australia planning to triple production capacity over the next five years. In July 2017, Petronas announced it was cancelling its Pacific NorthWest LNG project at Prince Rupert, confirming that British Columbia is not yet in a position to develop LNG, given this global competition amid slack prices.

It is clear from this review that the countries signing on to the Paris Agreement are generally struggling to meet their stated commitments, in part because voters are not convinced that bold measures such as carbon taxes are yet required and because decisions on carbon-intensive infrastructure are still politically appealing for their economic stimulus and job creation potential, and in part due to the potential for the US administration to undermine responsible regulations for carbon reduction. Even under Obama, annual subsidies for fossil energy in the US in 2015 were estimated by the US Energy Information Agency to be $25-billion, five times greater than the federal commitment to renewables.

It is also clear that consumers are not ready to make the transition. There remains a growing demand for fossil energy, industrial-scale food production and cheap water supplies, which public policy and private investment alike are still bent on meeting. As will be discussed in the final chapter, the transformation of policies guiding the future of the nexus (as discussed in chapter 2) will require a coordinated effort by governments, the private sector and consumers to become more effective in meeting carbon neutrality by mid-century.

The other part of the UN 50–50–50 goal is to increase carbon sequestration by protecting and restoring natural capital.

INCREASING CARBON STORAGE IN NATURAL ECOSYSTEMS

The foundation for the low-carbon, resilience approach is the integration of carbon-reduction measures with restoration of natural capital in order to buffer the impacts of increasingly frequent and intense extreme-weather events. "Natural capital" is the term applied to ecosystems that provide a range of "goods" essentially free of charge, such as water for drinking, irrigation, flood control and recreation. This natural capital also provides "services," such as carbon sequestration, soils, water filtration, drought protection and aesthetic values. Both of these sets of goods and services have been undermined due to poor land use, deforestation, rapid urbanization and the general failure to value natural capital in making decisions on resource development. As a result, less than 25 per cent of global carbon emissions are currently stored in forests and other natural ecosystems, well short of the target of 50 per cent.

An Obama White House report called *United States Mid-Century Strategy for Deep Decarbonization* estimated that the natural capital of the US could store between 30 and 50 per cent of the nation's carbon emissions through increasing the extent of forests on public lands by between 16.2 and 20.2 million hectares; improving forest management; protecting and restoring wetlands; and reducing the risk of forest fires and other natural disturbances.

Another key strategy is to restore carbon in agricultural soils. The UN Food and Agriculture Organisation estimates that if the pace of soil degradation continues, all of the world's topsoil could be gone in 60 years. More than a third of this precious resource has already been lost to industrial agriculture, deforestation and use of heavy equipment. Improved soil management can be achieved through application of agro-ecology principles – planting nitrogen-fixing crops; protecting soil by limiting tillage; precise application of fertilizer to reduce waste; and development of heat-tolerant crops. Such measures could increase carbon sequestration by a further 10 per cent. Scientists at World Agroforestry Centres in China, Kenya and Indonesia and at the Royal Botanic Garden in Edinburgh, Scotland, now estimate that agricultural lands properly managed under agro-ecology principles can hold four times the amount of carbon than originally thought by the IPCC.

An independent group of experts undertook a detailed analysis on how current industrial agriculture might transition to agro-ecology in a 2016 report titled *From Uniformity to Diversity: A Paradigm Shift from Industrial Agriculture to Diversified Agroecological Systems.* The shift includes significant additional merits such as improving crop nutrition, with resulting benefits for public health; soils management and increased carbon sequestration; reduced water applications and more efficient energy use; and improvements in food security by focusing on regional production rather than food imports. An additional policy area affected includes changes in agricultural subsidies. Globally, countries spend US$560-billion a year on supporting excessive fertilizer use, resulting in pollution of streams and lakes and encouraging massive drainage schemes that increase flooding of adjacent lands. In developing countries, women make up the bulk of small farmers, yet they currently have no access to financing. Improvements in real-time monitoring and better science will build capacity for new entrants to agro-ecology to help achieve these transformations.

Another key strategy is a switch to vegetarian diets. Growing a kilogram of wheat takes about 1260 litres of water, while the same weight of beef consumes 12 times that amount. Put another way, a kilo of beef is equivalent to 24 kilos of wheat in terms of total costs, not only for water but also for energy, fertilizer, greenhouse gases emitted and croplands used. Globally, carbon-equivalent emissions from livestock – much of which is the powerful greenhouse gas methane – exceed total emissions from the entire transport sector.

China is the only country that has formally committed to reducing per capita meat consumption by 50 per cent by 2030. If this target is achieved, it could reduce China's carbon-equivalent emissions by a billion tonnes per year, which is more than the entire emissions from Canada. But unless other countries follow suit, emissions from the agricultural sector globally could exceed half of the total carbon budget set by the IPCC to keep global average temperatures from increasing by any more than 2°C. Other, similar policies would have enormous benefits. For example, simply adhering to health guidelines for avoiding obesity and diabetes could cut global food-related emissions by one-third by 2050, and widespread adoption of a vegetarian diet would reduce them by a further third. Crops currently devoted to fattening livestock could feed more than four billion people.

Avoiding food waste is another area for policy transformation. Currently, about 20 to 40 per cent of all food produced never reaches anyone's table, to say nothing of all the water and energy it took to produce it. The estimated direct cost of food waste in Canada alone is $31-billion per year, but when related costs of labour, water, energy, land and logistics are included, the total outlay escalates to $102-billion per year. The problem is systemic, both in consumer attitudes and in the vertical structure of food retailing and distribution systems, which allows consumers to pick off deals at the retail end but in turn creates waste further up the supply chain. A more integrated model for the whole agricultural sector, with waste elimination as an additional goal, is required.

There are recent indications that the carbon storage potential of terrestrial biomass – forests, wetlands, soils and vegetation – is increasing due to less deforestation of tropical regions, expansion of temperate and boreal forests in Russia and China and naturally increasing rainfall in savannah regions. Some countries are introducing policies to increase capacity for carbon storage and restore ecosystem health. Europe has converted over 600,000 hectares of marginal agricultural lands back to forests through repayment schemes. In China the "Great Green Wall" program is designed to replant trees across 400 million hectares in the northern part of the country. Brazil has reduced by 80 per cent the loss of its tropical forests to agriculture through satellite monitoring, stronger enforcement and restricting rural credit for soy production.

WATER AND BIODIVERSITY

As discussed in chapter 2, water resources lie at the heart of the "climate nexus." Water for food production consumes over 70 per cent of total global demand. Power generation is also a thirsty business. Over 40 per cent of water withdrawals in the US are used for cooling in thermal power stations.

The World Economic Forum has said that water insecurity is now one of the major risk factors to the global economy. The World Resources Institute identified 33 out of 167 countries facing increased water insecurity by 2040 as a result of the changing climate and increasing water demand for food and energy production. Crop yields in parts of Africa could drop by 20 per cent due to both water stress and extreme heat.

Water management, like agriculture and energy, must address a suite of transformative policies if the climate nexus is to survive the coming decades. Some of these changes include universal pricing and tradable water rights; groundwater regulations; protection of aquatic ecosystems; reuse of treated waste water; and the adoption of precision agriculture in order to reduce the quantities of resources used for irrigation, pesticides and fertilizers and thereby restore water quantity and quality. Some current best practices for undertaking such strategies are covered in the next section. However, such examples would have to become universally applied if the water ethic outlined in the previous chapter is to come about.

Australia has developed the most advanced policies in water pricing and tradable water rights. The state of South Australia, for example, currently charges A$2.32 per cubic metre (plus a flat service fee of nearly A$300 per year) for average residential users, whereas water-short South Africa still provides water for free.

Governance systems will also have to change. In most of water-scarce western North America, the resource is allocated permanently on a first in time, first in right basis. In law, farmers with older licences, even if they use the water very inefficiently, can force out more junior licensees, even ones that use more efficient technology. Though this can become particularly problematic in times of drought, in practice some form of common sense usually prevails so that the water is shared. Australia has experimented with tradable water rights, where a base requirement is protected but additional amounts can be auctioned for the highest and best use, encouraging conservation. Expansion of this policy to other countries is unlikely, due to complicated existing rights and to concerns that trading water smacks of privatization that will lead to price gouging, a no-go policy in many countries.

Groundwater is considered to be the safety net of watershed management. In banking terms, groundwater is the capital asset, while surface water is the daily interest and transactions. Yet groundwater use is on the increase due to depleting surface water supplies. In India groundwater consumption has increased from 50 billion cubic metres annually in 1970 to 250 billion cubic metres in 2010. Not many countries have a well-managed groundwater system. California introduced groundwater legislation in 2014 as a direct result of its prolonged drought. The legislation requires water utilities to develop groundwater sustainability plans in higher risk areas by 2022. Such plans will enable utilities to restrict water extraction to the sustainable yield of the aquifer. It is worth noting that although the heavy snow and rainfall over the winter of 2016 in California eliminated the drought in surface waters, it did not eliminate the drawdown in groundwater. Scientists estimate that it would take five more years of heavy winter rains to fully recharge the depleted aquifers.

British Columbia too introduced groundwater regulations, under its Water Sustainability Act, in 2016. These regulations apply only to larger water users and not to domestic use and come into force in 2019 once the aquifers have been mapped. With over 25 per cent of total water use in the province based on groundwater sources, it was high time the resource was regulated. Experience from California and British Columbia could be invaluable in transferring innovative practices in groundwater management to other countries, such as India.

Protecting flows in streams for sustaining aquatic ecosystem health is another cornerstone of a new water ethic. Yet few countries have developed legislation and regulations to ensure that such flows are protected. British Columbia's Water Sustainability Act, however, does require that government decision makers on new water licence applications take into account environmental flow needs in deciding whether to allow withdrawal from streams. The regulations required for implementing this provision are currently being drafted and will likely include a sliding scale whereby rivers with large flows will receive only limited scrutiny, while watersheds where flows are limited will require more detailed analysis. One challenge will be to continuously adjust the low-flow requirements to accommodate the non-stationarity of hydrologic systems noted in the opening chapters. Again, British Columbian and Californian experience could prove invaluable to other provinces and countries.

The underlying tenet of the proposed new water ethic is to maintain water security for a whole range of uses. The fragility of water security was starkly stated by Cornell University scientist Toby Ault and colleagues in a recent paper, "Relative impacts of mitigation, temperature, and precipitation on 21st-century megadrought risk in the American Southwest." Mega-droughts are defined as those that last for more than 35 years. The researchers conclude that if the globe warms by more than 3.5°C, the risk of such droughts will increase by as much as 90 per cent by the end of the century. To indicate the magnitude of such events, the recent drought in California, which lasted only six years, killed over 66 million trees together with their concomitant ability to store carbon. Such megadroughts would stimulate a complete transformation of water management. Currently, for example, farmers continue to grow almonds for export, at the expense of domestic food crops, even though almonds consume four litres of water per nut. Further, cities can appropriate water from agricultural allocations in times of severe drought. In the event of megadroughts, the entire system of water infrastructure, water allocation, groundwater management and financing would have to be completely restructured.

A more integrated approach to reducing flood risk throughout watersheds is beginning to be applied in various countries. In the UK, the winter of 2015 – 2016 was the wettest on record. One storm alone resulted in more than £5-billion in damages. The problem was compounded by poor land use practices whereby farmers were encouraged to remove water from their lands through subsidized drainage systems, thereby increasing flooding of adjacent lowlands. A House of Commons committee established to review the flooding recommended a holistic catchment approach where benefits for flood control, improved water quality, habitat for wildlife and water conservation could be combined. There was an emphasis on protecting natural capital assets such as wetlands and swales to "slow the flow," rather than relying solely on engineered systems. So extensive were the committee's recommendations that they supported a wholesale change in governance structure to implement integrated watershed management.

The Department for Environment, Food & Rural Affairs is tasked to report back to the Commons committee with an action plan in 2017.

In summary, the plans proposed by Western developed countries to contribute substantially to the UN's 50–50–50 targets face implementation challenges due to a combination of policy differences with the Trump administration, practical politics requiring short-term investment in fossil fuel infrastructure in advance of provincial elections, and lack of financing capacity for transitioning to clean technology. However, as mentioned in chapter 1, Canada still has the opportunity to demonstrate to the world that with a well-informed population, sound and functioning democratic institutions, a commitment to evidence-based policy development, and independent research capacity in its universities, it can lead the way through these potential obstacles and create a more transformative approach than many of the other developed countries.

FOUR

The New Economy and Disruptive Technologies

Although the transition of the global economy to achieve the 50–50–50 goals of the UN Paris Agreement is too slow and timid, there are some encouraging signs of progress. Renewables are providing for a larger proportion of total energy use than ever before. Various jurisdictions have enacted some encouraging new legislation governing groundwater use and retaining ecological flows in streams. More and more farmers are recognizing the value of restoring natural capital on their lands, and the degradation of terrestrial biomass has been slowed. However, these shifts are still insufficient to reduce total global carbon emissions or significantly increase carbon storage in the Earth's ecosystems. There needs to be more-fundamental change, which in the past has been brought about by new economic conditions resulting from the advent of "disruptive technologies" – those that transform whole markets by making existing technologies or products obsolete.

Throughout history there are many examples of disruptive technologies. In the late 18th century and first half of the 19th century, whale oil was universally used for lubrication, heating and soap making and in processing textiles and rope. But its major value was as a fuel for lighting. By the mid-1850s, sperm whale oil was used in lighthouses, street lamps and public buildings. Over 236,000 whales were killed in the early 19th century alone, not only threatening the survival of that species but also increasing oil prices, as ships had to travel ever farther afield to find viable whale populations. By 1846 the industry reached peak sperm whale oil production.

The main reason for decline thereafter was not to save the whales but because a new, disruptive technology was brought onto the market. In 1846 a Canadian geologist named Abraham Gesner figured out a way to distill kerosene from petroleum. Kerosene was cheaper than whale oil, could be stored and did not produce an offensive odour. Within a decade there were 40 kerosene plants in the US, resulting in the Rockefellers establishing Standard Oil. The petroleum age was launched. By 1876, the once mighty whaling fleet had shrunk from 736 ships to 39. The price of sperm oil dropped from $1.77 to 40 cents, while the commercialization of kerosene dropped the price from 59 cents a gallon in 1865 to 7 cents in 1895.

The impact of disruptive technologies has created an economic theory based on "long waves" and explored by Paul Mason in his book *Post Capitalism: A Guide to Our Future.* Long waves last over 50 years and usually start with the market storing capital while it looks for new investments. Once a disruptive technology is invented, some of this pent-up capital is used to expand the technology, drive down prices and stimulate market demand. Markets grow and real wages increase, driving up consumption until the cycle peaks. At some point economic growth begins to slow, innovation dwindles and capital moves back into financial markets, waiting for a new wave of disruptive technology to spark another long wave.

Mason describes how industrial capitalism has experienced four such long-wave cycles over the last couple of centuries. In the first half of the 19th century, the steam engine, canals for long-distance freight transportation, textile machinery and agricultural tillage led innovation. The social and military revolutions of the mid-19th century brought an end to this cycle. The second long wave started in the second half of the 19th century with the universal development of railways, the telegraph, ocean-going steamers and advanced machinery. The third long wave occurred in the first half of the 20th century with the automobile, the telephone and the application of scientific management principles to stimulate mass production. Finally, the fourth long wave was ushered in following the Second World War with the advent of transistors, synthetic materials, mass consumer goods and the beginning of the information age with the development of early computers.

We are now entering a fifth long-wave economic cycle based on personal computer technology, artificial intelligence, application of quantum mechanics to commercial production and smart technology. Jeremy Rifkin, in his book *The Zero Marginal Cost Society: The Internet of Things, The Collaborative Commons and the Eclipse of Capitalism*, envisions a changing economic structure that has to adjust to decreasing marginal cost of production as the full effects of the information age take hold. Free access to information is changing the market's ability to set prices. This is because markets base their pricing models on scarcity, whereas information is abundant. Not only are costs of information technology tumbling because of Moore's Law, which states that product costs fall by half every two years because of improvements in technology. But it is now possible to replicate information-based systems for very low costs, as some entrepreneurs are loading their ideas on open-source Internet systems. For example, one of the conditions for Canada's funding a pilot carbon capture and storage project in Alberta was that the proponent, Shell Oil, share its blueprints, data and other documents publicly to reduce future costs of undertaking carbon capture and storage technology around the world.

Rifkin's thesis about zero marginal cost is based on information technology that is self-replicating and capable of creating its own upgrades so it can last forever. He forecasts that these systems will be powered by solar electricity, which in turn will be provided at close to zero marginal cost once solar technology, as outlined in the next section, has taken hold. The intelligent machines will be linked into an "Internet of Things," a self-organizing network of energy systems, logistics and communication that will enable transformation in sectors such as transportation, urban design, energy systems and food production. The Internet of Things is based on sensors that can produce billions of pieces of real-time information. Between 2007 and 2014, the number of sensors increased from ten million to ten billion, while unit costs for data storage decreased a thousandfold. Such smart technology could, for example, reduce truck transportation costs by 50 per cent without affecting total loads. Other spinoffs from the Internet of Things include online education courses at close to zero cost, thus reshaping accessibility to lifelong learning, an essential ingredient for surviving in a rapidly changing work environment. New 3D printing technology enables replication of a wide range of products at low marginal cost once the original program has been developed. In India a fully operational electric car has been created on a 3D printer and can now be reproduced for $10,000.

Another outcome of the Internet of Things is the "sharing economy," demonstrated by the Uber ride-hailing service and Airbnb for travel accommodation. Like them or not, these and other sharing systems have already disrupted the traditional capital model. Rifkin believes it will take only between 10 per cent and 30 per cent of a particular market shift to these self-help networks before similar disruptions occur in other sectors of the economy.

The sharing economy is also encouraging a resurgence of co-operatives. In the US there are already over 29,000 co-ops, with over two million workers and $500-billion in annual revenue. In the agricultural sector, over 30 per cent of farm products are marketed by co-operatives. The Rural Electric Cooperative Association in the US now generates 25 per cent of its electricity from renewable sources. In Germany there are 167 green co-operatives delivering 23 per cent of total renewable energy production. Most of the co-ops are located in cities and are beginning to branch out to provide a wider range of services.

Rifkin's third major prediction is that an "integrated horizontal network" based on distributed renewable energy will replace the existing vertically integrated energy system based on fossil fuels. It will be run by what he terms "customer–producers," people who both produce and use their own power, all networked together through a smart grid that routes power to where it is needed. Already the reduced cost of solar power technology installed on rooftops has created a class of customer–producers through two-way net metering, where surplus energy produced by a household is returned to the utility at a set price. Through the Internet of Things, these customer–producers can be networked together through a smart grid and create their own market.

One such network, under development in New York, is called the Brooklyn Microgrid. Dozens of solar panels are networked to enable producers to sell excess electricity credits to buyers in the network. The project is in its early stages as, at the time of this writing, there are only 50 participants. But the application of secure financial software will increase the reliability of transactions and eventually encourage participants to bypass the electric utility entirely. It is a classic example of the Internet of Things, where real-time monitoring allows participants to identify each other's needs and willingness to buy and sell. Already in development is a smart phone app that would enable customer–producers to manage their electricity generation and state how much they are willing to pay for purchases. Similar experiments in networked solar power systems are underway in Australia, Bangladesh and Germany.

DISRUPTIVE TECHNOLOGIES IN ENERGY AND TRANSPORTATION

Battery storage is the key to increasing the reliability of inherently intermittent supplies of wind and solar power. Experts believe that if homeowners can store electric power for up to four hours a day, it will give them flexibility to use renewable energy all day and also save money by using electricity in off-peak hours. Using smart phone technology and battery storage, consumers can save up to 50 per cent on their electricity bills by avoiding price spikes. Costs of batteries have been diminishing by 14 per cent per year recently. Tesla is building a $5-billion factory in Nevada to produce batteries for cars and homes, which could reduce the cost of battery storage by a further 30 to 50 per cent. By the early 2020s, battery storage should be affordable for most households to become part of the disruptive technologies supporting electricity generation from renewable sources.

Solar power costs have decreased more than 200-fold since 1970. Solar is now competitive with natural gas cogeneration plants at 5 cents per kilowatt-hour. Wind turbines have experienced similar price decreases such that, in 2016, more than 50 per cent of the incremental demand for electricity was met from renewable sources. As noted above, with the advent of networked solar panel systems, entrepreneurs are experimenting with various business models to gain market traction. In California and Nevada, solar companies are offering customers free installation of solar panels, financed through monthly charges on electricity bills, which are lower than for fossil-based power even with this amortization of the capital costs of the solar panels. The tipping point for solar power generation occurs when unsubsidized rooftop solar becomes cheaper than the basic costs of transmission over a long-distance grid, known as grid parity. In Australia such grid parity is already happening and is expected to occur in other countries during the 2020s.

A number of US jurisdictions are trying quite different approaches to encouraging the adoption of solar technology.

California installs solar panels for free to low-income families by using funds from its cap and trade program. It plans to invest $300-million by 2020. Colorado and New York are planning similar programs. In July 2015, the US government launched a "solar access for all" program to install 300 megawatts of renewable energy in federally subsidized housing, plus providing a tool kit to facilitate installation of solar panels. This is a multi-departmental initiative with support from the private sector.

In Nevada, when the MGM conglomerate installed solar technology on all its buildings, it affected 7 per cent of the total market for the major power utility in the state, NV Energy. The utility was forced to decrease the net metering prices it paid its other solar customers for surplus energy, thus creating a backlash. Eventually, the governor stepped in and brokered an agreement whereby solar customers with panels installed prior to December 2015 would be granted the original net metering rates. MGM paid an exit fee of $87-million to cancel its contract with NV Energy. But, in November 2016, Nevada voters overwhelmingly approved a ballot measure aimed at breaking up the NV Energy monopoly and enabling competition. Such a move will require a constitutional amendment through another vote in 2018, and, if approved, the state would deregulate its electricity market by 2023. Meanwhile, MGM was left free to innovate with solar power and has since expanded its capacity. The main solar energy providers – SolarCity and Sunrun – are gearing up to develop new technologies for the new renewable energy regulatory regime.

Traditional power utilities have invested billions of dollars in regional grids associated with large-scale power plants and need to recoup their investment by locking consumers into multi-year contracts. Meanwhile, renewable energy is becoming available at lower costs. This is a classic response when disruptive technologies take over and traditional business models of public utilities are threatened. Nevada is a case study of how utilities, regulatory bodies and governments will have to work out a new model to enable them to coexist with renewable energy.

Electric vehicles will be another disruptive technology. The energy efficiency factor for such cars is 90 per cent, compared with 20 per cent for gasoline-powered engines. EVs also have lower maintenance costs, as they are made with only 6,000 to 7,000 parts, compared with over 30,000 in cars with internal combustion engines. Fuel expenses are ten times lower than gasoline and diesel. Once EVs attain a range of at least 320 km, and their price becomes competitive with internal-combustion vehicles, consumers will take full advantage of these savings. Most of the large manufacturers are already investing in electric vehicle technology. In January 2017, Ford announced plans to put $4.5-billion into EV production by 2020 involving 13 new models. One of the planned cars is an SUV with an estimated range of 480 km. The company also plans an autonomous vehicle designed for commercial ride hailing in North America by 2021. Similarly, Daimler announced an $11-billion program for ten electric vehicle models. Volvo has committed to phasing out combustion engines in new models after 2019. France and the UK have announced policies that would phase out all combustion engine cars by 2040. In March 2017, the Canadian government announced a $7-million agreement with a Quebec firm to install 1,000 new charging stations across Canada and develop the next generation of faster-charging facilities. The program is part of a $62-million commitment in the 2016 federal budget to develop infrastructure for providing alternative transportation fuels such as natural gas and hydrogen, as well as charging stations for EVs.

In July 2017, Tesla released its Model 3 sedan, with a range of over 320 km per charge and a price of US$35,000. As such, it is directly competitive with luxury German cars such as BMW and Mercedes. But the latter models are diesel-powered, and after the scandal of Volkswagen tampering with emission controls, the public have lost faith in diesel. The German car industry is facing a highly disruptive few years as it switches from diesel to EV.

China too is investing huge amounts in electric cars. Between 2013 and 2015, the government spent $9.5-billion in subsidies to establish the industry. It also provides non-financial incentives such as open access to licence plates, which are heavily restricted for gasoline-powered cars. Two hundred Chinese companies are now designing thousands of models of electric cars, of which more than a thousand have entered the market. One company is building supermarkets for electric cars, each location selling multiple brands. China has also developed a smart phone app that enables drivers to not only locate the nearest plug-in but also pay for the recharge. China already has 49,000 charging stations (the US has 32,000), with the prospect of almost five million by 2020.

An umbrella group of 16 countries using electric vehicles has created the Electric Vehicles Initiative, with a pledge to get 20 million vehicles on the road by 2020. However, as noted in chapter 3, the International Energy Agency has forecast that to meet the deep decarbonization targets by mid-century, there will need to be some 700 million electric vehicles on the road by 2040.

The key to successful penetration of the electric car is improved battery performance. The Argonne National Laboratory in the US is working on the next generation of storage beyond the lithium-ion battery: the lithium-sulphur battery. The aim is to develop a battery priced at under $100 per kWh, which would lower the cost of an electric vehicle to less than $20,000. With the currently depressed price of oil, battery technology has to become even more efficient to compete. There remains a concern that the availability of lithium worldwide will constrain battery production, so this is another area on which scientists are now focusing their attention.

Conventional car companies are also rapidly investing in self-driving cars. So are governments: Ontario began testing its first self-driving cars in November 2016. Technology costs are decreasing exponentially. For example, the technology that senses the car's surroundings in order to avoid collisions has dropped from $150,000 per vehicle in 2012 to $250 in 2016 and is expected to sell for $90 by 2018. Cars have long since become computers on wheels. However, the interest by consumers to have more time to undertake other tasks while driving has to be tempered with the perceived greater risk to public safety.

The technology disruption from the combination of self-driving cars and ride hailing is making an automobile more of a service than a capital asset to be owned. Car sharing is also on a growth curve. In San Francisco, 15 per cent of Uber rides are already shared rides due to reduced costs. And since a personal vehicle is parked about 96 per cent of the time anyway, demand for parking will likely plummet as ride hailing, car sharing and self-driving vehicles become mainstream. This will create enormous opportunities for greening cities by using vacant parking space for net-zero-energy buildings, green space and urban agriculture.

The combination of improved batteries for residential energy storage and electric vehicles, along with self-driving technology, ride hailing and car sharing, is an example of the horizontally integrated networking model, which is rapidly replacing the vertically integrated model of traditional automobile corporations, urban transportation arrangements and electric utilities. The impetus for most of these potentially disruptive technologies is coming from individual entrepreneurs with access to private venture capital, with some support from governments in the form of start-up subsidies.

Follow the money. A real sign of progress in renewable technologies occurs when investment shifts in their favour. And there is mounting evidence that equity investments in companies that have high environmental, social and governance (ESG) standards are outperforming those that don't, by a significant margin. The MSCI Emerging Markets ESG Leaders Index, which includes over 400 companies that score highly on ESG, was showing a record outperformance gap over traditional companies in 24 emerging-market countries as of August 2017.

But the push for the transformative policies and technologies necessary to achieve the 50–50–50 goal cannot come from government alone, or from individuals and the private sector alone. It must be a combination of all these forces, which will be the focus of the final chapter of this book.

RESTORING NATURAL CAPITAL

Exciting new policies and technologies are also happening in the protection of natural capital to increase carbon capacity – and in restoring it to health wherever it is impaired. Innovators are turning to smart technology and the Internet of Things to reverse the degradation of topsoil, reduce water pollution from excessive use of pesticides and fertilizers, and switch to agro-ecology-based farming practices. There is a growing use of robots to plant and harvest food more efficiently. Financial firm Goldman Sachs predicts investment in robotics in the agricultural sector will have increased from its 2013 total of $1-billion to $240-billion by 2022. The main areas of innovation include self-driving tractors; precision agriculture using smart digital technology to control the application of water, fertilizers, pesticides and seed; and testing and growing increasingly drought-resilient seed strains. These techniques not only reduce costs and waste but also result in less pollution of both fresh and marine waters.

Technology can also support development of agricultural production in urban areas, closer to consumers. Japan is experimenting with an automated vegetable factory in Kyoto that can deliver 30,000 heads of lettuce per day. Production is stacked, as land is scarce, and after seeding, everything is done by machines, from watering to trimming to harvesting. LED lighting is now being designed to reduce energy consumption, and 98 per cent of the water used will be recycled. Transportation between producer and consumer is also greatly reduced, again eliminating carbon emissions and costs. As parking lots are gradually freed up by the advent of self-driving and car sharing, some could be converted to this new wave of urban farming technology.

Drone aircraft provide yet another dimension to automation, by monitoring crop growth, spotting early the onset of disease and water stress, even spraying crops with pesticides and herbicides. Improved technology for geopositioning enables drones to operate in remote rural areas. At one experimental farm in central England, the entire food growth cycle is automated, with no direct farm labour required. These trends will greatly improve the health of agriculture-based ecosystems but will challenge the future of farm workers. Over the past several decades, the agricultural workforce has declined from 81 per cent of total employment to 48 per cent in developing countries and from 35 per cent to 4 per cent in developed countries, at great cost to rural communities. The shift to robotics will likely accelerate this trend.

But it will take more than just a technological fix to transform agriculture from the currently dominant industrial model, which creates over 20 per cent of the global carbon emissions, to an agro-ecological model with its potential to restore carbon storage in agricultural soils. For example, US-based Tyson Foods, which supplies McDonald's and Walmart, slaughters 35 million chickens and 125,000 head of cattle every week, requiring more than two million hectares of corn a year for feed. There will have to be a wholesale shift in a wide area of policies from financing, research into crop types more resilient to extreme weather events, a new ethic on water use and conservation, and education of a new generation of farmers in agro-ecology techniques, together with the application of smart technology.

There already exist a few glimpses of what this future might look like. India's National Agroforestry Policy, as part of its plan to meet its commitments in the Paris Agreement, includes planting trees and forest cover on marginal agricultural lands to capture up to three billion tonnes of carbon annually. One example indicates such a policy could result in economic, social and ecological benefits. Over the past 19 years, Ramu Gaviti's 2.4 hectares in western India has been transformed from a marginal farm into a thriving biosystem with 1,000 fruit and nut trees and an income of US$1,200 a year. The agroforestry project was pioneered by a non-profit agency that supports climate-resilient agriculture. The initiative will have multiple benefits, as 50 per cent of India's land is degraded and 86 per cent of the degraded land is agricultural. Payment for ecological services by supporting farmers to grow trees and earn an income could become a major innovative policy instrument.

The United Nations has calculated that if all human waste were collected for bio-gas generation it could supply enough renewable electricity for 138 million households. But there remains a cultural stigma to recovering resources from human waste, which has held back its technological development. There are some signs that new technology will change this. Northumbrian Water, a private utility in northern England, has built two anaerobic digesters since 1996, saving $25-million a year in energy costs. In the US, Janicki Bioenergy of Sedro-Woolley, Washington, has pioneered what essentially is a small power plant that converts wet sludge into drinking water and electricity. The sludge, which comes from a sewage treatment plant, is dried and the resultant steam is used to generate electricity. The steam can also be condensed and reused as drinking water. With funding from the Bill & Melinda Gates Foundation, the firm has piloted a project in Dakar, Senegal, that treats the waste of 50,000 to 100,000 people. A second pilot was scheduled for the spring of 2017 and additional plants will soon be considered for US cities. The optimal size is for communities of 200,000, with drinking water produced for up to 35,000 people. Such technology, when fully developed, could become an essential ingredient for a low-carbon, resilient future in developing countries.

Digestion technology has been further refined in Aarhus, Denmark, which has become the first city in the world to provide its 200,000 residents with fresh drinking water using energy created from household waste water and sewage. According to the International Energy Agency's *World Energy Outlook 2016*, the amount of energy used in the water sector could more than double over the next 25 years due to expansion of urban populations, especially in developing countries. Application of the technologies used in Denmark, Senegal and the UK will greatly assist in managing this challenge.

DEVISING A NEW METRIC FOR HUMAN WELL-BEING

In 1936 British economist John Maynard Keynes wrote in his magnum opus, *The General Theory of Employment, Interest and Money*, that "most ... of our decisions to do something positive ... can only be taken as the result of animal spirits – a spontaneous urge to action rather than inaction and not as the outcome of a weighted average of quantitative benefits multiplied by quantitative probabilities." Our focus in the present book is that the global consequences of the Anthropocene, as sketched out in chapter 2, will become ever more severe as world responses to the hard problems of reducing carbon, restoring natural capital and increasing water security fall farther behind the pace necessary for meeting the 50–50–50 goals of the UN. It will require the unprecedented external pressure of the Anthropocene to stir the "animal spirits" that will urge us on to muster the technological ingenuity to make the required wholesale transition. One essential ingredient of this transformation is the creation of a different measure for human well-being than the current metric of gross domestic product.

Since the end of the Second World War, GDP (initially called GNP, "gross national product") has been the principal indicator for assessing progress in nations' well-being. But GDP was never designed to measure a broader base for *human* well-being, which would include ecological and social values. The flaws of the measure are well documented, especially perverse aspects whereby the costs of cleaning up and rebuilding damaged infrastructure after a major extreme weather event are actually considered to be adding to gross domestic product rather than as a sign of diminished well-being. In transforming the globe to a low-carbon economy with healthy ecosystems, there has to be a more comprehensive replacement for gross domestic product to measure progress in the transition.

There have been numerous attempts to devise a broader index of human well-being. One such was a commission set up in 2008 by the president of France at the time, Nicolas Sarkozy. Chaired by economist Joseph Stiglitz, the commission provided sound advice on how to broaden measures of economic performance to include consumption, wealth and inequality, though it failed to include measures of natural or social capital. At the 2016 meeting of the World Economic Forum, that organization's chief economist posed the question whether "we are living at the expense of tomorrow." The International Institute for Sustainable Development, located in Winnipeg, took up the challenge to broaden the measure of well-being from income to "comprehensive wealth."

Comprehensive wealth includes measures of natural capital (land, forests, waters, fauna, minerals and fossil fuels); human capital (skills and capability of the workforce); and social capital (trust, engagement in civic duties, shared norms). Indeed, many of these measures are contained in the UN Sustainable Development Goals. More consumption today at the expense of consumption tomorrow is unsustainable, yet this is the consequence of the continued decisions to support fossil fuel infrastructure in Canada and elsewhere. What is required is to understand how comprehensive wealth evolves over time, not just a measure of growth in gross domestic product. The International Institute for Sustainable Development set out to undertake such an approach for Canada.

To be clear, comprehensive wealth complements gross domestic product measures; it does not replace them. It deals with the *resourcefulness* of a nation, not just its resources; it tracks future trends, not just past performance. The IISD study, *Comprehensive Wealth in Canada: Measuring What Matters in the Long Run*, analyzed Canada's performance in this regard for the period of 1980–2013. They found that, overall, comprehensive wealth grew by only 7 per cent per person over those 33 years, or at a rate of only 0.19 per cent per annum. Though consumption of economic goods and services grew by 54 per cent, natural capital was drawn down by 25 per cent. Human capital did not grow at all, despite improved levels of education, meaning that in real terms the average Canadian worker had the same lifetime earning potential in 2013 that they had in 1980. Similarly, social capital was flat over the period.

For the first time, the institute ascribed non-monetary values to a number of indicators for natural capital: forest health, wetlands area, surface water, grasslands, glacier mass and indices of climate change. All these measures were trending downward, indicating that, from a comprehensive wealth perspective, Canada is not operating sustainably. If carbon storage is to be increased by 50 per cent, effectively eliminating net emissions from lands and natural ecosystems by mid-century, the value of such natural assets has to be measured and brought directly into all decisions on infrastructure.

Getting natural assets onto the balance sheet

Recently, five municipalities across Canada have been participating as pilots in what's called the Municipal Natural Assets Initiative. The five participants will investigate the extent to which natural assets – wetlands, rivers, lakes, aquifers, forests and soils – deliver core municipal services such as stormwater management, drinking water protection and flood control; quantify what the costs would be if these services had to be procured by means of an engineered alternative instead; and develop costed operational plans to continue delivering core municipal services. Such operational plans may require actions ranging from preventive maintenance and monitoring, to restoration and various management strategies.

As the climate changes, municipalities will be faced with increasing pressure to invest in infrastructure to provide services of all kinds, including, for example, water management. Traditionally, these functions have been provided by engineered systems – dams, dikes, pipes and pumps. And though natural assets can also provide these services, such assets have not been valued in accordance with generally accepted accounting principles, nor deliberately managed to provide service in a reliable manner, nor included in core financial and asset management systems. The objective of the Municipal Natural Assets Initiative is to provide support and tools to enable local governments to include natural assets – for example, forests, wetlands and creeks – into the overall suite of a city's financial and operational systems and thereby maintain and enhance the services it provides.

The five pilot participants in this initiative, located in British Columbia and Ontario, are focusing initially on stormwater management services, to test evaluation techniques and adapt traditional accounting practices to include ecosystem-based municipal services. The initiative plans to undertake another five pilots starting in 2017–2018, possibly clustered geographically so as to understand the potential power of several municipalities combining natural-asset management into a single watershed jurisdiction. The initiative's project team also plans to expand the methodology in 2018 to consider other services beyond stormwater management. The goal is to establish a substantial enough body of knowledge, as well as strong enabling conditions and methodologies, that the measurement and management of natural assets becomes a mainstream practice for municipalities across Canada. In a related move, the federal government and several provincial and territorial governments have published the *Ecosystem Services Toolkit* as a practical, step-by-step guide for all levels of government to incorporate ecosystem services management into a range of functions, such as planning, environmental assessment and management of watersheds and recreation areas.

Chartered Professional Accountants Canada, which oversees the role and function of all professional accountants practising in Canada, is also considering how ecosystem services valuation can be formally incorporated into accounting practices. Accordingly, the accountants organization has partnered with the Natural Step Canada to create the Natural Capital Lab to undertake a range of analyses to meet this objective. The lab convened groups of practitioners and experts in valuing natural capital in November 2016 as the first step in developing a new metric for formally including natural capital asset values in professional accounting practice across Canada. The Municipal Natural Assets Initiative project team has collaborated closely in the work of the lab.

One of the projects supported by the Natural Capital Lab is to work with Statistics Canada to include values for natural capital in Canadian national balance sheets. StatsCan and Environment and Climate Change Canada have developed an experimental ecosystem accounting approach to assess ecosystem goods and services using monetary, non-monetary and non-market values. The approach aligns with best international practices and has initially been developed for land uses including boreal forest, biomass extraction and marine coastal and wetland ecosystems. Ontario has provisionally indicated it is prepared to be an innovator for developing a full chart of accounts for its ecosystem values under the direction of the lab. An initial assessment of land accounts and forests has been completed.

The shadow price of carbon

Michael Mann and his colleagues have demonstrated scientifically that greater amounts of carbon in the atmosphere have resulted in increasing frequency of extreme weather events such as flooding, windstorms, wildfires and droughts and their associated costs. These costs provide another indicator of loss of human well-being, known as the "shadow price of carbon."

This figure represents the total cost to society as a whole that results from carbon emissions collectively reducing "comprehensive wealth" due to the changing climate. Economists favour adding this shadow price into the overall price of carbon, as it internalizes the full range of costs. The Canadian government estimated the shadow price to be around $28 per tonne in 2012, rising to $58 by 2050. These figures were used to evaluate the potential benefits of about $4-billion from implementing the proposed North America-wide fuel economy standards for cars by 2025 (now under review by the Trump administration). In 2015 the US Environmental Protection Agency estimated in its report, *Climate Change in the United States: Benefits of Global Action*, that if warming were held to the UN-proposed limit of 2°C, the US would save $7-billion in infrastructure repair costs and potentially $3-billion in avoided damage associated with sea level rise. Clearly, if a shadow price metric were included as a measure of human well-being, it would greatly speed up the transformation of energy policy outlined earlier in this chapter.

GETTING TO A "SAFE AND JUST PLACE"

The two international energy agencies mentioned earlier, which promote rapid transformation to low-carbon energy, have both pointed out that if the global response is slower, more investment will have to be put into capturing carbon before it enters the atmosphere if we are to avoid breaching the 2°C temperature rise ceiling. The IEA has estimated that five billion tonnes might have to be stored by 2050. At present this is not feasible. One engineered storage technology combines bio-energy with carbon capture and storage: crops are grown to capture atmospheric carbon, their biomass is converted into energy through biodigestion technology and the carbon emissions are stored underground. However, there is concern that cultivating biomass specifically for storing carbon might reduce capacity for producing much-needed food for the growing global population.

Shell Oil is piloting a carbon capture storage plant near Edmonton, but the facility stored only 1.3 million tonnes over 13 months. It also cost $1.3-billion, with large subsidies from both the federal and Alberta governments. Carbon prices would have to rise above $50 per tonne to encourage greater investment in such projects. Statoil, the Norwegian state energy company, is developing two pilot plants designed to store 28 million tonnes of carbon underwater. India has built a plant that is capturing all of its 60,000 tonnes of CO_2 emissions to make valuable chemicals. Carbon capture will have to become part of the overall strategy to achieve the 50–50–50 target, but the slower the transition to renewable energy is, the greater will be the need for costly investments in this as yet unproven technology.

We've mentioned a range of existing and projected disruptive technologies in this chapter to indicate there are avenues available to undertake transformative measures in water, energy, food production and restoration of carbon storage potential in natural ecosystems. Generally, the pace of change created by disruptive technologies has been faster than experts have predicted. So even though the international energy agencies are still forecasting that demand for fossil fuel energy may not peak until 2040, rather than by 2020 as recommended by both agencies to avoid overshooting the 2°C global temperature constraint, there is still a possibility the combined ingenuity of disruptive technologies plus supportive policies by governments may eventually lead to a reduction in global carbon emissions. But present attitudes of consumers of nexus resources will also have to be tempered as people come to realize such consumption habits are simply not sustainable. To deal with the impending effects of the Anthropocene, we will need to employ a combination of top-down policies driving change, supported by bottom-up technological creativity and a realization by consumers that we all have to live within our means.

Economist Kate Raworth has advocated a strikingly new approach to economic theory that brings many of the themes of the present book together. In her book, *Doughnut Economics*, she posits a metaphorical doughnut, the outer edge of which represents the nine planetary boundaries defined by Johan Rockström and his colleagues in 2009, beyond which lie unacceptable environmental degradation and potential tipping points of Earth's natural systems. Several of these boundaries, such as atmospheric carbon and biodiversity loss, have already been breached. The inner edge of the doughnut represents the social foundation of everyone on Earth having access to water, food energy and the other internationally agreed standards set out in the UN's 2030 *Transforming Our World* global sustainable development agenda. In-between is a "safe and just place" for humanity in which everyone has their basic human needs fulfilled and possesses a political voice but without impacting on the planetary boundaries. To attain this state will require a fundamental realignment of the world's economies, a shift from pursuit of endless economic growth, with its severe consequences for social equity, climate and biodiversity, to an economy that reallocates wealth to the underprivileged, reduces the global carbon footprint by 50 per cent and restores healthy ecosystems. Raworth argues such a transformation can be achieved with relatively small changes to current systems. For example, shifting just 3 per cent of global food production – most of which, in developed economies, is wasted anyway – to developing economies would achieve food security for all of humanity.

Since the dawn of the Industrial Revolution, nations have pursued a one-dimensional model of economic growth that has led to destruction of the biosphere, deprivation of a large proportion of humanity and massive inequality between rich and poor. We are now reaching the tipping point. Fortunately, there are disruptive technologies on the horizon that have the potential to change direction for the carbon footprint, and metrics have been introduced that will quantify natural assets in order to get them included in future decision making. The increasing impact of the Anthropocene will become the driver for changes such as these in order to create a "safe and just space" for all of humanity.

FIVE

The Hard Work of Hope: Ending a Titanic March of Folly

In her 1984 book, *The March of Folly*, historian Barbara Tuchman defined the work's title as "the pursuit by governments of policies contrary to their own interests despite the availability of feasible alternatives." The book details four such examples: the Trojans' decision to allow the Greek "horse" into their city; the Renaissance popes' failure to address the factors fanning the Protestant Reformation; the English taxation policies that exacerbated tensions with American colonists; and the US mishandling of the Vietnam War. She also noted that those responsible for these errors were provided with clear evidence that contradicted their ingrained beliefs and so was ignored. This flawed decision paradigm can also apply to the corporate world, as witnessed by the *Titanic* shipping disaster in 1912.

The tale of the *Titanic* begins with corporate hubris. A great deal of it, actually. The White Star Line was fiercely determined to dominate transatlantic shipping by destroying its archrival, Cunard. The race to build the largest and fastest ship in the world was symbolic. The *Titanic* was to be the largest human-made moving object in history.

Such hypercompetitiveness among large enterprises is hardly startling, of course. Some of the most senior oil executives in the world have variously confided privately to one of the authors of this book that one reason why major energy companies only pay lip service to climate change is that they are locked in mortal competition with one another to dominate the global energy market. The larger outfits, they will tell you, will do anything to gain advantage. Energy projects just keep getting bigger and bigger and more expensive. The Alberta oil sands, for example, are the largest mining operations on Earth and, as such, are vulnerable because of their very scale.

The sinking of the *Titanic* was the consequence of a long series of short-sighted decisions based on greed, arrogance and wilful denial. The inferior materials and workmanship in the construction of many parts of the ship, including its hull, were glossed over in favour of amenities for first-class passengers. A corporate decision was made, against the advice of the shipbuilder, to reduce the number of lifeboats to avoid clutter on the decks.

As we have seen, there is a certain amount of greed, arrogance and wilful denial in the climate story also. It remains a hallmark of our society that once we start investing in something big, we have to see the investment justified at any cost, even if we discover we are doing the wrong thing.

The infrastructure related to the use of fossil fuels is a highly sophisticated global system into which great sums of money have been and continue to be invested. Even though there are metaphorical icebergs of growing danger all around us, the *Titanic* that is the energy business continues to move full speed ahead as if nothing has changed. How could you even think of challenging the worth of the oil and gas industry? The fossil fuel sector is too big to fail, so there is no need to clutter the decks with lifeboats.

Another contributing factor to the sinking of the *Titanic* was the failure to undertake sea trials. The captain had never been on the ship until the fateful day it sailed. Nor did the crew undertake lifeboat drills. Similarly today, we have gotten into the habit of approving large resource projects that incur considerable environmental damage, such as oil sands, on mere promises that environmental and other problems will, in time, be somehow overcome. And we carry on even when they are not overcome.

For more than a century it has been widely held it was an undetected iceberg that caused the sinking of the *Titanic*. That is simply untrue. The *Titanic* received nine warnings about the presence of icebergs. Evidence demonstrates that the wireless operator was too busy transmitting chatty messages from passengers to take the warnings seriously. A century on from the *Titanic*, we are still blaming natural processes for human-caused disasters. Although there are some notable exceptions, the global media have been similarly distracted with gossip, chatty messages and meaningless entertainment to do their real job of warning us of greater danger. Obsessed with ourselves, we carry on ignoring the warnings.

Two of the warnings about the presence of icebergs were not even delivered to the bridge. So strong was his belief in the unsinkability of the *Titanic* that its captain, not unlike President Trump, ordered full speed ahead despite the warnings. Two inexperienced boys who were keeping watch in the crowsnest sent their alert to the bridge, but their version of today's intergenerational climate warnings came too late.

Even when significant alerts do "make it to the bridge," including the five assessment reviews undertaken to date by the Intergovernmental Panel on Climate Change, many of our politicians, not unlike the captain of the *Titanic*, have for decades been ignoring them. Just like the captain of the doomed ocean liner, they preferred to believe their own spin. Don't worry, Stephen Harper told Canadians in 2008, we can adapt to climate disruption; with a strong resource sector we are unsinkable. Like Donald Trump eight years later, Mr. Harper felt no need whatsoever to listen to the scientists waving their arms and pointing from the crowsnest. Don't worry, it's all a natural cycle, the climate is not changing.

In the aftermath of the *Titanic* disaster, much was made in the popular press about great human courage and heroism aboard the ship, but in fact there was little of either. The mental image of the lights staying on and the band continuing to play is haunting, of course. But it is sobering to note that after the vessel gashed its hull on the iceberg, third-class passengers were not allowed on deck until first- and second-class ones were off the ship. Though the British inquiry into the sinking argued there was no discrimination against third-class passengers, more of them died than any other passengers other than crew. The truth is that if the crew had been properly equipped with enough lifeboats, and with the knowledge and experience to launch them, everyone might have survived.

It is conventional wisdom that the wealthy will usually find a way off any sinking ship. If nothing else, they can buy a lifeboat. But the maiden voyage of the *Titanic* was not fully booked, and even at only two-thirds of capacity there were simply not enough boats for all passengers and crew. And just as with that historic disaster, there is hardly anything courageous or heroic about creating a climate disaster one clearly could have avoided.

For we are on a similar needless and heedless course today with the global water crisis, Earth system damage and climate disruption. There are not enough lifeboats for the seven billion people presently on Earth, and there will never be enough for the nine to 11 billion there may be by the end of this century. As with the *Titanic*, some wealthy nations and individuals will likely want to get off the ship first, leaving behind few resources and no lifeboats for the poor who will suffer most.

There were huge liability issues associated with the sinking of the *Titanic* that went back and forth in the courts for years. In this respect the world did learn something from the sinking, in that its aftermath brought a whole new series of regulations and a new institution to monitor the movement of ice in the shipping lanes of the North Atlantic. There is, however, an important difference between identifying immediately obvious lessons and actually learning more broadly from them.

All the practical lessons that were identified with respect to shipping regulations notwithstanding, we appear not to heed the larger lessons we might have learned from the *Titanic* fiasco. The moral, ethical, legal and political consequences of ignoring climate disruption remain incalculable. While laudable, the regulations that have been put into place in Canada to protect us from a climate disaster of *Titanic* proportions are tentative at best and remain largely aspirational. Even if fully implemented, they may not make the climate problem go away.

Sea ice and icebergs remain matters of great concern to all of humanity, though for different reasons than in 1912. While the owner of the conglomerate that in turn owned White Star, US financier J.P. Morgan, died the year after the disaster, various successor companies eventually got out of transatlantic shipping, and J.P. Morgan & Co. and related firms went on to make fortunes in financial markets.

The name J.P. Morgan (JPMorgan Chase & Co., now worldwide with assets of about US$2.6-trillion) would surface again a century later in the context of the global financial crisis of 2007–2008, which essentially did to the global economy what the sinking of the *Titanic* did to the White Star Line. Once again we learned that unbridled corporate greed and impunity in the absence of effective governmental regulation can create very large-scale disasters in which mostly innocent people suffer or die.

We are now beginning to realize that climate disruption threatens to do far more damage to our economic, social and political stability than all of the economic upheavals of the 19th and 20th centuries and to date in the 21st combined. Climate disruption could bankrupt us all.

While few believe in unsinkable ships anymore, or that anything is too big to fail, with enough hubris a civilization even as sophisticated as ours remains capable of sinking itself. That is something we should not forget. Nor should we forget that the sinking of the *Titanic* foreshadowed violent, irreversible change in the years immediately following. Dramatic disruptions included the first modern, mechanized war, which resulted in one of the greatest death tolls ever recorded until that time in the history of human conflict; the fall of entire empires and the monarchies that governed them; and the call for a new economic system that included broader forms of economic co-operation.

The Anthropocene similarly foreshadows dramatic, irreversible change and opportunity. Dark clouds on the Anthropocene horizon include yet another disruption in the world order; challenges to liberal democracy as a political ideal; and the recognized need for a better measure of human progress than gross domestic product. Opportunities include the urgent need to understand and act upon the direct link between hydro-climatic change and its current and projected impacts on the global economy and human well-being. Now, just as then, prior to the Great War, there are going to be winners and losers as we enter the Anthropocene. It doesn't have to be that way, however. It is not that we don't have the resources to effect positive change for all; it is that we don't have the will. And until we summon that will, we can expect more and more losers and fewer winners in the future.

STEERING THE *TITANIC* THROUGH THE ANTHROPOCENE

This book is about hope. But, as noted above, this hope must be embedded in realism, meaning that it will be very hard to achieve. We believe that the human race potentially has the resources, ingenuity, compassion and governance capacity to transition toward a carbon-neutral state. The means to overcome the challenge of living within the Earth system's bounds while still providing basic social justice for all in what Kate Raworth has called a "safe and just space" may well be within the capabilities and resources of the global economy. But the transition will not be without pain; indeed, pain will be a driver of the transition. The Anthropocene represents a permanent change in the global environment, in terms of both climate and fundamental ecosystem boundaries. These pressures impinge on the key Earth systems – its carbon, nitrogen and water cycles – in overlapping ways. The buildup of atmospheric carbon contributes to acidification of oceans; overuse of nitrogen fertilizers in industrial agriculture causes pervasive pollution of rivers, lakes and coastal waters; and the millions of dams throughout the world have disrupted natural hydrologic systems, making them less resilient to changes in hydrologic conditions. In addition, massive loss of biodiversity caused by deforestation, industrial agriculture, rapid urbanization, loss of wetlands and the mining of groundwater has exacerbated the effects of the changing climate on the nexus. On top of this is the continued growth in human population, which increases by 75–80 million per year and could reach 10 or even 11 billion by the end of the century. Realistically, it will be the harmful universal effects of global combustion, consumption and conception on the prospects of all of humanity that will drive the needed change.

All nations are facing significant challenges as they try to reduce their carbon footprints and maintain or restore healthy ecosystems to increase natural storage of carbon. Built-in consumer demand for energy, food and water resources encourages government subsidies and investments in fossil fuel infrastructure; consumers resist increases in taxes associated with putting a price on carbon, while the low price of water worldwide encourages overconsumption and pollution of often limited resources. The innovative and disruptive technologies required for changing this pattern either are not yet established or are inadequately supported by governments to make a significant reduction in the global carbon footprint.

Globalization of the economy has led to lack of equity and fairness. Income gaps between rich and poor countries and within populations in most countries have widened. Stagnant wages and lack of employment have led to increased public unrest and changed the face of politics, as witnessed in the United States and Europe. Long-standing discrimination against Indigenous peoples is entrenched, and the resulting legal challenges continue to impede orderly resource development.

The central thesis of this book is that these challenges require a complete transformation in global economic, ecological and social systems, and that the pervasive impacts of the Anthropocene on the Earth's critical functions will provide the necessary impetus to drive this transformation. The 2°C temperature increase threshold will be breached well before the globe becomes carbon neutral, and vast investments will be required in adaptation measures as well as decarbonization. The longer the hard decisions are delayed, the greater will be the effects of the Anthropocene and the costs of making the transition. The good ship *Titanic* is still steaming straight for the iceberg.

The UN has already charted two initiatives to support this transition, and both deal directly with managing the two respective boundaries illustrated by Kate Raworth in her depiction of "doughnut economics."

The first of these, the 2030 *Transforming Our World* global sustainable development agenda, is aimed at the inner boundary of the doughnut, dealing with social justice issues to end poverty, protect the planet and ensure prosperity for everyone. There are 17 goals in all, which will require the concerted action of all segments of society: governments, the private sector, civil society and individuals. This initiative has been accompanied by a further UN resolution, approved in December 2016, the International Decade for Action "Water for Sustainable Development" 2018–2028, as discussed in chapter 2.

The second initiative is the Paris Agreement, which deals with the outer boundary of the doughnut to limit overshoot of Earth's planetary capacities. Although currently not structured to avoid a greater than 2°C increase in global temperatures, the pact is at least the first one on climate to be ratified by essentially all nations. Its weaknesses have been canvassed in previous chapters, but there is a general feeling that, unlike its predecessor, the Kyoto Agreement, this time most nations will stay the course to meet their stated commitments as best they can, even though the United States has decided to end its participation. The stakes are much higher now than when Kyoto was signed, and renewable energy technologies are becoming competitive with fossil fuels around the world.

The impacts of the Anthropocene will affect all of humanity, but just as was the case with the first-class and third-class passengers on the *Titanic*, richer countries will be better able to adapt than poorer ones. This differential was precisely the impetus behind the 2009 creation of a stand-alone Green Climate Fund to operate under the auspices of the Conference of the Parties to the UN Framework Convention on Climate Change to help developing countries adapt. Although the fund was ostensibly intended to establish international financing of $100-billion a year by 2020, from both public and private sources, only $10.3-billion had been committed as of March 2017, with the US making the largest promise at $3-billion. Of that original $3-billion commitment by the US, only one-third has been transferred so far, with President Obama having delivered a second tranche of US$500-million just before leaving office. The remaining $2-billion is now up to the

Trump administration, with its stated intention to reduce funding to UN-sponsored agencies. It seems the rich governments are now not even investing in lifeboats.

THE NEW RENAISSANCE

This is not the first time the advent of globalization and rapid increase in communications and creativity have outpaced the ability of established institutions to respond to growing inequities. The European Renaissance that flourished between 1450 and 1550 brought the same breathless pace of innovation in communications (the printing press), navigation, trade, migration, arts, religious transformation and the foundations of early capitalism. It too created social friction leading to riots, populism, ideological extremism and a "bonfire of the vanities." At present, we face a very old challenge in this respect. Dante Alighieri's 1320 epic narrative poem, *The Divine Comedy*, for example, makes an observation that still haunts us 700 years later: that hell has a special place for those who remain neutral in times of moral crisis.

In their book, *Age of Discovery*, Ian Goldin and Chris Kutarna of the Oxford Martin School at Oxford University stressed that eventually societies adapt to these new ideas but that governments and individuals need to be smart and well organized to make this adaptation possible. Risk taking and openness to accepting new ideas, however strange they might feel at first, have to increase significantly, but the authors warn that "in an age when we must act, we hesitate instead." Unfortunately, this is our current reality.

John Ashton, a leading thinker and diplomat, made it clear that the market alone will not be sufficient to drive the transformation called for in the present book. He cited the

Renaissance book-burning incident in an address to a group of Swiss bankers and scientists in January 2014, titled "The Book and the Bonfire: Climate Change and the Reawakening of a Lost Continent." Among his remarks, Ashton said that "the market left to itself will not reconfigure the energy system and transform the economy within a generation.... Unless governments own and drive the transformation, acting on behalf of taxpayers and voters ... the transformation simply will not happen."

In conclusion, it is useful to recall the transformations in energy, water use efficiency and food production that will be required to meet the 50–50–50 goal set by the UN. Across the energy sector the carbon footprint has to be reduced by 70 per cent by 2050 through a complete shift from gasoline to electric vehicles; an 80 per cent reduction in carbon emissions in the entire building stock through massive investment in energy efficiency; conversion of all electricity generation to renewable sources; and significant investment in engineered carbon storage capacity. In water management there has to be universal pricing, worldwide groundwater regulations and protection of aquatic ecosystems through maintenance of ecological flows by regulations and by reuse of treated waste water and biosolids for energy and water. In addition to improving engineered solutions, investment must also be directed toward restoration of upland watersheds, with the goal of being able to use the forces of nature itself to help build more efficiently integrated water infrastructure that, as much as possible, operates and maintains itself. In the agricultural sector there has to be a transformation from industrial techniques to agro-ecology, with elimination of carbon emissions from farming and land use; increased carbon sequestration in soils and forests through use of nitrogen-fixing crops; limiting tillage; use of precision techniques to eliminate waste in the use of water, fertilizers and pesticides; and development of heat- and drought-tolerant crops. In the oceans there has to be regulated control of fishing in international waters through UN agreements; designation of marine protected areas; and restoration of natural vegetation along coasts to guard against increasingly frequent and damaging typhoons and hurricanes.

THE ROLE OF GOVERNMENTS

There are four key actions that must be led by governments across the globe and be implemented by 2030 in order to start to turn the corner. The first goal is to reduce the threat of water insecurity. Water lies at the heart of the nexus, yet it faces the most severe risks due to changing hydrology. The International Decade for Action "Water for Sustainable Development," 2018–2028 must result in meaningful progress for ensuring safe access to sustainable water management and sanitation globally; basic infrastructure – both engineered systems and using nature's assets – for managing the increased risk of flooding; application of emerging smart technology for more efficient water use; long-term drought management plans to deal with potential megadroughts; and a complete rethink of regulations for recovering resources from waste water treatment. Canada's current financial policies actually restrict integrated resource recovery from waste treatment. The Greater Victoria region on Vancouver Island was directed by federal–provincial regulations to construct a sewage treatment system, as its untreated waste water flows directly into the marine environment. But, in the end, despite setting goals for resource recovery, the two senior governments approved a traditional, large-scale treatment plant with limited opportunities for such recovery. The current cost-sharing formula splits the tab equally across the three levels of government – federal, provincial and local – with no specific requirements for recovering water and energy. Future waste treatment plant infrastructure must include specific targets for resource recovery scaled to levels of financing: the larger the recovery, the larger the grants.

Second, once all governments universally accept that impacts of the changing climate and loss of biodiversity must be immediately addressed, new governance systems will be required. Key actions include establishing a comprehensive set of indicators for human welfare to supplement the traditional use of only GDP. This must include valuing natural capital as an asset; setting a global price on carbon; eliminating all subsidies for fossil fuel development and replacing them with targeted subsidies to encourage innovation in clean technology; financing infrastructure in public transit, biking and walking; regulating standards in energy and water use in all building stock; and supporting investment in carbon storage through restoring ecosystems, afforestation and agro-ecology.

The crisis in the nexus is on the verge of becoming so universal that it will force governments to co-operate rather than compete with and challenge each other. One aspect of such co-operation must be in trade policy. One of the major weaknesses of the Paris Agreement is the lack of any sanction should a member country not meet its stated target for carbon reduction. However, there were binding trade conditions built into the 1987 Montreal Protocol on Substances that Deplete the Ozone Layer, to both stimulate and discourage international trade. An example of the former was a provision to encourage funding and transfer of ozone-depleting technologies between industrialized and developing countries. An example of the latter included provisions to prohibit offending industries from moving to non-participating countries, to remove any competitive advantage that might be gained from non-compliance. Though such trade sanctions would be challenged if they were to be included in the Paris Agreement today, when there is universal acceptance of the need for global action on the 50–50–50 goal, as the effects of the Anthropocene deepen, such co-operation will become not just acceptable but necessary. The unthinkable has to become the norm in this coming age of taking charge to steer the *Titanic*.

Third, these transitions will also have to be fuelled by a massive redirection of international financing. These resources are available but currently are allocated to national security, armaments and building national infrastructure rather than tackling global inequalities. The timid response to the Green Climate Fund will have to be replaced by an accelerated and more universal agreement to assist developing countries as the impacts of the changing climate and hydrology force migration and depredation on those unable to adapt. More overseas aid will also have to be directed to supporting the 17 goals in the *Transforming Our World* agenda as the crisis deepens.

Fourth, the role of corporations will have to change in order to respond to the growing need to include social and ecological values in decision making rather than the single metric of bottom-line profits. The G20's Financial Stability Board, for example, chaired by Bank of England governor Mark Carney, has established a Task Force on Climate-related Financial

Disclosures to look into opportunities for green bond financing of clean technology and infrastructure. "Green finance is a major opportunity," says Carney, but one that will have to be geared to a collective gain mentality, rather than private gain, and be focused on achieving the sustainable development goals. The Paris Agreement will require an unprecedented mobilization of both public and private financing, amounting to over $90-trillion by 2030.

In the face of the impacts of the Anthropocene, it only makes sense to at least examine alternative scenarios to more global co-operation in the context of an individual country's national security perspective. One option for states concerned about their future is to retreat within their own borders and use their resources to protect their own citizens. Such a strategy was favoured, at the outset at least, by the Trump administration, and to a lesser extent by some European nations. Indeed, isolationist approaches may be among the initial national responses as the crisis deepens, but in the end, if we are to survive the Anthropocene, humanity will have to come together or else fall apart. Sooner or later, however, we will have to face the fact that it is not all about where and how we live today. There are moral and ethical dimensions to our responsibility to the future that must also be considered.

In her 2016 book, *Great Tide Rising*, ethicist Kathleen Dean Moore explores the moral, as well as the environmental and economic, urgency of responding immediately to climate disruption. She is asking questions we will all ultimately have to answer. Moore begins her book by noting that culpable wrongdoing is defined as knowing and intentionally doing unjustifiable harm. Natural systems react to external disruptions. We humans are an internal disruption. What we are doing, we are doing knowingly, wilfully and recklessly. In ransacking the world, Moore says, we are failing in our moral and legal duty of restraint. It is immoral, she asserts, to think and act as though all ecosystem services exist to benefit only one species at only one moment in time.

Moore argues that, unlike the dinosaurs, we have a choice. At present, she contends, all our laws do is regulate the times, places and circumstances where environmental destruction can take place. Moore is of the view that we have to align with evolutionary processes rather than trying to bring them to an end. We cannot address the climate threat without granting natural objects, other species, ecosystems and natural cycles the rights of personhood, including rights of protection and restoration. We have to challenge the morally unsupportable claim that it is possible to mitigate the destruction of a natural place by creating a new one elsewhere. We need to construct a business model that respects the real value of ecosystem services rather than simply creating markets for repairing the uncalculated and often incalculable damage we do to them as a matter of prescribed course.

Moore also attacks the sloppy manner in which we fall back on optimism. The hard work of hope, she argues, is not tied to wishful thinking; it is linked to right action. Short-term, self-interested adaptation strategies do little to reduce rates of extinction – the price exacted from other beings by humans' failure to count other lives in the narrow calculation of their own self-interest. Nor do such strategies address inequality globally.

Reducing the rate of global warming is going to be difficult, Moore agrees. Plans to take the necessary steps will face powerful opposition. We will be left with no option but to turn to related responses – like adaptation – to improve our chances of success. But Moore questions whether adaptation as we presently conceive it is morally defensible.

It is widely held that there are two kinds of adaptation: linear versus transformational. Moore's view, which is one increasingly shared by poorer nations around the world, is that linear adaptation in the form of a single-minded accommodation to climate change is a moral failure. The danger is that "adaptation" is becoming a smokescreen that hides or minimizes the real consequences of global warming and so delays action, which allows people in developed countries to continue to live as they have for generations while the world misses its last chance to stop runaway climate change. By investing in linear "adaptation" projects, the privileged can use their power and money to try to shield themselves from the worst consequences of their own excess while imposing the costs of climate change on the disenfranchised and displaced. From this perspective this kind of adaptation is being seen as not just imprudent but unjust. What truly moral society, Moore asks, responds to the suffering it has caused to other peoples and cultures by investing billions to make sure the same thing doesn't happen to its own culture, to say nothing about broader global ecosystem impacts?

Moore believes we have a moral obligation not just to fix things for ourselves but to restore the system for everyone. To do that, we need a new narrative, one that includes respect for the rest of the life we share this planet with and whose presence we rely on for the stability of the Earth system that makes our presence even possible. As part of this new narrative, we also need to create a new sense of time that extends forward to include future generations.

THE ROLE OF INDIVIDUALS

A top-down approach, by governments alone, cannot achieve the transition outlined in this book. Indeed, the necessary effort can be successful only through the synchronization of government policy, with a network of energized and aware citizens operating through universities, businesses, governments and households that can influence and implement such policy from the bottom up. This support requires an informed citizenry that understands the potential roles they can play in contributing to solutions and engages in deliberations on the transition. In short, the public needs to become educated on how the nexus of water, energy, food and biodiversity functions, on the links to healthy ecosystems and how people can adapt their everyday decisions about consumption, travel and accommodation to reduce their individual and collective carbon footprint.

It is becoming clear that consumers are thirsty for clear facts about how they can change their habits to play their part. A survey in the Canadian city of Vancouver, for example, recorded 85 per cent of respondents as saying they would purchase an electric vehicle once they become available at a reasonable price with a reliable driving distance before recharge. Citizen streamkeepers oversee the health of many watercourses in British Columbia, actively engaged in restoring fisheries and protecting riparian vegetation. These are ordinary people undertaking roles that used to be played by governments before budgets were cut. They want to make a difference and are impatient with governments' slowness to take the necessary action.

This pent-up desire to make a difference would be greatly enhanced if all citizens had access to facts that would enable them to make informed decisions. Schools and universities should offer courses on the critical importance of healthy ecosystems, on how they work and can be restored. Such instruction should include ways to reduce waste in food consumption; the consequences of shifting toward a vegetarian diet in terms of savings in water, energy and carbon; application of new technologies and information systems to save on water and energy; and ways to grow one's own food in an urban environment. Experiential learning at a young age also can greatly enhance the capacity of the public to become actively engaged.

People are busy, so they need access to simple facts to aid their decisions. Internet sites are being developed to help consumers weigh the benefits and costs of installing solar panels or buying electric vehicles, to inform people of ways to reduce their carbon footprint associated with long-distance travel, or even deciding whether to buy new clothing or used. Recently, a think tank called the One Planet Economy Network has developed an integrated computer tracking system that will enable individual consumers to monitor how much carbon, water and energy are associated with their everyday tasks and to discover ways to reduce these factors by taking adaptive measures. The technology associated with the Internet of Things will enable cheap and simple tracking systems in the near future. As the public becomes more motivated to reduce the collective impacts of their consumption on nexus resources as a result of the growing effects of the Anthropocene, it will become possible to assess these impacts on individual items such as clothing, transportation, real estate and food in terms of water, energy and carbon, just as it is now possible to compare sugar and calorie content in foods. The maxim that what gets measured gets done (which will have to include natural-capital valuation) is a motivator, especially for a more engaged and concerned public with a heightened awareness for "doing the right thing." Hopefully, reducing one's overall footprint will become a badge of honour, a talking point in future conversations, where individuals share information on better practices and stimulate continuous improvement.

Another way of aligning everyday decisions toward a carbon-neutral world is to enhance the role that professionals play in either making decisions or advising governments on policy. Practitioners such as engineers, geoscientists, agrologists, biologists and planners are guided by a code of ethics and can be held accountable for their decisions by an association of their peers. As all levels of government have downsized over the past decades, more and more of the daily decisions on a wide range of activities are now made by such professionals practising outside government and funded by mostly development interests. In light of the tremendous changes that are now required in energy systems, food production, urban design, building efficiency and water management for flood and drought control due to the increased variability in hydrologic events outside the norm, professionals of all types need to be prepared to take a more active role in advising their clients on ways to support the shift to the 50–50–50 goal. This will involve taking more risks and encouraging the adoption of innovative practices to increase the learning experience. These professionals can also become more relevant by applying emerging information on the value of protecting and restoring natural capital in business-case assessments for their clients.

Transformative change of the scope outlined in this book

will be greatly enhanced by linking government policies and regulations with systematic improvements in the role individuals can play. The time to act is now, and as governments implement policies on setting carbon prices or establishing regulations for energy efficiency, they must provide parallel steps for better access to education and knowledge, ensuring that professionals have the decision tools and risk profiles to improve their advice on a wide range of services, and have information systems available to track the day-to-day footprint of consumers. The public wants to play their part to contribute to a carbon-neutral world, but they need the knowledge and the tools to do that. It is high time we make the right moves now to support this pent-up desire.

THE ROLE OF CANADA

Canada should be a leader in this transition. It has created a bold plan for reducing emissions, based on a universal carbon tax; enacted regulations to reduce methane emissions from the oil and gas industry (though now delayed by three years); advocated for a total, nationwide shift to renewable electric energy by 2030; and, in the latest federal budget, increased investment in clean technology. The government's mid-century strategy, though aspirational, supports significant investment in green public infrastructure and attention to restoring ecological function through reforestation, afforestation and some migration toward agro-ecology. Former Bank of Canada governor, Mark Carney, has been championing green financing during his current posting as governor of the Bank of England, and he will likely continue to promote such funding models after he returns home to Canada. Elements from the corporate sector have established a Council for Clean

Capitalism, which champions a new economic system where prices incorporate social and environmental benefits and costs in addition to purely economic factors. The Smart Prosperity Institute, supported by 27 of the most senior people across corporate Canada, has advised the prime minister to stay the course on his government's pan-Canadian framework climate plan, despite decisions of the Trump administration, and has a professional secretariat providing a range of advice on adaptation measures and clean technology. The Natural Capital Lab discussed in chapter 4 is breaking new ground on valuing natural capital as assets in municipal infrastructure and for restoring ecosystems services.

Canada also has a unique opportunity to become a leader in clean technology in North America. With the US administration committed to continuing to invest in obsolete fossil-based energy systems, together with their expressed disdain for science generally, the field is wide open for Canada to vigorously pursue the disruptive technologies that will be required in order to end the march of climate folly. In April 2017, a panel of distinguished scientists called Canada's Fundamental Science Review issued advice to the federal government on a long-term science strategy in a report titled *Investing in Canada's Future: Strengthening the Foundations of Canadian Research*. The panel concluded that investment in research and innovation had declined over the past decade relative to other developed countries, and they recommended establishing an independent National Advisory Council on Research and Innovation, plus a streamlining of existing research institutions to focus resources on applied and original science. If approved, such a council could provide momentum to Canada's potential role of not only being a leader in clean technology but also capitalizing on the economic opportunity presented by the trillions of dollars in future investment that will be required to make the transition.

But Canada is not immune to those forces that want to keep steering the *Titanic* into the icebergs. Opposition parties decry the carbon tax; corporate interests still push for more fossil fuel subsidies and investment in oil sands; and budget constraints limit needed investment in green infrastructure and natural capital. Hopefully, the good ship *Titanic* will only have to glance off a few growlers, some field ice and some small bergs as the effects of the Anthropocene deepen before the hardship of these impacts finally creates a universal resolve, even in Canada, to steer the ship of state toward a carbon-neutral future.

Time is of the essence, however. All realistic analyses of the shift to the 50–50–50 goals state emphatically that hard decisions must be made now, as it is already too late to achieve the 2°C goal, and the effects of the Anthropocene will just become insurmountable the longer the status quo prevails. Unless the policy and technology transformations outlined in this book can be implemented by 2030, we will have acted out yet another example of our collective inability to learn from past mistakes, and thus would the march of folly continue.

In a sense it really is April 14, 1912, all over again, and we are steaming full speed into risks we clearly understand and a potential disaster we know how to avoid. But this time there is no need for hubris or habit to sink us. Rather, our future depends on our heeding and acting on the warnings that are once again being clearly issued by Arctic ice. Our goal should be to navigate through the Anthropocene toward an exhilarating and exciting transformation of our current economic and social systems into ones in complete harmony with Earth's natural cycles.

ACKNOWLEDGEMENTS

Special thanks are due to Rocky Mountain Books, in particular to publisher Don Gorman for his commitment to books on water and climate; to editor Joe Wilderson for his diligence, intelligence, patience and good humour; and to art director Chyla Cardinal, whose mindful designs make books like this shine.

This book also acknowledges the commitment that Her Honour Judith Guichon, Lieutenant-Governor of British Columbia, has displayed in bring the crisis in the climate nexus to the attention of British Columbians and to a wider Canadian audience through her contacts with government and community leaders across the country. Her Honour has also dedicated her role as Lieutenant-Governor to engaging with the province's youth in Stewardship for the Future conferences. This initiative brings together youth from all across the province to report on practical projects they have undertaken with the encouragement of Her Honour on ways to improve the stewardship of the province's water, land and natural resources. The conference has been held in three successive years now and has directly engaged a large number of students and their teachers in seeking innovative approaches to environmental stewardship.

At the most recent conference, in June 2017, the students developed a vision of what the province should look like in 2067, the bicentennial of Canada's confederation.

Raising public awareness of the challenges of making the transition to a sustainable future for Canada is the highest priority over the coming 50 years. The authors are proud to have been associated with hard work and vision of Her Honour Judith Guichon in advancing this awareness amongst the province's youth throughout her time in office.

And of course, the authors accept sole responsibility for any errors, omissions, misrepresentations or misunderstandings.

BOOKSHELF

"Amsterdam Declaration on Global Change." In *Challenges of a Changing Earth*, edited by W. Steffen et al., 207–208. Proceedings of the Global Change Open Science Conference, Amsterdam, The Netherlands, July 10–13, 2001. International Geosphere–Biosphere Programme (IGBP) Science Series. Berlin: Springer, 2002. Accessed 2017-07-05 (pdf) at colorado.edu/AmStudies/lewis/ecology/gaiadeclar.pdf. See also Steffen, W., et al. below (two titles).

Ashton, John. "The Book and the Bonfire: Climate Change and the Reawakening of a Lost Continent." Address to the Tenth Swiss Biennial on Science, Technics and Aesthetics, Swiss Museum of Transport, Lucerne, January 19, 2014. Accessed 2017-07-05 (pdf) at e3g.org/docs/The_Book_and_the_Bonfire_speech_by_John_Ashton_190114.pdf.

Atherton, Emmaline, et al. "Mobile measurement of methane emissions from natural gas developments in Northeastern British Columbia, Canada." *Atmospheric Chemistry & Physics Discussion*, doi:10.5194/acp-2017-109 (April 7, 2017): 1–28. Accessed 2017-07-05 (pdf preprint) via *The Tyee* at thetyee.ca/News/2017/04/26/acp-2017-109.pdf.

Ault, Toby, et al. "Relative impacts of mitigation, temperature and precipitation on 21st century megadrought risk in the American Southwest." *Science Advances* 2, no. 10 (October 5, 2016). Accessed 2017-07-05 (pdf) at advances.sciencemag.org/content/advances/2/10/e1600873.full.pdf.

Australian Government Climate Change Authority. *Towards a Climate Policy Toolkit: Special Review of Australia's Climate*

Goals and Policies. Report 3 to the Minister for the Environment. Canberra: Climate Change Authority, August 31, 2016. Accessed 2017-07-05 (pdf, summary, data underlying charts, etc.) at climatechangeauthority.gov.au/reviews/special-review/towards-climate-policy-toolkit-special-review-australias-climate-goals-and.

Bloomberg New Energy Finance. New Energy Outlook 2017. New York: Bloomberg, 2017. Accessed 2017-08-05 (web summary only) at about.bnef.com/new-energy-outlook.

Canada's Ecofiscal Commission. "Similarities and differences between carbon tax and cap-and-trade systems." Blogpost, August 10, 2015. Accessed 2017-07-05 at ecofiscal.ca/2015/08/10/differences-carbon-tax-vs-cap-and-trade.

Canada's Fundamental Science Review. *Investing in Canada's Future: Strengthening the Foundations of Canada's Research*. Ottawa: Advisory Panel on Federal Support for Fundamental Science, April 2017. Accessed 2017-07-05 (pdf) at sciencereview.ca/eic/site/059.nsf/vwapj/ScienceReview_April2017.pdf/$file/ScienceReview_April2017.pdf.

Christensen, Randy, and Oliver Brandes. *California's Oranges and B.C.'s Apples? Lessons for B.C. from California Groundwater Reform*. Victoria: Polis Project on Ecological Governance, and Vancouver: Ecojustice, June 2015. Accessed 2017-07-05 (pdf) from poliswaterproject.org/orangesapples.

Conference Board of Canada. "Despite higher prices, Canadian oil extraction industry still in survival mode." News release 17-76, March 13, 2017. Accessed 2017-07-05 at conferenceboard.ca/press/newsrelease/17-03-13/despite_higher_prices_canadian_oil_extraction_industry_still_in_survival_mode.aspx.

Cooperative Institute for Research in Environmental Sciences (CIRES). "Methane leaks from three large US natural gas fields in line with federal estimates." NOAA (US National Oceanic and Atmospheric Administration) news archive, February 18, 2015. Accessed 2017-07-05 at is.gd/K6S8Tz.

Crowther, T.W., et al. "Quantifying global soil carbon losses in response to warming." *Nature* 540 (December 1, 2016): 104–108.

Accessed 2017-07-05 (abstract only) at nature.com/nature/journal/v540/n7631/full/nature20150.html.

Ellison, Garret. "Trump budget eliminates Great Lakes cleanup funds." MLive: Michigan News (website), May 23, 2017. Accessed 2017-07-05 at mlive.com/news/index.ssf/2017/03/trump_budget_eliminates_great.html.

Food and Agriculture Organization of the United Nations. *The State of Food and Agriculture: Climate Change, Agriculture and Food Security*. Rome: FAO, 2016. Accessed 2017-07-05 (pdf) at fao.org/3/a-i6030e.pdf.

Fukuyama, Francis. *The End of History and the Last Man*. London, New York, Toronto: Free Press, 1992.

Goldin, Ian, and Chris Kutarna. *Age of Discovery: Navigating the Risks and Rewards of Our New Renaissance*. New York: St. Martin's Press, 2016.

Gooch, Martin V., and Abdel Felfel. "$27-Billion" *Revisited: The Cost of Canada's Annual Food Waste*. Oakville, ON: Value Chain Management International, December 2014. Accessed 2017-07-05 (pdf) at vcm-international.com/wp-content/uploads/2014/12/Food-Waste-in-Canada-27-Billion-Revisited-Dec-10-2014.pdf.

Government of British Columbia. *Climate Leadership Plan*. Victoria, August 2016. Accessed 2017-07-05 (pdf) at pembina.org/reports/bc-climate-leadership-plan-august-2016.pdf.

Government of Canada. *Canada's Mid-Century Long-Term Low-Greenhouse-Gas Development Strategy*. Ottawa: Environment and Climate Change Canada, 2016. Accessed 2017-07-05 (pdf) at unfccc.int/files/focus/long-term_strategies/application/pdf/canadas_mid-century_long-term_strategy.pdf.

Governments of Canada et al. *Ecosystem Services Toolkit: Completing and Using Ecosystem Service Assessment for Decision-making*. Ottawa: Value of Nature to Canadians Study Taskforce, 2017. Accessed 2017-07-05 (pdf) at publications.gc.ca/collections/collection_2017/eccc/En4-295-2016-eng.pdf.

———. *Pan-Canadian Framework on Clean Growth and Climate Change: Canada's Plan to Address Climate Change and Grow the Economy*. December 9, 2016. Accessed 2017-07-05 (pdf) at canada.ca/content/dam/themes/environment/documents/weather1/20170125-en.pdf.

Hamilton, Clive. *Defiant Earth: The Fate of Humans in the Anthropocene*. New York: Wiley, 2017.

House of Commons (UK) Environment, Food and Rural Affairs Committee. *Future Flood Prevention*. Second Report of Session 2016–17, HC 115, October 26, 2016. Accessed 2017-07-05 (pdf) at publications.parliament.uk/pa/cm201617/cmselect/cmenvfru/115/115.pdf.

International Energy Agency. *World Energy Outlook 2016 Executive Summary*. Paris: IEA, November 2016. Accessed 2017-07-05 (pdf) at iea.org/publications/freepublications/publication/WorldEnergyOutlook2016ExecutiveSummaryEnglish.pdf.

International Energy Agency and International Renewable Energy Agency. *Perspectives for the Energy Transition: Investment Needs for a Low-Carbon Energy System*. Paris and Abu Dhabi: IEA and IRENA, March 2017. Accessed 2017-07-05 (pdf) at energiewende2017.com/wp-content/uploads/2017/03/Perspectives-for-the-Energy-Transition_WEB.pdf.

International Geosphere–Biosphere Programme. "Global Change and the Earth System: A Planet Under Pressure." IGBP Science Series no. 4. Stockholm: IGBP Secretariat, 2001. Accessed 2017-07-05 (pdf) at is.gd/jn1Yji. See also Steffen et al. below for series synthesis subsequently published in book form.

International Institute for Sustainable Development. *Comprehensive Wealth in Canada: Measuring What Matters in the Long Run*. Winnipeg: IISD, December 2016. Accessed 2017-07-05 (pdf) at iisd.org/sites/default/files/publications/comprehensive-wealth-full-report-web.pdf.

International Panel of Experts on Sustainable Food Systems. *From Uniformity to Diversity: A Paradigm Shift from Industrial Agriculture to Diversified Agroecological Systems*. Brussels: IPES Food, June 2016. Accessed 2017-07-05 (pdf) at ipes-food.org/images/Reports/UniformityToDiversity_Full Report.pdf.

Johnson, Tracy. "Just how many jobs have been cut in the oilpatch?" CBC News, July 6, 2016. Accessed 2017-07-05 at cbc.ca/news /canada/calgary/oil-patch-layoffs-how-many-1.3665250.

Judt, Tony. *Ill Fares the Land*. New York: Penguin, 2010.

Keynes, John Maynard. *The General Theory of Employment, Interest and Money*. London: Macmillan, 1936.

Liu, Yi Y., et al. "Recent reversal in loss of global terrestrial biomass." *Nature Climate Change* 5 (May 2015): 470–474. Accessed 2017-07-05 (abstract only) at nature.com/nclimate/journal/v5/n5/abs /nclimate2581.html.

Mann, Michael E., et al. "Influence of anthropogenic climate change on planetary wave resonance and extreme weather events." *Scientific Reports* 7, art. no. 45242 (March 27, 2017, with link to corrigendum May 26, 2017). Accessed 2017-07-05 (full-text html) at nature.com /articles/srep45242.

Marsh, Terry J., et al. *The Winter Floods of 2015/2016 in the UK: A Review*. Wallingford, UK: Centre for Ecology & Hydrology, 2016.

Mason, Paul. *Postcapitalism: A Guide to Our Future*. New York: Farrar, Straus & Giroux, 2015.

McKibben, Bill. *The End of Nature*. New York: Random House, 1989.

Milman, Oliver. "Meat industry blamed for largest-ever 'dead zone' in Gulf of Mexico." *The Guardian*, August 1, 2017. Accessed 2017-07-05 at is.gd/a5lYtN.

Moore, Kathleen Dean. *Great Tide Rising: Towards Clarity and Moral Courage in a Time of Planetary Change*. Berkeley: Counterpoint, 2016.

MSCI Emerging Markets ESG Leaders Index. "Cumulative Index Performance – Net Returns (USD) September 2007 to August 2017 and Annual Performance (%) 2008–2016." Brochure. New York: MSCI, 2017. Accessed 2017-09-05 (pdf) at is.gd/nqfRxP.

Municipal Natural Assets Initiative. "Making Nature Count." Brochure, n.d. Accessed 2017-07-05 (pdf) at institute.smart prosperity.ca/sites/default/files/mnai_flyer_web_version.pdf.

Navius Research Inc. *Modelling the Impact of the Climate Leadership Plan and Federal Carbon Price on British Columbia's Greenhouse Gas Emissions*. Report submitted to Clean Energy Canada, Pembina Institute and Pacific Institute for Climate Solutions, December 2016. Accessed 2017-07-05 (pdf) at pembina.org/reports/bc-clp-model-report-2016.pdf.

Nichols, Tom. "How America lost faith in expertise and why that's a giant problem." *Foreign Affairs* 96, no. 2 (March/April 2017). Accessed 2017-07-05 at foreignaffairs.com/articles/united-states /2017-02-13/how-america-lost-faith-expertise.

Nikiforuk, Andrew. "Warning for BC: Methane emissions still increasing in big US shale play." *The Tyee*, February 16, 2017. Accessed 2017-07-05 at thetyee.ca/News/2017/02/16/Warning-US-Shale-Play -Methane-Emissions-Still-Increasing.

One Planet Economy Network. "The Eureapa Tool: Applying Science to Policy Development." Brochure. Godalming, Surrey, UK: n.d. Accessed 2017-07-05 (pdf) at oneplaneteconomynetwork.org /resources/GreenWeek/Eureapa_Flier.pdf.

O'Riordan, Jon, and Robert William Sandford. *The Climate Nexus: Water, Food, Energy and Biodiversity in a Changing World*. Calgary: Rocky Mountain Books, 2015.

Pierce, Charles P. *Idiot America: How Stupidity Became a Virtue in the Land of the Free*. New York: Doubleday, 2009.

Raworth, Kate. *Doughnut Economics: Seven Ways to Think Like a 21st Century Economist*. White River Junction, VT: Chelsea Green Publishing, 2017.

Rifkin, Jeremy. *The Zero Marginal Cost Society: The Internet of Things, the Collaborative Commons and the Eclipse of Capitalism*. New York: Palgrave Macmillan, 2014.

Rockström, Johan, et al. "Planetary boundaries: Exploring the safe operating space for humanity." *Ecology and*

Society 14, no. 2 (2009): 32. Accessed 2017-07-05 (html) at ecologyandsociety.org/vol14/iss2/art32.

Sagan, Carl. *The Demon-Haunted World: Science as a Candle in the Dark*. New York: Ballantine Books, 1995.

Sandford, Robert William. *Cold Matters: The State and Fate of Canada's Fresh Water*. Calgary: Rocky Mountain Books, 2012.

———. *The Columbia Icefield*. 3rd ed. Calgary: Rocky Mountain Books, 2016.

———. *Storm Warning: Water and Climate Security in a Changing World*. Calgary: Rocky Mountain Books, 2015.

Saxifrage, Barry. "These 'missing charts' may change the way you think about fossil fuel addiction." *National Observer*, July 13, 2017. Accessed 2017-08-05 at nationalobserver.com/2017/07/13/analysis/these-missing-charts-may-change-way-you-think-about-fossil-fuel-addiction.

Schuster-Wallace, C.J., and R.W. Sandford. *Water in the World We Want: Catalysing National Water-related Sustainable Development*. Hamilton, ON: United Nations University Institute for Water, Environment and Health, 2015. Accessed 2017-07-05 (pdf) at inweh.unu.edu/wp-content/uploads/2015/02/Water-in-the-World-We-Want.pdf.

Schuster-Wallace, C.J., Chris Wild, and Chris Metcalfe. *Valuing Human Waste as an Energy Resource: A Research Brief Assessing the Global Wealth in Waste*. Hamilton, ON: United Nations University Institute for Water, Environment and Health, 2015. Accessed 2017-07-05 (pdf) at inweh.unu.edu/wp-content/uploads/2016/01/Valuing-Human-Waste-an-as-Energy-Resource-Web.pdf.

Smart Prosperity Initiative. *New Thinking: Canada's Roadmap to Smart Prosperity*. Ottawa: Author, February 2016. Accessed 2017-07-05 (pdf) from smartprosperity.ca/thinking/newthinking.

Steffen, W., et al., eds. *Challenges of a Changing Earth*. Proceedings of the Global Change Open Science Conference, Amsterdam, The Netherlands, July 10–13, 2001. International Geosphere–Biosphere Programme (IGBP) Science Series. Berlin: Springer, 2002. Table of contents and contributing authors accessed 2017-07-05 at https://link.springer.com/book/10.1007/978-3-642-19016-2.

———. *Global Change and the Earth System: A Planet Under Pressure.* International Geosphere–Biosphere Programme (IGBP) Science Series. Berlin: Springer, 2004. Accessed 2017-07-05 (346pp, pdf) from is.gd/eAoQYz.

Stiglitz, Joseph, Amartya Sen, and Jean-Paul Fitoussi. *Report by the Commission on the Measurement of Economic Performance and Social Progress.* Paris: INSEE (Institut national de la statistique et des études économiques), 2009. Accessed 2017-07-05 (pdf) from communityindicators.net/publications/show/9.

Sustainable Prosperity. *British Columbia Carbon Tax Review.* Submission to the British Columbia Government, September 2012. University of Ottawa, 2012. Accessed 2017-07-05 (pdf) from institute.smartprosperity.ca/library/publications/british-columbia-carbon-tax-review.

———. *The Value of Carbon in Decision-Making: The Social Cost of Carbon and the Marginal Abatement Cost.* University of Ottawa, November 2011. Accessed 2017-07-05 (pdf) at institute.smartprosperity.ca/sites/default/files/value-carbon-decision-making.pdf.

Task Force on Climate-related Financial Disclosures. *Recommendations.* December 2016. Accessed 2017-07-05 (pdf) at fsb-tcfd.org/wp-content/uploads/2016/12/16_1221_TCFD_Report_Letter.pdf.

Town of Gibsons. *Towards an Eco-Asset Strategy in the Town of Gibsons.* Gibsons, BC: n.d. Accessed 2017-07-05 (pdf) at gibsons.ca/include/get.php?nodeid=1000.

Tuchman, Barbara W. *The March of Folly: From Troy to Vietnam.* New York: Knopf, 1984.

United Nations. Framework Convention on Climate Change, Conference of the Parties no. 21, Annex: *The Paris Agreement* (FCCC/CP/2015/L.9 at p20). New York: United Nations, 2015. Accessed 2017-07-05 (pdf) at unfccc.int/resource/docs/2015/cop21/eng/l09.pdf.

———. *Transforming Our World: The 2030 Agenda for Sustainable Development* (A/RES/70/1). New York: United Nations, 2015. Accessed 2017-07-05 (pdf) from sustainabledevelopment.un.org/post2015/transformingourworld.

———. Vienna Convention for the Protection of the Ozone Layer, Final Act: *Montreal Protocol on Substances that Deplete the Ozone Layer*. Montreal, September 16, 1987 [English version begins at pdf p101, version française au p119]. Accessed 2017-07-05 (pdf) at treaties.un.org/doc/Treaties/1989/01/19890101 03-25 AM/Ch_XXVII_02_ap.pdf.

United Nations University Institute for Water, Environment and Health (UNU-INWEH). See Schuster-Wallace, C.J., et al. (two titles).

United States Environmental Protection Agency. *Climate Change in the United States: Benefits of Global Action*. Washington, DC: EPA Office of Atmospheric Programs report EPA 430-R-15-001, 2015. Accessed 2017-07-05 (pdf) at epa.gov/sites/production/files/2015-06/documents/cirareport.pdf.

University of Cambridge Institute for Sustainability Leadership (CISL). *Investing for Resilience*. ClimateWise Societal Resilience Programme report. Cambridge, UK: CISL, December 2016. Accessed 2017-07-05 (pdf) from www.cisl.cam.ac.uk/publications/sustainable-finance-publications/investing-for-resilience.

Watson, Cathy. "Indian farmers fight against climate change using trees as a weapon." *The Guardian*, October 29, 2016. Accessed 2017-07-05 at is.gd/Z1GWCG.

Welsh, Jennifer M. *The Return of History: Conflict, Migration and Geopolitics in the Twenty-First Century*. CBC Massey Lectures series. Toronto: House of Anansi Press, 2016.

White House, The. *United States Mid-Century Strategy for Deep Decarbonization*. Washington, DC: November 2016. Accessed 2017-07-05 (pdf) at obamawhitehouse.archives.gov/sites/default/files/docs/mid_century_strategy_report-final.pdf.

Wollenberg, E., et al. "Reducing emissions from agriculture to meet the 2°C target." *Global Change Biology* 22, no. 12 (December 22, 2016).

Accessed 2017-07-05 (html abstract only) at ncbi.nlm.nih.gov/pubmed/27185416.

World Agroforestry Centre. "India's National Agroforestry Policy." *Trees for Life: Annual Report 2013–2014*, pp. 5–6. Nairobi: World Agroforestry Centre, 2014. Accessed 2017-07-05 (html) at worldagroforestry.org/ar2013/index.php/indias-national-agroforestry-policy.

World Energy Council. *World Energy Scenarios 2016: The Grand Transition*. London: World Energy Council, October 2016. Accessed 2017-07-05 (pdf) from worldenergy.org/publications/2016/world-energy-scenarios-2016-the-grand-transition.

Zomer, Robert J., et al. "Global Tree Cover and Biomass Carbon on Agricultural Land: The contribution of agroforestry to global and national carbon budgets." *Scientific Reports* 6 (July 20, 2016): 29987. Accessed 2017-07-05 at dx.doi.org/10.1038/srep29987.

ABOUT THE AUTHORS

BOB SANDFORD

Bob Sandford holds the EPCOR Chair in Water and Climate Security at the United Nations University Institute for Water Environment and Health. In this capacity, Bob was the co-author of the UN *Water in the World We Want* report on post-2015 global sustainable development goals relating to water. Bob is also a senior adviser on water issues for the Interaction Council, a global public policy forum composed of more than 30 former heads of state or government, including Jean Chrétien, Bill Clinton and Gro Harlem Brundtland. Bob is a fellow of the Centre for Hydrology at the University of Saskatchewan and of the Biogeoscience Institute at the University of Calgary. He is a senior policy adviser for the Adaptation to Climate Change team at Simon Fraser University and a member of the Forum for Leadership on Water (FLOW), a national water policy research group located in Toronto. Bob is also the author or co-author of a number of books, including *Cold Matters: The State and Fate of Canada's Fresh Water*; *The Climate Nexus: Water, Food, Energy and Biodiversity in a Changing World* (with Jon O'Riordan); and, most recently, *North America in the Anthropocene* and *Our Vanishing*

Glaciers: The Snows of Yesteryear and the Future Climate of the Mountain West, all published by Rocky Mountain Books.

JON O'RIORDAN

Jon O'Riordan completed post-secondary studies in water resource management at the universities of Edinburgh and British Columbia. His professional career spanned 36 years, in both the federal and British Columbia governments, in environment and resource management. He is an adjunct professor in the School of Community and Regional Planning at UBC, where he taught for ten years. He is currently research adviser to Simon Fraser University's Climate Change Adaptation Team and to the Polis Institute for Ecological Governance in the Centre for Global Studies at the University of Victoria.

RMB saved the following resources by printing the pages of this book on chlorine-free paper made with 100% post-consumer waste:

Trees · 9, fully grown
Water · 4,297 gallons
Energy · 4 million BTUs
Solid Waste · 288 pounds
Greenhouse Gases · 792 pounds

CALCULATIONS BASED ON RESEARCH BY ENVIRONMENTAL DEFENSE AND THE PAPER TASK FORCE. MANUFACTURED AT FRIESENS CORPORATION.

CONTENTS

Jack O' Connell
Seattle
20 April 1982

The Discovery of Dura-Europos

Clark Hopkins.

The Discovery of Dura-Europos

CLARK HOPKINS
Edited by BERNARD GOLDMAN

NEW HAVEN AND LONDON
YALE UNIVERSITY PRESS 1979

Designed by Thos. Whitridge and set in Garamond type.
Printed in the United States of America by The Murray Printing Company, Westford, Mass.

Published in Great Britain, Europe, Africa, and Asia (except Japan) by Yale University Press, Ltd., London. Distributed in Australia and New Zealand by Book & Film Services, Artarmon, N.S.W., Australia; and in Japan by Harper & Row, Publishers, Tokyo Office.

Library of Congress Cataloging in Publication Data

Hopkins, Clark, 1895–
The Discovery of Dura-Europos

Bibliography: p.
Includes index.
1. Dura, Syria. 2. Excavations (Archaeology)—Syria—Dura. I. Goldman, Bernard, 1922– II. Title.
DS99.D8H66 935 78-31193
ISBN 0-300-02288-3

ILLUSTRATION CREDITS

Maurice Le Palud, Mission de Doura Europos, Syria, 1932–33: Pages 13, 30, 31, 33, 45, 51, 58, 70, 118, 127, 128, 134, 137, 154, 170, 191, 241, 250.
Vasari-Rome: Page 198 (bottom).
Yale University, Dura-Europos Excavations, 1933–34: Pages 74, 196 (top), 198 (top), 205 (photo nos. G1124A, G 939B, G 852A, G 884A); 1934–35: pages 40, 129, 179 (bottom), 204, 220, 236 (photo nos. H 154A, H 24A, H 139A, H 72A, H 142A, H 76A).
Yale University, Gallery of Fine Arts: Pages 111, 113, 141, 221.
All other illustrations are from private collections.

After the siege and victory by the Persians in AD. *256, the record is blank. The mute testimony that remained was of a site desolate and forlorn, where the lonely and level sands covered the bones of the city and stretched away across the desert.*

Contents

Following pages: Aerial view of Dura-Europos, December 14, 1932.

14 dic. 1932

Illustrations

Foreword

Dura-Europos first stood as a Hellenistic outpost built by the Macedonian Greeks. The desert city overlooking the Euphrates River in Syria changed hands to serve for more than four centuries as a heavily fortified western extension of the Iranian Parthians and, finally, as a town encampment for Roman troops guarding the eastern *limes* of the empire. It fell to the invading Persian Sasanians in mid third century A.D.; depopulated and deserted, it was gradually buried under the wind-blown sands and crumbling brick walls, a forgotten, nameless ruin sometimes sheltering the wandering Arab herdsmen behind its drifted fortifications. The chance discovery by British troops in 1920 of fragments of mural painting brought it to the attention of French and American archaeologists, who converted the desolate town in the course of twelve campaigns, from 1920 to 1937, into one of the famous archaeological recoveries of the twentieth century.

Today hardly a book that touches on the ancient Eastern Mediterranean sphere, on the history of religion or art in the Orient, is without reference to the spectacular find of the painted Christian Chapel of Dura, or to the astonishing revelation of a standing Jewish synagogue whose four walls had been covered with paintings drawn from the Old Testament. Both buildings remain unique discoveries, and both were found and cleared under the field direction of Clark Hopkins. The importance of the other discoveries made at Dura is somewhat dimmed by the brilliance of these two monuments, but they are of little less significance: fortifications, temples, palaces, baths, houses, and shops; sculpture, painting, and drawings; papyri, parchments, graffiti, and coins; clothing, armor, weapons, ornaments, and vessels. For a while it seemed that Dura would prove an inexhaustible source of historical and artistic treasure, and, indeed, we still have not completely measured its importance nor studied all its remains.

Clark Hopkins joined in the excavation and study of Dura-Europos on the invitation of Michael I. Rostovtzeff. He served first as assistant director and then field director at Dura until his field participation came to an end in 1935, when he left Yale University to assume his appointment at the University of Michigan and the directorship of its excavations at Seleucia-on-the-Tigris in Iraq. His retirement from

the University in 1965 meant that he could now give all his time to writing, among other studies, the final topographical and architectural report on Seleucia, to planning and taping a television series on ancient history and religions, and to restudying the archaeology of Dura-Europos.

Hopkins had planned a review of the excavations at Dura which would include a characterization of the people and events that restored ancient Dura for the modern world of scholarship, a detailed analysis of the arts (particularly the complex paintings of the Christian Chapel and the Synagogue), an appreciation of the religions practiced in the city, and a survey of the contributions of Dura to our understanding of the Hellenic-Semitic-Iranian complex in the twilight of ancient history. When he relinquished his pen at the age of eighty, on May 21, 1976, he had completed and partially revised a major portion of the writing. Much yet remained in first, preliminary draft, some in notes, and some sections only as items in the proposed table of contents. The richness of the information he had committed to paper—the narrative of his amazing discoveries, the keen insights that are the product of rigorous study and discipline, and the firsthand accounts of the processes and conditions of discovery—demanded publication rather than shelving because the work was incomplete.

With the consent and advice of his daughter, Mrs. Mary Sue Coates, and his son, Dr. Cyrus Hopkins, I have edited the manuscript. Left intact, except for minor changes, are the primary emphases of Hopkins's manuscript, his narrative of the several seasons of digging at Dura and his conclusions on the art and architecture, to which over forty years of teaching, research, and excavation have given authority and brilliance that must command scholarly respect. The narrative of the dig at Dura is captivating and forms a book-length entity in itself, but its importance lies in what the dig produced, an historical treasure trove. For the sake of brevity and to make the book available to the nonspecialist, I have excised much of the detailed descriptive analysis, particularly of the Synagogue paintings, and the patient, lengthy accumulation of data and references that provide the sound basis for his conclusions. Hence, some of Hopkins's summary remarks, which may appear here to be overbold and without substantial foundation, were raised on a scrupulously built documentation in the first draft. I felt it best not to repeat, unless absolutely necessary, the basic information on the Dura finds, which is readily available in the continuing series of volumes forming the *Final Report* on Dura-Europos. The specialists in

history, archaeology, papyrology, religion, numismatics, and so on will of course have easy access to these handsomely illustrated volumes.

I have made no substantive alterations of conclusions in the manuscript, and, hence, I have taken the liberty of not bracketing in the usual editorial fashion my modest contributions. I found in the voluminous correspondence of Clark and Susan Hopkins much good material which I have been sorely tempted to include, but have not. To have done so would not have been consonant with Clark Hopkins's innate modesty and his good humor, which forbade any hint of meanness in others.

Unfortunately, I cannot acknowledge the many people and institutions the author would have been glad to salute. John Griffiths Pedley and Louise Shier of the Kelsey Archaeological Museum must be mentioned for their assistance, as also Stephanie Weinreich, Barbara Dorr, and Carol Bromberg, who contributed their skills. I am also indebted to Harald Ingholt, a long-time friend of Hopkins, and to Frank E. Brown, who figures largely in the excavations of Dura-Europos. I am grateful to Clark Hopkins's son and daughter for entrusting me with the project. But most of all I must give full thanks on behalf of Clark Hopkins and myself to the lady who made the Dura excavations a reality, the Tyche of Dura-Europos.

That Clark Hopkins came to the task of summarizing the excavations at Dura-Europos with zest and enthusiasm, with the humanity and humor that characterized his career as archaeologist, scholar, teacher, and friend is everywhere apparent in the pages that follow.

B. G.

Preface

This account has to be a personal one, but I hope it will present a picture of the Yale excavations at Dura-Europos which will supplement the *Preliminary* and *Final Reports*. My purpose is not to review all the incidents of the seasons and the finds of Dura, nor even to detail all the events while I was a member of the team there. But excavations are journeys of discovery—and at Dura we had a particularly rewarding one—and such journeys demand a chronicle. The journey has its ups and downs, its periods of discouragement, and its sudden delights and triumphs. The significance of finds often does not appear at once, nor does the importance of the small detail and the circumstances of the finding. The final reports of an excavation, with the complete, detailed descriptions, often fail as estimates of significance to do justice to their subjects, because they are directed to the specialist and too often become lost in the particular archives devoted to the special phase.

At the same time there is in every expedition the personal side, the hopes, aims, and reflections of the members of the staff, the special contributions and the interparty relationships worthy of note, just as are those of a mountaineering team or of an Arctic expedition. Here again I have no intention of recalling all or even of mentioning all.

As the seasons of excavation progress, one sees gradually unfolded the history of the site, its place in the history of the region, and the special contribution it makes to our understanding. Although it may light one very tiny corner of the vast panorama of man's ancient history, it serves to help illumine a much greater area. "How far that little candle throws his beams."

Here I attempt to bring together in review the successive stages in the Dura excavations, and by recalling the circumstances of the discoveries and at the same time glancing forward to the reports of their special contributions and significance, to give both the general reader and the archaeologist a broader understanding of Yale's expeditions to Dura-Europos.

C. H.

Map of Dura-Europos Showing the Placement of the Buildings.

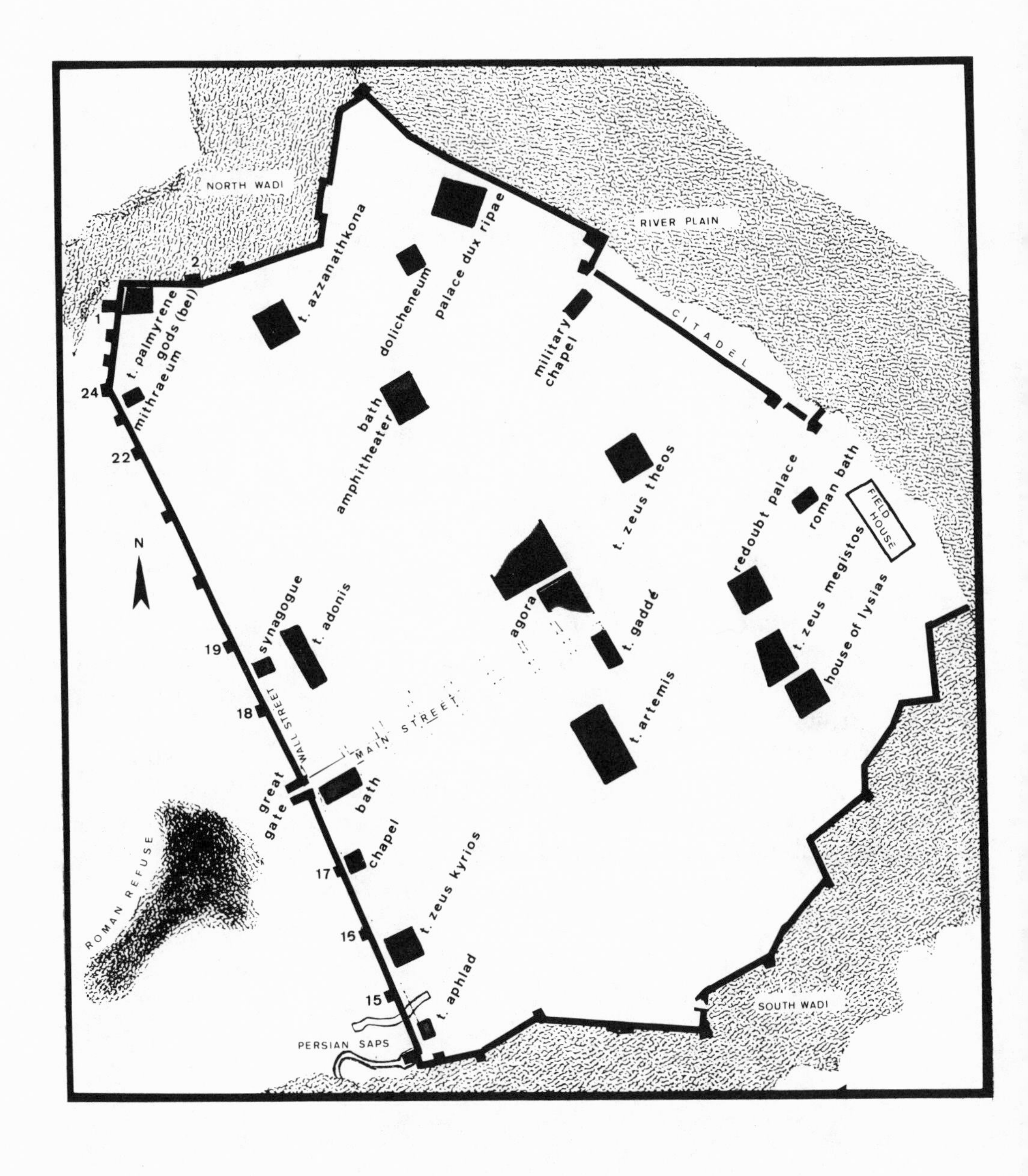

The Seasons at Dura

1898		Site visited by Schulz and Sarre.
1912		Sarre and Herzfeld explore site, find Greek inscriptions and evidence of wall painting.
1920	Mar 30	British troops discover paintings (Temple of Palmyrene Gods).
	Apr 23	Breasted asked to investigate fortress.
	Apr 28	Breasted leaves Baghdad for site.
	Apr 29	Site becomes part of French-Syrian protectorate (Treaty of San Remo).
	May 3	Breasted uncovers Konon paintings and finds name of city, "Dura."
1922	Jul 7	Breasted makes first report on Dura to French Academy.
	Oct	Cumont commissioned by Academy to excavate Dura.
	Nov 7–18	Cumont begins clearing Temple of the Palmyrene Gods, Tower of the Archers, Redoubt, center of city, some tombs, Temple of Artemis-Nanaia. Discovers name of "Europos" to identify Dura.
1923	Oct 3–Nov 7	Cumont's second season continues work of first; find soldier's shield with itinerary.
1924		No formal excavations, but soldiers prospect over area.
1925		Rostovtzeff at Yale, begins to plan for a major excavation.
1926		Druse revolt in Syria ends; area safe for excavating.
1927		Syrian government grants permission for joint Yale-French Academy expedition. Pillet sent to inspect site.

1928	
First Season	
Feb 28	Pillet, field director of expedition, arrives in Beirut.
Apr 13	Pillet opens first season at Dura.
Apr 20	Begin clearing of Great Gate, Redoubt, Citadel. Major finds: Nemesis relief, leather-covered shield.
May 6	Close of first season.
Second Season	
Oct 1	Hopkins arrives in Beirut to serve as Pillet's assistant.
Oct 21	Excavation team arrives at Dura.
Oct 25	Trenches opened; second season begins.
1929	Clearing continues on Citadel (skeleton of the strangled man), Roman bath, towers, Great Gate, Temple of the Palmyrene Gods; find large hoard of coins.
Feb 22	Official opening of new expedition house.
Mar 10	Hopkins leaves Dura.
Mar 31	Close of season.
Third Season	
Oct 30	Pillet opens excavations.
1930 Jan 28–Mar 27	Pillet hospitalized in Aleppo; supervision of work in hands of Naudy, Rowell, Walter.
	Clearing of Temple of Artemis, Temple of Atargatis, triumphal arch, city walls, Roman bath near Great Gate; discovery of relief of Hadad and Atargatis.
Apr 9	Close of season.
Fourth Season	
Oct 31	Pillet returns as director and opens season.
	Continue clearing of triumphal arch, desert wall, Redoubt Palace, House of Merchant Nebuchelus, Temple of Aphlad, House of the Large Atrium, House of the Archives; find Sasanian mural.
1931 Mar 31	Season ends, but part of staff remains.
Apr 29	Staff removing paintings from Temple of the Palmyrene Gods departs.

Fifth Season		
	Oct 4	Hopkins arrives in Beirut to assume field directorship.
	Oct 26	Season begins. Clear House of Archives, Praetorium, hall with inscribed steps, Temple of Artemis-Azzanathkona, Agora; find relief of Azzanathkona.
1932	Jan 17	Traces of wall painting found in house (Christian chapel).
	Jan 18	Excavation of wall paintings reveals baptismal font; wall paintings begin to appear.
	Jan 19	Paintings inside baptismal niche revealed and Three Marys at tomb.
	Feb 12	Find small Roman bath.
		Bits of wall painting found in embankment north of Great Gate (Synagogue).
	Feb 28	Fifth campaign officially ends but work continues through March and into April.
Sixth Season		
	Oct	Hopkins opens excavations in last week of month.
		Continue clearing of Chapel, Roman quarter of city, Citadel, Temple of Artemis, Parthian Bath, House of Scribe; major finds of scale armor, skeletons in sap, painted shield, Victory panel.
	Nov	Synagogue paintings begin to appear last week of month.
	Dec	Murals of hunting and banqueting found in house.
1933	Mar 5	Discovery of Diatessaron.
		Season closes last week of March.
Seventh season		
	Oct	Hopkins opens season end of month.
		Clearing of Wall Street, Agora area, Temple of Zeus Kyrios, Temple of Adonis, Temple of Zeus Theos, tombs in Necropolis; find painted shields, painting of Venus and Cupid.
1934	Jan 26	Von Gerkan begins study of circuit walls.
	Feb	Discovery of Mithraeum.
	Feb 23	Rainstorm causes slight damage in Synagogue.
	Mar	Season closes at end of month.

Eighth Season	
Oct 30	Hopkins opens his last season at Dura. Clearing of earlier Synagogue remains, Temple of Zeus Kyrios, House of Lysias, tombs in Necropolis, Temple of Bel and Iarhibol; find three painted shields.
1935 Feb 6	Discovery of Temple of Gaddé.
Feb 20	Hopkins closes season, leaves Dura to assume directorship of excavations at Seleucia-on-the-Tigris.
Ninth Season	
	Brown assumes field directorship.
1935–36	Continue clearing of Agora, Necropolis, House of Lysias; find Dolicheneum, Palace of Dux Ripae.
Tenth Season	
1936–37	Work continues in Necropolis, Temple of Zeus Megistos, Temple of Atargatis, Agora, Redoubt Palace.

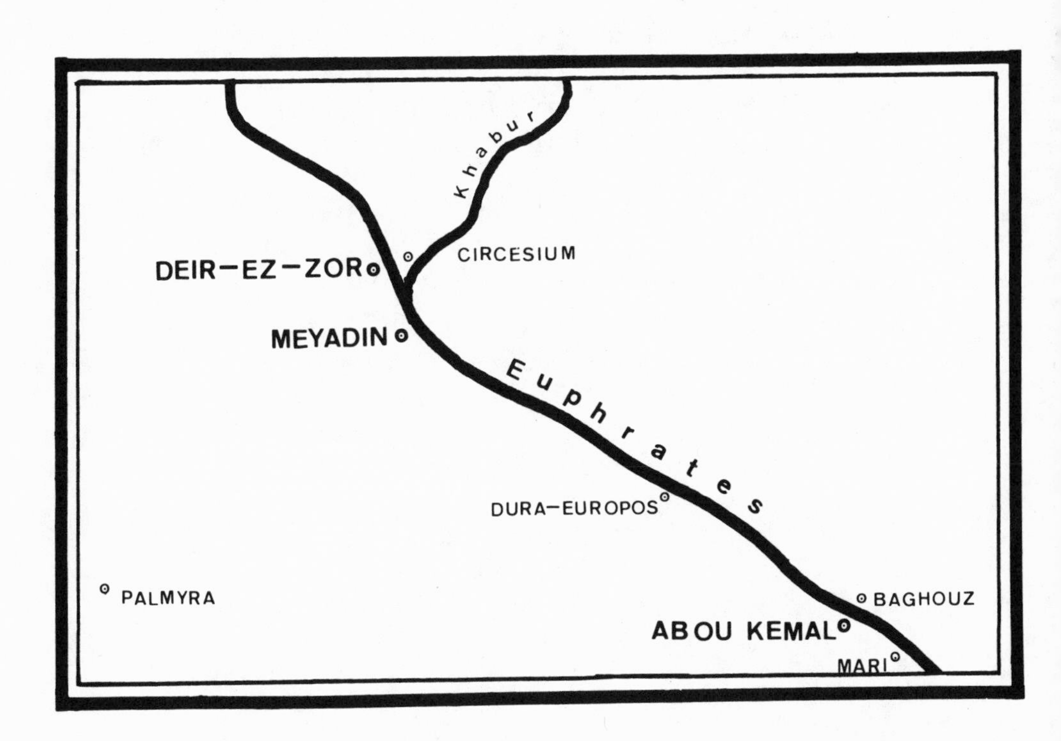

1 Discovery

The collapse of the Turkish Empire at the close of World War I left the Near East a political shambles, with the British and French foreign offices striving to impose order. In the years immediately following the close of hostilities, rival Arab factions, eager to control their own destinies, broke into open revolt. By 1920 the British army maintained only a tenuous hold on the parched Syrian desert along the middle Euphrates River. In March of that year, Captain M. C. Murphy and a company of British soldiers bivouacked in a ruined desert fortress near the village of Salihiyeh, overlooking the Euphrates. From British headquarters at Abou Kemal, a day's march downriver, Murphy sent a letter to his commanding officer, Lt. Colonel G. Leachman.

> While at Salihiyah I discovered on the 30th inst. some ancient wall paintings in a wonderful state of preservation. The paintings are in the west corner of the fort and consist of life-size figures of three men, one woman, and three other figures partly obliterated. The colours are mainly reds, yellows and black. There is also some writing which I have tried to reproduce below.
>
> I should be glad if you would forward this to the proper quarter.[1]

Leachman dispatched Murphy's report to the civil commissioner, Colonel A. T. Wilson, with a covering note.

> As a result of our occupation of the old fort at Salihiyah and the digging of trenches, a certain amount of finds have been made. The paintings to which the attached refers are most interesting and should, I think, be seen by an expert. If your American archaeologist is still about, it would well repay him to come and see this. The films enclosed are of the pictures. Could you please have them developed. If anyone comes up, it should be soon for obvious reasons.[2]

A rough sketch of the wall decorations was also sent to headquarters in Baghdad by General Cunningham, who was in command of the upper Euphrates.

The "American archaeologist" mentioned by Leachman was James Henry Breasted, heading an archaeological team for the new Oriental Institute of the University of Chicago. When he returned on April 23, 1920, from his expedition to the upper Tigris, where he had been exploring mounds for future excavations, he was asked by Wilson and Major General Percy Hambro, quartermaster general of the British army in Mesopotamia, if he could make the trip to the middle Euphrates. "They would be glad," Breasted later recalled, "if we could undertake to ascend the Euphrates . . . on their behalf."[3]

In the office of the civil commissioner, Gertrude Bell, the English traveler, Orientalist, and author, spearheaded operations and warmly supported the invitation. In her prewar explorations, Miss Bell had noted this "castle," as she called the ruins, "on the Syrian side of the Euphrates . . . hemmed in here by hills."[4] She reported that Wilson would provide Breasted with transport if he could be convinced to go. But Breasted needed no urging, particularly because the trip would permit him to go up the Euphrates well beyond the limit previously set by British headquarters, for the middle Euphrates was very much a war zone.

Breasted had been reconnoitering ancient sites in Egypt and buying antiquities for the University of Chicago. He had planned on visiting the ancient mounds of the Near East, but conditions had forced him to proceed to Mesopotamia indirectly by way of Bombay. Before he left Egypt, High Commissioner Allenby had given him a secret document to read which detailed the turmoil of the "whole middle section of the Fertile Crescent from Baghdad to Aleppo and Damascus."[5] Disappointed, he had shifted his attention to a survey of archaeological sites in Persia. Now the picture was radically altered. He must go almost immediately, "for obvious reasons": the civil commissioner told him in strictest confidence that a withdrawal, known only to the High Command, was about to be undertaken, an operation that would shift the British frontier on the Euphrates about one hundred miles farther downriver. If he wanted to seize this singular opportunity to dash north, Persia would have to wait.

It was nearly the end of April before Breasted's small caravan began its long journey.

> On April 28 in seven automobiles kindly furnished us by the British Army and Civil Government, we left Baghdad for the Upper Euphrates. The accidents and delays of desert travel by automobile were such that the nearly 300-mile trip from Baghdad

to the British Frontier on the middle Euphrates occupied an entire week.[6]

On the afternoon of May 3, General Cunningham, in command at Abou Kemal, drove Breasted and his assistants the twenty-seven miles from headquarters to Salihiyeh, past the sites of recent battles with the still hostile Arabs, for a preliminary examination of the paintings. Breasted was astonished by the enormous size of the ruined fortress and amazed by the magnificence of the painted figures.

> Descending from the car, General Cunningham led the way over the rubbish piles commonly found in such ruins and around a jutting corner of massive masonry. Suddenly there rose before us a high wall covered with an imposing painting in many colors depicting a life-size group of eleven persons engaged in worship. . . . It was a startling revelation of the fact that in this deserted stronghold we were standing in a home of ancient Syrian civilization completely lost to the western world for sixteen centuries.[7]

After obtaining permission to have several squads of East Indian troops available to clear the foundations of the building and to expose the paintings he could see in an adjoining hall, Breasted withdrew for the night to Abou Kemal. He returned next morning to find the troops already at work.

> We saw to our surprise a small scene in which a Roman tribune was depicted at the head of his troops, engaged in the worship of what looked like three statues of Roman emperors painted on the wall. There was the tribune's name written beside him in Roman letters: "Julius Terentius, tribune," and before him was the red battle flag of Roman troops. . . . Here were those tokens of Roman occupation full 35 miles outside of the well-known Roman frontier at Circesium by the mouth of the Khâbûr, which we were not expecting to see until we had put a two days' march behind us. We had before us the easternmost Romans ever found on the Euphrates or anywhere else for that matter.[8]

One of the paintings carried a depiction, carefully labeled, of the Good Fortune of Dura—the Tyche of the city—providing Breasted with the ancient name of the city.[9]

Photographs were taken by Daniel Luckenbill, detailed notes on

the composition were jotted down, and notes made for the later insertion of approximate colors. As the Indian troops cleared the ground, sketch plans were made, and the younger members of the party toured the city walls to determine the size of the fortress. Darkness fell, and the last photograph was made by the dying light. The next morning, after a cold breakfast, the party left in their Turkish *arabanahs* (horse-drawn springless carriages) with an escort of mounted Arab rifles provided by Colonel Leachman.

Breasted compared the style of the paintings he had uncovered with two mosaics from the Basilica of San Vitale in Ravenna, illustrative of the Byzantine style in the time of Justinian, the middle of the sixth century. Both paintings and mosaics exhibit the conventions of strict frontality of the figures, stylized robes, staring eyes, and rich costumes and jewelry. The Byzantine style was a development from the Roman with strong Oriental influence. At Dura, in that sudden revelation of 1920, a first link was revealed between early Eastern Hellenistic–Roman art and sixth-century Byzantine.

Two factors had been chiefly responsible for keeping the city of Dura from prior archaeological examination. First, the extensive ruins lay on the right (west) bank of the Euphrates, and the cliff on which they stood blocked the view from the opposite (east) bank, of all but a fraction of the old walls. Since the main caravan road lay on the east bank, the ruins had caused little comment, even from so astute a traveler as Gertrude Bell. Second, according to the historians, the southern limit of the Roman frontier lay forty miles up the river at the confluence of the Khabur River with the Euphrates, where Roman Circesium was founded to guard this important junction for travel and commerce. The ruins of Dura, therefore, had generally been consigned to the Arab period and so were beyond the interest of Classical and Ancient Oriental archaeologists.

The only geographic mention of Dura in ancient records is found in the *Parthian Stations* of the Greek Isidorus of Charax, written in the first century A.D. This reference was to a "Dura, called Europos by the Greeks," the first stop as one went down the river from Circesium. Although not recognized as Dura, the ruins had caught the eye of late-nineteenth-century travelers in the region.[10] Bell mistakenly identified the Dura of Isidorus as "the very striking tell Abu'l Hassan, the biggest mound upon this part of the river."[11] The site referred to by Isidorus, however, was known as Salihiyeh when Bruno Schulz and Friedrich Sarre visited it in the spring of 1898, and when Sarre re-

turned with Ernst Herzfeld in 1912. They reported on some Greek inscriptions found there, noted the Hellenistic style of the masonry, identified fragments of Parthian remains, and mentioned traces of wall painting in the west corner of the city on a building made of rough gypsum ashlar. They published drawings and striking photos of this "nameless city" in their monumental four-volume *Archäologische Reise im Euphrat- und Tigris-Gebiet* and correctly identified it as part of the caravan route between Palmyra and Seleucia.[12] Such a report would certainly have led to further investigation, but World War I held up both archaeological work and the distribution of German scientific volumes. Thus, the site lay neglected until Breasted spent his day there, and to him belongs the honor of finding in one of the paintings the Greek inscription that identified the site as Dura.[13]

The end of World War I brought two fundamental changes to the Near East. First, the Syrian coast and all of Mesopotamia were mandated to European powers, which encouraged archaeological research. Second, improved cars, less hampered by the absence of roads than earlier models and their range extended by the introduction of five-gallon gas cans, facilitated exploration and communication. Before and during the war, the horse, the camel, and the arabanah had carried desert travelers and freight. They were slow and difficult, and travel was often dangerous. After the war, the automobile, the autobus, and then the armored car claimed the desert as their own.

A better-qualified man than James Henry Breasted could not have been found to make the preliminary examination of the freshly uncovered paintings and to publish the results. He had begun teaching at the University of Chicago in 1894, serving as professor of Egyptology and Oriental history from 1905 until 1933. He also acted as director of the Haskell Oriental Museum from 1895 until 1901 and became director of the Oriental Institute in 1919. Although the outbreak of fighting in Mesopotamia had seriously curtailed the areas he could explore, the accidental discovery of the murals brought the opportunity to pierce the military frontier. Equally fortuitous was the presence of Gertrude Bell in Baghdad in a sufficiently influential position both to enlist the military and to induce Breasted to undertake the dangerous and arduous mission.

After the war with the Arabs, the district of Deir-ez-Zor (Dayr az Zawr), which included Dura, fell within the new boundaries of the French mandate. Breasted was cordially invited by the French Academy to continue the work so auspiciously begun, but obligations

in Egypt prevented his acceptance. As soon as possible, however, he called attention to the amazing new finds. He was invited to give an account to the French Academy of Inscriptions and Belles-lettres on July 7, 1922. His exposition of the result of that single day's work at Dura made up a slim volume, *The Oriental Forerunners of Byzantine Painting,* which included the exciting account of his journey to Dura and how it had all begun. A tour de force and an immense contribution to Near Eastern archaeology, it formed the first volume of the great series of reports and studies issued in subsequent years by the Oriental Institute.

From the vantage point of later expeditions and discoveries, I am tempted to call Dura the doorway to a new understanding of the great change in Oriental art that began with the Christian era. The Temple of the Palmyrene Gods, as we now know the building that Breasted recovered in part, brought to the fore the problem of Parthian art. Sarre and Herzfeld had recognized Parthian characteristics in some minor finds, but Herzfeld despised the Asiatic Parthians as interlopers (ca. 250 B.C.–A.D. 224) between the old and the new Persian empires of Cyrus and of Shapur, between the Achaemenian dynasty and the Sasanian dynasty. The Sasanians had not claimed inheritance from the Parthian kings, whose royal house they had overthrown in A.D. 224, but from the empire of Cyrus, Darius, and Xerxes. The Romans, early adversaries of the Parthians when their territorial interests collided, had feared and hated them. In general, history has ill-treated the Parthians. George Rawlinson recognized their importance and included them in his nineteenth-century four-volume study of the great Oriental monarchies, but a lack of preserved documents had seriously hampered scholars' knowledge of Parthian times and culture.

The paintings preserved in the Temple of the Palmyrene Gods at Dura promised a glimpse at long last of the art, religion, and institutions of an empire that rivaled the Roman for five hundred years. Nor was the promise in vain. Successive expeditions brought to light still more impressive and valuable discoveries, far beyond anyone's dreams.

2 Franz Cumont at Dura

Two centuries of Achaemenian Persian control of Syria were followed by a century and a half of Hellenistic rule. The Parthians and the Romans then contended for the lands of the ancient Near East, while the Nabataean Arabs carved out a separate state of their own in northern Arabia between the Euphrates and the Palestinian littoral, with capitals at Petra and at Charax at the head of the Persian Gulf. The new frontiers of Rome and Parthia were fixed in 20 B.C. by a treaty, and captured Romans and their standards were returned by the Parthians.

When the Sasanian Persians succeeded the Parthians in Syria in A.D. 224, and as the military presence of Rome waned, the fortunes of the independent Syrian city of Palmyra rose. For ten years (261–271) she ruled as queen of all the middle Euphrates district and extended her influence as far as Egypt and Asia Minor. Her brief period of glory ended when Aurelian captured Palmyra and carried Queen Zenobia captive to Rome. The underlying strength of the native Syrians remained, however, and after Rome had withdrawn from the East, after Sasanians and Byzantines had fought over the territory for three hundred years, the Arabs, at first under Mohammed and then under the banners of Islam, carried their faith and culture from Mecca and Medina as far as India, through northern Africa, and northwest up to the Pyrenees.

History does not repeat itself, but the nature of man remains much the same, and sometimes new circumstances parallel those of the past. After the Arab victories came the invasions of the Crusaders and Mongols; then the whole of the Near East was taken over by the Ottoman Turks, and the capital shifted from Baghdad to Constantinople. During the First World War, as Turkish power dwindled, Russia, England, and France sought to control the Near East. At the war's end, the Arabs, under Faisal, tried to make Damascus the capital of a new Arab empire. The attempt failed, and mandates over the whole territory were given to the European powers. Today the Arabs are once more asserting their independence in separate kingdoms and building their strength with the newfound wealth from oil.

Under Hellenistic, Parthian, and Roman rulers, the basic population of Dura, the Semitic Arabs, remained constant: conservative in

religion and social structure, retaining its language, and exerting an ever-growing influence on the cultural development of their Greek, Parthian, Roman, Semitic city. Arab workmen at Dura and the merchants of Deir-ez-Zor and Abou Kemal assisted in our excavations, just as the combined French-American staff and the French army officers supervised and safeguarded the work. The division of mandates determined the shift from British to French control, yet the Arab workmen represented the same strong Semitic influence under modern French and British rule that they had some two millennia earlier under Greek, Parthian, and Roman.

The excavations at Dura, conducted in the period between the two great wars, were vitally affected by the rivalry of the great Western powers and by the strength and goodwill of our Arab workmen and Armenian foremen. By the Treaty of San Remo, signed on April 29, 1920, the frontier between Iraq and Syria crossed the Euphrates immediately below Abou Kemal and so placed Salihiyeh and the ruins of Dura-Europos in the French mandate. The anticipated change in the border was, of course, the reason that British headquarters hurried Breasted on his journey, which began the day before the signing on April 28. Arab discontent with the French mandate eventually burst into open revolt, first in Jebel-ed-Druz, then all over southern Syria (1925–26).

Archaeology prospered in the 1920s, with the French supporting new museums in Damascus and Aleppo to care for the Classical and pre-Classical antiquities. Excavations were begun in coastal Byblos and Minet el-Beida; the great precinct in the desert capital of Palmyra was cleared of its crowded native houses and the population was moved away from the ruins. Restorations of precarious ruins were begun at Baalbek. The splendid archaeological journal *Syria,* the major vehicle for reporting the magnificent discoveries of the excavators, published its first issue in 1920, and the French Antiquities Service began a series of volumes to dramatize the archaeological wealth of the country.

Dura, on the Euphrates just north of the new Syrian border, came under the jurisdiction of the French army headquarters in Deir-ez-Zor. Breasted's report on the Dura murals to the French Academy in the summer of 1922 aroused great interest, and it was proposed that the excavations at Dura be extended. General Gouraud in Beirut offered to send troops to the fortress to aid in the dig and also to protect the archaeologists in their study of the site. The Belgian scholar Franz Cumont was commissioned by the French Academy to proceed to

Dura. The military situation permitted him only two short periods of work, in November of 1922, and in October and November of 1923.

Arab success in Iraq against the English strengthened Arab opposition to the French in Syria. This stubborn resistance brought ever stricter French control: the international press was muzzled and the legislature was kept on a tight rein. The country was under a form of martial law, and Cumont excavated by means of the soldiers of the Foreign Legion and under their protection. In 1924, after Cumont's two campaigns, no formal work was undertaken, but soldiers protecting the site employed their time in excavating a bit further. The following year, 1925, the Druse revolt against French control burst across the desert lands. Two powerful French columns sent to suppress it were destroyed, and the insurgents advanced from the south as far as Damascus. They failed to take the city but held most of the surrounding country. In September the Druse leaders were joined by nationalist leaders in other parts of Syria and the rebellion spread. At long last the French were forced to recognize that their attempt to rule the Levant by martial law was a failure. The war dragged on into 1926, with the environs of Damascus under fire and villages shelled by field guns and burned by French mercenaries. Finally, in August of 1926, peace was restored with the suspension of martial law, followed by the appointment of a new French high commissioner, with civilian administrative experience, Henri Ponsot, who was eager to reach a real agreement with the Arabs and their Syrian leaders.

Meanwhile the French archaeologists had extended their operations in sites along the coast. Claude Schaeffer disclosed startling Bronze Age remains in Ras Shamra and Minet el-Beida; Maurice Dunand was finding examples of early alphabetical writing in Byblos. The French Archaeological Service in Syria continued exploring and repairing the colossal remains of Baalbek. Palmyra now needed additional care and restoration, while the extensive ruins of the Hauran—ancient, Islamic, and Christian—required protection.

The political history of Syria between the two world wars provides the background for the practical work of excavation in Dura-Europos from 1920 to 1937. There was an interesting parallel between the modern scene and that of earlier times, as foreign powers fought over, controlled, and lost to each other this desert land. Centuries before, the Arabs had been attacked by the Crusaders from the West (1195–1300), and then their Near Eastern empire had been savagely destroyed by the Mongols from the East, first under Genghis Khan and

then by Timur (Tamerlane). The resulting decline in power gave the Ottoman Turks their opportunity, and under Selim I, the Grim, (1512–20), Syria, Palestine, and Egypt became subject to the rule of the sultan. The Ottomans were of a different race and spoke a different language from the native Semites, but by assuming the caliphate, Selim made himself and his successors the spiritual as well as the temporal heads of the empire and gained control over the holy cities of Mecca and Medina.

For four hundred years the Turks completely controlled Syria and the entire middle Euphrates. Ruling from Constantinople, they held the land largely as absentee landlords, controlling the districts with troops, suppressing local autonomy and stifling private initiative and the acquisition of independent wealth. At the end of this bondage, French and British forces seized control and stamped their authority on the land and its people. During and after the Second World War, the French gave control of Syria to the Syrians, and the British turned old Palestine over to the United Nations.

Our Arab workmen at Dura came to us from the desert, with their own language—the Arabic of the Koran—and their primitive way of life, which went back directly to the first Semitic kingdoms in Mesopotamia, the Babylonians and Assyrians, and far beyond them to the very beginning of civilization on the great plateau of Arabia. With the possible exception of some parts of China, the Near East—especially the desert of Arabia and the broad region between the Euphrates and the coastal mountains—is the only district in the world where land, people, and language have remained fixed from time immemorial, experiencing only the slow process of evolution.

In ancient times also the foreigners came to rule, first Greeks, then Parthians, Romans, and Sasanians. The local people of Dura, then as now, came out of the desert with their primal desert ways and accepted the technical culture of the foreigners and wondered at it, much as the contemporary Arab views the extraordinary achievements of European cultures. The Syrian of ancient Dura was accustomed to the old Oriental expression of religion and royalty stemming from Babylonia and Assyria. The introduction of Greeks, Romans, and Parthians induced amalgamations of the old concepts with new fashions, even as happens today. The modern Arab renaissance doubtless will derive tremendous advantages from the European impact, but the old conservative language, religion, and tradition still will dominate. In the Greek and Parthian periods the amalgam of the foreign and the native differed in

each particular district and place, only to be recomposed when the next conquering army moved in. One never finds an unadulterated culture in an occupied territory such as this.

Cumont's expedition under the auspices of the French Academy of Inscriptions and Belles-Lettres had been designed to explore, copy, and report on the astonishing paintings first found by the British troops. In his two short seasons working with army troops, Cumont completed work in the Temple of the Palmyrene Gods, the building that contained the paintings. He also excavated an adjacent tower which was part of the city wall, excavated some tombs, disclosed a temple in the middle of the city, and explored the Redoubt, a fortified building whose Hellenistic facade indicated its antiquity. To explore and excavate the ruins of Dura adequately, a full-scale effort was required.

Cumont, however, while a great authority on Oriental religions of the Roman Empire, was not primarily a field archaeologist. Also, French and Syrian funds for excavation were fully allocated elsewhere. Here, then, was a tempting opening for some outside institution. Yale University, under the urging of Michael I. Rostovtzeff, who had seen the rich opportunity, asked permission to continue the work with a combined French-American effort. But before recounting this major archaeological endeavor, I must review the astounding finds made by Cumont in 1922 and 1923.

The Syrians were hostile to the French mandate, and the Arabs were restless under French occupation. The middle Euphrates was, therefore, heavily invested with French troops, and Cumont reflected the military atmosphere in his account of the journey in November of 1922, which he presented to the Academy on January 12, 1923.

> If this archaeological mission has attained fortunate results, it is owing most of all to the able support granted me by the army of the Levant. If French forces had not occupied the ancient fortress of Salihiyeh, we should have found in this desolate and remote spot neither safety nor subsistence nor workmen. Thanks to the armed forces, instead of rapidly crossing the lonely ruins, as all travelers who have visited the site had done, we were able to stop and with their assistance begin a scientific investigation. We owe special thanks to the officers who gave permission to further our plans. General Gouraud, graciously accepting the suggestion of the Academy, lent his strong support to the execution of our project. General de Lamothe, who welcomed us most cordially at Aleppo,

> employed his wide authority in assisting us to reach our goal in complete safety. Finally, at Deir-ez-Zor, Colonel Bigault du Granut, who had foreseen the wishes of the Academy and had already sent some soldiers to Salihiyeh, facilitated with immense kindness the execution of the mission with which we had been intrusted.[1]

Deir-ez-Zor, a small commerical town established less than sixty years, bristled with prosperity thanks to the traffic attracted by its bridge across the Euphrates. Cumont's description of the sixty miles between Deir-ez-Zor and Dura is interesting, because the new era of the automobile had only just begun, and the riverbank he saw exhibited much the same atmosphere it had retained for more than a thousand years.

> Departing from Deir-ez-Zor for Salihiyeh, one crosses first of all a vast alluvial plain which reaches a breadth of twelve kilometers between the bed of the river and the abrupt height whose escarpment has been cut in ancient time by the force of the stream. The plain is not cultivated at all today except in the immediate vicinity of the little town of Mayadine and around the poor scattered villages, where irrigation is obtained by the most primitive means: the great skin buckets attached to ropes over pulleys are constantly pulled up by horses or donkeys and discharge their water in the *seghias,* raised irrigation trenches which bring the water to neighboring fields. Everywhere else the countryside is now a vast waste, but one also finds everywhere traces of ditches which formerly helped make it fertile. In antiquity and in the Middle Ages, the wide plain must have supported an extensive population, but irrigation, then as now, was the essential element.
>
> South of Mayadine, the cliff that edges the eroded valley slowly approaches the bed of the river, and the old caravan trail twists up to a bare and rocky plateau. Formerly, the Euphrates swirled its floodwaters against the base of the white cliffs that rise above it more than fifty meters, dislocating the crystalline gypsum and causing the crumbling of the friable stone, and so also a part of a chateau which stood upon it. However, no trace remains at the base of the cliff of the thick walls, which must have been engulfed by the swift current and covered with alluvial deposits. At Dura today, the river has swung slightly to the east and no longer washes

Hauling up water at Meyadin.

the face of the escarpment. The escarpment itself is cut by two deep *wadis* which rise to the west and so defend on north and south the summit of the height which they enclose. So this spur was open to easy attack only on one side, that which is level with the desert plateau, and its protection required only a transverse wall linking the ends of the two ravines. From this elevated observation point the plain of Mesopotamia, extending beyond the range of vision, can be surveyed from a distance, and no position is more favorable to control the passage of the Euphrates and guard the routes which follow each of its banks. Toward the south the river again winds away from the heights, and the valley leading to Abou Kemal forms a plain quite comparable to that of Mayadine.[2]

Cumont wrote Breasted on November 23, a few days after leaving Dura, explaining the damage suffered by the painting Breasted had uncovered, "the wall of Bithnanaia," which he named after the elaborately costumed woman prominently represented, and named, in the panel.

> On my arrival there the 7th of November. . . . I found the great sacrificial scene [wall of Bithnanaia] published by you already terribly injured. The sand with which the Indian soldiers had covered it had not withstood the desert winds and the rains. The officers told me that on their arrival they found all the faces mutilated by the Bedouin and the rest of the scene had faded considerably in two years under the action of the sun and the rains. We found practically intact only the small figures in the lower part of the scene.[3]

His postscript added that he had had the murals covered before he left with a layer of sand retained by a stone wall.

Cumont had arrived late on the seventh of November since the trip from Deir-ez-Zor was slow, and he had departed early on the eighteenth, when the first rain of the season stopped the excavations. He reported the opening of his first campaign, assisted by Commandant Eugène Renard,

> who showed an eager intelligence in conducting the archaeological work whose direction had been entrusted to him and was carried out through sound instructions to guide the excavators in their work. At Salihiyeh we found the ruins occupied by more than 200 men, Algerian riflemen, Foreign Legionaires and Moroccan spahis. Excavations had been started at the beginning of October under the direction of the commandant, Georges Hamel, seconded by Captain Thénard of the 19th Infantry and by Captain Beyer of the Fourth Regiment, Foreign Legion. They had unearthed before our arrival the temple [of the Palmyrene Gods] decorated with paintings and a part of its immediate surroundings, explored part of the fortifications, excavated several tombs in the necropolis, and, in the city itself, had brought to light beside several parts of private dwellings, the steps or seats and the entrance of a small [hemispherical theater].[4]

C. Virolleaud, French director of the Antiquities Department in Syria, was unable to accompany Cumont on his trip, but Brossé, the architect attached to the Archaeological Service, took color pictures of the paintings, made a watercolor sketch of the new mural, sketched an overall plan of the ancient city, and drew the pottery that had been excavated. A group of soldiers were put at the disposal of Cumont to dig wherever he desired. His own time, naturally, was taken up largely with examination of new murals, the deciphering and copying of inscriptions, the study of the numerous barely legible graffiti, and an overall study of the ruins and their excavation.

Results far exceeded expectations! That single day of work by Breasted had revealed only the lower register of the south wall of the the inner room—the shrine, or *naos.* On the west side, opposite the entrance, stood the semicircular base of the niche, or aedicule, for the statue of the divinity, and behind it remained the bottom section of some paintings, only the feet of large-scale figures. The north wall, as well as some of the pavement of the naos, was completely gone. However, the north wall of the outer room, the *pronaos,* had been preserved, covered in the embankment that sloped up against the great fortress wall of the city. It was there that the English soldiers had discovered the bottom register of the well-preserved paintings, showing the sacrifice of the Roman tribune and less recognizable scenes above and to the east.

Now the French soldiers cleared the room and the pronaos. Opposite the scene of the tribune they found a painting composed of a series of full-length portraits represented as standing between spirally fluted columns and flanking painted alcoves decorated with double ceiling tiles acutely foreshortened to provide a clear illusion of depth. The new discoveries significantly increased the total of paintings, and they revealed new and unusual features: the individual portraits, the Hellenistic illusionism of the ceiling tiles, and the imitation of spirally fluted columns.

The paintings had yielded to Breasted the name of the city, Dura, and the name of the benefactor in the sacrifice, Konon (I), son of Nikostratos. Under Cumont the site provided more dates and names. First, he found an inscription commemorating the building of a house or hall dedicated to Zeus-Bel in the Temple of the Palmyrene Gods. Dated A.D. 115, it was inscribed by Lysias, son of Konon, son of Patroklos, which suggested that the Konon (II) of the murals, the grandson of Konon I, belonged to the generation before A.D. 115. Thus, the great painting of Konon I could be dated to the last half of

the first century of our era, late in the life of Konon I, for he is represented with his grown and married grandson, who was named after him. One must be grateful also for the fact that the counting of years to record dates, which began with the Hellenistic Seleucid kings in 312/311 B.C., was continued by the Parthians as well as by the Romans at Dura. Therefore, we could easily determine dates, as can rarely be done in the study of the ancient world.

Konon I, son of Nikostratos, was the chief figure in the murals. "Beyond the two priests," Breasted had noted, "are Konon I's four children, a daughter and three sons. The third son, Patroklos, has his son Konon II (grandson of Konon I) standing next to him."[5] Cumont's argument, therefore, that the Konon I of the mural, son of Nikostratos, was great grandfather of the Lysias who erected the *oikos* (house) of the temple in A.D. 115 seems sound. The common Greek names in the prominent families occurred and reoccurred as first sons were named for grandfathers. A parchment of ca. A.D. 180 contains the name of Konon, son of Nikostratos, perhaps the same Konon mentioned in a parchment that Cumont said was found in 1921 in the gravel heaped against the city wall near the Temple of the Palmyrene Gods (but he did not mention who found it).[6] We can then reconstruct the family of Konon I with a good degree of probability:

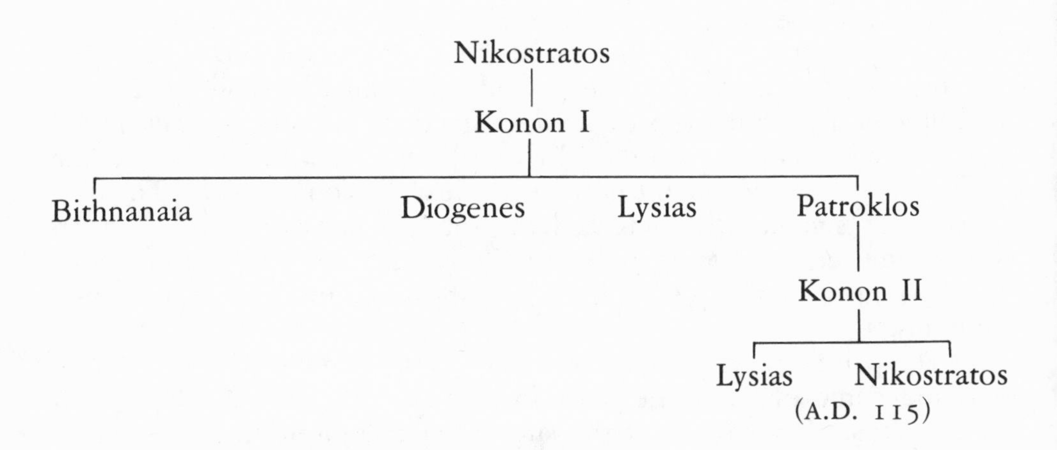

Further members of the family are portrayed in the Konon mural, but we do not have their names.

Ann Perkins recently suggested that the elaborate jewelry worn by Bithnanaia, the daughter of Konon I, in the mural, was more appro-

priate to the second half of the second century than to the first.[7] In the first century A.D., however, Dura seems to have been more prosperous than Palmyra. It had long been part of the Parthian Empire, and the inscriptions in the temple kept the names of one family close together. The balance of evidence in the temple, thus, points to the first century rather than to the second.

Fortunately, many years later, additional evidence for the earlier date came to light in the Temple of Zeus Theos, not far from the center of the city. Its wall paintings bore an extraordinarily close relationship to those of the Temple of the Palmyrene Gods. Zeus's temple had been erected in A.D. 114, and even if no other evidence for the Palmyrene paintings were extant, one would be strongly inclined to put both designs in the early decades of the second century.

The most striking resemblance in these two temples is found in the scenes on the wall in the naos. In both cases the remaining legs of a once colossal figure are clad in trousers, the feet wearing shoes with turned-up ends. Nearby are traces of smaller figures, a chariot wheel, and horses' hooves. The large figure is obviously a god, and the trousers indicate he is Iranian-Parthian. Could he be other than Zeus-Ahura Mazda, with his chariot on which no mortal could step?[8]

In the Temple of the Palmyrene Gods, the identification of the deity as Zeus-Baal—that is, Baal-Shamin, the Semitic lord of the sky (or Zeus-Bel)—had been preferred, especially after the discovery of the mural of the tribune and the three male gods clad in the costume of Hellenistic warriors. Inscriptions gave the Semitic names of Aglibol and Iarhibol as gods of moon and sun.

The name of the artist, Ilasamsos, was painted beneath the first two portrait figures and provides evidence of a local craftsman with a Semitic name common in Palmyra, an artist, therefore, belonging to the middle Euphrates district.

Cumont took up the excavations begun by the French troops the month before he arrived. He opened a room of inscribed steps or seats belonging to a Temple of Artemis-Nanaia, the earliest walls of which went back to A.D. 31. A marble statue of Aphrodite with the tortoise—Aphrodite as Ouranios—was well preserved except for the missing head. The city as a whole showed only superficial evidence of burning. It appeared to have been abandoned suddenly, leaving its monuments largely in situ.

But the excitement caused by the new paintings and statue, by fragments of shoes and leather sandals, wooden shields, and arrows was

dimmed by the sensational discovery of three well-preserved parchments written in legible Greek. One gave the name "Europos" to confirm the identity of the city. Two of the documents (Cumont's nos. 2, 3, 5) dealt with a loan of money and the law of succession, while the third (Welles's no. 17) was the remains of a registry roll of copies made from the originals kept by the principles, which opened up a whole new field to historians for the investigation of private life within the city.

Greek papyrus found in Wall Street.

The amazing state of preservation of such fragile materials, almost never found except in Egypt, held great promise for future excavations. Breasted had been at Dura one day; Cumont had now completed his first campaign of digging in little more than one week before the winter rains had put an end to all work. He immediately began preparations for his second campaign in 1923.

> Less favorable circumstances seemed for a moment to have compromised the success of the undertaking, but the intervention of General Weygand, who continually gave expression to a lively interest in archaeological research in Syria, smoothed over all ob-

> stacles. General Billote, commander of French troops at Aleppo, took the steps necessary to accomplish our task, and Colonel Andréa, commander of the Euphrates district, succeeded in securing the dispatch to Salihiyeh of a company of the Foreign Legion [the 15th Company of the First Regiment]. So we were able to continue our excavations from the third of October to the seventh of November, 1923.[9]

The task was not easy, for Cumont acted as director, archaeologist, epigraphist, photographer, and carpenter. In addition, he had neither trained assistants nor tools and all the while was exposed to the dangers of the desert, "where one might consider oneself fortunate to find means of subsistence, strength, and tranquility, and to be troubled only by the bark of jackals or the visit by night of a hyena."

Starting on October 3, Cumont and the soldiers had the prospect of a full month of work before the rains might be expected. He wisely chose to continue digging the three areas already begun, rather than to explore new sections, however tempting. It was clear to him that though the naos and pronaos of the Temple of the Palmyrene Gods lay north of the fortification tower, the grounds or precinct of the temple stretched across the tower's eastern face. The heavy sand embankment heaped up against the tower presented an obstacle that called for slow, hard work. The embankment stretched from the temple to a partly cleared tower, which because of some arrows discovered there, had been dubbed the Tower of the Archers. The tower and the embankment had yielded the precious parchments. Thus, the southward extension of the excavations was chosen as the second goal, a task made easier because the excavated sand could be easily and quickly dumped into the adjacent ravine.

Cumont did not have time to clear the tower of the Temple of the Palmyrene Gods; it was not only filled with debris, but also buried in the embankment. The stone tower and fortress wall had supported the sand of the embankment and so preserved the mud-brick walls of the temple room next to them. On the wide south wall of the room, opposite the door, was another striking painting, showing a eunuch named Otes offering sacrifice to the Tyches of Palmyra and Dura and to three great gods preserved next to him: Baal-Shamin, lord of the sky; Iarhibol, sun god; and Aglibol, with the crescent moon beneath the circle halo of divinity clearly visible.

Meanwhile, work in the center of the city revealed the temple of the goddess Artemis-Nanaia. A large part of the temple precinct was cleared, but the full space, which in fact proved to be that of two separate temples, lay beyond the reach of one month's clearing. The year A.D. 31 still stood as the earliest date for the construction of the building.

The great contribution of Cumont's second campaign came from the Tower of the Archers. The upper part of the side of the tower had broken away into the ravine, so the work of clearing from the top was accelerated, even while another gang of workers cleared the embankment along the entrance. The third day of clearing was well underway when two parchments, one in Aramaic and one in Greek, were exposed in the debris. It seems miraculous that these parchments, which must have been lying close to the surface after the digging of the previous season, had survived the rains and damp of the winter. As long as the embankment was steep, the rain was thrown off and the surface moisture dried before it could sink deep. But as soon as excavations leveled off the topsoil, the debris became vulnerable to moisture. Fortunately for archaeology, the winter rains of 1922–23 had not been as heavy and concentrated as in other years.

Before the excavations were concluded, five more parchments were recovered from the tower. One (Cumont's no. 1) recorded a contract of sale dated 195 B.C. and mentioned the "cleruchs" of Dura. The city, then, had been established well before 195 B.C. by *cleruchs,* veteran soldiers who were granted land in a conquered territory but retained their Greek citizenship.

It is easy to see why a settlement of veterans should have been located at Dura, a point on the river midway between the eastern and western Hellenistic capitals at Antioch-on-the-Orontes and Seleucia-on-the-Tigris, where the route from the great oasis of Palmyra conveniently linked the Euphrates with the rich valley of the Orontes.

The parchments confirmed that Greek and Macedonian settlers had laid out the city and that Greek families continued to use their own language and system of dating even after Parthians and Romans had taken over the district.

These important treasures from the tower were supplemented by other, more perishable objects. One find was outstanding: a Roman wooden shield with a leather covering illustrated with an ancient ship. Next to the ship was inscribed a list of stations from the Black Sea to Syria, obviously the itinerary of the legionary soldier whose detach-

ment had made the march. A list of cities along the Euphrates was already known (the *Parthian Stations* of Isidorus), but here was the first recorded conception of a map whose limits represented the Black Sea at one end and the fortress of Dura at the other. The paint was dimmed and the names not always complete, but the scheme and purpose were clear. The soldier had kept a convenient record of his journey on his shield, which a legionary apparently could adorn to suit his fancy.

Aerial photos showed the city plan to be a grid of streets in the Hippodamean, Hellenistic system, the enclosing walls following the contours of the land to take advantage of natural features of defense. Within the city was a fourth century B.C. masonry-walled structure, the Redoubt, fronting the ravine that cut through the east-center of the city. A great walled section along the cliff, called the Citadel, obviously formed a sort of acropolis within the circuit walls.

The few coins found were from mints in Phoenicia, Syria, and Mesopotamia; the latest one identified was from the Antioch of Philip the Arab, which demonstrated the continued existence of Dura down to the middle of the third century A.D.[10]

In all of Cumont's findings one striking feature was the Parthian influence. The painting of Otes in the Temple of the Palmyrene Gods portrayed all the gods in Hellenistic-Roman costume, except the central god poised on the globe of the heavens.[11] His well-preserved legs, clad in Iranian-Parthian trousers, identified him as an Iranian god or lord of the sky with strong Parthian influence. Also Parthian was the frontal position of the figures and treatment of heads and feet. The chief god's Parthian costume is echoed in the reliefs from Nimrud Dagh in Turkey, but not in those from Palmyra, reminding us that in the Parthian period at least, Dura was a Parthian center, that Parthian influence was, after the collapse of Seleucid Greek power, very strong, and that the supreme god, in some temples, retained more of the Iranian divinity of sky than of Greek, Roman, or Semitic conceptions.

What do we know of this new Parthian-Iranian god established in a strategic corner of the Dura fortress to help protect it from Roman attack? The Greeks in Dura identified him with Zeus, the supreme figure in the Hellenic pantheon. He was closely connected with the sun and moon and other celestial bodies, but superior to them all. The Hellenistic king Antiochus I had already identified Zeus with Ahura Mazda (Ormazd), the supreme Iranian deity, god of light against the powers of darkness. In Palmyra this Parthian supreme lord of heaven was identified with Zeus-Bel, and in the city of Baalbek the lord of the

sky attained a new dimension of greatness. Although the Romans called the famous and rich sanctuary in Hatra the Temple of the Sun, one recognizes the strong, new Parthian influence in the band of sanctuaries stretching across northern Syria and middle Mesopotamia.

The chariot, so firmly fixed with the god at Dura, identifies him with the charioteer god who brings the light of dawn without being the sun itself. Always he is portrayed in the later Iranian period as driving directly toward the observer, the four horses parting to either side to show the driver with radiate head, full front.[12]

The charioteer's character I find best expressed in Psalms 24: 7–8, as the gates of the temple at Jerusalem are thrown open at dawn: "Lift up your heads, O ye gates; and be ye lift up, ye everlasting doors; and the king of glory shall come in. Who is this King of glory? The Lord, strong and mighty, the Lord mighty in battle."

At Dura the temple belonged to Zeus-Ahura Mazda, a temple established in the middle of the first century A.D., a century after the Roman-Parthian frontier had been established at Dura, and shortly after the construction of the Parthian walls changed Dura from a Hellenic walled city into a Persian fortress.

3 The First French-American Campaign, 1928

In the Preface to his masterly account of the excavations at Dura, Cumont assessed the site and its promise.

> I should like to express the fervent wish that, once peace is reestablished in Syria, the excavations at Dura may be resumed. In 1924, it was possible to assign only a very small number of soldiers. . . . in 1925 the excavations were given up entirely. However, there are few ancient sites that may be more fruitful. . . . By extraordinary chance, an old Macedonian colony has been preserved on the banks of the Euphrates, scarcely changed by the Roman conquest, without any Byzantine restoration or Moslem rebuilding having ever transformed it. The Greco-Semitic civilization is reflected there just as the inhabitants left it, and a climate unusually favorable has assured the preservation of delicate paintings, parchments, and destructible articles which have disappeared almost everywhere else. The combination of so many favorable circumstances should tempt archaeologists searching for a site that promises to be fruitful. Situated on the frontier of two great empires and at the juncture of two civilizations, Dura-Europos, in revealing its history, will throw a new light on the whole Greco-Roman Orient.[1]

The appointment in 1926 of Henri Ponsot as high commissioner in Syria brought the promise of peace to the middle Euphrates after the desert revolt of 1925–26. The French in Syria, the English in Palestine, and the mandate government of Iraq were all eager to encourage scientific expeditions to explore their region's history, and to offer employment to nomad Arabs. There was fresh interest, moreover, in the Macedonian conquests and the Roman Empire, in part, perhaps, because of extended European and American economic and political interests in the lands of the former Oriental empires. Greco-Roman sites had been neglected in favor of the Sumerian-Babylonian-Assyrian, not to speak of the Hittite, in all of which exciting clay tablets pro-

duced records of the ancient empires as startling as the hieroglyphic documents of Egypt.

Yale University's interest in the Near East had been stimulated before the war by Professor A. T. Clay and his extensive collection of cuneiform tablets. His death and the war interrupted this work. In the first years after the war no one was particularly interested in the Classical Near East. The teachers and scholars in the Classics Department at Yale were largely concerned with the great periods in Greece and Rome and their outstanding authors. The Hellenistic period in the middle Euphrates, even the Roman Empire that reached through Palestine and Syria in the first century B.C., was of special interest only to theologians. In Egypt, the discovery of Greek and Roman papyri had concentrated the attention of Classical epigraphists and papyrologists on the North African area of the ancient world.

By a happy chance, the emigré Russian scholar Michael I. Rostovtzeff was persuaded to leave the University of Wisconsin and come to Yale in 1925 as Sterling Professor of Ancient History and Classical Archaeology, just at the time when interest in Near Eastern archaeology had been freshly stimulated. In 1926, the year Cumont's book was published, the excavations at Dura-Europos were abandoned by the French, in part because the site was so remote. A number of rich and famous sites in Syria—Baalbek, Palmyra, and Byblos, to mention only three—were more accessible and occupied the French attention and interest. At the same time, Syria, Lebanon, and Iraq offered a new field in the neglected Greco-Roman period, a period ignored in the earlier search for cuneiform documents and for the great empires of Assyria, Babylonia, and Persia. Princeton University, with its special interest in early Christian remains, was concerned with excavations at Antioch; the University of Michigan was exploring the district of Opis and Seleucia in the Tigris area just below Baghdad.

Before migrating to the United States, Rostovtzeff had explored the rich Scythian-Greek tombs of South Russia, and in Italy he had been especially interested in the ruins of Pompeii. His special study, the focus of his many interests at Wisconsin, was the Zeno papyri from Philadelphia in Egypt. Thus, he was admirably equipped to investigate and excavate anywhere in the Near East within the wide range of the Classical period. I do not believe, however, that Rostovtzeff expected to conduct excavations himself when he came to Yale. Yale had not hitherto engaged in excavations and had no skilled field archaeologists on its staff.

Rostovtzeff's attention was particularly drawn to Dura by the paintings, with their fusion of Oriental and Classical elements. The art of South Russia, where the different strands of Scythian and Sarmatian art were woven together with Greek illusionism, had many of the same intriguing features. In the Roman world, as in the Hellenistic, a strong Oriental influence made itself felt in the art, religion, and philosophy of the Near East. This rich field of study was almost untouched. At the same time, the Dura parchments opened a field in which Rostovtzeff was pre-eminently qualified.

I think Rostovtzeff was also eager to make a truly significant contribution to American universities as well as to Classical learning. He always felt grateful to the University of Wisconsin for his appointment as full professor in a land in which he had thought himself unknown. His prior appointment at Oxford had been temporary, and he anxiously had awaited a call, hoping that an opportunity, any opportunity, might arise. William Linn Westerman at Wisconsin had admired his work in papyri, had met him abroad, and had recognized his tremendous ability and energy. To Rostovtzeff the appointment to Wisconsin seemed to come out of a clear sky, almost from heaven. He thoroughly enjoyed Wisconsin. Although the state capital, Madison was still a university town. Rostovtzeff was accepted at once in university circles, admired by the Classical faculty, and received enthusiastically by the students, even by those who had some difficulty understanding his strongly accented English. His name, difficult to pronounce, became "Rough Stuff," and Rostovtzeff loved it. His affection for Wisconsin was strong, deep, and lasting, and he removed to Yale only because it offered wider horizons and greater opportunity. His special professorship would allow him freer rein, a broader outlook. Certainly the large stipend was a consideration, and the relative closeness to Europe perhaps also had an appeal. Rostovtzeff wrote in Russian, German, French, Italian, and English, and travel and contact with foreign colleagues was important to him.

And so he came to Yale. I do not think he regretted the change, yet he always looked back wistfully to the happy years in Wisconsin when he had been freed from anxiety and introduced to the society of fresh and enthusiastic students.

In Dura he recognized a rare opportunity to explore two of the fields in which he was most interested: the synthesis of Eastern and Western art in the Classical period, and the contributions that parchments and papyri might make in a region where none had been previ-

ously found. Here he, and he alone, could make a signal contribution to Yale and American archaeological research.

When Rostovtzeff learned that neither France nor Syria could offer support to continue the work at Dura, he began, with the approval of Cumont, to plan a French-American expedition.[2] The Yale Committee on Excavations listened sympathetically and took the necessary steps to determine the feasibility of working in the Syrian desert. In the late spring of 1927 the Syrian government agreed to permit the excavating of Dura by a joint expedition from Yale and the French Academy of Inscriptions and Belles-lettres. President James R. Angell of Yale obtained sufficient funds for three years' work. Rostovtzeff, representing Yale, and Cumont, the French Academy, were to be the scientific directors, while Maurice Pillet, a French archaeologist and architect with many years of experience in Egypt, was named field director. Excavations were scheduled to begin in spring of 1928.[3]

January of 1928 found Pillet gathering the necessary equipment and supplies, which he then shipped to Lebanon. He arrived in Beirut on February 28, but was forced to wait several days for his materials. His next stop was Aleppo, where he was again delayed, this time by heavy rains. Finally, on April 3, he reached Salihiyeh. He officially opened the excavations on April 13, the day before Rostovtzeff, his wife, Sophie, and Cumont arrived. The small group was joined for several days by the renowned French archaeologist Henri Seyrig, then attached to the French School at Athens.[4]

The trip across the Syrian desert by the first Yale expedition is best recounted in Rostovtzeff's own words.

> It is not easy to reach the site either from Beirut or Aleppo; a long tiring and difficult journey must be accomplished, which takes at the very least three days, although the distance is not much more than 700 kilometers. It is best to devote four days and three nights to the journey, setting out from Beirut if one comes to Syria by sea, or from Aleppo if one approaches by the train from Constantinople. There is a railway connecting Beirut with Damascus, Homs and Aleppo, but few people traffic on the railway—any Syrian prefers to pay one pound (thirty francs) and be shaken in an overcrowded Ford to travelling for double the price in an equally overcrowded, dirty and stuffy third-class railway carriage.

From Damascus or Homs, a Ford or Chevrolet took the traveler across the desert past Palmyra and then to Deir-ez-Zor, ninety kilome-

ters from Dura. The desert became the road, with its snares of sand dunes and boulders.

> Nor is the journey entirely a safe one, for, though the Bedouins still remember the severe lesson which the French gave them after the Damascus rising, they are sometimes carried away by their innate lawlessness. Thus they recently captured in the English territory of Iraq (so the French maintain), an English girl, whom they only consented to free after this virginal miss had informed the sheikh who wished to place her in his harem that she was already the mother of three children. More dangerous even than the Bedouin is the desert, "the bled." If there is a breakdown of the engine, if the tire punctures, or if an axle gives way, one may have to sit for hours without either food or drink, in the hope that help will arrive soon.

Safe arrival at Deir-ez-Zor meant for Rostovtzeff a dismal night in a hotel room provided with a half a dozen infested beds without linen. Dirty and dusty was the life of an archaeologist, complained Rostovtzeff, full of hungry and tiring work in either excessive heat or excessive cold.[5] But he was enchanted with his first view of Dura, poised on the steep bank of the Euphrates, facing the trackless desert. The stone walls of Dura rose out of the desert like the bones of some long-lost, half-submerged dinosaur buried twenty and thirty feet in desert sand. The site stretched for a kilometer north and south and for some seven hundred meters east and west. Add the five hundred meters of Necropolis, the pock-marked field of underground tombs circling the city on the desert side, and one was faced with a site almost a mile square. It was enough to make even an enthusiast like Rostovtzeff pause!

Camp was established within the arch of the Great Gate, still half buried, until a permanent expedition headquarters could be built. The practical problems of camp life quickly became apparent, but first Cumont must read the site for the new team. Here, in the northwest corner, he demonstrated, was the Temple of the Palmyrene Gods with the faded paintings still preserved on every wall. Beside stood the Tower of the Archers, which had contributed the parchments and wooden shield. The tower, standing within the temple precinct, was unexcavated; clearing would have to be from the top. Excavations in the Necropolis and private houses had not been very productive, but

An arch of the Great Gate.

there in the city center was the partly cleared temple with statues and inscriptions still intact. Moreover, more than half a hundred city blocks marked on Cumont's grid plan still awaited the excavator's spade, and the Necropolis seemed to stretch indefinitely into the desert. The twin-towered Great Gate had not been dug, nor had the Redoubt, the building with a stone facade reminiscent of Hellenistic fortifications. A wadi cut through the city roughly parallel to the river, and across lay the Citadel, obviously a fortified center, the east side broken away by the river. Not too large in area, it cried out for excavation.

If the site held much promise, the physical location of Dura left much to be desired. Although it stood on the bank of the Euphrates, the steep cliff made access to water for drinking and bathing difficult. Once fetched in buckets and brought up the steep cliff to the city, the water from the Euphrates—the color of weak coffee, noted Rostovtzeff—had then to be filtered and boiled. Partridge, pigeon, gazelle, rabbit, and plover were available in the desert, but the Arabs were forbidden firearms by the French, and the Europeans and Americans had no time for hunting themselves. Local goat and sheep were produced, but not quite eatable; of milk there was none, and the bread was gummy. Worse still, Rostovtzeff saw the Bedouin as "weak, underfed and lazy . . . a thorough barbarian" who smashed or stole most of the finds. The Europeans of Syria were little better, "mostly thieves and drunkards."

I think Rostovtzeff's view of country and natives had been tinged with the rather jaundiced outlook of Pillet. Also, Rostovtzeff was more at home in armchair archaeology than with roughing it in the field. He was quickly infected, however, with the excitement of working at a site so rich in possibilities.

The staff for those three weeks composed an interesting group: Pillet was tall, gaunt, saturnine, extremely courteous and gracious when he chose to be, but rather secretive by nature, especially in the presence of his peers, and at heart distrustful and bitter. Cumont and Rostovtzeff were old friends, comrades in learning, both outstanding scholars. Rostovtzeff was short, stocky, full of energy, tireless in any endeavor, physical or intellectual. Cumont was tall, polished, a gentleman of the old Parisian school but a scholar of the new school of archaeology, and acknowledged to be the best. Sophie Rostovtzeff came away with a bitter distaste for Pillet. She was tall and large with an imperious manner and a very keen mind, determined to take charge whenever household arrangements for her husband were at stake, and

30 Nomads at Dura-Europos.

Workman at Dura-Europos.

with a predilection to say exactly what she thought. M. Duchange, Pillet's assistant and a former sergeant in the French army, acted as secretary and, I imagine, as rather silent audience to the dominant cast of characters.

In preparation for the arrival of the scholars, Pillet had made one egregious error: he had forgotten to prepare toilet facilities. The Arab workmen and the Syrian soldiers simply retired into the desert, the male staff could adapt to the situation. For Mrs. Rostovtzeff, however, the problem was different. It was difficult enough to reconcile herself to desert ways, but an additional obstacle was thrown into her path. Pillet, afraid lest some accident befall the wife of Yale's representative, assigned a soldier to act as her bodyguard, with strict orders never to let her out of sight. Mrs. Rostovtzeff spoke many languages, but not Arabic. In vain she protested to the Arab rifleman; to Pillet she expressed in no uncertain terms her opinion of her plight and his lack of understanding and preparation. Her command of English was excellent, of French still better. Pillet accepted the rebuff, but he was not one to forget an injury to his pride.

The most surprising find this season was a great shield made of round rods thrust through a large leather hide. It was discovered in the entrance to the south tower of the Great Gate. Coming from the last period of the city, it was another reminder of that long-distant, last, desperate defense of the walls when every available man must have been thrown into the battle. Too clumsy for a marching or mounted trooper, the shield served only as an unwieldy defense against arrows.

Three bas-reliefs were found on April 20 on the western side of the Great Gate. One, fragmentary, was of a nude Heracles, posed in the Greek tradition, but facing full front in the Parthian manner. The second relief was also of a Heracles, strangling the Nemean lion, while the third, of white limestone, showed a dedicant sacrificing to the goddess Nemesis, with her attributes of dragon and wheel, dated A.D. 228.[6] As Cumont had noted, the association of the Hellenic sun god with this western personification of retribution does not appear elsewhere in the classical world. The inscription, written in both Greek and Palmyrene, gave the name of the dedicant, Julius Aurelius Malochas, son of Sudaius, a Palmyrene.[7] Here then was represented in Classical terms the Mesopotamian view of the sun god as avenger of wrong and injustice, a force against evil in the world.

Pillet surmised that the old caravan route from Palmyra must have run across the desert straight through Dura, from the Great Gate (and,

hence, sometimes called the "Palmyrene Gate") on the west to a river gate now washed away with the crumbling of the cliff. In the desert beyond the Great Gate, a little more than a kilometer to the northwest, Pillet identified the ruins of a Roman triumphal arch, at once marking

Mother and child.

the ancient route and offering another and most important piece of evidence of Roman activity.[8]

Because of the shortness of the season, priority was given to exploring the most noteworthy to determine the primary targets for future seasons. First, the Redoubt, with its Hellenistic bossed masonry facade, must be cleared.[9] The Citadel and the Redoubt would claim attention after the Great Gate. Clearing of the inner fortress, to establish the extent and purpose of the building, was carried to a depth of four meters, revealing Hellenistic, Roman, and even Arabic house walls.[10] On the river side, rooms and houses clung one above the other to the rocky slope.

All in all, the finds made that season were very modest, and obstacles abundant. Pillet had found a great number of graffiti scratched on the Great Gate, but they would need long and careful study before their importance could be properly assessed. No paintings or parchments were found, but the large wooden shield with leather cover held out the promise that the Great Gate still contained in its unexcavated debris spectacular treasures. Nor was this promise lost on Pillet, as we were to discover in a most unsatisfactory manner in the course of the next season.

But now the targets for future campaigns were obvious: the Inner Redoubt, the Citadel, Cumont's partly excavated temples, and the ruined triumphal arch in the desert beyond the walls of the city. The deep fill of debris and mud brick along the circuit wall facing the desert and the buried towers spaced at regular intervals, including the two towers that were part of the Great Gate, offered a grim, silent challenge. With the season at an end, the forty soldiers of the Syrian Legion who had assisted in the dig returned to their barracks, the small Arab force decamped, and Rostovtzeff and Cumont made their reports, urging the continuation of the work the next year. Yale, led by President Angell, accepted their judgment.

Excavating in Wall Street.

4 The Second Campaign, 1928–29

In midwinter 1927–28, when I was a Sterling Fellow at Yale, I was appointed assistant director of the next campaign, to go out the next winter. To prepare ourselves for the Syrian venture, Rostovtzeff urged my wife, Susan, and myself to gain some firsthand field experience, perhaps in Greece. Susan was well prepared, with a strong Classical background, degrees from the University of Wisconsin, and teaching experience in Latin and Greek. My degree from Wisconsin in papyrology should stand me in good stead in deciphering inscriptions. My Yale fellowship offered me a year's study in Greece, where I had the opportunity to visit prehistoric and Classical sites and to learn something of the importance of find-spots, levels, and stratification. David Robinson kindly invited me to join his staff excavating at Olynthus in the Chalcidice region of Macedonia. Here Susan and I spent the month of April, amid fields of brilliant poppies, with a view of the distant snow-clad peaks of Olympus, Ossa, and Pelion across the bay.

Susan helped catalogue finds and took charge of the large and still growing collection of terra-cottas: cataloguing, fitting, and repairing. I made a sketch plan of the site, just measuring by strides and trying to show the contours by eye levels, and then took charge of the small area where prehistoric remains had been found. My finds, later published by George Mylonas, who is particularly known for his work at Mycenae, required careful and exact recording of depth and location, providing me with an excellent exercise in careful observation and recording.

Back in New Haven for the summer, I received a letter from Pillet suggesting that I bring a revolver in case of Arab attack. Because the import of weapons into Syria was forbidden, he suggested a small .22 caliber weapon, which could be concealed from customs inspectors. I examined the supply available in a secondhand store. In answer to the proprietor's query, I told him I wanted the gun to repel possible Arab attacks in the desert. He seemed astonished and replied that a .22 revolver would be serviceable for shooting rats in a barn, but not very effective in repelling charging Arabs. I knew nothing of customs, calibers, Arabs, or the Syrian desert, but I had my director's instructions, and I followed them.

The Rostovtzeffs, when they returned from Dura and a visit in Europe, made no mention of attacking Arabs, but they had a good deal

of advice about preparations for the excavations. My wife was to keep the catalogue in our new excavations. I was therefore to be sure to have the necessary materials, especially India ink, stickers, and tags for labeling all finds. Then came Rostovtzeff's warning: "Pillet is a difficult man to deal with, Hopkins, but you will just have to do the best you can."

We met Pillet for the first time in Paris and were pleasantly surprised with his courtly French manners. He took us around the city, drove us out to Chartres, and entertained us at his home for lunch. He seemed most eager to please, assuring us that whatever we thought necessary for the expedition, he would be delighted to supply. I mentioned toilet facilities; he assured me warmly that the conveniences for women had already been considered and all necessary material collected. Surely Rostovtzeff had been mistaken in his judgment of the man. But then I mentioned the need for ink, labels, and tags. He explained blandly but firmly that naturally I did not as yet know the desert. The lack of moisture caused India ink to dry up immediately. Labels glued to objects lost their adhesiveness, curled up, and fell off. "And you know how these little tags attached with string are, Monsieur Op-Kan. They are loose and get torn or fall off, and thus are useless."

We easily located a stationery store where we could obtain these supplies, but the incident was not a happy augury. He knew Susan was to be in charge of the catalogue. Did he object to having a woman on the expedition? Or was he opposed to catalogues as such?

Having just spent a year in Greece and a summer in Italy, Susan and I decided to travel to Beirut via Istanbul and then through Asia Minor to Syria. The splendid museum in Istanbul gave us a chance to see the famous collection of Hellenistic reliefs and pottery, and we were fortunate enough to meet the members of the Chicago Oriental Institute expedition to Boghaz Köy: Hans Henning von der Osten, Erich Schmidt, and Richard Martin. Because they had just concluded their campaign, they could not take us to their excavations, but instead drove us to the great tombs of Gordion, those impressive monuments of the old Phrygian empire, which would be opened by Rodney Young after World War II. Schmidt was to go to Persepolis in Iran, where he took charge of the Oriental Institute's recovery of that most famous monument of the Persian Empire.

The journey to Beirut was not easy. As there was no direct rail connection, we took the train to Caesarea, then rode east by bus, to entrain once more for the south coast. We stopped at the site of an-

cient Tarsus, where Hetty Goldman led an American mission and made us welcome. The train took us across the Taurus Mountains to Aleppo; then we had an all-day ride by narrow-gauge railway to Beirut. We arrived, as Pillet had requested, on October 1, reaching the Hotel Metropole late in the evening. Pillet was out for the evening, and we did not meet him and Serge Dairaines, his secretary, until next morning. They had come to Beirut by boat, along with Jotham Johnson, the final member of our staff, and reached the city that morning. There proved to be no need to hurry, since we remained in Beirut for ten days preparing for our expedition. As I later learned when I became field director, it is important to have the staff on hand for an early start, as well as to have a staff that can be relied upon.

We were delighted to tour the city, to visit the American University and meet the faculty, and to make an excursion to the Dog River, where the conquering armies of first the Assyrians and then the Greeks had left their inscriptions on the rock walls. We were made welcome by Charles Virolleaud, director of the Antiquities Service, the Emir Chéhab, who headed the Beirut museum, and the learned and kind authority on Roman Syria, Father René Mouterde of the University of St. Joseph.

Most of all these leisurely ten days enabled the members of the expedition to become better acquainted. I had known Jotham Johnson as an able graduate student in Athens, who had been selected for a second-year scholarship in order to study Greek inscriptions. He was to supervise work on the Great Gate at Dura, with its many graffiti inscriptions. Johnson was self-confident and cocky, with almost enough ability to justify his self-assurance. Dairaines, serving as secretary to his relative Pillet and as photographer for the expedition, had just graduated from the Sorbonne with a law degree. He welcomed the chance to spend a year in Syria at an archaeological dig. Short and dark, of slight Celtic build, Dairaines proved congenial, warm, and friendly with a keen interest in ancient and modern Syria. Both men had bright minds and eager smiles.

As Pillet made almost all the official visits, in addition to securing supplies, personnel, chauffeur, and cook, he and Dairaines were away a large part of the day. Johnson, Susan, and I were left to explore the city and surroundings, but we all gathered for dinner at night. I am sure that Pillet would have preferred to keep himself and Dairaines on a path apart from the Americans, but his secretary had none of his aloofness. We quickly became friends, sharing hopes and troubles, experiences and ideas.

It was Dairaines who first told us of Pillet's low opinion of Americans. When Pillet, Dairaines, and Johnson had arrived in Beirut to find that Susan and I had still not made our appearance by dinnertime, Pillet declared such behavior was typical of Americans. We were all to have arrived that day; one could not rely on the promises of Americans!

There were more exacerbations. Johnson delighted in thumping with more enthusiasm than skill on an old piano in the hotel lobby. "Ol' Man River," the popular song of the day, particularly suffered from his efforts, which Pillet thought unworthy of the dignity of a French-American expedition, in particular one he directed.

Finally our caravan of five Model T Fords headed out across the mountains, reaching Aleppo before dark. A day or two later we headed into the desert toward the river, with a gendarme as guard, complete with bandolier and rifle. Every few miles we were stopped for inspection, a scrutiny of the chauffeur's license and the car permit, and endless questions as to our destination. Whenever Susan left the car, the gendarme followed. This northern route from Aleppo to the river, according to an English officer we met, was now perfectly safe. To the south, between Damascus and Hit on the Euphrates, Arab raids were still a danger. The country as a whole was quieting down. We arrived without incident at Deir-ez-Zor, where we stayed in the army barracks. A little Sudanese soldier made up our rooms, and we took our meals in the officers' mess.

Every morning we awoke at 4:30 and prepared to leave for Dura promptly at five, as Pillet had ordered. Dairaines informed us later that Pillet had had no intention of leaving the first four days, but he thought it was good exercise for the Americans to be up and ready just the same. The early start was justified when at last we did set out October 21, for though the site was only sixty miles down the river, the trip took four hours, and at the end we still had to pitch the tents, unpack the beds, set up a kitchen, and complete many small, but necessary tasks before dark. Night was already descending early and swiftly on us.

We set our tents inside the Citadel, with the entrances looking over the cliff to the view beyond. An old arched entryway became the dining room. The staff had comfortable double-roofed tents, and there were smaller tents for bags and equipment and to house the cook. Pillet pointed out to me the toilet facilities: a length of canvas a meter wide with vertical red and white stripes, staked out on the highest point of the Citadel, which was also the highest elevation for miles around, a beacon and attraction in the barren waste. Of course, a meter-high

enclosure would conceal no one above the waist.

I told Pillet that this facility would not do. Gravely, with a perfectly straight face, he told me I could make any changes I saw fit. I am certain that he was only sorry that he could not have faced Mrs. Rostovtzeff with this contraption, but of course he would never have dared that impertinence.

The campsite itself was magnificent. The sun rose across the river plain as we roused ourselves to begin the day. The evenings were cool, often chilly. The moon acquired a new importance for us as the only source of light at night, except for the winking bonfires of the Arabs in the plain below the cliff. The lonely call of the wild dog, the answering Arab dogs, and the occasional shrill laugh of the hyena reminded us that the desert was still a wilderness. Over against this scene of wild beauty were the flies by day and the gnats by night; mosquito nets were almost useless.

The first four days were occupied with recruiting workmen. They came out of the desert and pitched their black tents, some across the river, some under the shelter of the cliff over which the ruins brooded. There was no pattern to their settlement, no village order, just a scattering of broad tents, the sides lifted high to catch the least breeze and pegged down when the chill of evening fell.

Some of the Arabs came from the little settlement of Salihiyeh, not much more than a few huts a mile away on the riverbank, but they were more interested in selling dates and figs to the workmen than in working themselves. Others walked the twenty miles to and from Abou Kemal on the day of rest after nine days of work, returning to Dura with their round loaves of unleavened bread, like great griddle cakes a foot across, stacked a foot or more in height. This comprised their basic diet for the next ten days, their only food, except for a few dates, figs,

Opposite: View from the Citadel walls overlooking the Euphrates.

Top: Dwellers of the black tents.

Bottom: The black tents.

or black olives purchased from the local merchant. Those workmen who brought their own tents and stayed with their families had chickens and supplemented their slender diets with eggs.

They worked hard, men and boys, beginning at the age of thirteen or fourteen, if they were tall enough. Each of the youngest boys carried a shovel or two of dirt from his shovelman to the dump or, later, to the Decauville railway cart, and the next oldest group pushed the cart to the railhead and dumped its load into the wadi. The best men, the strongest and most experienced, were assigned to break the ground with picks, while the next best shoveled. The pickman had the best chance, of course, to find antiquities, but only if he could *read* the earthy remains. He had to develop almost a sixth sense that warned him of the presence of a fragile jar or pot when only a fraction of rounded surface appeared in the hard earth.

The key men were the Armenian foremen, who were paid in Turkish gold rather than with the large, pre-World War I, silver coinage and the little silver *bargouts,* about the size of a dime. There was a general, and natural, antipathy between our Armenian Christians and our Moslem Arabs, but the Armenians were recognized as educated, for they were able to read. Their presence also solved a problem of the Arab headmen. The sheiks, the leaders of tribes or groups of related families, did not work, but they controlled and guided the men within their units. If we had had to appoint Arab foremen, men from different groups or families would be in the charge of other men who were not their leaders. Quarrels over authority would have been inevitable. Hence, the sheiks were glad to see the socially neutral Armenians in charge within the excavations, while they remained in charge outside.

Some sheiks wanted places beside the paymaster's table so loans they had made might be repaid immediately. But no outsiders were allowed inside the large courtyard, and they had to be satisfied with stationing themselves just outside the gates, where they waited for their debtors to appear. The sheiks gave the word when the tents should be pitched near Dura and when camp was to be broken. They took up with the director the broad problems of hours and wages and holidays. As far as we could tell, however, policy decisions were made by the general community. The sheiks did not dictate, but rather guided their family and groups.

We rarely saw the women. When tents were struck, the camels carried the heavy wool tents and the poles. The men rode horses, and the women walked, carrying the babies and digging roots along the line

of march for the evening fires. There was no hurry, for there was no set destination to be reached by nightfall. The sheep and goats grazed as they went and moved a mile or two a day if grass was sufficient, perhaps six or eight or ten if forage was more difficult to find. When a halt was called, the young boys watched over the flocks near the camp while the girls stayed by their tents, assisting in household chores—cooking, weaving, tending the babies—or gazing out with faces veiled below the eyes on the desert and the men's activities beyond the shadow of the black shelters.

The desert offered little opportunity for remunerative work beyond raising the flocks and herds. When times were hard with famine, the villages were crowded with desert Arabs seeking any means to earn subsistence. There were no fat men in the desert, while in town the Arab put on flesh easily enough. The desert Arab knew bitter cold in winter and unrelenting heat in summer; he fought the climate and indigenous diseases—particularly anemia, syphilis, and ringworm—and he subsisted on a minimum of nourishment. Rice and bread, dates, figs, and nuts were the unvarying staples, but meat remained a luxury. For a special feast, a sheep was killed and the flesh roasted. Milk and sour cream were added in season, eggs and chicken occasionally, and a desert bird would be served as a special treat when the shot from a whirling sling found its mark.

The Arabs displayed their pleasure with childlike simplicity when they came to the pay table, singing noisily and brandishing their shovels and picks as if they were guns and all were off on a hunting party. Their dance was the Turkish line of men with arms around each other's shoulders, with the end man performing fancy bends and turns. The music was high and wild or melancholy and elusive.

A man bought his bride from a family within the tribe, and he settled down in his own tent, keeping his sheep in the common herd. There was no opportunity for school, for the herdsmen were constantly on the move. When the French insisted on an elementary school in one village, the tribal units gave money for a building, a teacher, and books, but providing the children was a different matter. The boys could not be spared from the flocks they tended, and the girls had no need for education. Schooling, if there was to be any, was only possible when the tribes remained in one place: in the spring when they settled by the river, and in the winter when the thick desert mud made migrations impossible.

The movement of tribes was to the northern highlands for the

Meharists riding up the Euphrates past the Citadel. "Am enclosing a couple of snaps of the camel troops going through, which is always one of the best sights around here. . . . They hunt with both [dogs and falcons] and the falcon brings down partridges, outards, and rabbits, and the dogs always appearing thin as rails outrun the rabbits and the gazelles." (Letter from the author to his brother in New Haven, February 6, 1929.)

summer and to the south for the winter. Modern political frontiers were ignored, for the desert had always been free and movement unchecked, except as the large tribes demarcated their territories and the smaller tribes concluded agreements, usually by marriage alliances and for mutual support. The broad desert was the grazing grounds of the tribes, much as the forest and plains were the hunting grounds of the American Indians, and little sympathy was expected or given between unrelated groups. Recognition of and respect for landmarks were phenomenal, but the ownership of border districts was left vague and came into question only in times of famine or unusual expansion. Raids, however, remained the popular pursuit of the daring younger men on their Arabian ponies, and the raids often evolved into deadly feuds. A rough justice prevailed, and a strong claim of injury, even against a superior tribe, might be honored, but the debt of blood devolved not on the law court but on the next of kin, and in the desert the round of retaliation rarely stopped.

Passing through Palmyra during one of our later seasons, we found the French officer of the Meharists, the desert police, trying to settle a feud between local tribes. It had begun with a small dawn raid by some impetuous young men eager to show their strength. They had cut out a few sheep and goats and driven them back into their own tribal territory. In a retaliatory raid a few shots had been exchanged at long range, and by chance a man had been injured. This called for more serious

retaliation. By the time the Meharists intervened a nephew of one sheik had been wounded, one man had been killed, and to further complicate matters, it was discovered that a woman from one tribe had been carried off by her lover from the other tribe.

All might have been settled without too much trouble, except the death of one man called for the nearest of his kin to avenge the shooting by killing the murderer. In a raid, who fired a fatal shot is rarely

The Meharists at Dura.

known, but the blood debt could be satisfied if any member of the attacking tribe were dispatched. Of course, without proof that the random victim had committed the original crime, his next of kin was obliged to avenge his death. The cycle of murder and revenge went on and on; feuds of small tribes grew in intensity, involving larger allied tribes, until all too often desert wars erupted.

In the Palmyra case the sheiks of the two tribes were summoned, and the injuries were reckoned in cash: How much for the wounding of a sheik? How much for the abduction of a not unwilling girl? How much for the life of one man killed? The details escape me, but I remember the life of the Arab herdsman, not related by blood or marriage to the sheik, was valued at twenty-five dollars. The French insisted that the amount be paid in silver on the spot and that both sheiks swear that particular feud was over. Life was plain but not entirely simple in the open spaces of the desert.

Pillet dreaded a native Arab attack on his mission, and he built accordingly. He had the great, wide courtyard surrounded by a solid wall of rubble and cement higher than a man's head. A corridor ran along one side of the staff quarters, and the private rooms opened off it, each long and narrow and lighted by a single small window cut through the thick cement much like an archer's slot.

One might be amused by his overanxiety, but there was some justification, as we discovered in later campaigns. In the summer of 1932 the few soldiers comprising the French garrison at Abou Kemal were attacked and as the lieutenant was returning to his fort in an army truck, he was ambushed and killed. One night our foreman's house in the great courtyard became someone's target. On another occasion, an old and experienced French archaeologist traveling alone at night was signaled to a stop by an Arab, who approached the car with one hand held high, the other behind him. When the car door opened, he pulled up the rifle he had been dragging behind him by the barrel and shot the archaeologist. Whether robbery or a fancied grudge was the motive we never knew. Our own expedition on a special investigation across the Euphrates at Baghouz, close to the Iraq frontier, was saved from attack by a sandstorm that rose at the right moment. Apparently the would-be attackers thought archaeologists carried their wealth with them and that the robbery could be covered by a quick flight across the border.

On the other hand we drove into Deir-ez-Zor once a week, withdrawing from the bank a thousand dollars in silver and gold to pay the workmen. Our route was well known and we had to drive through one

Street scene in Deir-ez-Zor.

or two narrow, steep passes where a roadblock and attack would have been easy. Although a thousand dollars was a fortune in the desert, we were never molested, nor was there ever any hint that we might be.

The Arabs were devout Moslems; all observed Ramadan: no food to pass the lips from sunrise to sunset during the sacred month. To work all day without sustenance even in the moderate climate of winter was a serious hardship. Gradually rules were relaxed and some food was taken by some of the workmen at the noonday break, and this was recognized as reasonable even by the total abstainers. The long fast ended as soon as the new moon was sighted. At the appropriate time, therefore, the western wall of Dura was lined with spectators trying to see the first faint appearance of the new moon. Everyone knew the man with the sharpest sight, and if he said he had seen the crescent through the mist of the horizon, then his word was accepted and the feasting began. If he or one of his comrades did not see it, no feast could be held. It was useless to say that astronomers had predicted the new moon on that date; nothing else but visual proof would do. Once, we left Deir-ez-Zor on the day after the new moon had been sighted and the feasting had begun. As we travelled over the lonely trail we saw an Arab half a mile away running towards us, his white robe fluttering behind. He arrived breathless and wanted to know only if the new moon had been sighted in Deir-ez-Zor. When assured that it had been seen, he departed happy. He had been unable to see it from the flat open desert.

On October 25 the trenches were opened.[1] Few workmen were to be found at the beginning and end of our campaign because the seminomads stayed near the Euphrates only during the winter. There was little use in beginning the excavations before the Arabs returned for the winter with their flocks from the summer pasturage. I rather bitterly complained to Rostovtzeff by letter about the lack of workmen until I became aware of this seasonal pattern.

The first winter rain came with the first of November, and I had my first experience of being stuck firmly in the mud with night coming on and the camp two or three miles away. We had been visiting Abou Kemal that day, twenty miles downriver, and did not think the shower would interfere with our return. We finally reached camp through thick mud in a pelting downpour convinced otherwise. More heavy rains came just after the middle of the month, followed by three days of intermittent downpours. Thanksgiving saw the temperature drop to just above freezing at night, announcing that winter had descended in

earnest. At least the flies and gnats ceased to trouble us, but our thin mattresses were no match for the cold. We had now spent a month in the field and had accomplished almost nothing.

But the splendid opportunities presented by the site were obvious. We were free to dig and study where we wished to our heart's content in an ancient city completely buried in desert sand, undisturbed since the time of its occupation. Instead of a small parcel of land purchased with difficulty and at great expense or a ruin buried beneath modern constructions and several layers of foundations, a veritable feast of monuments was waiting to be revealed just a few feet below the desert surface.

On November 19, a detachment of Syrian troops, the Meharists, rode into camp on their racing camels. They were a striking band, with bright red saddle trappings contrasting with the olive-brown uniforms, service rifles slung across their saddles, bandoliers across their chests, and their white turbans, the Syrian *kafias,* held in place by double bands of black twisted wool. The French officers shared our mess for three days and then invited us to their campfire feast, where they entertained us with an Arab dance. The entire troop formed a procession escorting us back to our tents under the star-filled night, with shouts and rifle fire echoed by the rifles of our own gendarmes. Early the next morning the long slow strides of their camels took them off to the music of the Arab flute into the uncharted desert where they served as guardians of the frontiers, watchmen over the nomad tribes, and arbitrators of the tribal quarrels and feuds.

There were certain advantages in starting with a small force of workmen, especially when the staff was inexperienced: the work could be carefully supervised, and the slow recovery of finds offered time to lavish our full attention on every precious bit. Johnson immediately began deciphering the graffiti in the Great Gate. I was put in charge of a small group excavating the north tower of the Citadel. As the number of workmen slowly increased in late November, we set a second squad to clearing the ground south of the Citadel, where a new expedition house and courtyard would replace the tents.

Clearing the tower gave me my first taste of desert dust in a concentrated form. Digging in the open air was often difficult, because the ground exploded into clouds the instant the surface was broken. In a closed tower with high walls and no vents other than a loophole or two, the dust was blinding and suffocating. On a calm day it merely hung in the air; when the wind eddied in the deep recess, the dust

Basket boys at work.

invaded one's body. If I spat, I spat mud. Later I wondered if Pillet had understood the situation and assigned me the tower just to see what his first scientific assistant would do.

The tower provided some fragments of second century B.C. pottery, bits of cloth, and the complete skeleton of a tall man apparently cut off in the prime of life by a hemp rope still twisted around the bones of his neck. He lay in the middle of the tower with no sign of a coffin. Excavations showed later that the Parthian palace on the Citadel had toppled with the collapse of the cliff face, not too long after the construction. I wondered if the architect had been blamed for the catastrophe and strangled in his own tower.

After clearing the tower we moved outside onto the Citadel proper. Most of the building in Dura utilized thick mud brick with gypsum plaster coatings on rubble foundation stone set in plaster. The abandoned city had collapsed into a fairly flat desert surface two to six feet deep. At floor level the small household objects began to come to light, while the temples revealed their inscriptions and bas-reliefs. Thus, the first few days of clearing were frustrating, full of dust, and almost empty of finds except for small bits of pottery. The pickmen were set to loosening the hard-baked soil while the shovelmen put it into the baskets or robe-pockets of the boys, who walked slowly in line to the edge of the cliff to dump their load. Progress was agonizingly slow. I walked over the site—two-thirds of a mile in length, a half a mile in width—from desert wall to cliff and wondered how we would ever find profitable material in this bare and dusty desert.

We worked a ten-day schedule: nine days of excavations and one of rest to avoid the problem of Friday and Sunday, the Moslem and Christian holy days. The workmen were encouraged to sift the earth carefully by our giving baksheesh to those who found worthwhile objects. Pillet suggested we note on a piece of paper the object discovered and then have the workman present the slip at the pay table, where he would be suitably rewarded. This plan sounded sensible, since there were no copper coins in the desert and a very small find would scarcely justify a silver piece. Several slips could easily add up to a monetary reward. I was fairly free with my slips the first few days to encourage both men and boys. On payday, Pillet personally disbursed the wages of each workman as he was called up to the pay table. When one of my men presented his slips of paper, Pillet glanced at the descriptions—a coin, a fragment of bronze—glanced at me, tore the notes in half, and declared the finds worthless. So it went with all the slips I had given.

I thought Pillet simply had wanted to discredit me in the eyes of the workmen. At the end of the month, however, he called me aside and asked if I should like to see what had been found, and then led me into his tent, where he produced a row of cigarette boxes containing the more valuable coins, trinkets, and bronzes discovered. As he smoked a box of fifty Bastos cigarettes each day, he had one box for each day's finds. I could not believe what I saw, for I had assumed that we had catalogued all the finds. I soon realized that by tearing up my slips at the pay table and handing out silver coins to workmen on the spot, he was able to monopolize all the good finds. Pillet had gone over the diggings during the midmorning twenty-minute break and given his rewards.

I told him this procedure would not do, that all finds had to be catalogued by my wife, and that each find had to carry a tag identifying the date and exact location of its discovery. He appeared to recognize the soundness of the plan, and gradually all finds, as far as I knew, came into the catalogue. Pillet was an architect who had had extensive experience in excavations in both Persia and Egypt. He had been trained,

however, in the old school of archaeology in which the field director was also camp manager and sent all finds back home for study. The importance of find-spot and depth had not been realized; archaeology was a treasure hunt for him, while as architect he was only waiting until the buildings were cleared in order to draw his plans and number the rooms.

Pillet took a special dislike to Johnson. Since Jo's particular interest focused on the graffiti of the Great Gate, that was the last structure Pillet wanted to dig. So we cleared the terrain for the expedition house, which was reasonable, and explored the north tower of the Citadel and the ground in front, which was also reasonable. I asked Pillet several times to dig the gate, but there was always a good excuse and postponement for a few days more. Early in January Pillet approved our suggestion of a ten-day trip to give us the opportunity to see something of Mesopotamia. Susan, Johnson, and I set off with a car and chauffeur secured in Deir-ez-Zor and returned to Dura ten days later.

Pillet's staff in 1929. *Left to right:* unknown Syrian, Jotham Johnson, Pillet, Susan Hopkins, Clark Hopkins, Achille the houseboy (seated), in the doorway of the expedition house. "AM. et Mme. Clark Hopkins en bon souvenir de Salihiyeh . . . 22 Février 1929—M. Pillet."

We had arrived late in the evening and gone directly to our tents. When Johnson visited his site at the Great Gate next morning he was amazed, and then angered, to find that in our absence his excavations had been largely completed. When questioned, Dairaines reported that as soon as our car was out of sight, Pillet had transferred the entire labor force to the Great Gate. Pillet clearly had counted heavily on the towers of the Gate providing the season's most important finds, since the Tower of the Archers had been so productive. Thus, he wanted the clearing to be under his, not Johnson's, supervision. His efforts were rewarded with the little wood panel painted with the goddess of victory which forms the color frontispiece of the *Second Preliminary Report.* The find was small but spectacular, not new in theme but a splendid example of a Greek subject in Oriental trappings.

Pillet was also a tyrant in the management of the camp. Our Syrian cook from Beirut had but a single burner on which to cook the meals. If dinner was late, Pillet thundered at him, and when only a small leg of lamb was brought to the table, he flew into a rage, accusing the trembling man of stealing meat. As a city Arab, the cook was terrified that he might be thrown out into the desert to cope with hostile Arabs. Early in November the chauffeur was discharged for incompetence.

My frustration was complete in those early days, for I had not yet learned the philosophy, so useful in Syria, that the worse things are in the beginning, the better they turn out in the end. The site seemed impossibly vast and seemed to grow as we tramped over it day after day. However, I enjoyed the open-air life. I had time to keep track of my trench and also to wander over the site, studying the details of walls and buildings. An old army sheepskin coat kept me warm in the early mornings until the noon sun brought a summerlike warmth.

Susan's catalogue steadily grew; the ink and tags worked well despite Pillet's warnings in Paris. She often avoided the midday meal to escape Pillet's temper, while he in turn became more reconciled to the circumstances and, at times, was courteous and considerate. Dairaines did all in his power as an intermediary, an unenviable role which he carried off with unfailing good humor and enjoyment.

After the first rains in November, the pattern of walls beneath the surface appeared; the shallow deposit above the foundations, drying faster than the earth inside the rooms, took on a slightly different color. On the highest part of the Citadel, Pillet recognized the buried walls of palaces for which he had searched in vain during the dry spring and autumn.[2] The largest force of men was transferred there.

Inside the southeast corner of the Great Gate.

On December 5, just six weeks after we had begun, a hoard of 120 Roman coins was discovered. Now we had our first significant find, and there was great rejoicing. Rather ironically, the hoard was found, not in our trenches, but by the workmen clearing the ground for the future expedition house.

I was on hand for our next major discovery on December 9. From the ground west of the north end of the Citadel, in the middle of low mud-brick walls, which proved to be part of a small Roman military temple, just four feet above the solid rock, we brought out part of a little altar inscribed in Latin. It was my first significant find. I read the Latin cut into the stone seventeen hundred years ago, the first person to do so since the abandonment of the city. The message itself was interesting: a centurion of the Fourth Scythian Legion, commander of the detachment at Dura, at the time of the enlargement of the military *campus* had erected the temple and a statue along with the Second Ulpian Mounted Cohort of Roman archers. There was no indication to which god the altar and temple were dedicated, nor was a date given. An altar found at the Great Gate in the first season had been addressed to the Roman emperor Commodus (A.D. 180–192) by the Second Ulpian Mounted Cohort, very possibly the same group, making probable a date in the last quarter of the second century.

With the outlines of buildings and streets clarified somewhat after the rains, I made a careful check of the pattern published by Cumont. In some cases I could discern within the blocks the outlines of house walls and courtyards. My only real contribution, however, made after a long and elaborate search, was the conclusion that north of Main Street, which stretched from the Great Gate to the city center, in the area opposite the temple area excavated by Cumont, the pattern of streets had been definitely interrupted.

The last few days in December provided some interesting discoveries whose importance we could not immediately measure. I found an interruption in the circuit wall just south of the Tower of the Archers, with a stone projecting from the wall at a peculiar angle. With a few workmen I began to clear the area and found a small side gate in the fortifications and plaster steps leading down the wadi toward the river. Going to and fro across the site, I noticed one day a piece of painted plaster, the design still visible, belonging to a long narrow building whose outline in part cut into the pattern of streets. This chance find was worth noting; it was the first, but not the last, time that locations of important buildings were betrayed by bits of plaster lying

on the surface still retaining traces of color. Usually, of course, centuries of rain and moisture had destroyed all painted designs.

Also at the end of December the rooms of a Roman bath were excavated. The shallow dirt in this northeastern section revealed walls only three or four feet high, but plaster fragments with distinct paintings of a woman's head and hand were found in the debris of this small building, which was of the common type found in Roman cities.[3] No one asked why a Roman bath should have been established at that particular part of the city, for it was still too early to diagnose the urban pattern.

Johnson, having completed the preliminary study of the graffiti at the Great Gate, roamed the city, recovering scattered graffiti and, more important, a zodiac so carefully drawn that the date of the graffiti could be established to within three days, July 3–5, A.D. 176. This established the first scientific correspondence between the Julian and Macedonian (Seleucid) calendars.[4] A second graffito of a mounted Parthian bowman indicated that the draw loom had already been introduced from China.[5] At Dura the mounted bowman usually advanced left, pulling the bow with his right hand. Johnson's graffito showed the archer moving right, pulling the bow with his left hand, not because the archer was left-handed, but because the artist was copying patterns from the draw loom, in which designs were punched through cloth and the second design duplicated the first except that sides and arms were reversed in mirror image. The draw loom produced the heraldic designs so common later in Egyptian and Sasanian cloth. Here one may notice that this Far Eastern technique and stylization reached Dura through the Parthians.

Pillet had found small pieces of wood, leather, and even papyrus in clearing three of the towers on the circuit wall. But a major contribution from the Gate was reserved for Johnson. On February 12 he came up to me and asked what I should most like to find. He then produced a splendid parchment, tightly rolled, long, and obviously well preserved. It proved to be a loan contract, opening with the name of the Parthian king Arsaces and witnessed by the commander of the garrison, an honored friend and member of the bodyguard.[6] One Phraates, the eunuch, had lent 300 drachmas of silver from the Tyrian mint in the year A.D. 121, given in both the Greek and the Parthian reckoning. Invaluable at any time, the document proved that Dura had not only been the frontier fortress against the Romans but also, because it served as the residence for such an important administrator in the

Persian hierarchy, the chief Parthian center of the whole district along the Euphrates and of neighboring Mesopotamia. The Citadel, with its Parthian palace and circuit wall, had obviously been the seat of the governor.

The circuit wall of the Citadel was built of rough-cut blocks laid in courses of headers and stretchers bedded in plaster. Excavations on the Citadel top revealed the foundations of two buildings, one of well-cut large blocks bonded with a minimum of plaster, the second of smaller stones in generously laid plaster and with a column drum cut with many flat facets in the Parthian style. There seemed no doubt that the fortification wall, gates, and arches of the Citadel belonged to the Parthian period.

Potsherds found in the north tower were of poor black glaze and a Megarian ware of the late second century B.C. On the Citadel the splendid black-glazed plates of early third century type indicated that the early palace belonged to the early Hellenistic period, the fortification wall to the early Parthian.[7] This later Parthian wall probably followed the course of an earlier mud-brick fortification.

58

The north entrance to the Citadel; a small Roman temple lies in the foreground.

Final clearing of the tower of the Temple of the Palmyrene Gods brought to light on February 10 a small altar and an inscription dedicating it to "Greatest Zeus." According to the inscription, in October A.D. 160, "about the fourth hour of the day, an earthquake shook the region." The very careful specification of time brought a note of special poignancy. Pillet seized on this inscription to explain the collapse of the corners of towers in the city fortifications. He reported to Yale that his study of the circuit wall convinced him it had been shaken by an earthquake, the one mentioned in the inscription as occurring about the middle of the second century. A single coin proved otherwise.

In the very bottom of the northwest tower of the Great Gate, the tower supposedly damaged by the earthquake, Johnson recovered a coin of the Roman emperor Philip the Arab (A.D. 244–249), effectively refuting Pillet's claim that the tower had collapsed and buried the debris over eighty years before.

Johnson's collection of inscriptions and graffiti continued to fill in dates in the history of Dura. The earliest date found by Rostovtzeff had been A.D. 65–66. Johnson found three earlier—one composed only of the name "Artemidorus" and the date 183–82 B.C., one of 17–16 B.C., and a little altar dated 134–33 B.C.—which forcefully underlined the question of the actual date of construction, a problem more conveniently dealt with in the discussion of the fortification walls.[8]

Shortly after our return from Baghdad, Sir Aurel Stein, the great Far Eastern explorer and archaeologist, rode into camp, his tents perched on a second camel. He had come from Kashmir, on his way to the Mediterranean. We had been hearing about him continually from Deir-ez-Zor, where the colonel had word that he would be coming through. His well-earned fame preceded him; everyone wanted to know if we had seen him as yet. There were not many Western travelers in this region, and he probably had news of his coming sent on ahead to prepare the way. He had known my father (E. Washburn Hopkins) in Lahore in 1896–97 and was delighted to learn that he had now met a relative of his old friend. He stayed with us for four or five days before departing for Palmyra directly across the desert to see if he could locate traces of the old caravan route.[9]

Before he left we drove him out about thirty miles in our Ford to one of the chief wells, but we could find no signs of an ancient track. Our return to camp took much longer than we anticipated and gave me some insight into English perseverance. Sir Aurel had been taking observations along the way, keeping notes and taking readings with his

prismatic compass. After we had left the well and had lunch in a wadi, Sir Aurel said he would direct the auto back across the desert to Salihiyeh. So he sat on the back seat with his compass and shouted to the chauffeur, "A little right . . . a little left," and this way and that every few yards. I could see from the horizon that we were going about the same anyway.

Finally we came to a rise and stopped for observations. Sir Aurel got out his compass and studied the angles, then took out his field glasses and scanned the horizon. Finally he sighted a large flock of sheep.

"Ah, just as I thought," he said. "There's Salihiyeh."

Pillet suggested we head that way, but Sir Aurel said he had been keeping very careful account of angles and distances and he could keep Salihiyeh and its flocks in sight, but he preferred to navigate by compass. So we went on again. We were no sooner seated again than his compass began to be affected by the metal in the car. We could not see the sheep with the naked eye so he had to direct, but I could see the point on the horizon about where they were.

"South a bit," he directed, and a minute later: "To the right again." Then: "More to the right."

It was plain by the sun behind our backs that we were turning in a large circle, so Pillet asked, "Don't you mean left?"

"No, no," a rather irritated Sir Aurel responded. "Right again."

"Are you sure your compass isn't affected?" I said.

"Oh, no, I'm holding it," and he had it in both hands and was bending over, peering at it to catch a reading when it was not joggling too violently from the bumping of the car.

So we rambled on some more until we got to another mound. This time he sighted some high cliffs in the distance, the tops just lighted up by the sun so that they did look rather like walls.

"Ah," he said, "Salihiyeh, Salihiyeh!"

Well, on we went again, but luckily it began to get dark and we could not see much of the horizon. We asked a couple of shepherds the way, headed north, and after some ten kilometers, swung into the village. Susan had said the trip with him would be most useful to observe his methods.

Sir Aurel was indeed a very nice fellow with a great store of information and useful ideas. We enjoyed his stay in more ways than one. But no real signs of a road or of stopping places were discovered, nor were any indications visible from the air. Here was another problem raised but not solved.

January 26 rewarded us with the largest hoard of coins yet recovered, found in the embankment just inside the Main Gate. February 12 provided the discovery of the Parthian parchment. Susan, cleaning the most legible of the coins, found they dated from early in the third century until after the middle. Coins of the emperors Valerian and Gallienus were represented. Their rules began in A.D. 253. We could now bracket the history of Dura between the early Hellenistic period, when it was founded by Seleucus Nicanor about 300 B.C., to just after the middle of the third century A.D.

The real harvest of valuable finds seemed to have just begun as we were ending the season. We reduced the number of workmen rapidly at the end as Ramadan began on February 11, and we left Dura on March 10. Although the season was closed, lingering hostilities remained. Our trip to Beirut was clouded by a stop at Palmyra, where we heard that Pillet had spread the rumor that Johnson had carved inscriptions on the Main Gate to enhance his reputation. I wrote a very sharp and indignant letter to Pillet denying his unfounded and malicious charges and pointed out that such conduct would have reflected on his own directorship as much as it would have on Johnson's reputation.

True, Johnson had cut one graffito at Dura. When the day of departure came, I made a final tour of our rooms in the new buildings to be sure nothing had been forgotten. In the still wet plaster beside the narrow window in Johnson's room, he had scratched with a sharp point, "Suffered under pompous Pillet." I rubbed it out of the still soft plaster with my thumb. There was no use in arousing the dogs of bitter feeling at that moment.

So ended the second season at Dura.

5 The Third and Fourth Campaigns, 1929–31

Since I remained at Yale during the third and fourth seasons, I am unable to give a personal account of discoveries. The 1929–30 season was unfavorable for excavations because of heavy rains; Pillet was not well, and the labor force was comparatively small. Pillet recovered for the fourth season, and the weather was dry and fine. The two seasons taken together allow a more complete review of some of the discoveries, particularly in the two temples of Artemis-Nanaia and of Hadad-Atargatis. The staff in 1929–30 consisted of Pillet as field director and Henry Rowell as scientific assistant. For both seasons, André Naudy served as secretary, and Antoine Walter was accountant and photographer. Alfred Bellinger joined the expedition for two months, starting early in November, as the Yale representative. Rowell stayed on for the first half of the fourth season to work with Alan Little, scientific representative for the whole season. At the close of the fourth season, C. Bradford Welles came out from Yale for a month to make a special study of the graffiti.

Henry Rowell had an exceptionally attractive personality and a fluent command of French. I thought perhaps the atmosphere in Dura during my stay had been clouded by the difficulty of frank and easy communication. Rowell was young and eager, always enthusiastic, a most promising student with an ability that carried him later to the chairmanship of the Classics Department at Johns Hopkins University for twenty-five years, after serving as chief of the Education Division of the Allied Control Commission during the war. For seven years he was professor in charge of the summer session of the American Academy in Rome, for two years the professor in charge of the Academy's School of Classical Studies, and director of the Academy from 1972 until January 1974. He was but twenty-five when he came to Dura to work with and under Pillet.

Little was a graduate student, able, eager, interested in both the linguistic and the artistic sides of archaeology. One could anticipate, therefore, that Rowell and Little would form an ideal team: the first a veteran in the second season, the second a novice, but eager to learn. Despite his illness in the third season, Pillet remained on the site until

the end of January, returning only briefly the end of March to supervise the closing of the campaign. Both Little and Rowell came away from Dura with an abiding resentment against him. As my experience had indicated, Pillet's only interest and competence lay in architecture, but he tried to cover his inability in other areas by a pretense of superiority and a disdainful attitude. A pose is very difficult to maintain in an isolated excavation with a small staff, close quarters, hard work, and adequate but uncomfortable rooms.

He could have learned. He was not a Classical scholar, but, though useful, Classical languages are not essential for the field archaeologist. Five languages were used in ancient Dura—Greek, Pahlavi, Latin, Hebrew, and Palmyrene, or Aramaic—and no one was expected to know them all. In Egypt, I suspect, he merely recorded the finds and was engaged largely in drawing. At Dura he was out of his depth as field director, and he knew it, but he was determined that no one else should discover the fact. He thought the best method was to handle all discoveries himself and keep assistants in the dark as much as possible. He would have preferred, I am sure, just secretaries and house assistants to serve largely his personal needs, and he tried his best to shape archaeological assistants into that pattern. It was not a happy situation.

Targets for excavation in that third season were obvious. Cumont had discovered the Temple of Artemis-Nanaia in the middle of the city, but had had no time to complete the work. The Great Gate had been excavated completely, but just inside the gate in a building on the north side of the street a jar with jewelry and more than eight hundred coins had been recovered from beneath the skirt of the embankment. Only preliminary work had been done in the Redoubt. The remains of the triumphal arch outside the city called for further investigation. Finally, the southwest corner of the fortifications, with its high tower, resembled the northwest corner, the location of the Temple of Palmyrene Gods, and offered therefore exceptional promise.

The winter rains were again a problem. They not only precluded work during the daylight hours but filled the trenches with water and mud and consequently seriously affected work after the clouds had disappeared. Pillet well described the difficulty.

> The winter [January and February] was exceptionally rainy, and the roads, already much impaired by the high floods in spring, were rendered impassable for several months, so that only the desert roads on the plateau could be used. The principal reason for this

> was the absence of the strong wind which ordinarily follows rains and quickly dries the ground. The excavations of the temple of Artemis and Atargatis, which are on a low level, were more than once inundated and I was forced to construct a strong embankment to the southwest to protect them. This was twice carried away, but finally stood, and the water which it thus held in covered a great stretch of the ruins lying between the temples and the walls along the desert. Work was thus interrupted several times on this site and was transferred to the south and west ramparts, where the slope of the terrain makes it possible always to work in dry ground. The roof and sides of the Excavation House were themselves much injured by the violence of the rain.[1]

In the second season, local Arabs had been recruited for the first time. The force increased very slowly but reached 150 in December and January. In the third season the 50 men and boys at the beginning of the season increased to 135 in December and to 200 in January and February. There was a rapid decline in March.

It quickly became apparent that the two temples in the center of the city lay side by side in two separate blocks. Cumont's excavations had uncovered parts of each, and the small section of street uncovered between the two had been taken as an intrinsic part of the temple. The new excavations disclosed that the Temple of Artemis-Nanaia had occupied one entire block; the Temple of Hadad-Atargatis across the street occupied one half a block, the other half having been allocated to a private house, presumably that of the priest. Even the third season was insufficient to complete the two blocks, half of the priest's house being left for the fourth campaign.

The two temples had a most unusual feature: the room immediately in front of the sanctuary was arranged as a little theater, and the benches or seats to either side of the entryway to the naos were made of gypsum and inscribed with the names of the donors or occupants. It gradually became clear that in the larger temple, that of Artemis-Nanaia, the room of inscribed seats had been given up in favor of the little rubble amphitheater erected in the courtyard to the south of the sanctuary rooms.

This arrangement—the anteroom of the sanctuary dedicated to a theater and the ceremony in part a display of sacred objects brought from the inner room—placed no emphasis on the paintings. Thus, we have a distinct difference in type between the temples of the gods and

goddesses at Dura. Those of the gods were decorated with paintings; those of the goddesses were furnished with inscribed seats in the pronaos, and sometimes with statues, but not with paintings. The temple of Hadad-Atargatis fell in the second category, making manifest the pre-eminence of the goddess Atargatis, which was so clear in the relief of the divine pair.[2] The sanctuary can be dated to A.D. 31.[3]

The cult relief of Hadad and Atargatis rounds out a brief but succinct picture of the temple. The two Semitic gods were the special concern of women of prominent families, and in the cult the great mother goddess predominated over the lesser Hadad, god of storm and rain as well as of the resultant grain. The cult reflects strongly the influence of native wives and the daughters of mixed parentage, since both cult and relief are Oriental in character. Only in the family names inscribed on the temple seats is the Greek element maintained, and even here it is amalgamated with the local Semitic.

The Temple of Artemis-Nanaia presented much more of a problem. The inscribed seats of the temple, removed by the Romans but used to repave the room, dated from 6 B.C. to A.D. 39–40, and another group from A.D. 102–103 to 141.[4] The Romans enlarged the temple to include the entire block and replaced the amphitheater of the inscribed seats with the *boule,* or *curia*—the senate house—of the city.[5] One chapel was dedicated by Aurelioi, Semitic families who adopted the name of Caracalla to show their Roman citizenship. Near the entrance to the pronaos of the temple, a beautifully cut inscription gave the name of Seleucos, son of Lysias, "strategos and genearchos" of the city. The title of *strategos* as military governor is well known; the *genearchos* appears to be the administrative head of surrounding Arab tribes. Several years after the discovery of this inscription, letters were noticed cut above the name of the dedicant. These provided the earliest date in the temple, 33–32 B.C., some years before the Roman-Parthian frontier was established in 20 B.C.

Placed in the center of the city, dedicated by the strategos and genearchos of Dura, older than that of Atargatis, and continuing to be enlarged until it became the largest temple in the city, this sanctuary must have been the most important locus for worship in the Parthian and Roman periods. Of the cult of the goddess at Dura we know little other than the name, Nanaia-Artemis, the first a deity distinct from, but related to, the Semitic Ishtar, and the second the goddess of wildlife in Greece, allied with celestial divinities through association with the moon.

Excavating next to the expedition house.

The placement of the cults of the two great goddesses side by side suggests complementary rather than parallel spheres of influence. The excavation of these two structures was completed in the third season, while the house of the priest—or priests—which occupied the south end of the block of the Hadad-Atargatis temple was completed in the fourth season. As at Hatra, the voussoirs from the archivolts of the courtyard of the house were decorated in the Parthian manner, with frontal human heads.

The primary purpose of pushing on the excavations inside the Great Gate was to find the origin of the hoard of coins and jewelry found at the end of the second season. Just inside the gate a Roman bath was uncovered with an inscription to the Tyche, or good fortune, of the bath. The weary traveler just entering the city after his long and dusty trip across the desert found here relaxation and refreshment.

Opposite the bath lay a small building, almost square, with four interior massive pillars of which only the rubble foundations remained. Pillet thought the building a customhouse, while Rostovtzeff called attention to the pillars in the square building, suggesting they supported a central dome and tunnel vaults between pillars and walls.[6] To me, the most striking element was the close similarity to the Roman temple west of the Citadel, which was dedicated by the Second Ulpian Cohort and the Fourth Scythian Legion. Both buildings were slightly longer than wide; in both the roof was supported by four heavy piers. The closest analogy is the ruin of a building in Parthian Assur where four heavy piers carry transverse arches that support the roof of a building just a little longer than broad.[7] I am reminded of the old Iranian room form, so common in ancient Persia: the square room with four columns supporting the flat roof. The Parthians retained the format but changed the shape a bit and introduced the transverse arches.

Rostovtzeff suggested that we recognize in the square room "the sanctuary of the city goddess of Dura which existed on the same spot before the Romans, and was rebuilt by them after the occupation of the place by the troops of Lucius Verus."[8] I think his identification was quite sound. The number of graffiti dedicated to Tyche in the Great Gate and the panel painting of Tyche found in the tower strongly suggest a shrine of the goddess close by. In our excavations, I liked to believe that discovery of a dedication to Tyche foretold a most successful season. This was a most useful superstition, because such a graffito was inevitably found no matter where we began to dig, and inevitably, we then had a most successful season.

A rear door gave entrance from the temple to a series of rooms, apparently the house of the priest. The exact site of the hoard of coins and jewelry was not clear, because the rooms had not been defined in the second season, but the presumption was that they came from the house of the priest of the temple itself and represented the sacred offerings and ornaments for the deity.

Pillet's survey of the mound in the desert beyond the walls had revealed some blocks of stone with large Latin letters, and the fourth campaign brought to light an inscription that identified the fallen triumphal arch as that of Trajan in A.D. 116, commemorating the expedition that captured both Seleucia and Ctesiphon on the Tigris. The Parthian parchment found in the second season made clear that Dura was in Parthian hands once more shortly afterward, in A.D. 121. Apparently Trajan had passed by and probably accepted the surrender of the city, hence the triumphal arch. The town presumably was not occupied for any length of time, for after Trajan's death in 117, Hadrian almost immediately restored the old frontiers. The Romans retreated almost at once to the old line at the Khabur.

Work along the ramparts above the south wadi exposed an altar with shallow reliefs on three sides. On the front stood a nude Heracles, on the left a camel, perhaps carrying the local camel god, Arzu or Azizou, with a dedicant before him.[9] Susan Downey called attention to the apotropaic nature of Heracles at Dura and to his close association with other gods, especially with the deities of Palmyra.[10] The altar seems to be a prime example of these two characteristics. Heracles occupied the prominent front face of the altar; the camel god of Palmyra and Dura, the left side; and the Tyche of Dura is presented on the right face in the guise, perhaps, of Tyche-Urania. All three are protectors from evil and donors of good fortune.

The weather during the fourth season was fine and dry, permitting full-scale operations for the first time with a work force of three hundred in December and January. Work in the desert clearing the ruins of the triumphal arch required a small, patient force rather than a large team. The debris could be conveniently dumped just outside the ruin area. The interior palace in the Redoubt was located, with its stone foundations and well-marked floors. One sizable team of workmen was employed all season in excavating city blocks; a still larger force was engaged along the desert wall on the west; and for the first time the great embankment along the circuit was seriously attacked. The embankment itself, still considered largely the accumulation of sand from

desert storms, held less interest for the excavators than the towers it covered. Therefore, Pillet had narrow trenches cut through the embankment toward the face of the towers to find the lower entrances, while other squads dug down inside the towers from above, protected from the mass of the embankment by the strong stone walls of the towers.

Much of the embankment was loose earth which, in front of one tower, caved in suddenly and without warning, burying two workmen. In front of one tower (No. 17) the trench encountered walls of a house preserved eighteen to twenty feet high on the west side. The plaster in some spots remained intact and in one place showed a most interesting graffito of a charging cataphractii, the heavily armed, mounted warrior with his horse also encased in armor.[11]

Finds in the towers were disappointing: the scales from *lorica squamata* (breast and back plates), parts of a wooden shield, and fragments of wooden arrows. Above Tower 15, just north of the corner bastion on the west wall, Pillet found evidence of a ramp reaching above the ramparts and even above the doorway of the top story of the tower from the parapet. A tremendous amount of work would be involved in clearing the walls, and the digging would be dangerous unless the whole embankment were attacked. But more discouraging, the rewards would probably be meager.

In the center of the city the debris was a shallow six to eight feet deep, sometimes rising to ten or sinking to four. The surface of the ground had gradually been levelled by wind and rain. The floors of the Roman occupation usually lay just on top of the rock floor of the plateau. A block halfway between the Temple of Artemis and the Palace of the Redoubt promised interesting results: a unique, large circular depression surrounded by a square plan of a building suggested a possible circular edifice. But excavation showed the ruins to be a deep cistern cut in the rock in the middle of an atrium.[12] The atrium attested a house far larger than the average at Dura, and the painted roof tiles showed the wealth of the owner. The block containing this large house demonstrated what future clearings confirmed: originally the blocks were divided longitudinally in half and the house plans originally called for four blocks on each side, making structures approximately square (17 × 17 meters), and cutting the block (about 72 × 39 meters) into eight sections.[13] The usual house plan was a central court surrounded by rooms, the chief room usually to the south of the central court. The house with the great atrium had been built out of

several smaller houses; enlarging a house only necessitated breaking doorways through the common walls that separated buildings.

In the fourth campaign, excavations along the Main Street disclosed a monumental arch spanning the street. It had been dedicated by the high priest of the Temple of Artemis-Nanaia and erected in honor of a philanthropist, a "savior" of the city, one Pacatianus.[14] It seemed logical to extend the excavations to either side of this arch. The main room of a house in the block on the south side provided a painting of a cavalry battle, called the Sasanian fresco because of the Pahlavi graffiti. Its date depends, of course, on when we place the first Sasanian occupation of Dura.

Alan Little believed that the Sasanians, after taking Dura in A.D. 256, occupied the city for a time and that some soldier or officer commemorated the victory with the painting. The graffiti support this view by identifying in Pahlavi the Sasanian cavalrymen vanquishing the unnamed Romans.[15] Rostovtzeff and Alfred Bellinger suggested rather that the earlier Sasanian attack of A.D. 253 had been successful and an occupation of the city immediately following had given occasion for the painting.

View down Main Street toward the Great Gate.

But it is best to postpone discussion of those last years until we come to the Roman period. Sufficient to say here that the evidence of the hoards of coins dating up to A.D. 253 attests to sudden deaths in the city, and a battle is the most reasonable cause. In early days, men buried their savings when robbing or looting threatened, against the time when the danger had passed and they could retrieve their treasure. Only the owner's death or his captivity and transport would explain a man's hoard not being recovered. But there is not an iota of evidence to suggest that the Roman garrison was withdrawn or evicted from the city then; I believe, therefore, the hoards of A.D. 253 are grim reminders of the price of a successful defense rather than of a defeat and occupation.

In any case, the paintings are specimens of Durene art in the period close to the fall of the city. Most interesting they are, partly because they are the work of a nonprofessional artist, sketching his vivid impressions and expressing himself in the current artistic tradition.

Beneath the scene of battle is portrayed a hunt with *sloughi* hound, hare, and jackal (or fox) posed with legs outstretched in the flying gallop (forelegs extended but both hoofs of the hind legs touch the ground).[16] In the battle scene itself a succession of single combats in two rows, one above the other, is represented, the victors with long lances overturning the fleeing enemy armed with reflex bows and small round shields. A parallel scene at Naqsh-i Rustam, near Persepolis, shows the Sasanian king Hormizd II unseating his opponent. At both Dura and Naqsh-i Rustam the defeated warrior is represented head down, feet in the air, but not yet fallen to the ground. It is a highly stylized form and most interesting, because it is repeated in the Han tombs of China in the second century A.D. In Persia the face of the overturned horseman is in profile; in China, he retains as far as one can see the typical three-quarters view of the Far East. But in Dura, in spite of being overthrown and falling, the defeated warrior manages awkwardly to turn full front to the viewer in typical Parthian style. The row of calm figures drawn above the battle scene undoubtedly represents deities watching the conflict.

The artist employed the well-known Mesopotamian-Assyrian device of framing the central scene by figures or objects shown in strict profile, not in vertical registers but in multiple view with successive levels shown from successive profile views. The British Museum relief of the Assyrian king Sennacherib fighting in the marshes shows the

Assyrian boats in the center, with successive series of reeds seen strictly in side view on the borders. The locale is indicated by the representation of crabs seen from above and of fish in profile.[17] At Dura this multiple view of space allowed the children in the Konon scene to be represented in front without interfering with the chief personages. This form of composition is customary, of course, in the Sasanian reliefs of the kings hunting in the midst of successive tiers of wild animals, all represented in strict profile through this multiple view of the artist.

The cavalry battle at Dura, though incomplete and sketchy, is most valuable, demonstrating the local Parthian-Mesopotamian artistic tradition as opposed to the Durene paintings in temples and synagogue, which might be attributed to the work of imported artists.

The excavations on the other side of the monumental arch disclosed the house of a merchant of the city, Nebuchelus, whose graffiti give a vivid picture of commercial dealing in the third century. The zodiac of December 11, 218 A.D., furnishes the earliest date among the graffiti and may well indicate the date of the erection of the house. Some ninety-four graffiti dealt with the business of Nebuchelus in barley, wool, textiles, boots, unguents, wine, and even loans of money, and Bellinger in his summary concluded they represented a rather gloomy period in the history of the city.

> We get few records of events, it is true, but we do get a reflection of the activities of a citizen, who, by virtue of the position of his dwelling, ought to have been prosperous if anyone was. But the record shows us a sorry time of uncertainty and pressure from the barbarian foe and consequent falling off of trade.[18]

It is true that the third century, particularly after the rise of the Sasanians in A.D. 224, was extremely difficult for Roman subjects on the eastern frontier. Trade and commerce, except very locally, must have been difficult, for Dura's economy leaned heavily on the international caravan routes. At the same time, it is easy to lay too much stress on the evidence of one individual. To substantiate Bellinger's conclusion, we would need evidence that Nebuchelus had extensive mercantile dealings in the previous period and was only now reduced to minor transactions. We have the record of a merchant located in the center of the city to be sure, but one who possessed a very modest house and engaged in rather petty trades. On the other hand, a few blocks away a

Excavating the buildings in the embankment along the city wall.

rich and affluent citizen was expanding his domicile and decorating it with painted ceiling tiles of baked brick.

One indication of events is the graffiti, dated in the spring of A.D. 238, stating, "The Persians descended upon us." This apparently referred to the great Persian raid of Syria in that year and the thrust to, or the taking of, Antioch. This time Dura escaped capture but presumably saw its lands overrun, caravans looted, and flocks and herds raided.

The two seasons' work had taken notable steps forward, completing the excavations of the temples of Artemis-Nanaia and Hadad-Atargatis, uncovering a small temple and a Roman bath just inside the Great Gate, exploring the Redoubt, and clearing a great part of the south wall along the wadi as well as the embankment in front of the southwest tower. The Sasanian Cavalry Battle had added to the catalogue of paintings from Dura; and the graffiti had given new insight into the history of trade in the city. A complete block of houses showed the pattern of the original city planning and the trend toward the concentration of wealth, with larger edifices incorporating adjacent rooms into one large structure.

Disappointing were the results of clearing the towers and digging the embankment in the southwest quarter of the city. There appeared to be promise of temple remains around the southwest tower, but there were no defined rooms, and the tower itself was in ruins. Other towers cleared between the southwest corner of the fortifications and the Great Gate yielded almost no important results, though some cloth and wood were added to the Dura collection.

The pessimistic view suggested we might have reached a dead end. Only the foundations of houses and the lowest parts of the walls remained in the city; the Citadel and Redoubt had been cleared; the most promising parts of the circuit wall had been cleared; and there was no other bright prospect. The campaigns were expensive and difficult; the results no longer spectacular. There was talk of making the fifth campaign the last.

The view north up Wall Street after the embankment had been removed. The West Wall of the city is on the left; the house walls are on the right.

6 The Fifth Campaign, 1931–32

By the spring of 1931 I had been at Yale for two years, finishing my contributions to the *Preliminary Report* of the second season and reviewing some finds of the third season. At that point Rostovtzeff invited me to take part in Yale's fifth season at Dura. The three-year contract Yale had made with Pillet was completed; Rostovtzeff hoped the position of field director might now be given to an American, while the first assistant would be a French archaeologist representing the French Academy of Inscriptions and Belles-lettres. If the French insisted on keeping Pillet, Rostovtzeff told me privately, I should go out as assistant; if they did not, I should be appointed field director. He felt also that though the last two campaigns had been productive, there was a good chance we would have to discontinue operations at the end of the 1931–32 season unless more striking results had been obtained. I took this to mean unless parchments and papyri or more paintings were found. The spectacular paintings had first drawn attention to the site, and then Cumont's finds of parchments had astonished the scholarly world.

With the permission of the French Academy, the position of field director was given to Yale, and I was appointed. As I set out, Rostovtzeff asked me to excavate one more of the fortification towers. The paintings of the Temple of the Palmyrene Gods had been found against one of the towers; Cumont had found the parchments, as well as wood and cloth, in the Tower of the Archers.

"Otherwise, Clark, dig wherever you see fit."

I have always been extremely grateful for this license. Rostovtzeff had studied the site with Cumont; he could easily have directed me to one spot or another; it is most tempting to guess likely sites for digging from the air photos, and then see how well one has succeeded. The man on the site, however, must search constantly for leads and should not have to wait for permission.

I had a very pleasant conversation with Pillet in Paris. He had no suggestions as to future digging. I had heard the French had taken new aerial photos, and I asked Pillet for copies. He replied he had none, though he understood the photos taken had been lost, both negatives and prints, in a fire.

Naudy and Walter, Pillet's French assistants, were free and willing to return to Dura. I went out early to study spoken Arabic before the season started and reached Beirut on October 4. Susan and our infant daughter, Mary Sue, remained in France.

In September 1931 I had celebrated my thirty-sixth birthday. As I look back I see myself a youngster, eager but with comparatively little experience, determined to find the treasures of Dura if they were to be found, embarking on what seemed to me—and was—a great adventure.

I needed help, and fortunately I found it. At the American University in Beirut a tutor in Arabic was recommended, and though the oral learning was extremely difficult for me, I made good progress in the two weeks I was in the city. I visited the museum to report my arrival to the director, the Emir Chéhab. By chance, Daniel Schlumberger, assistant director of the French Antiquities Service, was there as acting director in the absence of Henri Seyrig. I had met Schlumberger when, on a journey, he had come through Dura three years before. He invited me to lunch and offered to help me bring our supplies through customs. Next morning I brought him the list with a great sense of relief, for customs was ordinarily a long and arduous process in Syria. When I mentioned the army and our hope that we might obtain some supplies from the French military depot in Deir-ez-Zor, asking to whom I should apply, he promptly rang the captain in army supply and brought me over next morning to see him. The captain produced the permit given to the Dura expedition the previous year and promised to make out the necessary orders at once and send me a copy, which he did. Sometimes there are all kinds of red tape; sometimes progress seems simplicity itself.

Schlumberger mentioned that an old acquaintance and friend from the University of Chicago expedition whom we had met on our way through Turkey three years before was in town. I went to his hotel and left a note inviting him to lunch. Richard Martin was in charge of the new Chicago dig at Rihaniyeh, near Antioch, which had just ordered a Ford station wagon for their expedition. Such a vehicle seemed to be just what we needed, and I said so. If that was the case, I must hurry, he said, since new station wagons had to be ordered from Istanbul and come down by boat. At the Ford agency we discussed types of cars and prices. Next day I had a cablegram sent to Istanbul.

Three thousand nine hundred Syrian pounds, close to $3,900, remained in the Pillet-Naudy account for our mission, sufficient for the purchase. The Ford people in Beirut reported the station wagon was

due about October 20 from Istanbul. I visited Schlumberger to arrange that the car might come in without customs duties, and met at his house M. Proust, who told of the splendid Hellenistic-Roman mosaics found by the Princeton team at Antioch.

Walter was the first of our staff to arrive, on October 16. I had seen Naudy in Paris, but missed Walter, and I was most interested to meet him. I was delighted to find him a small, vivacious Frenchman bustling with energy. I had not recognized him at the boat, finally located him at the hotel, and found that he had brought in all his photographic equipment without difficulty as part of the scientific equipment of the expedition. He had served as photographer and accountant at Dura for two years; hence, he knew the customs officers, and they remembered him. He suggested he spend one day in Beirut to obtain both supplies and a camion, and then take off the second day, the eighteenth, for Dura, insuring the opening, cleaning, and organization of the camp before our arrival. All he needed was the key to the house and necessary funds.

But I had neglected to ask Pillet about this small matter of a key. Schlumberger remembered nothing, and Seyrig had not mentioned it, but my anxiety was quieted when Schlumberger finally found it in the Antiquity Services office. Walter needed an extra two days for his tasks and then left early on October 20.

I called on Harald Ingholt, the Classical archaeologist of the American University in Beirut (who was later to go to Yale), and on Father Mouterde, the amiable scholar at the French Jesuit University of St. Joseph, whom I had previously met in 1929. He urged me to use their archaeological library whenever I had the opportunity.

Ingholt, Martin, and Sherman (Martin's assistant) came to dinner, and we spent the evening speaking of archaeological news, especially the recent and continuing excavations in Syria: the French expedition at Byblos, Chicago at Rihaniyeh, Princeton at Antioch, Yale at Dura, and the French-Syrian service working primarily at Baalbek and Palmyra. The country had suddenly become available for research, and the future seemed bright.

On the twentieth, Deigert and Dave Clark of our staff arrived with Henry Pearson, from the Yale School of Fine Arts, and his mother. She hoped room might be found for her at the site as a guest assistant, working where and when she could, for a couple of weeks. Naudy, the last of our staff, arrived from France on the same day the car came from Istanbul. He spent many hours bringing in his equipment and the

plaster for removal of the paintings. The car could not be unloaded until the afternoon of October 22; once ashore it had to be assembled and the license obtained in Damascus, since Dura lay in that district.

There was a final rush on the twenty-third; I spent all day trying to hasten the assembly of the car in order to have a short trial run before taking the long run to Dura, while Naudy sought to hire a camion. The rest of the party took an excursion to the magnificent ruins of the old god of sky and sun at Baalbek.

We were up early on Saturday, October 24, and off in the car over the mountains ninety kilometers to reach Damascus at ten. I always liked to think our desert season began with our last glimpse of the Mediterranean. The sea slowly grew smaller as our car wound up the slopes, until, blue and beautiful, it disappeared in the distance. The Lebanons showed snow on the crests even in October as we drove down into the valley of the Orontes, the river flowing north, while to the south the valley reached to the headwaters of the Jordan. We crossed the slopes of the Anti-Lebanons, barer and not as high as the Lebanons, and then on to Damascus. Luckily the Moslem holiday is Friday, so the offices were open, and we secured our auto license without difficulty.

The next day we had a long, hot ride across the flat desert to Palmyra, with six of us, the chauffeur, and the baggage jammed tightly into the car. But the road was clear and well marked, and we had time to admire the magnificent ruins, standing clear now of the Arab houses and huts that had occupied the great temple precinct.

The road from Palmyra was less clearly marked, but because the chauffeur knew the desert routes, we felt no concern. The track did seem curiously uneven after we had gone some distance, never providing a glimpse of another car. I pointed this out to the driver. He remained confident; we kept on. Finally I told him to turn south across the open desert and, grumbling protests, he did so. After ten miles driving in this new direction, the straight, flat road to Deir-ez-Zor crossed our path.

The great advantage to Walter's advance party was that we did not have to stop in Deir but could continue on at once the ninety kilometers to Dura, secure in the knowledge that the house would be ready when we drove in on the evening of October 26.

In camp, Walter had cleaned the house, installed Hannah, our Armenian cook, procured provisions, engaged workmen, and had begun already the excavations. I was delighted to reenter the house

The expedition house with the Euphrates River in the background, 1929. "Souvenir de Washington Day à Salihiyeh 22 Février 1929—M. Pillet."

Pillet had built, a solid contribution to our work and comfort. A good-sized living room formed the center, with a fireplace for cold winter evenings. On one side were dining room and kitchen; on the other, a narrow unlighted corridor with cell-like dormitory rooms opening out, each with a narrow, deep-set window. The thickness of the rubble walls accounted for the deep sashes of the windows while Pillet's distrust of the Arabs was responsible for their narrowness. Yet they looked out across the valley and swift-flowing river to the eastern horizon. I found Pillet's room equipped with a real bed and mattress. His walking stick I appropriated as a badge of office. Deigert made plans and arrangements to add to the house an enclosed shower room and toilet. In contrast to the small house was the roomy court, enclosed by seven-foot walls, except to the east where the cliff offered ample protection. It was big enough to hold a garage, a bunkhouse for the foreman, and the three hundred workmen on payday, with space remaining for the storage of

large antiquities and for chairs and tables for the staff. Pillet had planned large in space, and his accommodations served us well.

Outside the camp lay the ancient city, little changed from the bare foundations and the walls I remembered. The house lay on the southern plateau and looked across a low depression toward the high-walled Citadel of the Parthians. The more or less level ground of the city was separated from house, precinct, and Citadel by a low area almost like a moat. In part, the Citadel wall had been built from stones quarried from the bedrock just outside, the quarry thus forming an additional obstacle to attackers. The depression between Citadel and expedition house had obviously led down to a river gate. The wadi running west from the depression was flanked on the south by the striking wall of the Redoubt, on the other by the high walls of the Citadel terminating in the river cliff.

On October 29 Naudy and I drove to Deir with the chauffeur and spent the day on errands. The bank officials were unacquainted with letters of credit, and though they recognized the French-American expedition, there was a long slow process of typing orders, signing papers, and talking to one official after another. Naudy renewed acquaintance with the commissary and arranged the formalities for making purchases. Fortunately the commissary had received the necessary orders from Beirut, and Naudy was able to purchase supplies at once. Colonel Goudouneix, the commanding officer of the military center, invited us to lunch, greeted us most cordially, and urged us to come to him whenever he might be helpful. I feel sure if we had been simply an American expedition we would have received full cooperation from him, but having the French Academy as partner was an added advantage. He was an outstanding officer, one who understood our difficulties and was most interested in excavations.

Naudy, serving as assistant director, photographer, supply officer in charge of the commissary, and supervisor of the kitchen, represented the French side of our combined operation. Pearson and Deigert of Yale were young architects recently graduated from the University of Michigan School of Architecture, both inexperienced in excavations. Deigert's first task was a survey map of the city as a whole; Pearson's was to study and draw the architectural remains as they appeared. Dave Clark, a young high-school graduate eager for travel and archaeological experience before entering the university, came as special assistant. Our guest, Mrs. Pearson, assisted in the cleaning of objects as well as in the search for graffiti. She also contributed most welcome American

dishes to the menus, which otherwise were based either on Arab eggs and chickens or canned goods and wine from the French commissary.

As we prepared to begin the new campaign, I reviewed the rough rectangle within the walls, 900 by 600 meters. Beyond the ruined walls stretched the desert, pitted with the depressions marking the sand-filled entrances to ancient tombs. The low mound marking the foundations of the Roman triumphal arch lay some distance northwest of the Great Gate, and scattered through the Necropolis was the occasional base of a tower tomb.

Fortunately the first steps were obvious, and work on these gave us opportunity to review other possibilities. Pillet had completed the excavations of private houses in the block of the Large Atrium (D5) and had cleared most of C7, the block of the Sasanian murals, as well as the temple blocks in the center of the city. The House of Nebuchelus, with its graffiti, had been completed, but a quarter of that block (B8) was still unexplored.

My survey of the site three years before had shown a break in the regular block plan just north of Main Street opposite the temples of Artemis and Atargatis, suggesting a district of central importance.

Excavations had approached the southwest corner of the circuit walls, where one might hope for finds comparable to those from the Temple of the Palmyrene Gods. Rostovtzeff had asked me to excavate at least one tower along the circuit wall; excavation of the southwest tower would fulfill that request and perhaps yield parchments or other finds comparable to those retrieved from the Tower of the Archers.

To Naudy as assistant I assigned the largest group of workmen for excavations in the group of eight blocks in Section G, which was gradually identified, as expected, as the market center, the agora of the Hellenistic city, the forum in Roman times. I took Pearson with me to explore and excavate the southwest corner and tower. The tower had been damaged, by an earthquake it was thought, and great cracks were visible on the exterior projecting above the sands.

Deigert was busy with the additions to the expedition house and the general survey of the city, a task that required almost half the season. I asked Naudy if he knew anything about the new air photos that Pillet had said were destroyed. He disappeared for a moment and returned with a detailed and complete set. The French air force, he said, had taken them at the end of the previous season and presented one set to Pillet and one to himself. He had heard nothing of their being burnt in a fire.

View of the expedition house with the Euphrates behind. On the far left above the excavations rise the walls of the Citadel.

I assigned to Dave Clark a small group of workmen (the numbers gradually rose during the first couple of months) to clean the rooms in Block C7 and to clear the rest of Block B8, particularly the rooms adjoining the House of Nebuchelus. I had been very sceptical of taking along a high-school boy with no special interest in archaeology, but was at once very much impressed with Dave Clark. He was mature, very intelligent, interested, and eager.

Those first ten days were hot, dazzling sun by day with omnivorous and omnipresent flies, followed at dusk by gnats that penetrated double mosquito netting. But the magnificence of the sunrise in the cool of the dawn and the blazing sunset at evening were almost compensation. Once night had come, a serene moon lighted a desert troubled no longer by flies and gnats, its silence broken only by the howling jackal and hyena. Dura was desolated, but it was dramatically beautiful.

On the sixth day of November the cold descended. The changes came quickly and unexpectedly, for there were no charts or daily weather reports. We had had unbroken heat from October 26 until November 5, except for a passing shower or two. Then, suddenly, despite bright morning sun, a cold wind came up that chilled the bones, killed the flies and gnats, and made us hug the stove and pile on clothing. By the time we finished breakfast, the day would have turned warm and summery until the sun reached the western horizon. In that vast space life seemed to move very quietly and calmly.

November 26, Thanksgiving, marked our first month at Dura. The first few weeks in an excavation are like a shakedown cruise, especially when members of the expedition are new to each other. One adjusts to the company and the work, fits into the harness, and becomes part of a team. The newness and freshness wear off, and a deeper interest grows.

By the end of November the houses in Block C7 had been completely cleared; the last two rooms of a building given the name House of the Archives, because it contained cabinets probably used to store manuscript records, had been dug in B8, and good progress was made in the shallow soil of the Agora. Finds of the first month were overshadowed by those of the second, just as the second month was obscured by the startling discoveries of January and February. Nevertheless, there was quite a number of significant additions to the Dura collections. From the Agora came an over-life-size head which echoed the early Hittite style and gave promise of impressive statues. In House G (Block C7) a representation of Hadad seated between bulls was found in two pieces. The upper part in shallow relief showed the god

Remains of the Agora. The shops face onto a colonnade.

with the double axe in the right hand and probably a bundle of wheat in the left. Above the left shoulder the thunderbolt was represented in the form of four undulating incised lines, two on each side of a straight, double-pointed staff, the whole bound together with a crescent-shaped band. The small bulls presented in three dimensions made it difficult for us to recognize at first that the sculpture fitted together. There was no inscription and, rare at Dura, no date. The frontal position, conventional drapery folds, and stiff posture belonged to the Palmyrene Orient.

No finds worthy of special mention were found in the southwest corner of the city, where work in the embankment and tower proceeded slowly. A well-cut graffito with a name and a clear date—A.D. 26—provided the earliest date found in these stone towers.[1]

Not far from the Temple of the Palmyrene Gods and close beside the road winding through the ruins, stood a thick-walled tower thirty feet or more in height, enclosing an arched doorway and built of baked brick. The brick at Dura suggested Roman or Byzantine work; the site beside the road suggested it might be of later Arab construction, but the tower itself was a challenge that could not be ignored. Early in December, I asked Dave Clark to transfer his team of workmen to this new and interesting area.

There was a little difficulty at first. The upper part of the tower had been badly cut away by desert sand, leaving a pinnacle of brick

which seemed likely to totter in any high wind or with serious digging. On one of the rest days Deigert and I shot at it with the .22 rifle, which had been added to the revolver and the twelve-gauge shotgun in the camp's arsenal, but the bullets bounced off harmlessly. We thought of lassooing the top and pulling it down, but the top was too high and there were no adequate ropes. Finally, one of the small Arab boys gathered to watch our futile efforts asked what we were trying to do. When told, one of the boys put his toes on the brick corner and in a moment had scampered up to the top, where one push released the top bricks. Successive pushes dislodged the other loose bricks and mortar, and the boy descended as lightly and easily as he had mounted.

The soil in front of the tower was shallow and largely sterile from an archaeological point of view, so clearing proceeded rapidly. A large court was revealed, bordered on three sides by columns, giving access to a broad room with five entrances from the court. Our tower enclosed not the central entrance, but the entrance to the right of center. Inside the building we rescued from under piles of earth and brick two platforms at either end of a great hall of the broad room type.

On December 10 a six-foot-square stone with part of a very large inscription was recovered from beneath the central doorway and gave us the name of Caracalla, and thus a date in the early part of the third century A.D. (211–217), and helped identify the building as the Praetorium of the Roman camp. If any doubt remained, fragments from another doorway provided part of a Latin inscription, and from a third doorway came the name of the Third Cyrenaican Legion, with the title "Antoniniana"—thus, belonging to Caracalla. Christmas Day, not a holiday at Dura, brought a present in the form of an extraordinary piece of plaster with a handsome inked inscription S.P.Q.R. (Senatus Populusque Romanus) and a dedication to Septimus Severus and his wife, Julia Domna, mother of Caracalla.

A Roman camp within the walls of a city is not common in Europe, and few have been excavated along the eastern frontier. The Praetorium, of course, formed the headquarters building, and the scene of Romans in the Temple of the Palmyrene Gods was now explained, since the temple lay close to, or was contained in, the Roman camp. In the biblical account of the trial before Pontius Pilate, Jesus was judged in the praetorium, and one may imagine an arrangement similar to that at Dura, with the Roman governor on the high dais.

Just before Christmas also, a bronze fragment, the toes of a statue more than life size, was found in the solid foundations of well-cut stones of the Agora. The bronze toes, which probably protruded from

beneath a plaster or stone sculpture, again argued for a marketplace adorned with large statuary.

The southwest tower, with its thick clouds of dust, hung together, though the gaping cracks became more and more apparent as the excavations went deeper. Fragments of cloth, arrows, and charred wood were found, but no trace of parchments or papyri.

My mid-December report, announcing the solid achievement of the identification of the Roman camp, brought joy to Rostovtzeff. Yale's Department of Classics had been mobilized, students and faculty, to study and publish the Dura materials. Baur took over the sculpture, Bellinger the coins, the advanced graduate seminar the triumphal arch of Trajan, Welles the graffiti in the House of the Archives, and Little the Sasanian paintings. Now Henry Rowell began his careful analysis of the Praetorium and the army camp.

Just before Christmas final clearing of the House of Nebuchelus brought to light a jar containing three to four hundred coins and, at the bottom, a gold necklace. The day before Christmas the temperature sank to 25 degrees Fahrenheit, with a freezing north wind so furious with sand that the workmen quit at midday. Christmas Day marked the end of two months' work, and the midpoint in our season of excavations.

Before the first of the new year we recovered under the great pile of debris in front of the southwest tower a room with rubble-plaster walls. The embankment mound sloped down over the room, preserving the rear (west) wall to a height of almost ten feet, of which about half was composed of mud brick over a rubble foundation. Only the rubble base of the front wall of the room remained.

The single large chamber was the broad-room type, with low wide benches running along both sides and back. Directly opposite the entrance, an empty arched niche of rubble projected in front of the mud-brick wall, and before that a plaster pedestal formed a narrow shelf. A small, uninscribed altar still stood beneath the niche. On one of the benches a faint graffito gave the date A.D. 75.

It became obvious as we dug out the back of the hall that the high wall as well as the niche leaned dangerously. Hence, we first turned to the heavy earth embankment behind to take the pressure off the wall. After a few days we returned to clearing the southern side of the room, which revealed the plaster still clinging to the mud brick and preserving a sketch in red and black paint of a great bird perched within an arched niche and a horned altar in front of which a dedicant drawn in

strict frontality made an offering on a small flaming *thymiaterion,* or incense burner. The whole picture was very sketchy. Although we were fortunate that the pieces of plaster had clung to the mud brick, we also were disappointed: here was an important room, in front of the southwest tower and thus corresponding to the rooms in the Temple of the Palmyrene Gods, yet the painting was so sketchy and carelessly done.

The next morning I asked Pearson to come and help me with the careful cleaning of the wall to the north of the niche. Meanwhile, one of the graffiti on the face of the niche gave a name to be remembered, the god Aphlad. We immediately started work on the wall and found to our continuing disappointment that the plaster above the rubble had entirely given way. About six feet from the niche we came on a break in the wall—a doorway?—and the rubble seemed to end abruptly. Soon, however, we saw we had a niche, not a door, with a block of stone across the top. Then, beneath, appeared the corner of another stone block. We dug down a bit more until we could see under the corner of the lower block. I peered underneath and saw relief carving. Pearson then looked and reported a big bird with feet reaching to the end of the block. We started working faster, attacking the block above to get it out of the way. Suddenly it occurred to me that that stone also might have something to do with the sanctuary, and I asked Pearson to feel under it for any inscription. He reached his fingers underneath and reported that he seemed to feel something carved. He worked his way around so that he could peer under and announced with great satisfaction that we had a large, well-cut inscription. The men working for us were as excited as we; they wielded pick and shovel faster than ever. Finally, the top block was cleared. I lifted it out and brushed off a well-made inscription. Then Pearson moved the underblock, turned it over, and exposed a magnificent relief of a god standing on two griffins. On the one side of the god stood a dedicant making an offering on an altar, and between the figures was the inscription: "This statue of the sacred god called Aphlad of the village of the Euphrates Anath, Adadiabos, son of Zadibolos, son of Silloi, erected on behalf of the safety of himself, his children and his household." The inscription block carried the equivalent date A.D. 54.

Careful measuring showed that the relief fitted the central niche. Apparently the inscription belonged to the second niche, where we had found the two. That single room had provided us, then, with a small niche with inscription, the central niche with its bas-relief, and on the left wall the little painting. The only damage was to the head of one of

Bas-relief of Aphlad.

the griffins and a few missing chips. We carried the stones down in triumph.

The inscription, we discovered, recorded the erection of the *andron,* this room, by members of an association dedicated to the god Aphlad on behalf of the safety of Seleucus, the general (*strategos*), and the members themselves, and their children. It is not uncommon to have dedications to gods on behalf of individuals and city leaders, but the year A.D. 54 may have had special significance particularly since the strategos would apparently represent the military leader, as did the strategos of the bowmen in early Roman times at Dura.

The years around A.D. 54 were difficult times. The long civil war in Parthia between Gotarzes II and Vardanes (A.D. 39–47) had come to a close before the middle of the century, but the ambitious Vologases I, ascending the throne in 51/52, had advanced into Armenia to establish his brother Tiridates on that throne; and Nero, beginning his reign in 54, had sent Corbulo to the East to oppose the Parthian advance (A.D. 63). No one could foresee the long-drawn-out war, but a major struggle between the two great powers in Asia Minor and Syria seemed assured. The Parthians defeated the Roman general Paetus in A.D. 62,

but Vologases was forced to evacuate Armenia when the able Roman general Corbulo defeated him the following year.

If troops from Dura were involved in the campaigns of Vologases, then perhaps the little painting of the eagle on an altar may be better understood. It seems curious that so sketchy a design should have been depicted on the wall of the andron. It stands alone, and though one might find a random painted sketch (dipinto) in such a place, it seems that it should have a more than passing significance.

In his careful study of the Sacred Portal, the arched entranceway that represented both the heavenly portal and the portal of the dead, Goldman interprets the scene at Dura as part of the iconography in which the eagle with palm branch signifies the carrying of the dead prince to the world of the gods.[2] At Dura it is not clear what the eagle carries in his mouth, but just behind, a winged figure lifts the crown of victory, and the arched canopy almost certainly represents the entrance to the next world. One may say, at least, that if Seleucus, the general, had been lost in a campaign after the room had been built on behalf of his safety, the suggestion of triumph in resurrection would have been most appropriate.

Just before January 11 we laid a line of track for our dump cart just inside the walls from the Great Gate to the south wadi. There was still a little clearing to be done around the Great Gate to facilitate detailed drawing. The track would run conveniently close to the house, already largely excavated, which lay in front of Tower 17. Little was expected of this cleaning-clearing operation. However, the unexpected soon happened.

I was checking the trenches before breakfast on the seventeenth when Abdul Messiah, our foreman, came up from the house at Tower 17 to announce they were uncovering a painting. The workmen had just cut through a doorway left blocked at the conclusion of the previous season's work, and at the very inside corner of the door there appeared a red-and-black painted design with a geometric pattern beneath. The find was particularly striking, because we had given up all expectation of any unusual finds in that building. Pillet had dug out all but the two back walls, which had thick curtain walls of mud brick in front. Two walled-up side doorways obviously belonged to one or two side chambers. Behind the mud brick on the west walls I had seen a fragment of plaster still in place above the ground surface, so we dug there first, thinking there might be a painting. The plaster was blank,

except for a graffito dated A.D. 232. In the second room no plaster at all remained on the back wall. We had dug the final little side room simply to complete the house plan.

Now, however, I eagerly returned to this building. We dug forward to the end of the room and then started to clear down. Next morning I arrived about seven, and Abdul Messiah had just uncovered the front of a vaulted canopy that had appeared suddenly in front of the rear (west) wall. Fortunately the workmen had not dug down through the top but, in clearing immediately in front of the arch, had discovered the open space beneath it. The vaulted area was full of plaster and debris, but on the soffit we could see rosettes and stars. Even the small space opened allowed me to see fragments of painting on the back wall and ornamental bands on the sides. As we dug down in front of the vault and its supporting columns, part of the earth against the north wall gave way to reveal four small figures in half a boat (the other half was broken away), and two figures in the foreground apparently standing in the water. Only later did we recognize Jesus walking on the water to meet Peter. Nearer to the canopy there seemed to be a continuation of the same scene, a man lying on a bed in the foreground, a god approaching on a cloud above, and a third figure in the rear hastening along with another bed. Once more a section of earth fell away to show a second register of painting beneath those already exposed. But here we stopped, leaving still unexplored the paintings on the south walls of the little room and the vaulted canopy.

The day of rest for the workmen gave Deigert and myself opportunity to clear the canopy and uncover the scene on its back wall, a shepherd with his flock and, in the lower lefthand corner, two people nude except for white loincloths, picking fruit from a tree in the presence of a large serpent—obviously Adam and Eve. Next day Pearson and I shifted our efforts to the adjacent wall below the panel already uncovered. Patiently we exposed two figures carrying bowls and wands, advancing left, toward a large white unmarked building displaying a large star over each corner. Part of a third figure, posed like the first two, was later uncovered farther along the wall. Of course we had no idea what type of room we were excavating; small wonder that in the excitement of this spectacular discovery we speculated rather freely, and sometimes in error, as to the subjects of the painting. It certainly seemed as if we were gazing at two of the three wise men in the nativity story. On the south side of the room the wall was much destroyed and we did not dare to dig far, but we could discern a painted figure holding a sword or club in its upraised right hand; along the

forearm in clear Greek letters I could read DAOUID and above an immense prostrate figure, GOLITHA. On the same, south wall we also found the painting of a woman bending over a wellhead.

An inscription carefully composed of square Greek letters painted in the decorative band between the niche and the painting of David and Goliath read TON CH(RISTO)N IN UMEIN MNESKESTHE *PRO*KLOU (The Christ Jesus is yours, remember Proclus). I feel sure that this was the only dedication belonging to the original chapel; it confirmed the conclusion, if further confirmation was necessary, that we were standing in a Christian chapel. Whether the dedication referred to some offering in the niche or to the chapel as a whole was not clear, but the niche seemed more appropriate for a lighted lamp in a room in which the lowered ceiling would cut out all light from above. Certainly the paintings would require more illumination than that given by the low doors. It seemed reasonable in any case that Proclus was a benefactor. I wrote in the *Fifth Preliminary Report* that the IN was intended for EN with the meaning *Christ among you*.[3] Welles's interpretation in the final report on the Chapel of IN as an abbreviation of the name Jesus is obviously better, especially since the words and letters are all carefully made.[4] Discussion of the special significance of the words will be reserved for the next chapter.

We needed several days to dig out the Chapel and some additional days to identify all the scenes. The scene of David and Goliath was easy, as was the healing of the paralytic; then came the scene of Jesus walking on the water. The woman at the well was more difficult, because it was not clear what she was doing and there was no representation of Christ. Henri Seyrig suggested the three Marys coming to the tomb of Christ better fitted the large panel on the north wall than the three wise men. Then the sketchy gate with the feet of other women beyond would represent other women accompanying the Marys before dawn to the tomb (Luke 24: 1–10). Above the panel of the woman at the well, fragments of painting showing only the foliage of trees and shrubs suggested a garden scene, that of paradise. That the room was a baptistery rather than a martyrium was clear from the absence of any signs of bones or covering for the basin we found beneath the canopy.

Our camp was awestruck by the extraordinary preservation of Christian murals dated more than three-quarters of a century before Constantine had recognized Christianity in 312. The scenes were small, but they were unmistakable. It is true that compared with the paintings in the Temple of the Palmyrene Gods they were sketchy and amateurish, but that little mattered, for they were Christian!

Cumont had recalled in his historical summary the account of a monk, a recluse, who had inhabited the ruins of Dura in the ninth century, and naturally there was some discussion about ascribing the construction of the chapel to him. Only very gradually was the real significance of this small, remote Christian center recognized.

The Christians had already been expelled from the synagogues by the third century and were meeting in private houses, but a building openly dedicated to Christian use was a rarity indeed. There was no widespread and national persecution in the first two and a half centuries but beginning with Nero, who blamed the Christians for the fire at Rome, the new sect became the target whenever disaster struck. The Christians refused to worship the old gods of Rome, who had brought victory and grandeur to the Roman state; this refusal was seen as the cause of local calamity and national disaster.

Under Marcus Aurelius in 177 there were riots against the Christians in Vienna because of local droughts. Irenaeus was saved from the mob in Lyons (177) due to his providential absence on a visit to Rome. Persecutions under Septimius Severus in 202 cost the life of Origen's father, and in 215 the Emperor Caracalla drove all teachers of philosophy from Alexandria, where Christianity and Greek philosophy found mutual foundations of belief.

Christian communities, whose central places of worship we know only from literary sources, spread and grew even in the third century, this dark period in the history of Rome. Pictorial remains had been preserved only in the Christian catacombs of Rome or in small, often cryptic chance finds.

In the middle of the century under Decius came the first universal and systematic persecution of Christians: "to compel Christians to sacrifice to the old gods under whose aegis Rome has grown great." The attack was renewed with increased ferocity under Valerian in 257–58. In the Decian persecution the great Origen, distinguished pupil of Clement of Alexandria, was tortured and died in Caesarea or Tyre; under the Valerian attack Cyprian was executed in Carthage.

How the Dura chapel had escaped unscathed we may review in our chapter on the Roman occupation. It is sufficient at the moment to note that the destruction of the city occurred some six years after the decrees of Decius but before the renewed and more intense attacks of Valerian. In any case, it seems almost a miracle that the Dura meeting-house was preserved to give us a glimpse of a Christian community in

the mid-third century. We have contemporary Christian painting only in the catacombs. While the private house as meeting place and church as such was continually mentioned in the early Church Fathers, the house at Dura is our sole archaeological representative for three centuries of houses dedicated to Christian use.[5]

The Church before Constantine was a mosaic of bishoprics, with leadership and guidance in the hands of the local bishops, leaders often at odds with one another, and only gradually accepting the Four Gospels as a scripture equal in sanctity to the records of the Old Testament. In the middle of the third century Dura could bear witness only to the beliefs of Syria and, with the discovery at Dura of a fragment of Tatian's Diatessaron, probably only to Tatian's interpretation of the Gospels. What one would like for comparison are examples of communities, the meeting places, the paintings, the texts of districts in different parts of the empire.

Only very gradually were the Four Gospels recognized by the various divisions of the Church, the Creed formalized, and the way of Christian belief and living fixed. In the very center of this critical period, the Dura chapel appears. The fragment of the Diatessaron, containing phrases from all Four Gospels, decisively demonstrates that the four already were accepted as orthodox. Tatian belonged to Rome and the coast of Syria. His harmony of the Gospels found at Dura indicates a widespread dissemination of his interpretation in Syria.

Kraeling summed up well the significance of the Dura finds "Among all the examples of the pre-Constantinian *domus ecclesiae* [assembly house] of which probable or possible traces have been found anywhere . . . ." Dura is the best, dated between 232/3 and 256, as early or earlier than any comparable Christian structure. It illustrates vividly the position and stage of development of the sect under Severan rule. The best preserved, it offers the first view of a Christian chapel as part of the Christian community, and the first cycle of murals in the setting of a baptistery.[6]

We were quite surprised by the publication in 1934 of the plan of this early Christian building by Armin von Gerkan, the distinguished German archaeologist who had made a special study of the Dura fortifications in 1933.[7] I am confident that von Gerkan felt he could make significant contributions to the history of the Christian building before the final report was published by us. Why he consulted no one at Dura, nor announced to anyone interested in Dura his intention of

publishing, is not clear. His article did inspire a still more searching review of the building than was possible in our preliminary report and that was incorporated in the account of Kraeling, and the plan of Pearson published in the *Final Report.*

The history of the Christian building is divided into two parts: the period when it served as a private residence, and the time after its dedication to Christian use and purposes. There was no change during these two phases in the dimensions of the building nor in the basic arrangement, the usual house plan of a series of rooms around a large open court. When the building was devoted to Christian use, two rooms on the south side were thrown together to make an assembly hall, and the northwest room was transformed into the Chapel by the addition of the baptismal font and paintings of biblical scenes; the cesspool was paved over and minor changes were completed in the court and surrounding rooms. There is still question as to the years in which the house was built and the alterations transforming it to an exclusively Christian building were made. The most important date, that of the Christian murals, has been narrowed to the period between 232, the date of the graffito on the wall of Room 4b, marking the plastering or replastering of the room, and the destruction of the city in 256.

The building, of red clay brick on rubble foundations, was located immediately in front of Tower 17, but it was separated from the city wall by the street kept open along the fortifications (Wall Street). It shared common walls with the adjacent buildings.[8] The early Christian of Dura entered the open court by an entryway in the northeast corner, which shielded observation of the court from the street. On his left, as he entered, stood a portico of two columns across the length of the court, which, in the original house, was unpaved. The large south room was furnished with the wide low plaster benches that marked the Greek *triclinium* (dining room), and a decorative moulded Bacchic frieze, a part still in situ, at the height of the high door lintel.

Kraeling concluded that, except for the portico, "in all other respects the Dura house and the Christian Building of Dura perpetuate the traditional dwelling of Mesopotamia that goes back to at least Early Dynastic times and that survived down into Hellenistic, Parthian, and Roman days and still survives in the smaller towns and villages of the region."[9]

Certainly the builder of the house was an affluent citizen, as the portico, the decorated triclinium, and the enclosed staircase next to the entrance testify. These features are also distinctly Greek, while the

broad-room shape of the triclinium and the turn of the entrance to prevent observation of the court from outside belong more to the Oriental tradition. We now know the relationship and development of the Greek house in the later Roman; we still need, however, to trace the gradual change from the typical Greek-Oriental house into the basilica and baptistery of the early Eastern church.[10]

No study has been made of the comparative dates of red and gray mud bricks at Dura. Von Gerkan in his examination of the fortifications assigned the gray brick to the Hellenistic period and seemed to assign the red brick to the Parthian and Roman periods.[11] I remember well that the brick in the Parthian palace on the Citadel was gray and difficult to distinguish from the desert soil that enclosed it. That red brick only gradually replaced the gray seemed clear, and its introduction should provide a date post quem for the construction of the original Christian building. My impression is that the brick of the Christian building points to the Roman period.

Two graffiti of horsemen of the pre-Christian period were found on the wall of Room 4, one armed with a reflex bow, the other a fully armored cataphract. As Surena, the Parthian military leader at the Battle at Carrhae (53 B.C.), used the cataphractii, the graffito could be early, but the representations of the cataphractii became popular only gradually, and the drawing on an early coat of plaster suggests a late rather than early date for the wall.

The most important graffito in the assembly room, that on the west wall, supplies the date 232. I noted that it was on an inner coat of plaster, and it was my impression it was the innermost, but this is extremely difficult to determine until the plaster breaks off and one can study it in cross section. It is not uncommon at Dura for a graffito to mark the erection of a building; it would have been equally appropriate to have had it mark a complete renovation and rededication. Kraeling believed the date 232 represented the erection of the earlier building.[12] A coin was found in the plaster of the assembly room floor, an issue of Alexander Severus (A.D. 222–235).[13] It could thus belong to a floor plastered over in A.D. 232, or to one constructed a little later. A major question is whether the building was erected in 232, or whether that date marks the year renovations were undertaken to convert the house into a Christian structure? Because the rule of Alexander Severus seems especially appropriate for Christian construction, I believe that if the Christian alterations were not made in 232, they were made very shortly thereafter.

The year 232 falls within the reign of Alexander Severus, who

accepted Christ as one of the inspired and so allowed freedom of worship. That tolerant atmosphere would account for the fact that at Dura, though the Christian Chapel was in a closed building—as (we were to discover) were the Synagogue, the Temple of Mithra, and the other temples—no special precautions were taken to maintain secrecy. The presence of the Christians must have been widely accepted. I doubt the Christian community of Dura would have dared to convert a house into a religious edifice so visible to their pagan neighbors except under such circumstances.

The first great general persecution of the Christians was instituted by Decius in 250 and this was followed in 257 by a still more severe attack under Valerian. It would be strange, I believe, to have so conspicuous a Christian building erected in the years between the two persecutions. Although the Chapel was apparently looted before the embankment engulfed the building in 256, the motive must have been plunder rather than religious intolerance, for we found no evidence of an attempt to deface the paintings. Therefore, I feel the chapel must have been erected in 232 or later, up to 250.

Kraeling has pointed out that the essential features in the early churches were the assembly room and the baptistery. He considered the west room of the Dura building, between these two, the room for preparation for baptism, the *exorcisterium,* since exorcism played an important part in early ceremonies. This special ritual taking place in Room 5 would explain the construction of a rather ornamental doorway between the exorcisterium and the baptisterium, Room 6.

The single formal graffito that forms an intrinsic part of the original baptistery was written in black letters in the painted band beneath the niche and above the scene of David and Goliath. It is a dedication comparable to the painted graffito bricks that we were to find in the Synagogue, since it named a donor and bore a very special message, "Ch[rist] J[esus] is yours." The translation is easy, but the exact significance is difficult and will be taken up with the interpretation of the paintings. The general meaning, "Christ Jesus is yours, remember Proclus," suggests this as the cardinal point in the whole baptistery and, I think, strengthens Kraeling's supposition that after the baptism, the final act of the ceremony, called the *consignatio,* with the oil of confirmation marking the completion of the ceremony, took place directly above the table.[14]

The triumphant cable I sent to President Angell and Rostovtzeff ("Five beautiful Christian frescos and room with steps found") heralded

Top: The Christian building: the north wall of the baptistery with the painting of the three Marys before the tomb, Christ on the Sea of Galilee in the panel above. The scene emerged from the debris in January 1932.

Bottom: Temple of Azzanathkona. The pronaos, made in the shape of a small theater, carries a series of inscribed risers for the seats. Set into the wall next to the entrance, above an incense altar, is a relief of the goddess enthroned between lions, with a dedicant standing beside her.

not only the recognition of fine Christian paintings, but also the discovery of another temple room with inscribed steps providing the dates A.D. 25 to 104. We located the room behind the newly uncovered Roman Praetorium, about fifty yards south of the Tower of the Archers, a spot that seemed least likely to be productive, for it lay close to the circuit wall in the debris we thought too shallow to yield anything of importance. Unmentioned in the cablegram I sent was a splendid, amazingly preserved bas-relief still in its original position on the wall of the room above the inscribed steps. The relief pictured a temple front with fluted columns on either side supporting a pediment adorned at the corners with acroteria in the form of leaves. In the pediment posed a bird, and between the columns Azzanathkona was enthroned between lions. On one side a man led forward a bull, and on the other a much larger figure was poised with raised arm as if placing a crown on the head of the goddess.

The relief was extremely difficult to clean; we discovered the reason at the base of the wall. At the first-step level was a small altar for offerings (not a fire altar) and beside it a hollow box of plaster, open on one side and with a hole in the center of the top. The relief was covered with an oily soot that clung to the stone surface. When we collected a little of the soot and touched a match, we could still smell, or thought we could smell, the scent of frankincense. We visualized the ceremony that probably took place in the sanctuary as the worshippers sat in the theater-like pronaos: incense burning in the plaster box; the smoke rising in front of the relief; and on the little altar alongside offerings being made. Then, from the naos, the inner room, the priest brings forth the still more sacred objects of worship to display before the audience.

Through the inscriptions on the steps, Susan was able to trace the family relationships by means of the patronymics. The cult was one to which only women were admitted. The patronymics were almost entirely Greek, belonging to families prominent at Dura, but a number of the women's names were Semitic, showing the same gradual mixture of the population that Cumont had discovered in the Temple of Atargatis.

The relief of Aphlad, carefully stored beside its niche, and the representation of Azzanathkona, still in situ on the wall, together added one more important contribution to our study of Dura. They made perfectly clear that the embankment along the circuit walls which had encased both buildings had been man-made and considered as a temporary means of defense to reinforce the circuit wall of the city on the

west. The embankment must have been erected to meet a sudden emergency; it was to have been removed as soon as the danger had passed.

Then came the most frustrating day I had ever experienced as an archaeologist. I had just written that all we needed to make the season a complete success were some parchments and papyri, when, close to the surface and just south of the room with the inscribed steps, almost on the edge of the ravine, we found papyri, masses of papyri, papyri on which letters were strong and clear, but in which not a single whole word remained. The letters were preserved as print in the dust—printed, as it were, on the dust—and to dust they quickly returned. The documents had been tied in bundles; while the letters of ink remained intact, the papyri upon which they were written had long ago dissolved into dust. In that dust I could see thin layers or pages—six, eight, ten together—but when I tried to brush or read, they dissolved into nothingness. We tried to put some letters together, gently touching the dust to either side, but the least breeze would lift the dust and letters to leave visible a letter or two of an undersheet. It was an impossibly tantalizing experience. We had stumbled upon a room of stored documents, for there was no sign of box or bag, only documents gathered in packages. We tried to cut out small sections; the slightest movement shattered the entire segment. The immediate transformation was not even into dust—just powder. In desperation we cut out a brick-sized piece, sealed it in paraffin, and sent it back to Yale. All was futile: a dark powdery dust arrived. I remembered reading of an ancient tomb opened and a beautiful form seen reclining on a couch; then a breath of air collapsed the whole. Had the form miraculously held its color and shape until that moment? Or was it just in the eager imagination of the beholder?

I was discouraged, but Dave Clark was not. I met him next day before breakfast, carrying a cracker-box-size container toward his trench.

"What's that for?"

"Papyri," he answered confidently, and we both took the path to his dig. The room of inscribed steps, the pronaos, and the naos lay close to the circuit wall. Clark, clearing to the west in the direction of the Tower of the Archers, had come on a series of small rooms in the temple complex whose north wall was the circuit wall itself. The wall was not high at that point, and the debris was shallow. I looked in astonishment when a workman stooped down, picked up something,

shook it out, and held up a complete papyrus almost a full page in size. Work was immediately slowed in order to sift all dirt for fragments. Next the men turned up a complete parchment, very dark in color and written in a minute hand. Sifting produced only fragments in the very slow process of lowering the level. Just as work stopped for the pay hour and the day of rest, we had indications of a richer vein beyond, a dozen or so fragments in better condition than those found already.

A welcome building inscription gave us the full name of the goddess to whom the temple, built against the city's north wall, had been dedicated, Artemis called Azzanathkona. When work began again, we immediately found more fragments of papyrus. The outline of a large piece appeared. We grubbed around in the dirt until the length and width of the piece were clear before I worked my hands slowly underneath, lifted cautiously, and extracted a whole package, several pieces wrapped together and complete. Wormholes and gaps there were, to be sure, but there was also much well-preserved writing. Before breakfast we extracted many more packs, not of rolls, but of sheets stacked together. One sheet that still held its seal comprised a package with four or five good-sized sheets of well-preserved scripts. We had just dug by chance in the best corner of the room, for the complete clearance yielded only a few more fragments of any size.

The night we completed the room, February 1, the rain began and for the next three days alternated with terrific winds. I received a cable from Yale on February 8, "Sincere congratulations," signed by President Angell and Rostovtzeff. The rains provided two days of rest and then a holiday to allow the Arabs to celebrate Ramadan, festive despite the downpour.

Henri Seyrig, the able director of the French Archaeological Institute in Beirut as well as the Syrian representative for the partition of our finds, arrived on the afternoon of February 13 for a two-day stay. According to the terms of the excavation agreement, the field director was to divide into two sections all movable finds at the end of the season, and the Syrian representative would choose the section he preferred. I suggested that Yale take the Aphlad relief, Syria that of Azzanathkona, and that we divide the Christian paintings. After a moment of reflection, Seyrig responded that he wanted the Aphlad relief because all previous reliefs had gone to Yale. He suggested Syria receive the two large bas-reliefs and Yale take the Christian paintings.

Seyrig was most interested in the papyri and eager to see what we had found. The pieces were exceedingly brittle and most of them

crumpled. We needed excellent photographs in case the originals were lost or damaged in transit. Here Dave Clark, a collector of Civil War envelopes and skilled therefore in steaming open letters, volunteered the same careful steaming he used on his envelopes to flatten our papyri before placing them under glass. The operation was a success. Seyrig and I sat down before one beautifully written document, in the light of the acetylene lamp, and read a part of the military calendar for the Dura garrison. It was dated to the reign of Alexander Severus (A.D. 222–235).

The appropriate ceremonies and observances for part of April, May, and part of June were inscribed on the single preserved piece. In that first reading we could just recognize enough to establish the birth-date of Alexander Severus and a few other special days. One item stood out dramatically—the birth day of Rome! Dura, on the very edge of the far-flung Roman Empire, must have seemed infinitely distant from the capital in those days of marching on foot. But on April 21, the eleventh day before the kalends of May, according to that precious calendar, the Dura garrison celebrated the birthday of Rome and sacrificed a cow (*bos femina*). Obviously the troops had roast beef for dinner that night. Ever since that discovery at Dura, I have tried to remember April 21, celebrating with roast beef the birthday of Rome and recalling that wonderful evening when the calendar came suddenly, startlingly to life.

One other papyrus I remember well, partly because I was able to read it almost completely. It was short and clear and comparatively legible. Addressed to camp commanders, it announced the arrival of a Parthian envoy who was to be received, given lodging, and speeded on his way. There was just one word that I could not make out, and it was some time before the reading could be made at Yale. It was the Greek word *xenia,* but written in Latin letters, the word for "hospitality," the accord due an honored guest. At the bottom of the document was the list of posts along the river to which the message was to be delivered, in military parlance, the "distribution list." I scarcely needed to be told by our French officer visitors that it was almost exactly the message they received from time to time announcing the arrival of distinguished guests.

In those few days we recovered from that single room (W13) seventy-seven papyri and seven parchments (all that were found in the 1931–32 season), more than half the total recovered in all the campaigns at Dura: 109 papyri and forty-five parchments, a sum including

the eleven parchments found by Cumont. Compared to Egyptian finds, this total was insignificant; but except for Egypt, parchments and papyri from the Classical period are exceedingly rare, confined largely to seven from Avroman, the Old Testament parchments from Qumran in the Holy Land, a few from the desert of the Sinai, and the charred pieces from the House of the Philosopher in Pompeii. The papyri of Room W13 belonged to the records of the Twentieth Palmyrene Cohort, stationed at Dura. Military papyri from Egypt are rare, since the garrisons there were few.

In this summary of the contents of the documents, I am looking beyond the campaign years of discovery at Dura, anticipating the painstaking work of C. Bradford Welles and the staff at Yale published in 1959. The first reading in the field of papyri and parchments giving new glimpses into the days of long ago has a special appeal of its own, but on the day of discovery the triumph comes suddenly, dramatically, and after the two previous meager seasons at Dura, it was incredible.

The discoveries in the temples of Aphlad and Azzanathkona, with their reliefs and records, even those of the Roman camp with Praetorium and Roman inscriptions, were overshadowed by the startling finds of the Christian Chapel and the military papyri. And for a moment we almost lost sight of the interesting dipinti found in the room immediately in front, to the east, of Room W13, a series of little sketches and puzzles drawn in ink on the plaster walls.

The first sketch, reaching Yale before the announcement of the papyrological discovery, was considered too perfect, a counterfeit, a master hoax by the artist of the expedition's staff. The frontality, the Parthian gallop of the horse (all four legs off the ground and extended, the hind hoofs turned back), the balance between horse and wild boar, the trousered costume with long-sleeved jacket, the reflex bow—all fitted almost too perfectly into the Dura tradition. It was extraordinary also that such a drawing apparently should have been made by a Roman scribe.

The find of papyri in the adjoining room coincided with the discovery of three more ink sketches of life at Dura. The lion hunt seemed more hastily drawn, but the careful portrayal of the sacrifice to Iarhibol before a triumphant general was carefully, excellently drawn and came apparently from the same pen as that of the wild boar sketch.

Of course the papyrus records found in Room W13 presuppose a Roman scribe working there, and the scarcity of sketches in ink elsewhere suggests that scribe was also an artist who amused himself with sketches of the local scene. But whoever this careful artist, he was a

skillful observer of the tradition of Parthian influence, better expressed here at Dura than anywhere else. The artist gives us a glimpse of the local costumes, actions, and the artistic tradition of the time.

Far more important is the support the little sketches give to the authenticity of the more ambitious scene of a sacrifice made to the god Iarhibol to celebrate a victory. I assume the artist was commemorating a scene he had witnessed, adding only the divine manifestations inherent in the ceremony. The nature of the victory and how it fitted into the history of Dura will be discussed later. Here, however, is an episode in the Dura chronicle, recorded by a contemporary artist, an extraordinarily vivid picture of life in the frontier Roman camp.

Three examples of parts of the well-known acrostic come from the room of the ink sketches. Our artist seemed to be intrigued by this cryptic palindrome.

ROTAS
OPERA
TENET
AREPO
SATOR

In itself it is an epigraphical tour de force, with no clear meaning and no purpose except the fact that the letters may be read up and down and forward and backward to form the same words.

An attempt has been made to connect them with early Christianity through arranging the letters in the form of a cross, and placing the extra vowels above and below the cross bar as the alpha and omega, the beginning and the end.

P
A
A T O
E
R
P A T E R N O S T E R
O
O S A
T
E
R

A fatal flaw, I believe, is the need to translate the Latin letter O into the Greek omega (Ω), to form the Christian symbol of Alpha and Omega especially common in the Gnostic interpretation of the faith. It is true that in the middle of the third century, when persecutions of Christians began, there would be every reason for Romans, and particularly those in the military, to conceal their adherence to the new faith. The Christian interpretation is, however, only a most ingenious resolution to the puzzle and we should, therefore, interpret the three drawings of the palindrome at Dura, all found in the Temple of Azzanathkona, and so in the district of the Roman camp, as just another avocation of our scribe and his friends.[15]

I recall two other exciting incidents in that rich fifth season. The plaster of one wall of the Christian chapel was preserved up to the modern surface of the ground. I had noticed a similar appearance in another room where mud brick built in front had protected the wall and plaster behind. If the plaster gave telltale evidence of well-preserved walls there, it might also in other places. A careful search of the great stretch of mounds along the circuit wall north of the Great Gate revealed such scraps of plaster outlining two rooms, one close to the Gate, the second almost two blocks away. It was too late in the season to start extensive new excavations. The last thing we wanted to do was to suggest to the native workers that there might be valuable finds within easy reach, for it was difficult to protect against pilfering in our absence. Careful mental note was made, however. There was no telling, of course, the size or type of room and amount of decoration on plaster. There was only hope, and I so reported to Rostovtzeff when I next saw him.

A curving wall of baked brick visible on the surface between the Roman camp and the Citadel was cleared on February 12 until it became apparent it belonged to another small and undistinguished Roman bath. Baths had already been excavated in the Roman camp and the Great Gate. The season was almost at an end, and I left the one room excavated as it was.

Our four months of work ended with the end of February, and funds ran out rapidly as we contemplated the necessity of an additional month or two to allow Émile Bacquet, expert technician from the Louvre, to remove the Christian murals and the last of those in the Temple of the Palmyrene Gods. I asked both Pearson and Deigert to stay on with me and complete drawings and plans of the unexpectedly large series of finds.

Naudy was understandably a little disappointed. He shared in the common triumph, but his special section in the center of the city, the agora, where the larger group of workmen toiled, produced very little in portable remains or spectacular results. Beneath the fill and rubble were the foundation walls of many small shops, the well-worked ashlar masonry of large Hellenistic buildings. Along the streets the bases of columns indicated porticoes. A rubble wall with hundreds of diamond-shaped niches suggested the city record office, but no trace of records remained. There was the forepart of a bronze foot, obviously from an important akro-metallic statue whose extremities, at least, were made of metal, and a crown of silver: significant finds but overshadowed by discoveries elsewhere. Nevertheless excavations of the civic center made a major contribution to the city plan and the sequence of wall foundations contributed valuable evidence of the agora's historical development.

Unexpected burdens had fallen on Pearson and Deigert with the need for plans, the drawing of architectural and ceramic details, copying of paintings, and the study of buildings. Our staff was small and the demands were large. Pearson and Deigert performed more than adequately as students trained in architecture, and Dave Clark gave us an extra, greatly needed hand. I was delighted when Pearson agreed to return the next year. He was not only an architect and interested scholar of ancient architecture and architectural decoration, but extraordinarily able in practical problems and in overcoming the many difficulties and obstacles that confront every archaeologist.

My wife and daughter, not yet quite a year old, came out for that last month, April. The desert grass and flowers were beautiful as spring turned rapidly into summer. Still too many tasks remained for us to feel self-satisfied, but it was happy work to conclude a triumphant season.

7 The Paintings of the Christian Baptistery

One of the major problems concerning the Christian community at Dura, namely what Gospel tradition the paintings represented, was solved in part toward the close of the sixth season, the campaign following the discovery of the Christian building. In that amazing sixth season, we uncovered the first of the Synagogue paintings late in November and all during the winter explored more and more of the astounding series of panels and walls. In addition there were the murals with banquet and hunting scenes from a house a block away from the Christian chapel, the painting of Aphrodite and the ceiling tiles from the House of the Roman Scribes, a bath from the Parthian period, horse "housings" of cloth and metal astonishingly preserved; a Roman shield with painted decorations, some sculpture, a few first-class parchments, three hoards of coins; and the continuing minor finds in pottery, wood, and cloth.

In early March, during the sixth season, the work was slackening off as the trenches began to be blocked out for closing; the massive work of packing and crating frescoes began and the digging came to a close. Not much more, therefore, was expected from the dig when in one of the baskets of finds from the embankment, behind (west of) Block L8 and not far from Tower 18, a piece of parchment scarcely three inches square appeared. Susan, compiling the catalogue, entered it on the daily register and made the usual attempt to decipher and identify what she could. The little piece, not badly crumpled, was written in a clear, legible hand, as far as the complete letters were concerned.

Next day, Susan reported with triumph that the top line contained the name "Salome," clear because, though there was no division between words, the Greek connective *kai* (and) came both before and after. It was a day or two before the fragmentary letters following the connective were recognized as "the women" or "the wives"; and further down in the fourth line of the document, the word *Sabbaton* (the Sabbath) stood out. The Gospel of St. Luke mentions the "other women" (24:10) who came with Mary Magdalene, Johanna, and Mary the mother of James, to the empty tomb of Christ, and St. Mark mentions

The Christian building. The room at the extreme right is the baptistery; the arch of the baptismal font, flattened by the weight of the embankment in which it had been buried, is visible above the deep shadow of the niche it encloses.

Salome as coming to the tomb (16:1) with Mary Magdalene and Mary, the mother of James.

It was one of those chance finds, a fragment of parchment found two blocks away and on the other side of the Great Gate from the Christian building. How it got into the debris at that point remains a mystery, and how it happened to be preserved and then discovered is another. Since it was impossible to sift the great mass of the embankment, we depended on the sharp eyes of workmen. A small piece of parchment, dirt brown, appearing in the shoveled dirt and dust required good fortune as well as sharp eyes.

The find was made on March 5, 1933, and there was an enthusiastic but unsuccessful searching in the Bible to find the appropriate passage. We found readings close and tantalizing. Clearly, we had some sort of Gospel text, something indubitably connected with the Christian community. Susan made the transcription, and we took photographs and sent parchment and copies on to Yale, still not recognizing its extraordinary significance.

That opportunity fell to Carl Kraeling at Yale, who was able to

identify the piece as a part of the Diatessaron (or Harmony, a synthesis of the four Gospels into a single narrative) made by Tatian toward the close of the second century.[1] By another extraordinary chance, the fourteen short fragmentary lines on the little piece of parchment contained words and phrases from all four Gospels, which made its identity certain.

The Dura fragment begins in the middle of the list of disciples who followed from Galilee and witnessed the crucifixion and the earthquake. Zebedee is the first word preserved and the name "Salome" follows. In the Bible account, the mother of the sons of Zebedee is mentioned specifically in the Gospel of Matthew (27:56) and Salome in the Gospel of Mark (15:40). New in the Dura fragment is mention of "the wives" of those who followed from Galilee, instead of the more general mention of "other women."

The fragment speaks:

> [And the mother of the sons] of Zebedee, and Salome, and the wives of those who followed Him from Galilee, to see Him crucified. It was the day of preparation and the Sabbath grew light. While it was still twilight in the day of preparation,* there came a man, a councilor, a hyparch** from Arimathaea; a good man and just who was a disciple of Jesus; but secretly, for fear of the Jews and he looked forward to the kingdom of God and he was not in agreement with the [Jewish] council. . . .

Luke (23:50) speaks of Joseph of Arimathaea as a good man and just, but only John (19:38) mentions him as a disciple of Jesus in secret for fear of the Jews.

In addition to its dramatic account, the document is of special interest and unusual importance for three reasons: it shows the Greek version of the Diatessaron was already well known in Syria in the middle of the third century; it adds one more piece of evidence that the Four Gospels had already been recognized as the only true testaments of the life of Christ; and in Dura it bears witness to a recognized background for the paintings of the Chapel.

Tatian, according to accepted tradition, wrote the Diatessaron about 172.[2] There were versions in Syriac and Greek. The Dura fragment is the earliest in Greek, in fact the earliest ever found in any

*The Sabbath, strictly speaking, did not begin until dawn.
**A senior officer.

script. One might expect a Syriac version for Syria and the middle Euphrates, especially since Tatian was conversant in both tongues. At Dura in the middle Euphrates, however, the Greek version is current, suggesting that Greek was probably the common standard text in the Near East.

Besides the four accepted Gospels, there were extant in the third century the Acts of St. Thomas and St. Philip and the traditions of the Gnostics, who claimed an oral tradition of the life of Christ and visions beyond those of St. Paul. At Dura, however, the Four Gospels as preserved by Tatian remained triumphant.

Without the Diatessaron one could never be certain at Dura just what Christian tradition the paintings of the Chapel represented, and the fragment is our only witness! It forms a very precious document indeed, therefore, in the collection of rare manuscripts at Yale.

It seems almost as if the painter of the baptistery were making a special effort also to show the inclusion of each of the Gospels in the paintings. The Marys with other women belong only to Luke; only John relates the story of the woman at the well; the descent of Jesus from David, mentioned only by Matthew, seems indicated in the picture of David and Goliath; and Christ imbued with the Holy Ghost and so the performer of miracles as told in Mark seems presented in the healing of the paralytic. But we cannot be sure that this combination of the Four Gospels is not the result of happenstance, particularly since we have only a portion of the upper panel of paintings.

The paintings of the baptistery have been described and analyzed many times. The interpretation of the symbolic meaning of the selection and arrangement of the murals in the Chapel is not entirely clear, since the west end contains the baptismal font; the north side, as well as the east, presents the great paintings of the tomb and the Marys; but the south wall is distinguished by the important inscription of dedication. The understanding of the decoration depends on the identification of the individual scenes. The style of the paintings indicates the schools of art to which they belong, and their antecedents and relationships may be identified from comparisons with other paintings, both inside and outside Dura, pagan and Christian.

Unresolved is the question of whether the letters of Paul and especially the moral force of Paul's teaching were reflected in the Chapel.

It seems appropriate to begin the description with the entrance of the convert through the door between the exorcisterium and the baptistery. If Kraeling was correct, as I believe he was, the evil spirits had

Painting of the Samaritan woman at the well. (Courtesy of the Gallery of Fine Arts, Yale University.)

been successfully exorcised, the power of Satan had been broken, and the cleansed convert then entered the baptistery.

As the candidate turned toward the baptismal font, he saw on his left the figure of the woman at the well, sketchily drawn in the manuscript style. The wellhead is drawn with the illusion of modified depth and the woman holds the loose rope in both hands. Her body, largely in profile, although she turns toward the spectator, is bent slightly forward. Were she to straighten up, she would fill the whole panel, in the Greek style of presenting characters prominently in the foreground and occupying all the space. In Greek drawing, however, the lower border of the panel would ordinarily form the ground line.

At first one wonders at the absence of the figure of Christ. However, the scene stresses the message (John 4:14) "But whosoever drinketh of the water that I shall give him shall never thirst; but the water that I shall give him shall be in him a well of water, springing up into everlasting life." Not uncommon in catacomb painting is the single figure representing the word or miracle of Christ. Here the word and the suggested miracle of water leading to eternal life convey the significance of the scene.

Next, the convert would stand in front of the baldachin. The white stars in the blue field in the vault, the decorative columns, and the decorations of fruit and grapes on the face of the arch are the fruits of the earth with the stars of the sky to suggest the earth and the heavens. A rubble step in front of the basin allowed one to sit easily on the edge of the font and then turn to stand or kneel in the basin itself.

On the west wall, within the curved frame arch over the basin that formed the long niche, or arcosolium, the artist painted the good shepherd and his flock. In the left corner below stand Adam and Eve holding fig leaves or aprons in front of their bodies; a great serpent with angular coils advances across the front of the scene. Kraeling suggested that the scene of Adam and Eve was a later addition, and the fact that the branches of the tree reach up into the scene of the good shepherd above strengthens this position.

The styles of the two paintings are quite distinct.[3] The heraldic style of Adam and Eve is Oriental; Adam raises his right hand to the tree, Eve her left in a manner suggesting the patterns made on a draw loom but probably reflecting the strong oriental preference for symmetrical composition. The good shepherd, posed frontally in the typical style of Dura, is sketchily drawn and the sheep he carries is more in the heroic mold than the shepherd himself.[4] Commentators have pointed

out that the combination obviously represents the fall of man and the redeemer of mankind, although the complete message of the artist—loss of Eden, fall of mankind, the recognition of good and evil—is obscure.

Curious is the empty foreground and the absence of any ground line. But these features are neither accidental nor designed to keep the painting above the action of baptism or the splash of the baptismal water, for the same deep, empty foreground appears in the healing of the paralytic. Both paintings were designed in the sketchy style of the manuscript tradition, but the manuscript sketches, rather tightly enclosed in the texts, never exhibit a large and empty foreground.[5] Dura, however, does reflect in part the Mesopotamian tradition of centering important figures in a scene rather than placing them in the immediate foreground. If Adam and Eve and the Samaritan woman at the well reflect the Greek megalographia (large figure) style, the good shepherd and the healing of the paralytic reflect the Mesopotamian tendency to push the important figure away from the front frame and into the center of the canvas.

In the healing of the paralytic the emphasis is obviously on the miracle, for the artist centered the man lying on his couch, a representation taking up much of the center of the panel but a little to right of center and balanced on the left by the figure carrying his couch, walking toward rather than away from the center. Portrayed above and behind the smaller figure is Jesus, extending his right arm to signal the miracle. He is drawn in the usual manner of the third century, clad in simple tunic, standing on a ground line represented above the reclining figure. Infrared photos clearly show a series of small black rectangular (almost square) protrusions from the head of Jesus. These have been interpreted as curls, but their shape and manner of drawing is quite different from the drawing of hair on the head of Jesus in the walking on the water scene.[6] Possibly the protrusions represent divine inspiration as the miracle is performed. They do not appear in the walking on water scene for a reason that I shall suggest in a moment.

The perspective of the couch is the Greek modified depth, which gives the bed a tipped-up appearance, rather than the extended distance of the Roman artists. The painter has given no indication of a foreground; the broad space is flat, as probably is the terrain in the scene of the good shepherd.

Between the healing of the paralytic and the walking on the water there is little break; at first we thought it might be the scene of a

Jesus on the Sea of Galilee.
(Courtesy of the Gallery of Fine Arts, Yale University.)

rescue at sea continued with the figures on the land. The sequence belongs to the Hellenic manuscript tradition but may be traced back to the cyclic style, one episode in a story succeeding another without a marked break between.[7] Interesting is the delineation of the water that fills the foreground and provides support for the boat of the Apostles. The view is double: the little figures in the boat raising their arms in astonishment are shown frontally, as is the outline of the boat; in the foreground the figures of Peter and Jesus are also pictured from directly in front. The bright color of the apostles' robes compared with the sketchy outlines of Peter and Jesus, emphasize the astonishment of the disciples of Jesus rather than the miracle itself. Their astonishment signals the miracle and obviates the need for marks of divine inspiration on the head of Jesus.

The slope of the embankment across part of the walking on the water scene preserved only part of the painting. We have no clues as to what other miracles might have been depicted on the north wall and also perhaps on the upper part of the east wall. The lower panel of paintings ran the length of the north wall and appears to continue without interruption along the east wall. First on the north wall is shown the tomb with the three Marys behind, followed by five other women, around the corner on the north wall, who are coming to the tomb of Christ before dawn. The great white sarcophagus with its many-pointed stars, perhaps symbolizing the divine figures announcing the tomb is empty, is balanced visually by a paneled gate that separates the group of three women from the group of five behind.[8] Only the hems of the robes and the feet of this latter group are preserved.

I wrote in my first letters after the uncovering of the panels that there was sufficient space for one or two more figures to be painted between the three women at the tomb and the gate behind them. If such had been the case, then perhaps the group of five women especially mentioned in the Gospels, as opposed to the general term "other women" (Luke 24:10), was represented twice, once on the north wall and once on the east wall. However, I think the interpretation of but three figures standing before the tomb is better.[9]

The women on the north wall are portrayed, as far as one can see, in the usual processional manner, a succession of viewpoints that have turned each figure to the frontal position, the feet seen tiptoe in profile, though the women advance left. Where preserved, the right arms are seen carrying torches; the left hands, held across the waist, hold bowls of incense or ointment. The hairdo of the women belonged

to the period of Alexander Severus, but might go back to the early part of the third century.[10] They wear long-sleeved and flowing garments of rich material but unadorned. The entire panel, displaying a tradition quite different from the manuscript style of the other paintings, comes closest to the procession of martyrs in the Byzantine Church of S. Apollinare Nuovo (ca. 568) in Ravenna. Quite striking, with the light robes contrasting with the Pompeian red background and the heads outlined against white veils, the scene emphasizes the miracle of resurrection, its message conveyed by the divine personages represented by astral symbols.

The south wall would seem unimportant, compared with the baptismal font and the great scene of the Marys, were it not for two factors: first, one begins with the woman at the well upon entrance to the baptistery and continues around the room to the final scene between the doors; second, the only inscription belonging to the original construction of the Chapel was carefully placed just below the niche and above the scene of David and Goliath. The inscription provides the name of Jesus Christ and records the dedicant, Proclus.

One expects in the pagan temples at Dura a representation of the deity with the name of the dedicant to mark the altar, the focal point of adoration. In the Chapel there is naturally, in compliance with the biblical injunction, no representation of God and the image of Jesus was portrayed twice as well as being symbolized in the scene of the good shepherd. The name of Christ is written only twice, once in a little graffito beside the niche and, more important, in block letters beneath the niche, coupled with the name of a donor or *the* donor. The conclusion is almost inescapable that this portion of the Chapel—the part between the two doors on the south side of the room, the wall containing the story of David and Goliath below and the garden of Paradise above—represents the focal point of the entire baptistery.

The inscription itself is simple: "Christ Jesus is yours: remember Proclus." The little scratched graffito beside the niche reads: "Christ Jesus; remember the humble Siseos."

What did the niche contain and why the table below? Kraeling suggested the baptistery may have served also for the celebration of the eucharist.[11] The niche would have held the bowl of holy oil, and the table below would have been used for the communion bread and wine. However, he did not feel the celebration of the eucharist was particularly appropriate to the scene of David and Goliath. On the other hand, the holy oil used in the Syrian baptismal rite would be very relevant in

relation to the victory of David, the chosen of God, over Goliath. In the Syrian rite of baptism the holy oil of anointment was used in connection with the baptism itself. Elsewhere in the East the convert was anointed both before and after the immersion.

Kraeling asked the pertinent question why one whole room should have been devoted to the "infrequent rite of baptism and have been decorated so lavishly."[12] The answer, I think, becomes obvious when we recognize that the table of the eucharist stood in the center of the south side of the room, and that the celebration of the eucharist concluded the ceremony of baptism. I suggest that the scroll of the New Testament was installed in the niche, on the analogy with the Synagogue, where the Torah shrine is located in the center of the wall of the room. Perhaps the arched top of the niche reflected the vault of the Torah shrine. I feel that the only object of sanctity to take precedence over the bread and wine of the eucharist would be the writings of the New Testament. "This is my blood of the New Testament which is shed for many."

Thus, while the west end of the room is given over to the baptismal rite, all the rest concentrates on resurrection, the miracles of God, and the triumph of right over wrong, of immortality over death. There is no reason why, from this point of view, the sacraments and the holy oil should not have been kept together; and there is good reason for the arrangement at Dura if the south wall marked the center of the sanctuary. The scene of David and Goliath is not carefully drawn because the eye of the observer would be fastened on the vessels of the sacraments.

If one takes the dedication as the simple statement it is, directly expressed—"Christ Jesus is yours"—one thinks first of the sacraments spread on the low table below the painting and the words of Mark and Luke (Mark 14:22–24; Luke 22:17–20). The coming of the kingdom of God mentioned in Luke offers a new thought, the promise of resurrection and immortality. Then, the Garden of Paradise immediately above the niche and the inscription become relevant. Jesus is the "anointed one" in the original meaning of "Christ" and in the Hebrew the anointed one is the messiah. "Jesus is your messiah," the inscription might be interpreted to read, and *messiah* is the longed-for savior and deliverer.

Luke 17:20–21 makes the connection clear: "And when he was demanded of the Pharisees, when the kingdom of God should come, he answered them and said, The Kingdom of God cometh not with

observation. Neither shall they say, Lo here! or lo there! for, behold, the kingdom of God is within you." The phrase in Greek, *entos humon,* is taken sometimes to mean, "within the individual," sometimes "within the group," and a still better meaning has been suggested: "within your reach."[13] At any rate, with the painting of the Garden of Paradise immediately above the niche and the painting of David and Goliath just below, an interpretation of the savior being within the possession of the initiate strongly suggests that the convert will now belong to the elect and receive immortal life.

The paintings of the baptistery show at least two hands, two styles. One is seen in the careful, formal representation of the Marys, the other is found in the sketchy scenes of the good shepherd and the miracles of Jesus. Perhaps the very formal, balanced scene of Adam and Eve represents a more Oriental influence. The manuscript tradition and the Greco-Roman large-scale figures of processional scenes belong to the westernized Near East rather than to the Eastern Mediterranean region.

Here frontality plays a dominant role; there are no distinctly Parthian features, but rather deeply imbedded Babylonian and Assyrian concepts. The Oriental costume of trousers and elaborate robes is absent; the open foreground belongs to the Mesopotamian part of the Near East rather than to Syria. There is little in the drawing that would seem out of place in the early Christian paintings of the catacombs, except perhaps for the insistence on frontality and the broad, open foregrounds. Most striking is the contrast of the paintings in the Chapel with those we later uncovered from the Synagogue at Dura and the Temple of Mithra. The Synagogue illustrated the Iranian dress in part and the echo of the draw loom as well as elaborate costumes. Still, the Synagogue lacked such typical Parthian features as the Parthian and flying gallop, the quilted trousers, and the reflex bow, all found in paintings of Parthian design and in the common graffiti of Dura. Thus, it will be seen that the Temple of Mithra belonged entirely to the Iranian tradition; the Synagogue echoed quite clearly *some* of this tradition; the Christian chapel had very little of it.

Durene Christianity came from the West and remained in the Western tradition, not quite the tradition of the catacombs, but rather that of westernized Syria. I must underline the distinct difference of this westernized Syrian style of the baptistery from the Mesopotamian tradition of the Synagogue, and the Persian-Parthian tradition of the Temple of Mithra.

Hopkins and crew attempting to free the expedition's station wagon from the mud during the 1932–33 season.

8 The Sixth Campaign, 1932–33

Three weeks before the beginning of the 1932–33 season the Arabs attacked the French-Syrian garrison and the village of Abou Kemal twenty miles below Dura. Abou Kemal was a tempting target because it was located close to the Iraqi frontier. Hastening by camion to the assistance of the beleaguered outpost, André Chavelet, lieutenant in the Sixth Battalion of the Syrian Legion stationed at Deir-ez-Zor, was killed.

We never experienced any serious trouble ourselves during our ten years of excavating at Dura, although the money we carried from the bank to camp every now and then must have made us an easy mark. But I think the small detachment of Syrian soldiers assigned to guard our camp was useful as a deterrent in time of stress.

At the request of Colonel Goudouneix, our labor staff was increased to 360, sixty beyond our normal complement, to help give employment to the famine-stricken Arabs. The meager rainfall over the last two years had seriously threatened the existence of the Arabs living around Abou Kemal. With the enlarged labor force we worked on a twelve-day schedule; ten days of work and two of rest.

For this season, Le Comte Robert du Mesnil du Buisson was appointed to represent the French Academy as associate director. He had been a captain in the French army during the war and director of French excavations at Mishrifé-Qatna in Syria after the war, and therefore filled admirably the double role of able excavator and liaison officer between our American camp and the French military headquarters in Deir-ez-Zor. He was an able man of action, an enthusiastic excavator, and a sound organizer. He had not been extensively trained in research, but he was quick to recognize new leads and to follow them on his own initiative. His years in the army had given him confidence in command; he acted with energy and insight. Through his acumen, a number of most interesting leads were explored which might otherwise have been neglected. His knowledge of Hebrew and Aramaic enabled us to prepare transcriptions of graffiti and to read the Aramaic inscriptions in the Synagogue.

Frank E. Brown came as assistant from Yale. Well trained in his graduate years, he had spent two years at the American Academy in Rome, where he had been particularly interested in Classical architec-

ture and, most fortunate for us, in architectural drawing. He was—and remains—an exceptional man, both scholar and excavator, not exceptional in having both skills, but exceptional in the double excellence he displayed. Meticulous in his work, unflagging in energy, gifted with extraordinary insight, he was a bulwark on the work of the last three seasons of my stay at Dura. After my appointment at the University of Michigan, he became the director for the last two seasons. A more admirable selection could not have been made.

Architect this year was Van W. Knox of the Yale School of Fine Arts, who not only drew the plans of buildings excavated, but also made copies of the paintings found in a private house. He drew a reconstruction of the Torah shrine in the Synagogue and was always eager to help in fitting together the broken pieces of painted walls.

Toward the close of November, Émile Bacquet came out from the Louvre to remove the paintings of the Christian chapel. In mid February, Maurice le Palud joined our staff to assist in photographing the great series of new paintings, and in March, M. L. Cavro was persuaded to come from Beirut and make copies of them.

Margaret Crosby, from the Yale graduate school, came as assistant. Trained in classics and interested in archaeology, she was eager to obtain firsthand experience in fieldwork. Her participation was also an experiment, to see what part a woman who was there in her own right, and not because she was accompanying one of the men, might play in a Near Eastern excavation isolated in the desert. We had no trouble with the Arabs, and my experience has been that the desert is no threat if proper precautions against being lost are observed. This early experiment of a young woman in the camp was most successful. Margaret Crosby first made copies of the graffiti in the Palmyrene Tower; she took charge of the photographic catalogue, helped to put together the broken paintings in the Christian building, catalogued and studied the important series of ceiling tiles from the House of the Scribes, and, early in January, took over full supervision of the trench being cut through the area between our camp and the Citadel as well as of the buildings across the Deir road below the Redoubt.

My wife, with the experience of the 1928–29 campaign and the last month of 1932 behind her, took over supervision of the kitchen and direction of our Syrian cook. Once more she accepted charge of the catalogue, which meant the repair of broken pottery, the cleaning of coins, and the identification of papyri. The special study of the inscribed seats in the Temple of Azzanathkona I placed in her charge.

Above all she took care of our infant daughter, Mary Sue, now a year and a half old. It was a great comfort for all of us, isolated in the desert that surrounded Dura, to know that there was an American medical mission stationed at Deir-ez-Zor, and to have its competent director, Dr. Ellis Hudson, within one hundred miles of us in case of an emergency. The presence of our infant made this comfort a necessity.

Chief targets for the sixth campaign were obvious. First there was the embankment, particularly the block stretching between the Christian building and the Main Gate. Still more intriguing were the traces of plaster noticed at the close of the previous season that indicated buildings and rooms reaching the surface of the embankment north of the Gate and promising walls well preserved to unusual heights. The great expanse of the Roman Camp had been only partially explored. It was essential to explore the depression between the Citadel and our new house to determine whether Main Street extended to the river and if remains of a river gate could be found. In the previous season we had found indication of a possible temple area forming one corner of Block L5, and the area just north of the Citadel and the northern part of the Citadel itself required further investigation. Certainly we did not lack promising and challenging areas to excavate.

I confess we felt a pressing need to uncover striking finds in that period of the Great Depression. Even Rostovtzeff's enthusiasm and eager elucidation of the new Christian building at Dura had met with disappointing reactions among Classical scholars, who were more concerned with Greece and Rome than with early Christianity, and with only moderate response from American churchmen, more interested in the modern interpretation of the Bible than in older biblical representations and ancient meetinghouses. The discovery of a house dedicated to Christian worship in the period before Constantine, they admitted, was splendid and unusual—yes, even unique. But there it ended.

European clergy and the continental schools were much more interested in the history and development of the Church, but they were in no position to give aid to Dura. In a moment of sadness, Rostovtzeff wrote me that the response to the finding of the Chapel had been disappointing. The churchmen did not appreciate the significance, and the scholars who did could not give financial aid. The Jews, he wrote, were much more interested in their historic monuments; they appreciated much better the role of history; they continued and spread the study of Hebrew. "Find a synagogue," he suggested, "and we shall

gain the acclaim which such startling discoveries really deserve." The building whose walls showed above the embankment with bits of plaster still upright converted his wish into prophecy.

Pillet had begun the excavations of the Christian building, the structure he called the edifice of Tower 17 because its high back wall was just in front of the tower. The excavation of the tower, which was his objective, required a certain amount of clearing in front so that the entranceway might be opened and debris removed through the lower doorway.

The discovery of the Chapel suddenly demonstrated that it was not necessarily the towers and the walls as much as the buildings encased in the embankment that might offer the richest rewards. I still wonder, with the benefit of hindsight, why we had not thought of this before. There stretched before us the great sweep of the embankment from the Great Gate north along the desert, a third of a mile and more, broken only by shallow trenches made by British soldiers in their defense of the river road. The embankment also reached south from the Great Gate, past the Chapel to the southwest angle of the wall. South of the Chapel where towers had been dug, the embankment was penetrated, a long slow process, but with no notable results.

With good reason, then, at the end of the fifth season we paced the embankment afresh, closely scrutinized its sloping surfaces, and carefully checked every stray piece of plaster that might indicate the presence of buried walls. In the embankment north of the Gate, our survey not only disclosed fragments of plaster, but upright bits, and in the two places where these occurred, one within the first block from the Gate and one in the second, the pattern suggested the rooms of houses whose mud-brick walls had miraculously retained at least a part of their plaster coverings.

We had no good reason to announce these special indications at Yale. We were assured financial support for at least the next season, when the new leads could be investigated. There was no need to raise hopes that could easily lead to disappointment. We decided to wait and see. Expectations would be high enough at Yale in any case after the spectacular discoveries of 1931–32.

Although impatient, we did not begin the 1932–33 season with these new prospects. Work in the center of the city was discontinued for the moment, but three new trenches were begun. Du Mesnil attacked the district just south of the Citadel; Frank Brown continued on the Praetorium in the Roman quarter, while I worked around the

Chapel, clearing along the block of embankment toward the Great Gate.

Because Du Mesnil was experienced, I gave him the second-best foreman. Frank Brown's inexperience recommended giving him our best foreman. I took the third foreman, a well-meaning Armenian able to follow instructions but not very capable in organization and independent judgment. We started the first day's work with the laying of the Decauville railway tracks so that dirt could be hauled on the small cars and dumped into the south ravine. The workmen were sorted out and the tools allocated: picks for the best workmen, one or two shovelmen to each pickman according to the hardness and depth of the ground, a group of boys to carry dirt, and a responsible man or two on each wagon. The organization and instruction of one hundred men took time, and my foreman was far from expert. He knew the men better than I did, however, and it was his function to allocate tasks, at least for a beginning. There were some altercations, for the position with the pick is the special prize. The pickman worked where most finds were to be made, and so received the most baksheesh. There was some confusion, natural under the circumstances. I felt impatient, then frustrated, and finally rather dismayed. I thought, "If I have all this trouble and confusion, old and experienced excavator that I am, what must poor Frank Brown feel in his first experience in this desolate, difficult land?" I looked across the half mile of flat ground to his dig. There, billowing from the ground, was the biggest, most extraordinary cloud of dust I had ever seen. It looked as if every workman had thrown himself with almost fanatical devotion for at least half an hour to shoveling all the dirt he could into the air.

In one way it was a pity that the sensational paintings of the Synagogue were discovered in the sixth season, for the Synagogue and its astounding walls so overshadow the other finds of the season that one forgets the campaign would still have been an outstanding success without the revelation of the Synagogue.

The excavations south of the Citadel revealed the Main Street running east and west. The Roman bath, only partly excavated because of the modern road, followed the pattern of the baths found in the northwest corner of the city and just inside the Main Gate. The new find made almost certain the presence of a gate to the river at the end of the depression. But all remains of gate and wall had been swept away by the river's crumbling of the cliff. The bath suggested the gate was close at hand. Probably the slope of ground extended beyond the gate

to the level of the riverbank. A slope beyond the wall would have protected gate and wall in time of floods.

Contrasting sharply with the regular plan of the Roman baths was that of the Parthian bath, built along oriental lines, found beneath the ruins of a military theater, an *amphitheatre castrensis*, whose foundations had helped to preserve the earlier walls to an unusual height. The Parthian structure was adjacent to, or a part of, a palaestra complex but had the special Eastern features of dome with squinch supports and terra-cotta pipes and pottery within the framework of arches and domes, the earliest evidence of this method of reducing the weight of stone vaults. The bath and its domes were small, not lending themselves to easy expansion, and on this account perhaps were not continued in the Roman period. Thus, the two types of architecture, side by side, were separate and distinct.

Frank Brown discovered, beyond the bath in the northeast corner of the city, a great house of Parthian plan built in the third quarter of the first century and later remodelled for the Roman soldiers during their occupation.[1] When the military garrison that came to Dura under Verus in A.D. 164 had moved from the campus in front of the Citadel to the northeast corner of the city remained a problem. But the Roman headquarters, the Praetorium, was erected, according to the inscriptions, between February 211 and February 212.[2] A logical shift would have been under Caracalla (A.D. 211–217) when frontier garrisons in the Near East were heavily reinforced.

Most important for the history of Dura was the growth of the northeast corner of the city in Parthian times. Hellenistic mud-brick walls enclosing that corner of the city were partly replaced by stone near the beginning of our era. Early in the first century, the Temple of Azzanathkona was built just south of the corner, and the Temple of the Palmyrene Gods was laid out in the angle of the wall in the middle of the century. In the third quarter, the great House of the Parthian was built in Block E4.[3]

The open space beside the bath would first have served as a palaestra for the Hellenistic training of youths and for their military service; then it was replaced by the Parthian bath, which in turn gave way to the Roman Camp, Praetorium, and bath. The overall evidence seems fairly convincing, if not conclusive.

In the first days of December, paintings in a private house in Block M7 came to light in much the same way the Christian murals had been discovered. Starting from the Christian building we had dug north in

the embankment all through November toward the Great Gate, uncovering the walls and rooms of a house of ordinary plan. We broke through a blocked doorway to discover on the wall some painted plaster: a happy chance, for we were thus immediately warned to clear with caution, which insured the preservation of the paintings. One entire wall of the room had been painted. Half carried a rather stiff, conventional representation of a banquet, a type common in the reliefs of Palmyra; the other half showed a spirited hunt of onagers in the desert by a galloping horseman clad in a long-sleeved tunic and trousers, armed with a reflex bow. We could identify the wild asses by their long ears and hairless tails. The long unbroken lines sweeping from head to tail, a yellow beneath, then red and black, betrayed the hand of a master craftsman who drew with supreme confidence, correcting little errors in the outlines with the stronger colors.[4]

The banquet was remarkable for the presence of a woman at the end of the table and a winged Eros with inverted torch, a symbol of the funeral banquet. The name of the hunter, Bolageos, undoubtedly the deceased, was written beside him. A representation of the evil eye attacked by snake and scorpion, cock and ibis served as a guard against evil spirits. The painting was a remarkable combination of Parthian, Palmyrene, and Greek elements, but the onager hunt was typically Durene and Parthian.

A second banquet scene of three men reclining beneath festooned walls and attended by servants embellished the smaller west wall. We can only guess at the number of otherwise undistinguished houses such as this one which once held comparable paintings now completely destroyed.

At almost the same time as the murals were discovered, a fallen piece of plaster with a painted Victory in the *frigidarium* of the Parthian bath was recovered.[5] A later replastering had admirably preserved the colors, fresh and bright. Despite the fact that the Victory came from a Parthian bath, the painting was much closer to Hellenistic models than was the Victory on the wooden panel found in the second season.[6]

The first papyrus of the season was found in the embankment less than a month after our start. Two days later we came across our first parchment, a harbinger of the score of legible texts we were to recover. The last and most important, the fragment of the Diatessaron, was found on March 5, also in the embankment. Tower 19 yielded a complete Roman legionary shield of laminated wood covered with leather painted in bright oriental design and with the lion of the Third

Painted panel of the Victory as uncovered.

Cyrenaean Legion defiantly facing the foe.[7] Beneath the shield at the very bottom of the tower was the most astonishing treasure: two complete sets of scale armor for horses, one of bronze, one of iron, the plates still sewn to the cloth beneath. They were so well preserved that they could be lifted and placed on a horse's back.[8] Even the cuisse, the covering for the base of the tail, was intact.

Three hoards of coins, wooden doors, a stone carved on both sides in low relief, a bronze brazier decorated with griffins, various clay plaques, small sculpture, pottery, glass, and coins were recovered by the teams. But above all stands the Synagogue.

The traces we had found near the close of the previous season—tiny bits of upright plaster in the bare dull slope of the embankment just sufficient to mark straight lines—spoke of walls beneath. As soon as we were well established and had the three teams working, I took one of our best workmen and explained to him what I wanted. He was to follow with great care the line of plaster by cutting down just six or eight inches to see if the bits of plaster represented a continuous wall. Equally important was to determine whether the fragments of walls meant a series of rooms, as I expected, or one large hall.

Work was slow, careful, methodical. The Arab workman pursued his task with great skill; he had understood at once what was required. At the end of one day the outline of plaster turned into an unbroken line of wall, a wall of plaster on mud brick, which meant plaster very irregular in thickness. There were no cross walls and no breaks except

The Goddess of Victory from the Parthian bath. Émile Bacquet is doing field repairs on the painting.

The Synagogue before the great hall had been excavated. Du Mesnil stands in the doorway. Only the plaster-faced wall of the room has been lightly trenched to trace its outline in the embankment sloping up against the city's west wall.

one in the middle of the front (east) wall, well down the slope where the lines of plaster ended. No doubt this marked the door. One corner showed traces of red, black, and yellow colors without design or form. We were not certain at first that we had one large room, perhaps forty by twenty-five feet, but it *looked* as if it *might* be a single chamber, and I took a photograph of Du Mesnil standing in what we guessed was the entrance. The picture clearly shows the lines of walls in the slope of the embankment as well as the tremendous amount of earth that would need to be removed in the excavation. Behind the traces of the back wall can be seen the line of stones that formed the section of the circuit wall protruding from the embankment. The space between we later dubbed Wall Street.

What to do next? Fortunately we decided to give this structure the full archaeological treatment! The wall as far as we could see was mud brick coated with plaster. If we dug immediately in front, the weight of the earth behind, filling Wall Street, might cause the mud brick to cave in. The bits of plaster brought into view seemed vertical, but the wall as uncovered might slope dangerously forward.

Plan of the Synagogue. Entrance to the Synagogue complex was made from Street A through a doorway off the narrow alley (71). After walking through a series of rooms that formed the original house, one entered the colonnaded court, on the opposite side of which were the main and auxiliary doorways into the prayer hall that backed on Wall Street.

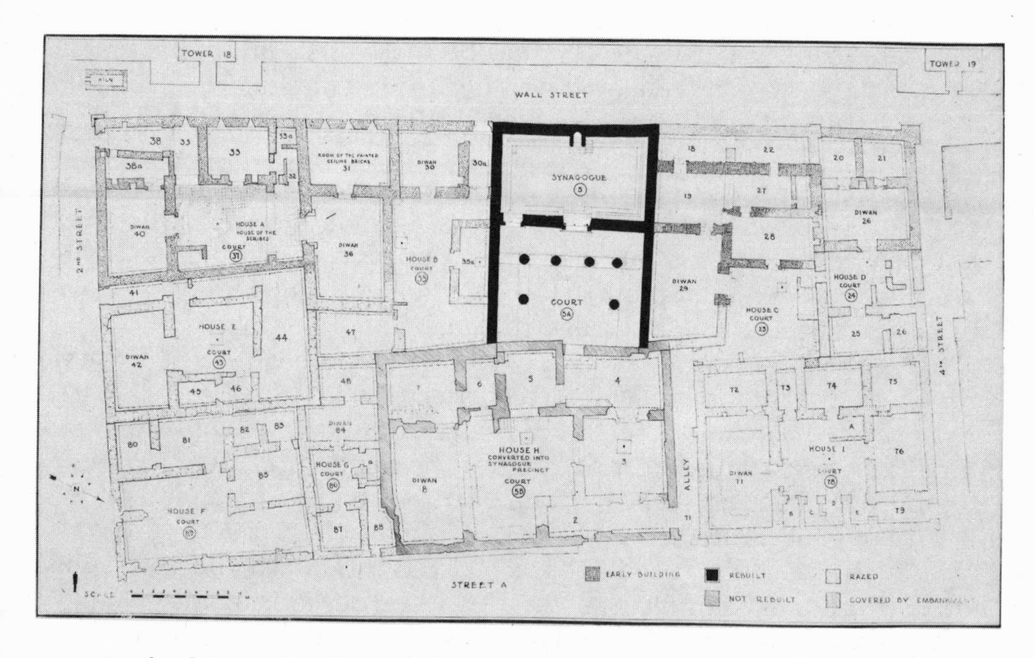

I asked Du Mesnil, who had the largest team of workmen, to start halfway down the embankment, excavating toward the back wall in such a way as to reach the west wall some six or eight feet below the top. At the same time a second group would dig between the wall and the stone fortifications, uncovering the outside of the wall, but not cutting too deeply until the pressure of dirt from the inside of the room was removed. Du Mesnil, an experienced excavator and engineer, understood perfectly what was required and personally supervised the work. Track was laid along the top of the embankment. My team was working just south of the gate; Du Mesnil's team would be a block and a half north of the Gate; hence, both teams could use the track to dump the debris in the south wadi. Work progressed slowly, but fortunately the season was unusually dry and clear.

Our workmen were eager and had to be restrained from cutting through the covering of untouched debris left against the side and back walls. I do not know if any of the men realized the possibilities, but curiosity as to the condition of the walls themselves was natural. The chance of finding paintings was a remote or near possibility, depending on the degree of optimism. If paintings were preserved, how much might remain intact? How much plaster had retained its hold on the fragile surface of mudbrick walls? We had left a foot of loose dirt against the side walls, a bit more in front of the back (west) wall, as the

level of digging gradually descended. By loose dirt I mean dirt that had been packed down by time and weather for fifteen hundred years; now it was a bank of solid earth, but originally it had been the debris of loose dirt.

As we came close to the back wall we saw that the whole area formed one great room, for there were no indications of cross walls, though a very sharp lookout had been kept. Pieces of mud brick were carefully outlined, undercut, cleared, then discarded only when we were certain there was no wall beneath.

On November 17 a small piece of papyrus, the writing still legible, appeared in the debris of Wall Street back of the block. Two days later a fragment of parchment surfaced. Important in themselves, these finds were of still greater interest because of their condition. Their preservation not far from the surface meant that the soil had remained dry and undisturbed over the centuries.

As Du Mesnil's trench approached the back wall it became apparent that the height of wall above the cut on the north side would be eight to ten feet. What if they held paintings and suddenly a hard rain or sandstorm should come up? Blowing sand would perhaps be even more destructive than rain. However, good fortune smiled: the sky shone clear and blue and vast. At the end of November the days were still magnificent, warm and sunny, with cold nights. We waited through a rest period so that the workmen would be available in force for any emergency.

At breakfast the day after the rest, about a week before the end of November, with the day dawning clear and calm, I announced we should have the unveiling immediately. All who cared to do so were welcome to come and see. My wife remained in camp with the baby; Frank Brown declared his trench in the Roman quarter also showed promising fragments of painting and his *chantier* (work area) would rival any finds we might discover in Block L7.

Only a few details of that day are still vivid. Du Mesnil was in charge of the work; Van Knox, as architect and artist, was assisting. Miss Crosby came in hopeful curiosity; Bacquet had not yet arrived. But I clearly remember when the foot of fill dirt still covering the back wall was undercut and fell away, exposing the most amazing succession of paintings! Whole scenes, figures, and objects burst into view, brilliant in color, magnificent in the sunshine. Though dwarfed against the vast backdrop of the sky and the tremendous mass of the embankment, they seemed more splendid than all else put together.

Once, when I was involved in a train wreck, I had no recollection of the moment between the shock when I was thrown from my seat and when I began to pick myself up from the bottom of the overturned car. So it was at Dura. All I can remember is the sudden shock and then the astonishment, the disbelief, as painting after painting came into view. The west wall faced the morning sun which had risen triumphantly behind us, revealing a strange phenomenon: in spite of having been encased in dry dust for centuries, the murals retained a vivid brightness that was little short of the miraculous.

As the full extent of the wall came into view, I sent a note to Susan to come at once. Frank Brown came running from his trench. He could not stand the strain of waiting when he saw the bright colors in the distance. We stood together in mute silence and complete astonishment. A casual passerby witnessing the paintings suddenly emerging from the earth would have been astonished. If he had been a Classical archaeologist, with the knowledge of how few paintings had survived from Classical times, he would have been that much more amazed. But if he were a biblical scholar or a student of ancient art and were told that the building was a synagogue and the paintings were scenes from the Old Testament, he simply would not have believed it. It could not be; there was absolutely no precedent, nor could there be any. The stern injunction in the Ten Commandments against the making of graven images would be sufficient to prove him right.

If, finally, this passerby had been in my shoes, the director of the excavations, responsible for the success of the expedition, and the one who would be credited most with its achievements, then the discovery of the Synagogue that day would be like a page from the Arabian Nights. Aladdin's lamp had been rubbed, and suddenly from the dry, brown, bare desert, had appeared paintings, not just one nor a panel nor a wall, but a whole building of scene after scene, all drawn from the Old Testament in a way never dreamed of before.

We asked ourselves, when the great series of pictures appeared so suddenly, what they all meant? Of course, we did not know what sort of building we had found. What was the meaning of this first great scene of the gate of a city in the far upper-right corner? And of the huge figure in front of a procession of people moving left? Between the feet of the great figure that dominated the highest panel, there was an inscription in Aramaic. Du Mesnil slowly read it, "Moses, when he went out from Egypt and cleft the sea." A pity we could not, by some magic, tell that ancient Dura writer how much his inscription meant to

Top: Discovery of the painted plaster at the top of the Synagogue wall. The author sits at the top of the embankment against the west wall of the city; Miss Crosby kneels "inside" the hall; Susan Hopkins is seated behind her.

Bottom: The Synagogue with its temporary corrugated iron roofing. Only the top two registers of paintings and the Torah shrine have been uncovered.

us. We would not have dared to dream that those paintings of people and soldiers, those vividly portrayed events, belonged to the story of the Old Testament unless strong epigraphical proof had been given.

We had the additional good fortune to discover that just at this point the embankment preserved part of the walls to their full height. The top of the northwest corner of the room was painted to represent a column supporting the roof and ceiling. The capital of the painted column was partly preserved, indicating that at least this corner showed the full height of the paintings if not of the room itself. The side (north and south) walls sloped down rapidly, following the angle of the embankment. The upper register of paintings remained in only that one corner, showing the dream of Jacob, with angels ascending to heaven by means of a ladder.

Those inauspicious little dabs of red and yellow that first drew our attention when the trenching began just six inches deep now took on their true meaning: they were the plagues of hail and fire pouring over Egypt in the scene of Moses leading his people in the Exodus. Now we could easily recognize and identify biblical scenes: the Egyptians drowning in the Red Sea, the sweetening of the waters of Marah, the numbering of the tribes, Moses before the burning bush, and so on.

Needless to say, we were not immediately aware of the full import of our find. The niche with its raised floor to hold the sacred scrolls was at first thought to be an important seat. Two or three days were needed to clear enough of the clinging dirt to recognize the little inscription and to be sure of its message. The paint was tempera and, when dry, could be smudged. The dirt had to be removed with extreme care. And, of course, we had no reason to believe at first that a muraled synagogue could exist here or anywhere else. Hence, we first guessed we were looking at Roman armies, refugees, troops fording rivers, etc. Once the inscription was established, however, the contents of the paintings were unmistakable.

At the south end of the room the wall as excavated was preserved to a height of six feet, sufficient to disclose the whole of the sectioned middle register of paintings. At that juncture in our clearing we had, running from one end of the room to the other, an unbroken panel of pictures slightly over five feet wide and forty feet long.

A single word inscribed in that stretch of pictures was of tremendous help. "Aaron," written in Greek, was inscribed beside the chief figure in a temple and courtyard scene. In that first season of discovery of the Synagogue, we dared not go below the second register. We

Opposite: The Synagogue painting of Exodus. Between the legs of a dominating figure appears the Aramaic inscription "Moses, when he went out from Egypt and cleft the sea," which identified the great hall as a Synagogue.

Below, left: The first cut through the floor level of the Synagogue on the west side of the hall reveals the Torah niche.

Below, right: Susan Hopkins seated in the niche before it had been identified and the walls on either side excavated. "Our discoveries have gone on at a remarkable rate. Clark now has a room in a private house with frescoes, and the big room is getting more and more impressive. We are a bit worried however about its falling, particularly because one in a room next door collapsed this morning narrowly missing the workmen. . . . Those frescoes really are wonderful and the whole effect with the big seat topped by the story of Abraham and Isaac is simply stupendous." (Susan Hopkins to her family, December 4, 1932.)

Aaron, the high priest, his name written in Greek, as seen on the walls of the Synagogue.

all, metaphorically, held our breath until the kindness of Colonel Goudouneix in Deir-ez-Zor, the cooperation of the French supply officers, and the architectural ability of Du Mesnil combined to provide a durable roof of corrugated iron on wooden posts to protect the paintings from wind and weather. The natural slope of the embankment and the side walls of the building helped enclose the paintings; the stone circuit wall on the west would break the winds from that quarter. A season that was cold and dry rather than rainy and windy added to our good fortune.

When we had the roof in place, we could not resist the temptation to dig a narrow trench through from the base of the embankment to the rear wall, and there we found the Torah shrine intact below the second register of paintings. We detected one further band of paintings to either side, but this could wait. Reward enough were the paintings of the sacrifice of Isaac by Abraham on the front of the Torah shrine, and the sacred symbols of the seven-branched candlestick (*menorah*), citrus fruit (*ethrog*), and palm branch (*lulab*), not to mention the columned arcosolium to contain the Torah itself, splendidly preserved and lacking only the curtain and treasured scrolls.

Not until the middle of the next season did we remove the upper paintings and dig out the rest of the room. The lower half of the room was itself impressive. Two low plaster benches circled the room at floor level, backed by a decorative dado above which rose the paintings in three bands, or registers, that encircled the room. For the first time we could grasp the full scope of wall space that had been covered with paintings in the original building.

We had one unfortunate accident. We had found fragments of painted plaster, as well as baked-brick ceiling tiles, in the debris. We tried to piece together the plaster on the floor of the building, now roofed over with the corrugated iron, but the light was not good and the space inadequate. We carefully moved outside in order to see better what joins might be made among the hundred and more pieces with recognizable details—a part of a face, a hand, part of a garment. But whatever significance they might have had was lost when a sudden and violent rain washed out all color. Only a single inscription on a baked brick, written in black, came back miraculously when an infrared photograph was taken. Our loss forcefully reminded us how fortunate we had been in weather and circumstance up to then. For a sudden squall or sandstorm in those first days could have destroyed utterly our newfound paintings!

ΑΡΩΝ

Top: Erecting the temporary shelter over the Synagogue.

Bottom: The west wall of the Synagogue emerges from the embankment. Susan Hopkins stands in front of the scene of the return of the ark from Dagon's temple; to her left is Solomon's Temple; above are the scenes of Exodus.

As foil and contrast to the artistic tradition of the Synagogue, one could not ask for better representations than the paintings found later that season in the house we labeled the House of the Scribes: a scene of Venus and Cupid and the hunt of the onagers mentioned earlier. The Venus and Cupid are clearly Greco-Roman; the hunt is Parthian. At the same time, the funeral banquet painted on the wall with the hunt, and some military portraits also found in the House of Scribes, belong to the native Greco-Semitic background of Palmyra and Syria.

This amazing season held yet another surprise: the discovery of skeletons, armor, and money belonging to a small party of men caught in the embankment during the last siege of the city and suffocated by the sudden collapse of the tunnel in which they were crawling. On the basis of the Roman coins found in the pockets of the unfortunate soldiers, Du Mesnil judged them to be part of the Roman defense, but I think they belonged to the attacking Persians.* The chain-mail breastpiece of the leader added one more important article to the extraordinary collection of Dura military equipment. The coins, which included one from Antioch of A.D. 256, provide the latest Roman coins from Dura and point to the destruction of the city in that year.

*Hopkins vacillated in his identification of the trapped soldiers as Romans or Sasanian Persians, as to whether the Roman coins in their pockets indicated they were Romans or the heavy armor spoke for Persians. He gathered that Frank Brown and Henry Detweiler, in studying the siege operations, thought the soldiers to be Persians.

9 The Paintings of the Synagogue

If the siege and destruction of Dura had occurred a dozen years earlier than they did, sometime before 243–44 instead of in 256, only an earlier, small synagogue would have been covered in the embankment. Its excavation would have occasioned very little excitement. The first assembly room, like the Christian hall, fitted into a conventional house plan, and the decorations were largely architectural rather than pictorial. A dado ornamented in the old Greek incrustation style imitated marble with diamond-shape geometric pattterns. The cove of the ceiling was marked off with red, black, and white bands, and the ceiling was brightly painted to represent coffers with gilded rosettes in the center.[1]

The designs belong to the old Greco-Roman tradition. There were no animal or human figures, and the repertoire fitted into patterns not uncommon in private houses in Dura. There was nothing to suggest any drastic change from the conservative Jewish artistic tradition. The building itself, its dimensions and decorations, would have been interesting as a monument of one center of the Diaspora in the third century A.D., its place of meeting, its preferred decoration, and its physical arrangements.

But in A.D. 243–44, the second year of Philip Caesar and the year 556 of the Seleucid era, the roof of an enlarged building was constructed over the remains of the earlier building. A ceiling tile, found in the embankment within the building, gave in Aramaic the date and fixed the time of the completion of the room with the installation of the ceiling.[2]

While the first assembly hall had been large for a room in a private house, the remodeled hall was huge for Dura. The new room had a central entrance in the middle of the east wall and a smaller doorway near its south end. In the middle of the west, opposite wall, stood the arched Torah shrine. The walls around the double row of benches were painted with a dado and three registers, or bands, of panels reaching from the dado almost to the ceiling. Decorated baked tiles set into the ceiling completed the decorative scheme. Therefore, the change both in size and decoration between the first and second structures in A.D. 243/44 was revolutionary.

Composite photograph of the paintings on the west wall of the Synagogue. (Courtesy of the Gallery of Fine Arts, Yale University.)

We found successive stages of development in the second building in the floor levels and in the paintings. The complete excavations showed two floor levels, the first laid when the building was erected, the second added some time later. A well-worn step in the main doorway led down to the earlier level.

Construction of the ceiling, with its cofferlike tiles, marked the drastic shift in artistic decoration from the conservative approach of the first building. The tiles carried Aramaic and Greek inscriptions identifying the building, its officers, and its benefactors, as well as a series of decorative motifs: animals, masks or human faces, fruits, and flowers. Two of the tiles represented the human eye, one evil, the second apparently benign.[3] The completely preserved dado echoed some of the animal and mask designs of the ceiling as well as repeated some of the older imitations of marble, and so presumably belonged to the first stage of decoration.

With its arcuated niche resting on plaster columns painted in imitation of veined marble, the Torah shrine probably belonged to the first phase, or was chronologically close to it. The design of the facade of the shrine contained not only symbols of the Jewish faith, but also the sacrifice of Isaac by Abraham. Here, then, was the shift to representation of human figures and of biblical scenes. Appearing as they do on the facade of the Torah shrine itself, they mark a very radical change indeed.

The symbol of the evil eye on the ceiling coffer reveals the prevalence of this superstition. Kraeling, noting that in some of the synagogue paintings the eyes were gouged out when the building was first buried, suggested that this indicated the workers "were by no means friendly toward the owners and worshippers."[4] I may call attention to the fact, however, that because only the eyes were harmed, and not the rest of the paintings, we have evidence instead of a special taboo and not of hostility to the worshippers.

A curious feature in the Synagogue is a series of Persian dipinti written on the paintings rather inconspicuously. These recorded the presence of visitors who viewed the pictures. The dipinti, several containing the day and the month in the year 253/54, seem to indicate the paintings in the building were complete at that time and had already attained considerable repute. The writing is Middle Persian, to be expected since the visitors came from across the Euphrates, from Mesopotamia, which had been under the Parthians for four hundred years before the Sasanians arrived in A.D. 226. All the visitors who left their mark behind were from the Parthian-Persian districts, not from the West.

If we suppose the wonder and fame of the Dura paintings were greatest when they were first completed and so drew these distinguished visitors, then there is a ten-year period—243/44 to 253/54—in which to place the completion of the successive stages of the Synagogue. These ten years fall during the decline in power and prestige of Rome, and the corresponding increase in the authority and expansion of Sasanian Persia. With the shift in power came Christian persecutions in Rome and religious freedom in Persia. Persia under the great Shapur I (A.D. 241–272) had adopted a very liberal policy toward non-Iranian religions, while the West saw the first imperial persecutions of Christians begun under Decius (249–251) and continued under his successors, particularly Valerian with the edict against Christians (257–258). In lower Mesopotamia the Jews had flourished under Greek and Parthian rule and now had two great theological centers, one at Sura on the canal near the Euphrates and the other at Nehardea on the Euphrates-Tigris canal.[5]

What we witness in those ten years of synagogue development between 243/44 and 253/54 is an increased liberal interpretation of the Jewish attitude toward artistic production. At Dura the temples, the Christian Chapel, and many of the private houses were decorated with

paintings. The Jews had never tried to compete in this field. Now, suddenly, the new synagogue surpassed them all in the size of the main room and the extent of the paintings. The Jews, proud of their history and religion, were eager to surpass all rivals in the richness of their pictorial decoration.

If the change in the Jewish community at Dura to an extremely liberal view toward paintings were due to the ascendence of Sasanian prestige and the loss of Roman power, we might expect to find in the paintings increased Sasanian artistic influence, particularly if prototypes belonged to the Babylonian schools. However, definite Sasanian characteristics are difficult to define. The synagogue betrays a trend away from the Greco-Roman tradition toward the Oriental, with the incorporation of such striking features as frontality and the Parthian riding trousers. The Greeks preferred to place the important figures prominently in the foreground; the Oriental artist expressed importance by size and would often place principal characters in the center of the painting. In later Sasanian reliefs of hunting and fighting, heroes magnified in size are placed in the center of vertical scenes of miniaturized men and animals above and below. In the Persian miniature, the features of the landscape are grouped around the significant figure of king or hero in the center. Physical features, such as trees and buildings, are viewed in profile, floating in the landscape portrayed in the Oriental bird's-eye view.

In the Synagogue depth is achieved by overlapping; scatter-patterning, with all objects viewed directly from the side, translates the horizontal of landscape or seascape into the vertical painting. We find in surveying the individual scenes the interplay of old Semitic, Greek, Parthian, Roman, and perhaps Sasanian motives and artistic traditions. The Synagogue paintings are the last and greatest of the long series at Dura.

The individual paintings that comprise the decoration of the assembly hall have been intensively studied and described, particularly in the two monumental volumes of Kraeling and Goodenough. The origin of the artistic styles represented in the paintings and the meaning, the iconography, of the several scenes have been carefully scrutinized, with varying answers. Here I attempt no more than to identify the panels and provide my tentative conclusions on the style and symbolism, particularly when they differ from those expressed by other students of the paintings.

The Facade of the Torah Shrine

The artist of the Synagogue, following Oriental patterning, formed the central themes of worship, the temple and the priesthood, in a vertical panel from which radiated the horizontal registers of the walls. The shrine itself, a columned niche with arched top and holes in the reredos, suggests the presence of a Torah curtain at one time. In the spandrel above the niche were painted three thematic units: a columned building facade that may represent the Temple; the seven-branched candelabrum (the *menorah*), the citrus fruit (the *ethrog*), and the palm branch (the *lulab*) which stand as the symbols of the Jewish faith; and the sacrifice of Isaac by Abraham (the *akedah*), symbolizing complete obedience to the will of God.

The painted facade clearly represents a supreme sanctuary, perhaps the ark or the Temple of Jerusalem. While the two other representations of the Temple in the Dura Synagogue show it to have a pitched roof, that represented on the Torah shrine is flat, either to fit the restricted space or to refer to the earlier temple in Jerusalem of 1 Kings 6:9: "So he built the house, and finished it and covered the house with beams and boards of cedar."

The scene of the *akedah* shows only the back of Abraham as he faces the altar on which Isaac has been placed. In the foreground is the ram and a small tree, and in the distance stands a tent with a small figure, also seen from the rear. Above all is shown the outstretched hand of God, recalling the story in Genesis 22:1–14. While the patriarch may have been shown only from the back because he is receiving the message that stays his hand, looking up toward the (invisible) angel of God, this view violates the strong tradition of frontality at Dura. As this scene was one of the earliest figural scenes painted in the hall, it is possible that the artist was still hesitant to show the face of Abraham, remembering the prohibition against images. But I suggest, rather, that Abraham is so represented to avoid frontality and the evil eye, reflecting the significance of the Jewish tradition.

The small figure in the tent is best interpreted as Isaac freed from his bonds, and not as one of the servants. In the Christian Chapel the healing of the paralytic was represented with Christ above, the paralytic on his couch, then also shown walking away with the couch on his back. This conflation brings the details together, presenting the successive steps in one picture. And so also in the Synagogue Isaac is shown twice,

once on the altar and then entering the tent of his father to indicate he had returned safely and was in a dwelling place again. This simple, straightforward exposition seems the characteristic and appropriate one in the Synagogue. It agrees with the format of other paintings at Dura, and I think an interpretation in terms of the Dura audience should always be considered first.

The Early Painting of the Reredos

Immediately above the Torah shrine two registers of paintings flanked by wing panels covered over an earlier design which apparently covered all the central area, even extending into the later wing panels.[6]

The first painting was of a great tree or vine beneath whose branches were posed rampant lions and a tall table with a circle beneath and a bolster-shape on top. We recovered broken fragments of gilded plaster rosettes at the foot of the painting, and a series of holes along the branches of the tree established that the tree had been adorned with flowers. This great tree-vine (to use Kraeling's term), interpreted as flowering with the sweet fragrance mentioned in the Jewish commentaries on Scriptures *(Midrashim)* from the tree of life, may represent the family, the continuity of generations, and beyond that the trees of paradise.[7] The "sacred tree" was a common motif in the Orient, and the ancient Syrian would have recognized the heraldic lions as associated with the tree and the heavens.[8] The table with the bolster and circle is more difficult of interpretation. On the basis of comparison with the mosaic floor in the early-sixth-century synagogue of Beth Alpha in the Galilee, I suggest the configuration has to do with the columns (the table legs) of the sky over the wheel of time (the zodiac).[9] Hence, the tree-vine becomes not only the tree of life and heaven, but the immortal tree with the great clock of the sky continuing forever beneath its branches.

The age-old Semitic symbol of the divine tree representing earth and sky is equaled in importance only by that of time and eternity, a conception inherent in the divine tree. This concept is ably and amply portrayed at Beth Alpha; its symbol is the round wheel of the zodiac. One expects, then, a wheel in the symbolism at Dura, and it appears beneath the table of the sky.

The Reredos Painting, Second Stage

A subsequent phase of decoration was painted over the vine with a series of three vertical panels depicting the generations in the family of Israel: Jacob and his sons and Joseph and his two sons in the lowest panel, above David alone, and in the upper part an enthroned king (probably David) among the tribes of Israel. I should see the tree in this second phase of painting translated into the generations of Israel. Whether the painting shows the seated King David reigning over the tribes of Israel or portrays the divine messianic king of Israel, he symbolizes eternal kingship over the generations.

In this or a later repainting of the panel, the tribes of Israel were added, clustering around the enthroned king. But two figures at the foot of the throne wear different costumes from the rest. They may represent Jews not belonging to the original twelve tribes, later converts but nonetheless loyal and devoted followers. The long Aramaic inscription on a ceiling tile of the Synagogue mentions a proselyte as a leading member of the Jewish community, for he is listed with two others who are in charge of the work.[10] Josephus records that the Nisibis district of Mesopotamia was ceded by Parthian king Artabanus (d. A.D. 60 to Izates of Adiabene, who with his mother was converted to the Jewish faith.[11] There must have been many converts in Mesopotamia, as well as at Dura, and recognition of this element in the divine kingdom seems appropriate. The Iranian (Eastern) costume of the tribes as opposed to the Greek (Western) costume of the two other figures may symbolize the foreign element.

The Wing Panels

The four wing panels of single, standing figures, strongly Greek in both costume and style, form a frame to the reredos. One figure even shows the Greek shadow line, although they all show some of the stiff Parthian frontality and share the Parthian position of the feet, the "profile-tiptoe."

The identification of Moses is easy: he stands with shoes beside him, and so in the presence of God, represented by the hand reaching toward him, and a burning bush behind his arm.

One of the figures is obliterated above the waist and beyond identification. In the panel below stands a figure with white hair and

The reader of the scroll.

covered hands, sometimes identified as Moses or Joshua or Jacob, but who may be Abraham, reported in the Babylonian Talmud as the first human to show the effects of age.[12] The frailty of mankind, expressed in the white hair, contrasts with the planets and stars about him, symbolizing the everlasting order of God and the eternal covenant made with Abraham.[13]

The fourth panel figure reads from an open scroll held in either hand, a receptacle covered by a red cloth at his feet. There are almost as many suggestions as to his identification as there are scholars who have studied the paintings: Jeremiah, Ezra, Moses, Josiah, Samuel, and a Samuel who is mentioned in a graffito as the elder of the local congregation.[14] So many interpretations supported by strong arguments suggest something basically wrong with the approach taken. These prominent, central panels with single figures should be easily intelligible to the viewer for whom they were intended, the members of the congregation at Dura. I think Sonne moved in the right direction when he said:

> Now it seems to me that the search for a Biblical personality is based altogether on an unwarranted generalization. Realizing that most of the scenes and figures of the paintings derive from Biblical stories, we assume that the portrait too must be connected with some Biblical personality, an entirely gratuitous assumption.[15]

But Sonne did not go far enough! Perhaps we have been misled by

the certainty of the identity of Moses and the burning bush; the Aramaic *titulus* inscribed in the panel provided the proof positive: "Moses, son of Levi." One thinks of the burning bush as identifying Moses; but it is equally possible for the burning bush to identify the word of God represented by the divine hand above it, and for the scene to represent the oral message of God. A very important message it was.

> And I am come down to deliver them out of the hand of the Egyptians, and to bring them up out of that land unto a good land and a large, unto a land flowing with milk and honey (Exod. 3:8).

If Moses and the burning bush represent the oral message of God, then the partially preserved panel would depict Moses receiving the tablets of the law, and the painting would represent the written communication. The reader of the scroll, then, may be any reader of the sacred scriptures, past or present.[16] Now, if three of the panels depict the covenant between God and man in three different aspects—the oral communication, the written word, and the reading of the Torah—then the fourth panel should illustrate some feature of the communication also. The eternal nature of the covenant seems a very logical and plausible solution.

These four wing panels are painted in a strikingly similar style, not completely Hellenistic in their rendering but still belonging to the school that places the chief figures prominently in the foreground, dresses them in the classical costume of himation and chiton, creates a modified depth, and shows skill in portraiture and variety of detail. Such details as the shadow line, the turn of a head, the overlapping to add depth, and the tipping of the scroll with the fingers clasping it are Hellenistic Greek features which distinguish these panels from other Durene work. They illustrate not passages related to the Bible, the common theme in the Synagogue, but abstractions concerned with man's communication with God.

But the enthroned king and the generations of Israel in the center panel they frame are strongly Iranian in style. The use of size to indicate rank particularly reflects ancient Persian reliefs and heralds the new Sasanian mode of depiction. Thus, we have combined in this section Mesopotamian-Iranian elements, Semitic symbolism, and Greek-Hellenistic artistic tradition. It is well to bear in mind that at Dura the combination of Hellenistic with old Semitic art may hark back to early

Hellenism in lower Mesopotamia, as well as look forward to the new Hellenistic-Roman.

*The Exodus Panel (WA 3)**

Joseph Gutmann has offered the excellent suggestion that the successive pictures arranged in three broad bands around the four walls, leading to that illustrating the completion of the Temple, marked the steps of the ceremony and procession of the congregation in the Synagogue.[17] I interpret the top register as belonging to the history of Israel, with the conquest of Canaan, and the bottom band as accounting for the delivery and prophecy, with the promise for the future. Thus, it is appropriate to begin with the scene of Exodus in the top register.

This panel is the easiest to interpret but, artistically, the most intriguing and complicated of all the paintings in the Synagogue. When we first uncovered the walls, the dipinto between the legs of Moses—"Moses, when he went out of Egypt and cleft the sea"—was a godsend in identifying for us the building as a synagogue. Once the identity of the building was established, the major details were obvious: the escape of the children of Israel from the walls of Egypt; the crossing of the Red Sea; the miracle of the spring of Marah; and the numbering of the tribes under the guidance and protection of God.

It is an excellent example of the cyclic style, the succession of incidents depicted one after another without formal breaks between. Moses appears three times; leading his people out of Egypt, causing the Red Sea to divide, and standing under the sheltering hands of God. In the top corner, beneath the hail and fire that symbolize all the plagues that descended on Egypt, are painted two Corinthian columns, one red, one black, placed side by side. They represent the pillars of fire and cloud which led the Israelites through the wilderness by night and day (Exod. 13:21–22).

We take the recognition for granted now, because the columns became the accepted signs of divine guidance in later, Christian paintings. As far as I know, this use of red and black columns to represent

*The letter and numeral designations are those commonly assigned to the wall paintings, given here for easy reference to illustrations in Kraeling's *Final Report* on the Dura Synagogue.

guiding posts by day and night belonged only to Hebrew symbolism. Their use at Dura is proof that the artistic interpretation of the biblical passage was already well established in the middle of the third century, and the congregation was expected to recognize it.

Moses is first shown striding toward the Red Sea, his raised staff explained by the dipinto beneath his feet: "Moses cleaves the sea." Once again, clad in Greek dress adorned with decorative strips, the Roman *clavi,* he stands on the edge of the sea, holding back the waters with his staff. Most impressive is the third and larger figure of Moses close beside the other, for the body is set off by the empty space of the sea behind, and his head is framed by the great hands and forearms of God stretching across the upper frame of the panel like a divine arch in the sky: Moses holds his staff down to touch the pool of Marah, made fresh by the miracle.

In the great panel as a whole, one is impressed with the Hellenistic emphasis on variety of detail contrasting with the Oriental tendency toward pattern. The panel adheres most closely to the Hellenistic-Semitic tradition, though overlaid with Parthian frontality. Semitic features of size and multiple view are retained; the trend toward tiers of figures is manifest. In spite of these, Greek genius shines through, and the establishment of the new Jewish symbolism and representation is obvious. The combination of all these factors makes this depiction of the Exodus not only outstanding in the Synagogue, but among all painting of the Roman third century.

Jacob in Bethel (NA 1)

The panel depicting Jacob's dream is almost two-thirds complete. I do not know whether it should be construed as belonging to the Exodus story, but the drama concerning the future conquest of Canaan is most appropriate (Gen. 28:10–15). Jacob, in plain Greek dress, appears to lie on a bed while angels cluster on a diagonally placed ladder. They stand in the Parthian tiptoe-profile position, clad in Persian trousers tucked into high boots, with the long-sleeved chiton, the robe thrown back from the shoulders, similar in dress to that of the magi in the Temple of Mithra. Old Testament angels, such as these are, have no wings, and in Christian art they did not attain the eagle plumage until the fourth and fifth centuries.[18] The scene has some basic Hellenistic Greek elements with a veneer of Parthian frontality and some striking

Oriental features, such as the tipping of the ladder and the position of the angels.

Solomon (WA 2)

We were extremely fortunate in finding in a partially destroyed panel the depiction of the three lower steps of a throne, on the topmost of which was written in Greek letters the name "Solomon." The throne as constructed by Solomon is described (I Kings 10:18–20), but Kraeling has pointed out that the representation at Dura of his throne reflects the Aramaic interpretation *(Targum Sheni)* to the Book of Esther: "And there stood upon it [the dais of the throne] twelve lions of gold and over against them twelve golden eagles, a lion opposite an eagle and an eagle opposite a lion, so that each golden lion's right paw was opposite each golden eagle's left wing."[19]

Beside the throne is the seat of the *synkathedros* (an officer of second rank) whose feet and lower legs only remain.[20] To the right of the throne is another seated figure, whose legs are almost twice the size of those of the other figure. Since size indicated the relative importance of the figures in the Orient, the man on the one side must be of distinctly higher rank.

The beginning of the fourth chapter of I Kings lists the princes and officials in the court of Solomon.

> So King Solomon was king over all Israel. And these were the princes which he had; Azariah the son of Zadok the priest; Elihoreph and Ahiah, the sons of Shisha, scribes; Jehoshaphat the son of Ahilud, the recorder, and Benaiah the son of Jehoiada was over the host; and Zadok and Abiathar were the priests; and Azariah the son of Nathan was over the officers; and Zabud the son of Nathan was principal officer, and the king's friend.

In the Greek, the princes are *archontes* (rulers), the scribes *grammateis,* the recorder *hupomimneskon,* the priests *hiereis;* Orneia, son of Nathan, is over the "officers" [*kathestamenoi*], and Zaboth, son of Nathan, is the companion [*hetairos*] of the king. The King James version translates that Azariah was over the "officers" and Zabud was the principal officer and friend of the king.

The Greek *kathestamenos* (officer) literally means "the one who

stands below (the throne or the king)." In the painting the two figures seated immediately below the throne would be described as "the ones seated below (the throne or the king)," that is, the *kathedros.* One would be *kathedros;* and the second, that is the assistant, the "one accompanying the *kathedros,*" would be the *synkathedros.* The right to sit in the presence of the king would indicate an officer higher in rank than one standing. Apparently, then, the principal officer and "friend" of the king was painted to the right of the throne and his assistant to the left.

Interesting is the parallel in Christian literature: the *cathedra* is the "seat below (the supreme pontiff)" and becomes the "seat of the bishop" and the cathedral is his "seat." Whether at Dura the artist was borrowing a Christian title or the Christians later took over the Hebrew designation is not clear. The painting indicates that in the Eastern-Iranian monarchies, the "ones seated below (the king or the throne)" were well established and easily recognized.

In the left-hand portion of the painting the feet and lower garments of three figures are seen, probably the queen of Sheba and her attendant and the courtier who presents the queen to Solomon. The king's massive throne stems from the Parthian-Sasanian period rather than from the Greek or even from the Old Persian.

The Battle of Eben Ezer and Hannah and Samuel in Shiloh (NB 1, NB 2)

The sequence of events in the middle band or register seems fairly clear, since their action moves toward the center, toward the Torah shrine. The scene of the capture of the ark in the Battle of Eben Ezer is followed by the return of the ark, and then the representation of the Temple, the ultimate resting place of the ark.

Next to the battle is a partially destroyed panel, probably of Hannah and Samuel before the walls of Shiloh (I Sam. 1:24).[21]

The scene of the Battle of Eben Ezer is complete except for the upper-right-hand corner.[22] The ark is carried on poles on the shoulders of Philistine captors, guarded by soldiers. In the right-hand half of the picture a spirited battle between two charging horsemen is framed above and below by soldiers fighting on foot. Their mounts are not in strict Parthian gallop, for the hooves are turned down and braced on

the ground as in Greco-Roman as well as in Old Persian and Babylonian styles, but exceptional at Dura. The viewpoint, however, exhibits the well-known Parthian shifts: the chest of the horses is seen from in front and a little above the legs, the flanks and shoulders are seen in profile, and the back legs are viewed from in front and slightly below so the far leg is outlined in front and below the near one. The horsemen, drawn strikingly larger than the soldiers in the foreground, exemplify the central focus that indicates the importance of the subject by size, a striking characteristic of Assyrian and Sasanian reliefs.

The painting is revealing from the point of view of warfare at Dura and the artistic tradition of battle. It is the only panel in the Synagogue, except for the Torah shrine facade, where frontality is not observed. While the horsemen fight with lances, the foot soldiers in armor *(lorica squamata)* use sword and shield, but wear no helmets. The scene is not typical of either early Hellenistic or Roman art. The equipment belongs to the so-called Gaulish type, an intermediate form in Mesopotamia between the phalanx form of Alexander's forces and the Parthian-Roman types.

From the artistic point of view the panel seems a patchwork, put together from parts of earlier compositions. The Philistines appear to be triumphantly carrying off the ark, but the civilian bearers seem quite out of place on the battlefield. To be sure, soldiers, stiffly drawn in the frontal position, guard the ark. The strict frontality and the static quality of one part of the panel contrast sharply with the desperate and diverse action in the other half. Whether the horsemen originally came from a separate panel, their interpolation among the tiers of foot soldiers heralds the later Sasanian practice used in bas-reliefs.

Clearly the parade of the ark belonged originally to quite a different composition. Here it is taken over as an adjunct to the battle, the soldiers introduced as appropriate guards. The pole-carriers clad in Hellenistic civilian costumes belong to the earlier tradition; the soldier guards, to a slightly later period.

The biblical account in I Samuel 4:1 makes no mention of cavalry nor of individual combat. The distinguishing feature that identified the battle in the panel was the carrying off of the ark from the scene of strife. With the adjacent scene, around the corner on the west wall, showing the return of the ark, we have the three events—Hannah and the child Samuel, the Battle of Eben Ezer, and the return of the ark—following one another in logical and biblical sequence.

The Return of the Ark (WB 4)

Outstanding among this amazing sequence of paintings is the return of the ark from the Philistines, for it skillfully combines all the essential elements of the story in the single scene, as well as adroitly blending both Greek and Oriental artistic traditions. The upper half of the scene on the right presents the facade of the Temple of Dagon with altars before it, along with scattered utensils and two broken statues of a trousered Dagon. The left half of the scene shows the ark carried in a cart pulled by two white milk kine and driven by two Philistines.

Above and behind, three elders of the Philistines, clad in chitons and himations, watch the departure of the ark.

The panel illustrates the key verses of I Samuel (5:2–5), which recount that when the Philistines put the ark into their temple, the statue of Dagon was destroyed despite their efforts. The priests of Dagon advised them to

> make a new cart, and take two milch kine, on which there hath come no yoke, and tie the kine to the cart, and bring their calves home from them: And take the ark of the Lord, and lay it upon the cart; and put the jewels of gold, which ye return him for a trespass offering, in a coffer by the side thereof; and send it away, that it may go (I Sam. 6:7–8).

The ark is the focus of attention: placed in the center, enlarged in size, frontally posed, with a suggestion of open space between it and the temple and the scattered utensils.

The Temple (WB 3)

The next, and the last, scene in the sequence before reaching the wing panels framing the reredos is of a temple placed in the upper center of the panel, balanced below by a great wall with three gates. Unique is the absence of all human or animal forms, and puzzling the appearance of seven crenelated walls which fill the panel from top to bottom. This is no earthly temple, but a divine or idealized structure, quite appropriate, since the Temple in Jerusalem had been destroyed by Titus a few centuries earlier in A.D. 70, never to be rebuilt.

Kraeling discussed the various possibilities in Hebrew history and vision for the building with its seven walls, and concluded that this must be the Temple in Jerusalem with the added details from the Midrash which described the heavenly Jerusalem surrounded by seven walls, each with its own particular color. He cited the Midrash Tanḥuma as pointing to Jerusalem as center of the world and to the Temple as the center of Jerusalem.[23]

That the entire panel is devoted to the temple and its walls and that the entrance to the court is greatly emphasized seems sufficient evidence that it is the Temple in Jerusalem, not that in Bethel, but I think it is represented only in the idealized form. The only visual relief

from the vertical and horizontal lines of walls, doors, and temple is the drawing of small winged Victories on each corner of the temple pediment and figural decorations in the panels of the central door. Both leaves of the central door are divided into three panels carrying motifs: a standing female, a tall nude male with twin nude children, and a reclining humped bull.

These are curiously pagan in appearance in this Jewish setting. The females carry a cornucopia, which identifies them as the symbol of fruitful abundance; halo and turreted crown indicate their position as city goddess of good fortune; the oar carried in the right hand represents sovereignty over the sailing season. With this identification in hand, the symbolic meaning of the figures above then becomes clear: after the spring equinox in the calendar of the zodiac comes Taurus, the bull, and then the Heavenly Twins; at the summer solstice, the midpoint in the solar year—Cancer, the crab—comes the season of abundance and navigation.

Goodenough's suggestion that the large male figure with the two children was the Iranian Gayomart with the twins Masya and Masare is possible, for the artist may have introduced Iranian figures into a Greek sequence. I think, however, that the nude figures are far more at home in Hellenistic than in Iranian iconography.

Purification and Sacrifice (WB 2)

The large, cloaked figure labeled in Greek *Aron* (Aaron) dominates this scene. He stands beside a templelike structure, the Tabernacle; at his feet is a stone wall, marking the courtyard, with three doors. Placed before the priest are the *menorah* and an altar and incense burners aflame. Standing on either side of this tableau are trumpeters, sacrificial animals, and attendants.

Kraeling labeled the scene "The Consecration of the Tabernacle and its Priests," and he remarked on the compositional shortcomings, "entirely static and frontal, telling no story . . . developing no action . . . failing to combine the figures and objects portrayed into a unified, well-organized composition, distributing them about instead in various parts of the space enclosed by the frame and at various levels from bottom to top."[24] His judgment was based, naturally, on Western artistic standards. From the point of view of the Eastern artist, the compo-

sition is equally faulty, with one figure standing in front of and obscuring part of the Tabernacle, with too much action, and showing an illusion of depth accentuated by the altar presented in three-quarter view, the billowing curtain over the main doorway of the wall, and the flaming of the incense burners.

But from the point of view of Durene painting, and this means the combination of Eastern and Western artistic tradition as well as biblical illustrations, the scene of the purification and sacrifice is a real treasure.

Aaron, dressed in his rich robes of office, dominates the tableau. To his left in a vertical space are the essentials appropriate for the sacrifice at the beginning of the month (Exod. 40:2–8): two men with silver horns, mentioned in Numbers 10:2–3, call the assembly to the Tabernacle for the sacrifice of a white, fat-tailed ram and a humped bullock, pictured without straps or halters. On the opposite side of the tableau, the trumpeters are drawn once again, standing above a red humped bullock (a heifer? the sex is not indicated) held by an attendant who carries a long-handled axe.

In the biblical account in Numbers there is no direct time sequence between the rite of purification and sacrifice, of Eliezer and Aaron, but the two would not unnaturally go together. It is obvious in the Dura painting that the two ceremonies have been combined into one panel, resulting in the duplication of the two trumpeters to the left and right of the Tabernacle, although the Bible explicitly states that two, and only two, were present. The other unusual feature is the decided change of style between the two sides of the picture; one in the Western style with modified depth, the other with the flat, vertical echelons of the East. Reasonable, it seems to me, is the supposition that the Dura artist combined two well-known scenes, one of purification, one of sacrifice. In both, the central figure of a priest and of an altar remained essentially the same.

Kraeling believed the scene represented the consecration of the Tabernacle and its priests and pointed to Exodus 29:1, as the best proof, while Goodenough suggested "an idealized generalization of the priesthood of Aaron."[25]

Moses Smites the Rock (WB 1)

One of the most dramatic episodes in the Old Testament, told in both Exodus and Numbers, is that of Moses smiting the rock and

Moses smites the rock (WB 1).

bringing forth fresh water for the parched and complaining Israelites. Here in the Synagogue, Moses stands with bare feet, holding his staff in the pool which has risen from the rock. It is the same figure of Moses as in the Exodus and the burning bush scenes: a lordly figure, dwarfing the tiny elders, each standing in front of his tent, hands raised in the regular gesture of worship and astonishment. The Tabernacle, mentioned in Numbers, finds its appropriate place in the top center of the picture, with a graceful *menorah* before it.

The striking detail in the painting, not mentioned in either account, is the twelve streams of water issuing from the spring and

winding to the tents. The spring of Moses, identified by the Tabernacle, is distinguished from the pool of Marah by the leaping fish that demonstrate the water is not a new, fresh miracle. At Be'er (Num. 21:16–18) the nobles of the people dug twelve wells, one for each tribe, and Israel sang a song of rejoicing. In the wilderness of Horeb the basic difference between sand and rock is amply demonstrated. The water lies beneath sand—how far depends on season and water level; the rock is hard, dry, uncompromising, and impenetrable. The miracle is that Moses strikes the *rock* with his staff and the water gushes out.

The problem for the artist was how to represent the audience, how to show the rock in the sight of all, and how to express the water in abundance. The traditional artistic representation of Moses seemed sufficient, particularly when the staff is displayed so prominently. The painter has raised the springhead well above the lower border of the picture to emphasize its prominence and clearly to show that the water flows not from a well dug in the sand, but from a rock raised well above desert level. The twelve tents with their twelve elders standing before them represent the twelve tribes, and the streams of water indicate an ample supply for everyone.

Setting Up of the Tabernacle (SB 1)

Kraeling identified this scene as the dedication of the Temple and Goodenough as the procession of the ark.[26] I prefer the setting up of the Tabernacle, believing that it illustrates the passage in Exodus 40:2–3:

> On the first day of the first month, shalt thou set up the tabernacle of the tent of the congregation. And thou shalt put therein the ark of the testimony, and cover the ark with the veil.

The painting, in the west corner of the south wall, is a most tantalizing fragment, just half or less of the original panel, the slope of the embankment having cut diagonally across the picture. The only clear indication of purpose among the several figures represented is a group of four men who carry poles on their shoulders as they move left. Kraeling suggested they were carrying the ark, just as in the scene of the Battle of Eben Ezer, and I agree that the same model was followed. Instead of portraying the accompanying soldiers, as in the battle scene, the figures of children are introduced.

The Bottom Band of Panels: Programmatic Composition

In my interpretation, the top band of paintings (Register A) recorded the history of the people of Israel, the action on each side moving toward the Torah shrine in the center of the wall. On one side the Exodus from Egypt is completed; on the other the conquest of Canaan ends with the kingdom of Solomon. The action of the middle band (Register B) is similar in movement: one side shows the progress of the ark culminating in the divine Temple; the other side was the connecting theme of the Tabernacle and the sacred utensils, culminating in the ceremonies of purification and sacrifice. So the central portion of each panel contained the Tabernacle and the most sacred utensils. The sequence of the lowest band (Register C) is quite different.

Here the paintings are preserved all the way around the room, completely on the west and north walls, partly on the south wall, and on the east wall to the extent of a large panel to two-thirds of its height and one between the doorways to less than half. The subjects vary, and there seems no common denominator that would parallel the sequences of the history of the people of Israel in the top band and the concentration on the Torah, the Temple, and the priesthood of the middle register.

The reredos and the wing panels present, however, the divine kingship of David and the promise to Abraham as well as the word of, and covenant with, God. These broad themes are obviously continued in the lowest register with striking illustrations of deliverances and prophecy in Old Testament history.

The episodes, however, go beyond these general hopes to reflect in part the special aspirations of Jews of the period, both at Dura and in the broad field of the Diaspora, even those of Palestine itself. The times were troubled! In the middle of the third century the Roman Empire was threatened along both the European and the Asian frontiers. The Temple in Jerusalem had been destroyed, the Jews had been driven from Jerusalem; orthodox Judaism itself was threatened by the rapid growth of Christianity. The independent movements of Essenes and the growth of Gnosticism claimed new revelations, with borrowings from both the Old and the New Testaments.

It was inevitable that the restoration of the Temple and the kingdom of Israel in Jerusalem should be the steadfast hope and vision of the orthodox Jews. The unity of the Jews and deliverance from apostacy, as expressed in David's kingship over all the tribes, would pre-

sumably play important roles. The question of immortality and resurrection became more and more relevant in the troubled times, particularly since both Christianity and Mithraism were gaining strength and prestige, partly through insistence on the afterlife of the soul. Christians, as they clearly proclaimed in deriving their name from *Christ,* the anointed one, identified Christianity as fulfilling the promise that the anointing of David had given to the Israelite kings and to the divine kingship of Judah.

The glorification of Moses as the great triumphant leader of Israel is the organizing principle of the paintings that underlies Goodenough's interpretations in the three volumes of *Jewish Symbols* he devoted to the Dura synagogue. Wischnitzer looked to the divine leader, as expressed in David and the restored Temple, as the guideline to her interpretations in her *Messianic Theme of the Dura Synagogue.*

It is very difficult to determine the central theme. I do not believe the Durene authors of the conceptions and the artists had such precise notions in mind. The two upper bands recorded the splendid history of the Jewish people and the development of their religion in Torah and Temple. The bottom series of paintings, the panels at eye level, depicted some of the aims, difficulties, and great achievements of the past, and some of the burning hopes for the future.

It is well to remember, perhaps, that there was no division between ancient and modern history. The Maccabean revolt occurred when Dura had already been in existence almost a century and a half; the old Persian empire had been reborn in the Sasanian; the Temple of Jerusalem had been destroyed about a century and a half before the Synagogue was erected. The visions of Ezekiel were imminent and pertinent.

The Anointing of David by Samuel (WC 3)

The painting of the anointing of David by Samuel, recorded in the first book of Samuel 16:11–13, is very clear and precise: Samuel towers over the seven sons of Jesse, holding the horn of oil over the head of David. David stands in front of his brothers, his hands concealed beneath his robe in sharp contrast to the bare hands of his brothers. There are clear and deliberate differences in the decorations of the robes, in the concealment of the bodies, in the positions of the legs, and in the grouping to break up the rigid pattern of Oriental design of the figures in formal rows.

Why should this scene of the anointing appear beside the Torah shrine? The dipinti of the ceiling tiles records a Samuel as the *presbyteros* of the Jews. David, the divine king, sits enthroned in the pictures of the reredos. The anointing is the symbolic act which makes him the messianic leader, the *Christos* (anointed one) of God. Thus, there are strong reasons for seeing the messianic theme as a very important one in the Synagogue at Dura.

But perhaps there is another, special reason. In the Christian Chapel, the victory of David over Goliath is placed above the altar, opposite the great scene of the coming of the three Marys to the empty tomb of Christ, signifying the resurrection. Above the David and Goliath tableau was painted the garden of paradise, the picture pointing to the overcoming of obstacles and the winning of paradise through Christ, the Messiah of the Christians.

The painting in the Synagogue answered in part the great new popularity of the Christians and their threat to the Jewish faith. The anointed one was David, the messiah, and he alone was king over Israel. In the Gospel of St. Matthew, Christ as he entered Jerusalem for the Passover feast (Matt. 21:9) was hailed as "Son of David." The Synagogue suggested that the real messiah, to the Jews the only Christ, was David.

Pharaoh and the Infancy of Moses (WC 4)

It seemed strange at first that the infancy of Moses should be illustrated in so large a panel and placed in so prominent a position in the Synagogue. The birth of Moses, however, and his rescue by the daughter of Pharaoh mark a new era in the history of Israel. If David was the anointed king, Moses was the great savior and leader, coming at a time when the condition of the Jews was desperate.

The beginning of the Exodus pointed forward to the later great deeds and achievements. So, too, the wide panel of Moses in its prominent position on the front wall anticipated expositions of his accomplishments. If the lower band of paintings was, as I believe, painted first, surely the infancy of Moses promised the great panels of his career in the upper registers. The programmatic scheme must at least have been planned in general before the top register (Register A) was completed.

In the Old Testament when Pharaoh's daughter saw the ark

among the flags, she sent her maid to fetch it. In the Aramaic version of this section of Exodus *(Targum Onkelos),* Pharaoh's daughter "stretched out her arm and seized it."[27] The painting shows Pharaoh's daughter standing in the water holding the infant. Her handmaidens wait on the river's bank with bowls of ointments, while to one side the infant's sister Miriam passes the child to his mother, to whose care it was confided. On the opposite side of the panel the two women are represented again, now standing beside Pharaoh's daughter, who lays the child at the feet of Pharaoh. The enthroned king is flanked by his scribe and chief officer.

Rather interesting in the painting is the employment of frontality. Pharaoh's daughter in the river, the handmaidens, and the king and his courtiers face full front. Before the seated Pharaoh, however, Pharaoh's daughter is shown almost in profile; Miriam and her mother are painted with a decided turn of the head. As in the wing panels, then, strict frontality is gradually superseding the Greek preference for the three-quarter view. The infant faces front, but the body shows the illusion of depth. The attendants, though frontal, are portrayed with their arms and hands held across the body or to the side in the Greek tradition.

In Hellenistic art of the mid third century B.C., robes were drawn projecting sharply in the front between the wearer's legs. In this new artistic fashion, which became increasingly popular on through to the end of the century, the strong vertical lines of the chiton were retained, particularly with the projection of the garments between the legs, and only partially obscured across the middle of the body by the diagonal folds of a light and sometimes semitransparent himation. The costuming of Miriam and her mother is drawn in imitation of this style and, hence, provides a clue to the sources of the Synagogue imagery.

The two women wear long-sleeved chitons with a separate broad drape or sash circling the waist and a third garment with heavy vertical folds extending from just below the hips to the ankles. The drawing is somewhat ambiguous, the combination of garments difficult to visualize, because the artist has misunderstood and misinterpreted his Hellenistic model. The semitransparent or very thin himation wrapped diagonally around the body, as drawn by the Hellenistic artist of the latter part of the third century, extended only as far as the knees, allowing the strong lines of the under-robe to emerge below the bottom of the himation, but those fold lines could also be seen extending up beneath the overgarment. The Synagogue artist interpreted the lower hem of the himation as a strong horizontal fold. Perhaps the

figure of the seated Esther in the scene with Mordecai suggested this interpretation, for her himation is gathered into folds over her lap.

The artistic interpretation here is supported by the same drawing of drapery in the scene of the widow's son. In both scenes, then, the Hellenistic fashion of the late third century B.C. is followed, but misunderstood in the drawing. Other Hellenistic drawing traits are retained, such as, for example, the illusion of spatial depth in the left sleeve of Miriam as she reveals her forearm raised to the child.

One may place the prototypes of these figures, then, at or just after the end of the third century B.C. and the synagogue artist isolated now from the regular Western progression in the treatment of drapery. It is, perhaps, the best proof that some of the Synagogue prototypes at least originated in Mesopotamia and date back probably close to the beginning of our era.

The Triumph of Esther and Mordecai (WC 2)

The ten short chapters of the Book of Esther read like a tale taken straight from the Arabian Nights. Ahasuerus, the great king, king in Susa and of 127 provinces from India to Egypt, set aside his disobedient queen, Vashti, and installed in her place Esther, the cousin of Mordecai. A conspiracy against the king by two of his chamberlains was discovered by Mordecai, reported to Esther, and so brought to the attention of the king. The guilty conspirators were hanged and Haman, the son of Hammedatha, was then promoted by the king. But Mordecai refused to bow down before the new lord, thus bringing down on him the hatred of Haman, which extended to all the Israelites, whom he sought to destroy.

Haman offered to pay ten thousand talents of silver to those in charge of the killing of the Jews, and letters in the name of the king were dispatched "by posts unto all the king's provinces, to destroy, to kill, and to cause to perish, all Jews, both young and old, little children and women in one day." When the news of impending disaster reached Mordecai, he went to Esther, who commanded three days of fasting and lamentation. While Haman had a gallows erected for Mordecai, Esther planned a feast for the king and his chief official. Meanwhile, a sleepless king asked for his chronicles to be read to him and in them discovered that Mordecai had never been rewarded for saving his life. Haman, on being asked by the king what honor should be bestowed on a man who had served him well, assumed the king meant him.

> And Haman answered the king, For the man whom the king delighteth to honour, let the royal apparel be brought which the king useth to wear, and the horse that the king rideth upon, and the crown royal which is set upon his head: and let this apparel and horse be delivered to the hand of one of the king's most noble princes, that they may array the man withal whom the king delighteth to honour, and bring him on horseback through the street of the city, and proclaim before him, Thus shall it be done to the man whom the king delighteth to honour. Then the king said to Haman, Make haste, and take the apparel and the horse, as thou hast said, and do even so to Mordecai the Jew, that sitteth at the king's gate: let nothing fail of all that thou hast spoken (Esther 6:1–10).

At Esther's banquet the king promised his queen to grant any petition she might make, "even to the half of the kingdom." Esther asked only for her life and the lives of her people. On being told that Haman threatened the life of his beloved, the king "in his wrath sent into the palace garden." Haman, having drunk too much wine, fell asleep on the queen's bed, where Ahasuerus discovered him. Thus Haman fell victim to the gallows he had prepared for Mordecai.

In the painting, Mordecai, seated on the horse and clad in the royal robe, with the royal crown in the shape of the pointed Persian cap with diadem on his head, is conducted by Haman, barefoot and clad in the short tunic of a slave, before a group of admiring citizens. Ahasuerus, seated on the royal throne, receives a parchment of petition from a chamberlain; Esther, the crown on her head, is seated in the royal chair, her feet on a stool. Mordecai wears the Parthian diadem, differing sharply from that of the Hellenistic and Sasanian, and carries the Parthian quiver and short bow. Since the Sasanians superseded the Parthians twenty years before the Synagogue was painted, the equipment of the Parthian bowman and the diadem with Parthian ribbons strongly support the conviction that the prototype was drawn in Parthian Mesopotamia.

The chamberlain with petition is presented in profile, the assembled citizens frontally, providing another example of the Oriental frontality gradually superseding the Greek three-quarter view and the Oriental profile.

From the time of the old Persian empire, the celebration of Purim annually commemorated the deliverance of the Jews, but the Persians also held a special place of honor in the history of the Jews because of

the first great Persian king, Cyrus, who freed the Jews from bondage in Asia, allowed them to return to Jerusalem, and authorized the rebuilding of their temple. Cyrus was himself hailed as the anointed one, the messiah, by Isaiah (Isaiah 45:1).

Now, in the middle of the third century at Dura, there was special reason to recall the benevolence of Cyrus and the rebuilding of the Temple. The Sasanids had replaced the Parthians as rulers of Asia and claimed the inheritance of the old Persians against the Parthian interlopers and the Roman invaders. The Sasanian Shapur I succeeded his father Ardashir I in 241 and vowed to restore the ancient frontiers of Persia, which would include not only Dura but all of the Holy Land.

The dipinti painted on the walls of the Synagogue and particularly on the scene of Esther show not only the interest of Iranians in the Synagogue but also the interest of the leaders of the Synagogue to have Iranians from across the Euphrates recognize a close kinship with the Dura Jews.

The triumph of Esther, while just one episode in the long history of Jewish experience, could have provided a very meaningful, though veiled hint in the situation at Dura. Its special prominence in the Synagogue pointed to the hope of a second restoration of the Jews by the new Persian rulers, and the restoration of the Temple after its destruction by the Romans.

Elijah Revives the Widow's Son (WC 1)

On one side of this panel a woman in the black clothes of mourning holds a lifeless child in her arms. In the center, Elijah, reclining on a couch, has received the child and reviving him causes him to sit up. He then passes the child to the woman on the right, now clad in the bright colors of happiness.

Any doubt that the healer is Elijah seems dispelled by the immediately adjacent panels of the south wall, which are given over entirely to Elijah in this register: the proclamation of a drought, Elijah at Zarephath, and Elijah's triumph over the priests of Baal on Mount Carmel.

There is a little question why the biblical sequence is interrupted and the restoration of the widow's son is placed after the triumph on Mount Carmel instead of before. The double panels of the triumph over the priests of Baal, however, belong together and fit conveniently

into the one wall rather than severely crowding or eliminating the panel of Esther on the next wall. Speculation on what the Dura artists might have wished and chosen to do is rarely profitable. Better it seems is the analysis of panels and arrangements as they stand.

In the Bible the widow of Zarephath and the healing of the child are tied closely together, for the child is the widow's son. The episode comes just before the triumph on Mount Carmel. The widow and her son are provided with meal and oil for many days; then suddenly her only son falls sick and dies (I Kings 17:17–24). Then comes the account of the victory over the priests of Baal. The sequence of paintings, placing the revival before the contest, is changed.

Details of drapery and drawing indicate that the prototype for this painting could scarcely date before the second century B.C. Probably it belonged to the period after the beginning of our era and, presumably, originated in Mesopotamia, at least in a school divorced from the regular greco-roman development.

Why this scene should hold such a prominent position in the Synagogue is still unanswered. It does follow the scenes on the adjacent wall dedicated to Elijah. In the middle of the third century, moreover, the question of resurrection and immortality became increasingly important. The dangers on the frontiers of the Roman empire only reflected the desperate situation in the empire as a whole. The growth of Christianity and Mithraism signaled the new emphasis on life after death. The story of the widow's son is one of restoration to this life rather than to the Garden of Paradise. The whole panorama on the north and east walls is that of restoration: of the dead bones brought to life again in this world, of the return of the Jews to Palestine, and finally of the cleansing of the Temple and the reinstitution of the rites, sacrifices, and ceremonies of the Jewish faith. The emphasis in Elijah and the widow is not simply on the restoration of life but in the final words of the woman: "And the woman said to Elijah, Now by this I know that thou art a man of God, and that the word of the Lord in thy mouth is truth."

Elijah Meets the Widow of Zarephath (SC2)

In the Old Testament the meeting of the widow and Elijah occurs as she gathers sticks for a fire (I Kings 17:10–16). The Synagogue artist represented in front of the city gate a woman bending over before a

standing male figure. Unfortunately the panel is partially destroyed, making difficult the identification of another figure whose existence I had not recorded in my field notes, but which was noted by Gute. This fragmentary figure may be Elisha, the later pupil of Elijah.[28]

The Victory of Elijah over the Priests of Baal (SC3 and 4)

Preserved complete is the victory of Elijah in two great panels. The first, showing the discomfiture of the pagan priests, shows four priests on either side of a great altar, on top of which lies a bullock adorned with the garlands of sacrifice. The fire requested from the pagan gods has not descended, and the little man posted beneath the altar, according to the Talmud, to light the fire at the appropriate time, is thwarted by a monstrous serpent approaching from the right.

The second scene presents a white bullock on a great altar being consumed by the fire, whose flames billow up behind it. Three priests stand to one side, and the one nearest the fire, Elijah, raises his hand in triumph. Youths carrying water jars have no effect on the flames. The priests of Baal had called on their gods in vain; Elijah had mocked them, and their appeals were fruitless. Then Elijah built an altar, and on his petition the Lord set it and his offerings aflame, despite the fact that all had been soaked with water (I Kings 18:29–39).

The biblical passage is graphic, and its illustration in the panel presents strong evidence for the type of artistic interpretation employed in the Synagogue paintings. Two ideas are expressed both in the paintings and in the biblical account: the first is the victory of God over the pagan priests of Canaan, and the second is the power of God even over the laws of nature. There is no need to pour water on the sacrifice and for a burnt offering it seems senseless. The fires, however, consuming the offerings despite the water, show the greater power of God over two elements.

The Vision of Ezekiel (NC 1)

The bottom register of paintings of the whole north wall is devoted to one long scene dominated by the vision of Ezekiel. The painting is the largest in the Synagogue and obviously conveys a very special message. A light-colored background on one side, dark red in the center, and

light background again on the other side divide the painting into three unequal parts.

According to the vision of Ezekiel, life will be restored to the dry bones of the dead and the children of Israel will be restored to Palestine. In the painting on the left, Ezekiel, touched by the hand of the Lord, stands among disjointed human remains. Above is shrubbery painted in the Iranian manner in solid clumps of large leaves with serrated edges. A mountain is shown split in two; a castle is overturned; above, the hands of God proclaim the miracle. Four female figures with the butterfly wings of the Greek Psyches, incorporating the breath or spirit of life, descend to revive the corpses. Centered in the panel is a group of ten small figures standing between two tall figures drawn in Greek megalographic style. To the right of this group are dismembered human bodies, and the hand of God once more appears above. The painting illustrates the famous passage in the thirty-seventh chapter of Ezekiel (1–14). "The hand of the Lord was upon me and carried me out in the spirit of the Lord, and set me down in the midst of the valley which was full of bones." The Lord commanded Ezekiel in his name to prophesy that the bones would live again, "and the bones came together, bone to his bone," and the breath of life was given to them, and they lived, "an exceedingly great army." And then Ezekiel was instructed to bring the people to the land of Israel.

The painting illustrates beautifully the hand of God touching the head of Ezekiel, placing him in the midst of the valley. The prophecy of Ezekiel comprises two thirds of the panel, a picture divided into two distinct parts by the change of color in the background, but the sections are linked by the right and left hands of God above.

The vision seems fairly clear as interpreted in the painting: there are the dry bones and the dismembered human parts coming together to form corpses; then the Psyches breathing life into the lifeless figures; and finally the restoration of the dead to life represented by the ten figures. Whether the two large figures on either side of the revived smaller ones are the new leaders of the people in Israel or represent Ezekiel is difficult to determine in both cases. I favor Ezekiel, appropriately repeated, as now bringing an end to his vision. Artistically, the two large figures, prominently to the fore, are excellent examples of the Greek megalographic style, and contrast with the smaller figures, equally frontal and pictured in tiers, representative of Mesopotamian art both in the contrasting sizes and the successive rows. The Greek Psyches, with their split skirts, spread wings, and fluttering dresses,

preserve best the Greek illusion of depth, even though three out of four turn to the front. The type of shrubbery and the translation of horizontal ground into vertical space are Iranian elements. The section might be called a Mesopotamian-Hellenistic interpretation with Parthian frontality.

The choice of broad or literal interpretation determines in part just what one sees in the last third of the great panel. I believe the vision of Ezekiel is followed on the northern wall by the revolt of the Maccabees, an event quite separate, but continuing the history of Israel after its return from Babylon. This is followed, in my estimation, by the victory of Judas Maccabaeus over Gorgias in the large panel on the front (eastern) wall. Finally, the cleansing of the Temple is portrayed between the doors on the wall opposite the Torah shrine at the back of the hall.

In the last third of the great panel, a Roman soldier is represented standing over a figure in Iranian costume kneeling at a pagan altar (identified by the presence of a small idol). To the right of the altar a civilian in Iranian costume grasps by the hair a man crouching before him and raises his sword to kill him, while four soldiers in Roman dress look on.

Explanation of the painting depends on the identity of the man at the altar, and the recognition of the man slain and his slayer. If, as I believe, the same figure who kneels at the altar is the man who is slain, then the painting obviously represents the beginning of the revolt of the Maccabees. The costumes of the kneeling man and the one slain are the same: long-sleeved tunics and dark trousers tucked into high boots. In the conflation style, the two successive steps in the story come together in the same scene, with thȩ center piece of the altar linking the two together.

The account in the First Book of Maccabees describes the incident in vivid detail as Mattathias reaffirms that he and his sons and brothers will follow the way of God but not that of the king if it would take them from His path. Then, when an apostate Jew responded to the king's command to worship at the pagan altar, Mattathias slew him and the king's lieutenant and destroyed the altar.

> And Mattathias cried out with a loud voice in the city, saying, "Let everyone that is zealous for the Law and that would Maintain the covenant come forth after me!" And he and his sons fled into the

mountains, and left all that they Possessed in the city (1 Maccabees 2:15–28).

The key to the correct identification of the painting is the representation of a civilian being slain by another civilian, the two placed prominently in the foreground, not far from a great altar, also prominently placed, and in the presence of a group of soldiers depicted in the background. Supporting evidence is provided by the figure of a civilian forced to kneel at the great altar by a soldier standing above him. If the civilian at the altar is the same man being killed by the civilian, it would be difficult to find any solution other than the account in Maccabees.

The Maccabeean revolt did not occur until 168 B.C. and because of the troubled years that followed, an artistic tradition for the depiction of any part of the story would not have been established until some time later. The costumes worn by the soldiers, the only Roman military dress included in the Synagogue, is of a type common in the second century A.D., a reasonable date for the formulation of the pictorial tradition upon which the Synagogue artist based his panel.

Final, and conclusive, evidence is furnished by the position of the panel between the vision of Ezekiel, which saw the Jews safely restored to the Land of Israel, and the painting of the decisive Maccabeean victory over Gorgias and Lycias, pictured in the adjacent register: a victory that led immediately to the purification and rededication of the Temple in Jerusalem.

Victory of Judas Maccabaeus over Gorgias (EC 1)

This great panel on the front wall of the Synagogue is tantalizingly abbreviated by the defacement of the upper part of the picture. Fortunately, however, enough remains, particularly in the north half, to show the chief features and a good many significant details.

The painting is divided into three parts: on the left, a body of armed horsemen advance; in the center, two male figures recline on couches among seated women; and on the right, unarmed men advance, carrying obviously valuable objects in their hands.

Clearly the band of horsemen come from the desert and approach unsuspecting inhabitants of a city. The horsemen, in Iranian riding dress, carry the bow case and quiver of the Parthians. The leader of the band is not only magnified in size, but also distinguished by the royal

kandys (cape) which billows out behind his back. Since his trousers are adorned with an ornamental vertical stripe, he very closely resembles the mounted Mordecai in the Esther scene. It is impossible to tell, because of the faded upper portion of the picture, if he wears the royal Parthian diadem. To indicate the setting, the dog of the desert, the sloghi, is represented.

The account of the surprise attack on the town of Emmaus by Judas Maccabaeus is related in both the first twenty-five sections of the fourth chapter of the First Book of Maccabees and in the *Jewish Antiquities* of Josephus (XII. 298–316). The accounts agree that the attack at dawn was a complete surprise, for Gorgias, the Greek commander of the garrison at Emmaus, had gone off with the greater part of his troops in search of Judas. Judas, however, had already decamped and fell upon the diminished Greek forces at Emmaus. His victory was complete, but he forestalled his troops from pausing to loot, for he feared the imminent return of Gorgias. When Gorgias and his men looked down from the heights to the plain that held Emmaus and beheld the Israelite host arrayed for battle among the smoking ruins, the Greeks "became frightened and turned to flee" (*Jewish Antiquities* XII. 311–312). Only then did Judas and his men carry off their spoils of war.

The painting represents the Maccabeean force of mounted troops taking the camp, the surprise of the dawn attack symbolized by the reclining men and seated women. The battle itself is not portrayed; instead, the artist chose to depict the victorious outcome, the carrying off of "much gold and silver and stuffs of purple and hyacinth." Both Josephus and the First Book of Maccabees then continue to relate the further victory of the Israelites over Lysias, the commander in chief of the Greeks, followed by the triumphant entry of Judas into Jerusalem, the cleansing of the Temple, the rededication of the Holy Place, and the rebuilding of the altar.

The very strong Iranian-Parthian character of the scene is natural in the artistic tradition, since the Maccabeean revolt took place in 168 B.C., only a few years before the Parthians captured the large and important Hellenistic city of Seleucia-on-the-Tigris in 141 B.C. The Greek artistic tradition persisted in placing the mounted military leader in the center and prominently in the foreground. The leader can be confidently identified as Judas Maccabaeus.

If the killing of the apostate by Mattathias Maccabaeus was a later addition to the vision of Ezekiel and added to form a transition to the

victory of the Maccabees, then there is a very interesting chronological sequence: first the vision of Ezekiel with the restoration of the dead; then the victory of the Maccabees; then the intermediate scene, which binds them together. As for the artistic tradition, the actors in the vision wear only Hellenistic costumes, but the scene does include Iranian shrubbery. The victory of the Maccabees is peculiarly Parthian in the use of costume, horsemen, and dogs; but there is a trace of Hellenistic megalographia. The transition scene would naturally reflect a still later tradition, and the appearance of Roman costumes, combined with the stronger feeling for spatial depth, supports this view. The helmets with peaked fronts, narrow chin guards, and short vertical plumes are the distinctive second-century Roman type. They would have been well known in Dura.

The Cleansing of the Temple (EC 2)

All that remains of this small panel painted between the Synagogue doors are two long-legged birds, one in flight and the other standing on the ground, and some utensils, among which is the Temple *shofar* (ram's horn). Following the scene of the victory of Judas, this painting represents the cleansing of the Temple with the rejection of pagan utensils and the pagan birds for sacrifice. The desolation of the Temple and the court, here symbolized by the discarded *shofar,* is stressed by both I Maccabees (4:36) and Josephus (XII.7.6).

Whatever may have been the rest of the scene, the birds remain a prominent feature, placed in the center foreground, although they are not very accurately drawn. Probably the artist was relatively unacquainted with the desert, but the birds belong in that setting, bustards recognizable by their long legs and necks and by their preference for running. When disturbed they fly only a short distance, then run again. They inhabit only deserted regions, and their presence exemplifies the desolation of the whole Temple area, an abandonment emphasized by the scattered vessels of the Temple.

This scene is a fitting conclusion indeed to the long messianic sequence before it. The visions of Ezekiel, the return to Israel, the revolt of the Maccabees, the victory over Gorgias, and finally the triumphant return to Jerusalem, and the reinstitution of the daily sacrifice recapitulate the history.

There are additional reasons why the Maccabees and the cleansing

of the Temple should be portrayed in the Synagogue at Dura. The revolt of the Maccabees occurred in 168 B.C., when Mattathias killed the apostate Jew. At that time Antiochus IV was engaged in civil war with Demetrius I and the Seleucid empire was the stage for that great struggle which continued for more than thirty years. Antiochus IV, in revolt against his brother, chose Zeus Megistos Ouranios as his patron god and tried to introduce his worship into the Temple at Jerusalem. In Dura the Heroon shrine of Zeus Megistos (found in the course of the tenth season) stood as a permanent reminder to the citizens of the devastating civil war, the war that allowed the Parthians successfully to attack and take the eastern provinces and Seleucia-on-the-Tigris.

The victory of the Maccabees over the Greeks and the dedication of the new altar in Jerusalem in 165 B.C., just three years after it had been profaned, were celebrated with the eight-day festival of Hannukah, instituted as a yearly celebration as soon as the Israelites had returned to Jerusalem. It does not seem pure accident that the great celebrations of the Jews should find place in the Synagogue, although it is perhaps accidental that the panels recording them were preserved: the Exodus from Egypt (WA 3) recalling the Passover; the scene of Esther (WC 2), the days of Purim; the tents of the desert (WB 1), the Feast of Tabernacles; and the return to Jerusalem with the cleansing of the Temple (EC 2), the Hannukah. Yom Kippur is perhaps less clear, as the year's end and beginning, but surely the purification and sacrifice (WB 2) at the beginning of each month bring it to mind.

What is the overall theme, then, of the bottom band (labeled C) of scenes, which were painted before the two bands above? One begins on the south (side) wall with the series of Elijah, the triumph over the priests of Baal, and his power over death itself. (Why is he not shown in his fiery chariot going up to heaven in a whirlwind, one might ask? But it is scarcely profitable to speculate why some biblical scenes are omitted, especially when only half the paintings are preserved. Yet, the victory over pagan priests and the visible conquest of death are of prime importance.) The deliverance of the Jews in the old Persian empire, represented in the story of Esther and Mordecai, is particularly relevant in the era of the new Persian Sasanids. David represents the great king of all the Isralites, and his anointment marks him as the messiah, the anointed one *(christos)*. The rescue of Moses looks forward to his great role in the history of Israel and perhaps suggests the miraculous escape with the aid of the Lord.

Then, finally, come Ezekiel, the Maccabeeans, and the rededica-

tion of the Temple. The Torah shrine contains the law, the wing panels show the direct communication with God. David enthroned and the Temple symbolize the divine background. The history of the people of Israel as far as the kingdom of Solomon appears in the upper register of paintings. The establishment of Tabernacle, Temple, and the priesthood is shown in the middle register. The bottom register recalls the triumphs of God over the enemies of His people, the restoration in Jerusalem, and the messiah.

I have one further suggestion: if the scene of Moses and Pharaoh is the prologue to and anticipation of Moses leading his people out of Egypt, then it is quite possible that the anointing David looks forward not only to his role as divine king, but also to David as the great leader in the establishment of the kingdom in Canaan. I suggest, therefore, that the top register of paintings to the right of the Torah shrine illustrate the early history of Israel and the exodus from Egypt while the panels on the other side of the shrine depict the conquest of Canaan, the institution of the kingdom, and the kingship of Solomon, the builder of the first temple.

If the top and bottom registers are linked by the history and visions of the laws, where does the middle band of paintings fit? I do not believe that in any synagogue one needs to explain special emphasis on the Torah and law, the Tabernacle, the ark, or the Temple. Joseph Gutmann has suggested these panels reflect certain rites and observations belonging to the congregation.[29] The entire focus in the sanctuary is on the worship of God; the history and the visions are simply reflections.

The fascinating panorama of paintings in the Synagogue desperately needs further investigation from the point of view of artistic development in Syria and Mesopotamia. A Jewish artistic tradition, like the Homeric in the West, should retain successive stages of development, and this is clearly indicated in the Synagogue paintings. We find there a gradual shift from the Hellenistic development in the illusion of depth to the Eastern use of tiers or echelons of figures in vertical depth with decreasing spatial illusion. The Iranian costumes reflect a middle period in the paintings at Dura. The Hellenistic phase is perhaps best found in the Exodus scene, with the massed helmets of the soldiers, and in the wing panels, where the shadow line is employed. The misunderstood fold lines of the drapery in the scene of Moses as a babe and in the healing of the widow's son signals a Mesopotamian tradition now divorced from the Hellenistic. The enthroned king, centered and

magnified, as well as shown frontally, is representative of an Eastern tradition, probably Parthian, as the Mithraic magi show the combination of Hellenistic depth and Oriental frontality.

One of the great contributions of the Synagogue panels is this display of the successive steps in the Eastern development from the early Hellenistic to the early Sasanian, a tradition in part of its development divorced from the Greek and accepting more and more of the Asiatic. So the combination of Greek and Parthian style appears in the Parthian gallop and the enthroned monarch, whereas the Orient introduces the tiers or echelons of figures in vertical space and the focus on the center element, magnifying in size the image of the leader. Later comes the Sasanian use of vertical space to frame the most important person or element in the scene.

Other incidental details are the use of overlapping and network patterns, the differences in Iranian costumes to conform with changing fashions, the Iranian shrubbery as opposed to the Greek, the strong frontality and profile-tiptoe step, the Eastern animal in profile carefully and expertly depicted, the new scatter patterns, and the arbitrary use of architectural facade profiles.

The cartoons and prototypes of the Synagogue paintings very probably came from western Syria, Palestine, and Egypt, as well as from Mesopotamia and the Jewish schools of lower Mesopotamia. But the artistic treatment of many points can be attributed to Mesopotamia, both in the Hellenistic and Parthian periods, and I am convinced that further study will relegate more and more of the paintings to Mesopotamian prototypes. The devout artist who designed the tremendous series of paintings at Dura was a genius who magnificently fulfilled the challenge presented by the great bare walls of the Synagogue.

10 The Seventh Campaign 1933–34

The seventh season of excavations began at the end of October, 1933, and almost immediately was rewarded with a painting of Venus and Cupid in the House of the Scribes. A few days later, on the eleventh of November, we recovered our first parchment, a well-preserved and clearly written document, and shortly after came the discovery of a graffito dedicated to the goddess of good fortune, Tyche, scratched by an unknown hand on the wall of a private house. An insignificant find, it is true, but I happily welcomed the graffito because of my continuing conviction that the appearance of Tyche augured well for the season.

A veteran and able staff returned for the campaign with me. Henry Pearson was an architect and technician of unusual skill. He had worked with Bacquet in the fifth season, taking down some of the paintings of the Christian Chapel; now he undertook the tremendous task of dismantling the Synagogue. Confident, skillful, combining all-around ability with an understanding of ancient architecture and archaeology, he filled a great need with admirable style. He drew the plans of buildings excavated, made special studies of architectural features, and both removed and preserved the painted walls.

Du Mesnil returned as associate director and representative of the French Academy. He took direction of the largest team of workmen, continuing the excavations in the great embankment and systematically clearing the wall north of the Synagogue. With Francis B. Comstock, a volunteer graduate from Yale, Du Mesnil was able not only to shoulder the major work of excavations, but to conduct individual investigations both inside and outside the site.

Comstock was neither classicist nor archaeologist, but interested in history and travel. A blithe spirit who undertook with enthusiasm whatever task was assigned, he not only added a much-needed extra hand to the work, but delighted in the few recreations we had to offer at Dura: hunting, horseshoes, and bridge. He brought his own car as a reserve and extra conveyance, a vital necessity when we had guests. Frank Brown came back, if not as associate director, at least as right-hand man. An outstanding student and meticulous excavator, he was

Top: Visitors in the Synagogue, March 1934. *Left to right:* Du Mesnil, Hans Lietzmann, Michael and Sophie Rostovtzeff, the author and his daughter, and Henry Pearson.

Bottom: Frank Brown and his crew in the Temple of the Gaddé. The head pickman stands to the left, and Abdul Messiah Taza, the foreman, on the right.

the ideal archaeologist, combining field expertise with armchair research.

Herbert Gute, a young and gifted graduate student in the Yale School of Fine Arts, came out to copy the paintings of the Synagogue before they were lifted from the walls. He was a painter of the first order, with a German precision and attention to minute detail. He worked hard, long, and most intelligently. An able athlete as well as a scholar, he added immensely to our recreational activities. He had been a baseball pitcher in high school and quickly excelled in horseshoes. A keen competitor, he kept daily scores of our doubles matches.

My wife, Susan, took charge of the kitchen, the cook, and the supplies. She kept the catalogue, cleaned the coins, repaired the pottery finds, continued the study of Greek inscriptions from the Temple of Azzanathkona, looked after Mary Sue, now approaching three years of age, and in her spare time kept up her correspondence and took an occasional hand at bridge.

We were working with three-hundred men on the twelve-day schedule—ten days of work and two of rest. Those leisure days were occupied with hunting pigeons along the cliffs and pheasants in the river valley, playing horseshoes, and the evening hand of bridge. Seasonal rains the previous spring had helped to bring the desert out of its long drought. Mail came down once or twice a workweek thanks to the kindness of the French military courier on his route between Deir-ez-Zor and Abou Kemal. We worked from dawn to dark, with two brief breaks during the day. Once a week one of us went into Deir for supplies and to withdraw from the bank the Turkish gold and the heavy bag of silver to pay workers and foremen. Visitors were few; no tourists, just the occasional military man and his family. Our routine was filled with outdoor activity at the digs and indoor struggle with notes, records, reports, letters, and financial calculations. We enjoyed the good life of sun and stars, broad desert, wind, and wide horizons, and best of all we had the close companionship of a congenial group engrossed in an immensely satisfying, if equally tiring, common project.

At the end of December, Nicholas Toll, the distinguished editor of the Byzantine periodical of the Kondakov Institute in Prague, joined our expedition as photographer. A student of Greek sculpture, he was already well known as a scholar of Greek art, but he lacked experience in field archaeology. He added a scholarly knowledge to our study of the paintings, reliefs, and architectural decorations. He particularly de-

Lt. de Barmond, in charge of the French garrison at Abou Kemal, in front of the cook's tent.

lighted in his photographic tasks and was eager to contribute wherever he might.

First objectives this season were obvious: continuation of the removal of the tremendous amount of dirt in the embankment on both sides of the gate, and expansion of our clearing in the center of the city. We had begun the season with a general feeling of confidence, partly because of the extraordinary successes of the past two seasons, partly because the excavations were in areas already partly explored and clearly promising. The veterans knew the site, knew each other, and were well acquainted with the work and the men. Gute and Comstock fitted in easily and comfortably.

Early in November, Frank Brown began the systematic excavations of the Temple of Adonis (in Block L5) not far from the Synagogue. In the course of the fifth campaign, Pearson had discovered from his review of the block plan that the southeast corner of the block did not correspond with the normal plan of private houses at Dura. I had thought, because of its location near the Great Gate, that it might be a caravanserai. A very small exploratory dig proved me wrong, but did show that the building was unusual. Two rooms placed together and facing west were separated by a wide bay, and the inner one was surrounded on three sides by a narrow corridor.[1] The finds of the fifth season were limited to fragments of painted plaster from the inner room and to two graffiti or dipinti in ink from the court: one of a warrior armed with a spear, and a second holding a bow. Although the fill in the block was shallow, the pressures of the sixth season did not allow continuation of the work until almost the end of the campaign,

when more of the court was excavated and an inscription to the god Adonis was recovered.

Almost immediately Frank Brown's systematic excavation of the area brought rich returns in antiquities. After two months' work the excavations of the temple were complete and enough of the remaining quarter of the block was uncovered to show it contained only two small private houses of the usual type. One of the two relief sculptures found was a portrait of the goddess Atargatis crowned with the turreted walls of the city goddess, seated in a high-backed chair, the tops of whose sides supported doves, symbols of Aphrodite. Her face was carefully carved with Greek proportions and features, but to the Orient belonged the snail curls, her wide staring eyes, and the heavy earrings.[2] The second relief presented the dromedary god seated on a camel before a flaming altar, with the symbols of the sun and moon over his shoulders and a great palm branch behind.[3] He was clad in Greek costume with very stiff folds and a scarf that seemed caught in a stiff wind. Among the half dozen inscriptions in the temple, one on the lintel of the doorway of the *pronaos* provided the date A.D. 152.

Early in December as I walked along the crest of the city plateau, not far from the steep slope west of the Citadel, I came upon a piece of painted plaster, the color still visible. This was the second time I had recognized painted plaster on the ground. The first had occurred in the second season in the Roman quarter and had led to the excavation of our first Roman bath. I still do not understand how plaster exposed for centuries could retain even a trace of its tempera color. Possibly it had lain face down and protected, and by some accident was turned up; perhaps the wind had uncovered a fragment at just the right time. We had found several fragments of wall plaster projecting above the fill, here and there, but once above ground all color had disappeared.

This find of painted plaster drew our attention to Block B4, and Frank Brown moved his team from the Temple of Adonis, cleared the new area, and brought to light not only the walls of the Temple of Zeus Theos, but also inscriptions to place its foundations in A.D. 114. Enough pieces of painted plaster were recovered, some still *in situ* not far below the surface, to allow the restoration on paper of the great back wall of the naos and the painted panels to either side. The god was represented standing beside his chariot, crowned with winged Victories, his four horses abreast, leaping away to the right.

While the paintings of the side walls in the Temple of the Palmyrene Gods had been preserved, there were only fragments left on

the northwest wall immediately behind the seat of the god, just enough to show the lower legs of a great trousered figure standing beside part of a chariot wheel and the hooves of horses. Now the Temple of Zeus Theos supplied the design common to both temples.

When Nicholas Toll arrived at the end of December and at once offered to undertake whatever tasks might aid our explorations in addition to his duties as photographer, I suggested he survey our work—both the care and cleaning of finds in the courtyard and the operations in the trenches—and see what might appeal to him most. He spent a few days wandering around the site, watching the excavations in action, exploring various parts of the city, and then asked if he might explore the Necropolis outside the city walls. I was delighted. There beyond the Great Gate lay the broad field of ancient cemetery, almost unexplored and completely neglected. Cumont had explored a dozen tombs in 1922–23, and after his departure the soldiers had cleared three more good-sized and well-constructed underground chambers.[4] The bases of funeral towers, reminiscent of the tower tombs of Palmyra, had been recognized by Cumont, and one was cleared sufficiently to establish its identity. The graves in every case had been systematically looted. Only broken pieces of pottery and an occasional coin remained.

The chamber tombs consisted of good-sized rooms cut out of the subsurface rock. The gypsum beneath the hard ground surface was easy to cut. The walls of the chambers were carved with *loculi,* shelflike niches just big enough to hold a wooden coffin with its corpse. A single chamber would hold twenty, sometimes fifty, but the wood had disintegrated, the bones were scattered, and every object of value had long since been removed. The entrances to grave chambers not excavated or explored were easy enough to locate, sunken depressions in the ground, sometimes giving access to just the upper meter or so of a chamber otherwise filled with sand and mud. The cemetery stretched for half a mile into the desert and across the wadi to the south of the city.

Du Mesnil had made a sounding in a small tomb at the edge of the northern wadi and excavated a second one nearer the gate, but found nothing of value. I had surveyed but found no interesting leads. Thus, I was delighted when Toll suggested he make a study and survey of his own. A very real difficulty was that the systematic exploration would be slow and laborious and would require the constant supervision of a staff member. There was too much other work in hand, supervising our

three hundred workmen, for me to spare a member of our regular team. A couple of good workmen was all that Toll required, and fortunately we now had many able, experienced workers. Those unskilled to begin with quickly learned under his tutelage. He was both a thoughtful scholar and a meticulous and thorough worker, a man of great personal strength with an eager, intelligent approach to practical problems.

One problem at Dura, from the point of view of the archaeological historian, was the shallowness of the fill that covered the greater part of the city. Certainly it facilitated rapid clearing, but it lacked the stratification of deposits that shows the gradual development of the city and records, so to speak, its historical development. Located on the rocky plateau, the city's foundations were of rubble and the walls of mud brick, but the practice had been to keep the streets free of accumulating debris, and when house walls fell, they were cleared away in order to build again on the old foundations, or to construct new wall foundations on the rock floor.

I had made a fruitless search for some deeper deposit in out-of-the-way corners of the city. There was none. With the prospect of a visit from Rostovtzeff, I redoubled my efforts to locate some stratified remains that would then give us a useful building chronology for Dura. I prospected outside the city walls and found one deposit not far to the west and south of the Main Gate: a mound, obviously a dump heap with ashes and Roman sherds on top. The promise was not great, since the mound was not more than ten to twelve feet deep, but in the absence of anything better, it was worth exploring.

As we dug down into the dump heap in January, it became clear that much, if not all, of the great mass was Roman refuse, chiefly ashes from the Roman baths, and that little or no stratification of value might be expected. Our disappointment deepened as we worked our way deeper into the mass of ashes and Roman pottery fragments.

Du Mesnil, meanwhile, discouraged after his unsuccessful exploration in the Dura Necropolis, suggested we explore the necropolis of Baghouz on the river ridge across from Abou Kemal, twenty miles downstream. He asked Henri Seyrig, then French director of Syrian excavations, if he might make a probing search as a preliminary to further investigations. The modern road ran down the right bank of the river, and few other than Arab herdsmen occupied or explored the left bank. Opposite Abou Kemal, however, ancient habitation was attested to by tower tombs of stone still standing above the low plateau, and a

survey had shown tomb depressions over a wide area. The necropolis had obviously belonged to a town of some size, now unnamed and forgotten. Permission was given to explore as an extension of our Dura excavations, and Du Mesnil took Toll and Pearson with him for a three-day foray.

With the help of a small team of Arab workmen, they opened a number of shallow graves, remarked with astonishment the size of the cemetery, and gathered some Parthian glazed pottery. The team decamped for Dura unexpectedly early when on their final day a red sun rose with a gathering wind and a black cloud of dust on the horizon. A sandstorm is not to be taken lightly, particularly when one is caught in an open camp. The tents were quickly taken down, the baggage packed, and the river crossed for the hurried retreat to Dura before the main force of the storm obliterated the countryside. We heard afterwards a persistent rumor, emanating from Abou Kemal, that the little expedition had been eyed as easy prey by some malcontents in Abou Kemal. They thought, erroneously, that the Americans would be carrying sums of money. The team was unescorted by guards, and escape across the Iraq border would have been temptingly easy for thieves. The robbery was planned for that last day but was abandoned when the sandstorm erupted, for escape—indeed, any movement—would have been impossible in the storm. So the rumor went, and it was plausible enough. Perhaps the Good Fortune of Dura was responsible for the plan having come to naught.

We judged from the scanty remains Du Mesnil and his team brought back that the site might have been the Parthian frontier center after the Romans had seized Dura in A.D. 164. But the team returned to Dura with one real prize. Pearson's workmen had come up with some bits of horn and wood, which Toll recognized as the fragments of a Parthian reflex bow, the first complete example ever found, at least in Syria. The horn center of the short bow allowed the bending back of the wooden ends to make a bow of unusual strength and one that required only a short pull and not very long arrows. It could be conveniently stored in a case (the *gorytus*) scarcely longer than the quiver. Sometimes arrows and bow were contained in one case. Thus, it was an ideal weapon for the mounted bowman and became the mainstay of Parthian mounted archers.

Although arrows were plentiful, no bows were found at Dura, and hence we were doubly thankful to find this example. For reasons unknown to us, some hunter or soldier had taken his bow with him to the

tomb and unwittingly provided us with the one major piece we needed to round out our picture of the equipment of war at Dura.

In the course of the fifth campaign we had found a heart-shaped piece of ivory with a little indentation in the top and with two curved bands, like wings, projecting from either side. The curved portions seemed too large to make a ring, but clearly were once a single band. The whole object was less than a half inch at the widest point, and we thought it might have been part of an inlaid table leg or have served as an ornamental pendant or trinket. We catalogued it and kept it for future reference.

It required a visiting archaeologist from further east correctly to identify the object as a thumb ring. He picked it out, recognizing it immediately. The curve of the band of ivory was too big for a finger but just right for the thumb, and the wide, pointed heart-shaped end, when turned up instead of down, covered the ball of the thumb. And of what use was that, we asked?

The Oriental pulled the bowstring with the thumb rather than with the fingers. He locked the bowstring in by curving the first finger over the end of the thumb, and the ring kept the cord from cutting into the flesh. We now had another bit of evidence to add to the panoply of war Dura provided.

Well-preserved shields were already counted with our outstanding finds, but even among them, Cumont's discovery of part of the oval shield on whose painted cover the stations between the Black Sea and Dura were listed was totally surprising. Apparently the soldiers recorded on their shields if they so desired the stops they made on their long marches. Was this a particular shield for display or for service in combat? The oval shield is that of a cohort, not of a legionaire, and the combination of good leather over wood marked the piece for defense. Pillet, in the course of the first Yale campaign, had found a shield composed of long wooden rods run through a strong leather hide to form a formidable protection against arrows. Too large for cavalry, inappropriate for the legionary or even the Roman in the cohort, the shield must have been for some light foot trooper, auxiliary to the Roman regulars, or else have belonged to the Parthians.

The elaborately painted shields that must have served for parade rather than combat raise the question whether troops posted in a small town on a distant frontier would own two shields, one just for parade, and whether they would have parades and ceremonies that called for decorative equipment. And if they did, would the shield not be much

Left: The Roman wooden shield as it was uncovered; its center boss is missing.

Right: Painted Roman shield found in Tower 19, in its original condition.

simpler, just the oval of wood covered with a gesso and painted with the required design. We found many fragments of oval shields, none that could be distinguished from the type actually employed in combat.

One fine example of chain-mail armor came from the skeleton in the sap of the embankment. The soldier may have been a Persian who had penetrated well within the city wall, because he carried an iron sword narrower than the usual Roman blade. Yet because some eighty-two Roman coins had spilled from his pocket and were found among his bones, I believe he was one of the Roman defenders rather than a Persian sapper. Perhaps the local defending militia was armed with the Parthian blade rather than the Roman.* With the skeletons were the round bosses of the oval shields belonging to Roman light-armed troops. Either the men wore no helmets or else were protected by leather headgear which had disappeared. The same year we found the warriors' chain mail provided the scale trappings for horses, one of iron and one of bronze, with scales still sewn to the cloth underneath.

The paintings and graffiti provided further and corroborative information on the military equipment of the Romans and Parthians, as well as bringing to life the agonizing, mortal struggles for which the

*See page 139 above concerning the possibility the remains were Sasanian.

scattered finds were tangible but mute testimony. They provide a glimpse of the ancient soldiery in action, the brilliant ranks of foot soldiers and the charging horsemen here in the Eastern plains. We can almost witness the eclipse of the Roman legionary, as the soldier on armored horse, the forerunner of the knights of the Middle Ages, begins to take military precedent. The heavy cavalry used the lance to unseat its opponent. The old-fashioned scale armor was beginning to be replaced by chain mail. And the graffiti of the quick Arabian horse with the rider drawing his reflex bow hints of the later Moslem armies. The so-called Sasanian Fresco shows the heavily armored cavalry, the cataphract, with spear or lance in the right hand but not couched for support or to take the shock of contact. The rider's feet are dropped, toes pointing to the ground clearly to indicate that he had no stirrup, nothing to brace the feet when he drove in to overturn his enemy. At Dura we found no traces of saddles or of anything that might conveniently fill the open space left in the middle of the back in both pieces of the horse armor. Probably the ancient warrior used only a soft pad or pillow to support his seat.

Individual combat between mounted riders is represented in the house at Dura and also in the Sasanian reliefs at Naqsh-i Rustam, where we find the Persian king unseating his opponent. Interestingly enough, the vanquished enemy is overturned in the saddle with legs in the air, head falling to the ground. Rostovtzeff thought the mode of representation quite primitive, but the same basic elements are present in the reliefs from Han tombs, and the same format was accepted in the Sasanian combats at Naqsh-i Rustam. I do not object to calling the house fresco at Dura Sasanian, because very likely it was painted some time after the Persian Sasanians came to power in A.D. 224. But I cannot accept the supposition that a fighting man in a conquering Persian army would have entered the city and decorated a private house. Rather, I think this art belongs to the Parthian tradition, which was so strong at Dura, and which was to become an integral part of the later Sasanian style.

To our continuing dismay, excavations at our own necropolis did not produce as spectacular a find as the compound bow brought back by Du Mesnil, Pearson, and Toll. By the middle of January the Arab team working through the middle of the dump heap outside the walls of Dura under Comstock's direction reached the bottom and proved, unhappily for our expectations, that the Roman refuse continued down to virgin soil. Toll's excavation of two small tombs also contained

The discovery of the horse armor. *Left to right behind the armor:* Marguerite van Berchan, Émile Bacquet, Henri Seyrig, and the author.

nothing of interest. Beneath the debris of Roman ash that formed the dump heap, however, the sunken orifices of more tombs were discovered. Obviously these tombs had fallen into disuse before being buried beneath the Roman debris, where they lay sealed from later intrusions. The vital question was whether they had been looted before the Romans had used the area for a dump. Toll, now with a bit of experience in excavating subterranean tombs, undertook their clearing.

Excavating the Necropolis presented some special problems. First the dirt had to be removed from the entrance tunnel, which opened into the underground chamber cut beneath the surface gypsum. Sometimes that chamber would be filled from floor to ceiling with dirt washed down by winter rains; sometimes a narrow space remained close to the ceiling if the entrance to the tomb had become blocked before the chamber was completely filled. The work underground was largely performed in the dark, except for the faint light that filtered down from the entrance tunnel. There was no wind, just an eerie silence and, with each thrust of the spade, the dust of ages billowing up to blind the eyes and clog the nose.

Fortunate for us indeed was the location of the Roman dump, close to the Great Gate, for that meant that it must cover some of the earlier graves, since the cemetery would have expanded gradually away from the city wall. The dump had obviously been placed in a portion of the cemetery long neglected, where the identity of the graves was no longer important. Hence, the graves must not only be considerably older than the dump heap, but have also belonged to families no longer active or remembered in Dura, but from a more distant time in the city's history. It quickly became apparent that the Roman dump, in covering the graves and their entrances, performed a real service for us: the ever-accumulating layer of debris dumped over the tombs protected them from the usual clogging by rain-washed dirt. Once the entrance was cleared, the tomb chamber itself might have only a foot or two of debris instead of being filled almost to the ceiling. Toll had reached this point in his excavations just as Rostovtzeff reached Syria, and indications were that though the graves had been stripped of jewelry and any other precious objects—probably by the undertakers after the funeral ceremonies were finished—the graves were otherwise intact.

In January my team of Arab workmen, who had been steadily clearing the embankment south of the Great Gate since the season began, hit pay dirt. They had been working on the face of Tower 16,

Horse wearing the scale armor. The workman holds the reed shield.

when they discovered set high up in the stone work a relief of Zeus, the Lord or Master—Zeus Kyrios. This tower was the second south of the Great Gate and at the south end of the block that contained the Christian Chapel. Carefully inscribed in both Greek and Aramaic was the name of the deity, as well as the date A.D. 31. Cut in broad but carelessly made letters in the face of the rough stone beneath was the inscription *Roumes epoiei,* "Roumes made this," and the date A.D. 28/29.

I am strongly inclined to believe that the date A.D. 28/29 represented the time when this section of the wall and the tower was constructed. Then, two centuries later when the embankment was built, the relief was carefully covered over and not removed from the wall. If the citizens of Dura had conceived of the embankment as a permanent feature, would they not have removed the relief and placed it in a new setting where Zeus Kyrios was to be worshiped? But the facts of our discovery of the relief are yet another indication that the embankment was undertaken as a temporary measure against a specific Persian attack.

On January 22 I set out to meet the Rostovtzeffs in Palmyra, but a heavy rain and the resulting deep mud forced me to turn back. Next day I was more successful, but the driving was very slow and difficult. By the twenty-fifth, when we returned, the desert roads, no more than tracks, were dry and safe. We could never tell when the rains might come, for winds at Dura sometimes came from the northeast, sometimes the south. Rains were expected the last half of November and during December; there should have been a lull in January, with the desert remaining moist and the grass growing. A second series of good rains in February assured a successful growing season. In that year of 1933–34 the rains were ample, but not excessive; the drought of 1932–33 was broken.

In Palmyra I could report to Rostovtzeff and Armin von Gerkan the probability of a new temple at the foot of the circuit wall at Tower 16, the discovery of the Temple of Zeus Theos, the clearing of the Temple of Adonis with its inscriptions and reliefs, and the painting of Venus and Cupid, as well as some excellent though sporadic finds of parchments and papyri. Two other items were well worth reporting, one anticipated, the other quite unexpected. The upper registers of paintings in the Synagogue had been removed, the pressure of dirt on the back of the Synagogue wall in Wall Street relieved by the clearance of the debris, and then the bottom register of panels and the dado had been cleared and removed. The dado we found to be intact, a fine

series of panels imitating marble sheathing with representations of masks and animals. The bottom register of panels above the dado continued completely around the four walls of the hall, interrupted only by the two doors in the east wall.

Rostovtzeff had seen in Dura the possibility of finding the early amalgamation of Eastern and Western art which led to the formation of the Byzantine. Breasted's book had stressed this aspect in the paintings he had uncovered in the Temple of the Palmyrene Gods, the paintings responsible for our being at Dura in the first place. Now the Synagogue exemplified this combination in extraordinary fashion. If the Synagogue paintings had not already become renowned as outstanding examples of early biblical illustrations, I think they would have been equally distinguished as landmarks in the history of painting, executed eighty years before the establishment of the Byzantine empire.

The completely unexpected find, however, was the discovery in the Necropolis of the tombs under the Roman dump, stripped by the ancient undertakers, but unmolested by desert diggers and later robbers. Just as Rostovtzeff arrived, Toll triumphantly uncovered the first of the subdump tombs and began to recover the pottery and minor trinkets. A small tomb it was, but it had been looted only in part. The faience vases were well preserved, the *loculi* contained the bones of the dead, albeit in some cases not in order, and there was no indication of later disturbance. The first precious tomb revealed that the tombs under the heap indeed were covered before systematic looting had taken place and gave rich promise of other tombs in the same condition. So an entire new field of fruitful investigation suddenly opened.

Because dirt had not accumulated in the tombs from the desert floor, they required only a fraction of the usual clearing time and effort. Toll led Rostovtzeff into one of the newly opened tombs and pointed out in the semidarkness the pottery still standing in front of the *loculi,* the fragments of bones, and the splintered remains of a wooden coffin. The dark, sealed chambers, silent and waiting, awakened suddenly from the sleep of two thousand years. It was an extraordinarily impressive experience.

Less than a dozen days after the arrival of the Rostovtzeffs, we recovered part of an inscription dedicated to Sol In[victus] from the embankment south of Tower 2. Two days later the base of a column still *in situ* was found sufficiently preserved to show a splendid dipinto dedicated to Mithra, and on that same day a small part of some paintings was uncovered.[5] By the time I left camp temporarily on the ninth

of February it was clear that the long-awaited Temple of Mithra had been discovered.

Rostovtzeff had promised Cumont, when first they had surveyed the site together, that he would find a Temple of Mithra. He now remembered making that promise while standing on the very place where we were to find the temple. Telepathy? Certainly Rostovtzeff had not indicated the spot before the discovery, but the promise to find a Temple of Mithra in the ruins of Dura was reasonable. Mithra was a favorite deity of the Roman soldiers in the third century, and it was reasonable to suspect that the Eastern god would be popular at Dura. The problem was where to find the god's home in the vast expanse of ruins. And, when found, would the remains throw new light on the fascinating subject of Mithra worship?

When I stopped in Paris on my way out to the East for the seventh campaign at Dura, I paid my usual courtesy call on Cumont, the first excavator of Dura and codirector with Rostovtzeff of the excavations. I warmly invited him to come out and see our latest discoveries. He said he would be delighted, but he had much to do in Paris and was getting rather old for desert travel. I asked if anything we might find would induce him to come.

"If you find a temple of Mithra," he answered, "I shall come."

And he came!

An unexpected twist of fate took Susan to Beirut for a week in February. The coin of fortune, which had only shown us its bright face of good luck and splendid health, for a moment turned up its darker side. Our small daughter, not yet three years old, was attacked by a skin disease, a fungus that kept her restless by night and uncomfortable all day. With the loss of appetite and sleep, she gradually lost weight. Dr. Hudson in Deir-ez-Zor recognized the symptoms of Monilia, a disease not uncommon in Syria, but for which there was no known cure. The only hope was that the patient could maintain sufficient strength to outlast the disease. "Perhaps the University in Beirut will have better information." The restless nights had also affected Susan's health, and headaches were added to the strain of peering through cracked lenses in her glasses. She had brought an extra pair of glasses, but they had disappeared along with the suitcase containing them on the journey through the desert.

So when Harald Ingholt and his wife, then at the University of Beirut, visited camp early in February and suggested taking Susan and

the child back with them to Beirut, the invitation was eagerly accepted. I was to come on a week later and bring them back to Dura.

In Beirut the doctors at the hospital had just found a formula that arrested and cured Monilia. They were at the point where they welcomed volunteers to take the inoculations. Mary Sue was one of the first given the treatment, and happily began to respond by the time I had arrived. We spent a week in Beirut to make sure the injections were effective. Susan obtained new lenses as well as an extra pair of glasses, and a thyroid difficulty that had cropped up subsided. Finally I was instructed how to give injections in the child's arm, and the three of us happily returned on the road to Dura.

During my ten days' absence, the crews had cleared the Temple of Mithra to floor level, and, though small, the temple was a most impressive jewel. The foundation walls were intact and the niche in the back wall and the wall itself were preserved almost completely. There were two bas-reliefs *in situ,* paintings on the niche facade, a great mass of painted plaster fragments, and innumerable dipinti and graffiti.

It is one thing to be present at the unveiling, to see and admire each detail as it comes into view, but is was quite another, exciting experience for me to see the revelation as a whole when I returned after my brief absence. There were two splendid paintings on the facing walls of the niche and two similar paintings of Mithra hunting on the side walls, one almost complete, the other showing enough to prove it was a close replica of the first. Set in the back of the niche, one above the other, the two reliefs represented Mithra killing the bull. Above the larger, upper one were the signs of the zodiac in relief. Still higher on the wall were painted sketches of the life and deeds of Mithra.

On the side pillars of the niche two enthroned magi holding the ebony staffs of their office were painted. The hunting scene provided a splendid example of both the Parthian and the flying gallop as well as a strong central focus. The hunted animals, all viewed from just in front and above, were arranged in the network pattern, an overlap of forms resulting from a combination of Western and Oriental art perception.

This was the outstanding month for experts at Dura. Franz Cumont was the world authority on Mithraism as well as the first excavator of Dura. Rostovtzeff, guiding spirit as well as moving force of the Dura excavations, was most interested in the amalgamation of Oriental and Western culture in the late Classical period, and the expert in papyri and the decipherment of graffiti and dipinti. Von Gerkan, who had

Top: The Mithraeum begins to emerge as the painted arch of the shrine becomes visible.

Bottom: Mithraeum paintings. Mithra the hunter is portrayed on the lateral wall; the seated magus is adjacent to him; the intrados of the arch show the signs of the zodiac around the niche, which holds two reliefs of Mithra slaying the bull.

come at Rostovtzeff's invitation, was expert in Hellenistic city plans and fortifications. Seyrig, serving as the director of the Syrian archaeological services, visited us briefly, adding his expert knowledge gained from visiting all the Syrian archaeological sites. And, lastly, Ingholt was the leading authority on Palmyrene art, particularly sculpture.

A seemingly constant flow of casual visitors also came and went almost daily. One day the colonel in Deir-ez-Zor announced at eleven o'clock that he was bringing three people for lunch and then arrived with six, who could hardly be fitted into the dining room along with our regular staff, Cumont, and the two Rostovtzeffs. Another day two people arrived for breakfast, and as soon as they left in the afternoon, four more arrived for dinner and the night. Never did we laugh as that night, my wife wrote her family. We had four unexpected guests and only three beds, and those without a single sheet among them. Fortunately the new arrivals were young and American; they were not in the least discomfited, although they were accustomed to the far more elegant accommodations of the Oriental Institute expedition, where the housing facilities included linen sheets. They must have found life at Dura a bit primitive.

The days between February 19 and March 18 remain in my memory as particularly delightful. The early spring weather brought warm days with brilliant sunshine and cool nights. Susan was now fully recovered, and Mary Sue once more took fresh delight in each day. No doubt my relief at the restoration of health colored those days for me.

Cumont was a scholar of the old school and one of the best: always courteous, always kindly, tremendously learned, although now physically quite frail. With his arrival on March 9 there was work for one and all. Rostovtzeff spent hours on end trying to detect and decipher the graffiti in the temple. The sun on the white plaster was dazzling, a constant strain on the eyes, but Rostovtzeff seemed untiring. He was joined by Cumont in a general review of the details and paintings of the cult.[6] Mrs. Rostovtzeff struggled to put together the broken pieces of painted plaster. With a small team of workmen, von Gerkan cleared corners, probed along the walls and into towers, making detailed sketches of the fortifications. His six weeks' stay was too short for the study of the circuit walls themselves, not to mention those of the Citadel and the Redoubt. The days went by with unusual speed, and to the outsider the work, though brimming with activity, might have seemed monotonous.

Only small incidents and the unusual occurrences stand out in memory. Just before Cumont arrived, Rostovtzeff felt ill and attributed

Top: Franz Cumont and Michael Rostovtzeff before the shrine of the Mithraeum during the season of 1933–34.

Bottom: The author stands behind Rostovtzeff and to the left of Cumont, Du Mesnil, and Pearson in the sanctuary of the Mithraeum.

his discomfort to contaminated water from the river. He asked Susan if she was sure she knew when water had reached the boiling point. Francis Comstock drove him to see Dr. Hudson, and he returned on the third day, "miraculously" cured, he declared, by the amazing genius of the physician. Later I asked Dr. Hudson what had been the trouble. A simple upset stomach, not uncommon with a change of diet, was the answer. A pill and a good night's rest had achieved the "miraculous" cure.

The desert was beginning to change with the development of the oil fields near Kirkuk in the middle Tigris area. The long pipeline was destined to stretch from the fields east of Mosul, across Mesopotamia, reach the Euphrates close to Abou Kemal, and then stretch on south of Palmyra to Haifa and Tripoli. A new all-weather road for the heavy trucks of the oil company was built. Oil men came from the States along with drivers and engineers, workers and directors. Syria itself was developing apart from the oil interests and the wealth they produced. The international traffic between Beirut and Baghdad went through Palmyra, then off to the south of Dura, but the traffic running from Aleppo to Baghdad and on to the Persian Gulf went through Dura itself.

We no longer heard the hyena at night, and the call of the jackal rose only occasionally. Wildlife, particularly what could be chased by daylight—gazelles and long-legged outards, the running birds of the desert—were fast disappearing. The night predators, the wild cat and the wild boar of the swamplands, were more slowly being eliminated.

Three members of a pipeline crew visited us one Sunday while Rostovtzeff was in camp. We were working through Sundays but raised the flags of the United States, France, and Yale to salute the first day of the week. One of the oil men introduced himself with, "I seen Old Glory flying at the mast and I knowed there would be some guys would speak our language." We made them welcome, and the two years I had spent in Texas forged a strong bond between us. Rostovtzeff very kindly offered to show them around, and I saw them at various intervals in corners of the walls and the excavations, with Rostovtzeff talking long and earnestly. I well knew that an opportunity to expound on the Hellenistic states, Parthian domination, and the Roman emperors and Sasanian kings would not be neglected. They came in for tea weary but much impressed. During a pause in the conversation, Rostovtzeff remarked, "I saw a hare in the desert today."

"What did you see?" one of the oil men asked.

"I saw a hare in the desert."

"What?"

The man next to him poked him with an elbow. "A rabbit!"

"Oh," said the first oil man, "a rabbit. I ain't seen. . . ." He paused to consider. "I haven't saw. . . ." He paused again. Then, triumphantly: "No rabbit has crossed my path in more than a year."

I asked Cumont before he left whether he was disappointed in what we had found at Dura. He thought a moment, obviously wishing to put the best face possible on the matter. "In one way I am disappointed," he declared, "because here at Dura I hoped to find a temple stemming from the East; but the temple here has come in with the Roman soldiers, so it is not as startling as I had hoped. However, I am very glad I came, for the paintings, if not the reliefs, are quite extraordinary and the finds at Dura have been amazing."

Had Cumont had more time to study the overall excavations, however, his disappointment might have lessened. In fact, it was not Roman soldiers that erected the early buildings, but the "Archers of Dura," presided over by a local *strategos* and not by a military commander *(praepositus).* The translation suggested the archers were Palmyrene, quite possible, since presumably the Roman force contained Palmyrene soldiers. The impression is strengthened by the mention of the Twentieth Palmyrene Mounted Archers in later inscriptions. The Parthian bath in the northeast section of the city and Temple of Bel, or the Palmyrene Gods, in the northeast corner of the walls suggest a concentration of troops in this area even before the Romans came to Dura. Under the earlier, Parthian regime, local troops and probably horsemen would very probably have supplemented Parthian forces.

I have no objection to seeing Palmyrenes and Durenes united in cavalry units, but I do not see the need for insisting on Palmyrenes or Romans in the early worship of Mithra at Dura. Mithra was known largely through the Roman camps and in Rome itself, but he represented the great Persian god Ahura Mazda, and Dura is the most eastern center in which Mithra's temple has been found. The temple with the murals and reliefs we uncovered was constructed in the Roman period, between A.D. 168 and 171.

It was logical for the Mithra temple at Dura to follow Roman models in size and shape. The Mithra fraternity consisted of small but close-knit groups with secret writings, ceremonies of baptism in the blood of the bull, the communal banquet, and successive steps in a religious hierarchy. So standard were these features all over the Roman world, from the Rhine to North Africa, from Rome to the farthest

Inspecting the excavations. *Left to right:* assistant foreman, foreman Abdul Messiah Taza, Du Mesnil, the author, Henri Seyrig, Michael Rostovtzeff, Henry Pearson.

frontiers, that the liturgy and religious program must have been set once and for all at the very origins. Yet within these limits, the Mithraeum at Dura has a very Eastern flavor.

The paintings on the walls showed the god on horseback, hunting the animals of the desert. He wears the Parthian riding costume, riding to the right but facing full front in the Parthian style, and draws a Parthian reflex bow, while the stocky desert horse portrayed is in the Parthian gallop. The animals he chases were portrayed in the typical network pattern of the East. The seated figures painted on the front of the pillars are equally Parthian in costume and pose. In their hands they held the ebony canes associated with the magi and may represent Zoroaster as originator of the Mithraic mysteries, and Osthanes his most famous pupil.[7] I have but one suggestion to make concerning this brilliant contribution of Rostovtzeff. Because the seated dignitaries are not distinguished in rank one from the other, as might be expected in a master-pupil relationship, I doubt that they are to be identified as Zoroaster and Osthanes. Rather, I suggest that they represent the unnamed leaders and expositors of the faith, the same figure rendered twice just as was done with the representation of the Mithraic hunter.

One of a pair of seated magi with an ebony staff from the Mithraeum.

The ebony cane is Persian, an import from the East. No such staff appears in Western Mithraeums.

Because the reliefs of Mithra killing the bull, the core of the cult, were so standardized, it is difficult to demonstrate that the Durene examples came from the East. One can only show that they originally belonged to Persia rather than to Asia Minor, and that, therefore, they may have come to Dura with the magi and their ebony canes. The question is whether a Roman model was changed at Dura to conform with local artistic practice or whether the Dura-Palmyrene version was earlier but transformed gradually as it penetrated the Roman world. I may note that Mithra slays the bull with the long dagger or short sword, the Persian *akinakes,* rather than with the Roman sword. But his costume is Parthian: trousers, boots, and pointed cap.

In summary then, until a more ancient or elaborate shrine of Mithra is found at Dura or Palmyra, I suggest that the Dura Mithraeum as it stands provides our best example of the intermediate, Parthian cult, brought to Dura under Parthian domination and continued in Roman times. We would expect the introduction of Mithra to come early in the Parthian period at Dura, especially as the popularity of Mithras in the Persian court is reflected in the adoption of throne names: Mithridates I, the Great, conquered the Seleucids, Mithridates II ruled Parthia from 123 to 88 B.C., and Mithridates III ruled 58–55 B.C. It seems difficult to believe Mithra's worship was not well established at Dura, if not before the treaty with the Romans on May 12, 20 B.C., then at least shortly afterward when Dura became the Parthian center in the middle Euphrates, when the great walls were erected, and when the garrison of the city was necessarily a combination of Parthian soldiers and local troops.

The Mithraeum seems an anticlimax until we realize that the Persian religion was the third great monotheistic religion at Dura. It was the great rival of Christianity in the third century, and it played a most important role as the background of Islam—growing up under Persian control, but linked with the West—and especially belonging to the Arabian peninsula, close to Palestine, under the Persians. Astounding at Dura in the third century is that we have in this outpost sacred buildings of the three great religions preserved almost side by side.

Apart from the new discoveries in the excavations, three outstanding achievements marked the season: the copying of the Synagogue paintings by Herbert Gute, the removal of all the paintings by Henry Pearson, and the study of the walls by Armin von Gerkan.

The Mithraeum after the excavations have been completed and the paintings and reliefs have been removed.

It was impossible to leave the Synagogue with its paintings standing in the desert. The roofing we had installed was only a temporary measure, far from secure or adequate protection over a long period. Unfortunately, Bacquet was fully occupied in the Louvre. Though available for a short time the year before, particularly to bring some of the paintings of the Palmyrene Temple back to the Louvre, he was now no longer able to get away, much less to spend the long term necessary for the complicated task of removing the Synagogue paintings. But we were inordinately lucky to have our own architect, Henry Pearson, able to step into the breach. He had assisted with the removal of the panels by Bacquet in the fifth season and was particularly interested in the techniques and problems associated with the removal of the mud-brick walls that formed the backing for the paintings.

Once he had carefully cut away the wall from portions of the painted plaster, he applied plaster of Paris reinforced with fiber strands; long cross pieces of squared wooden rods, one half inch thick, were attached to brace the panels and form a secure backing. The technique

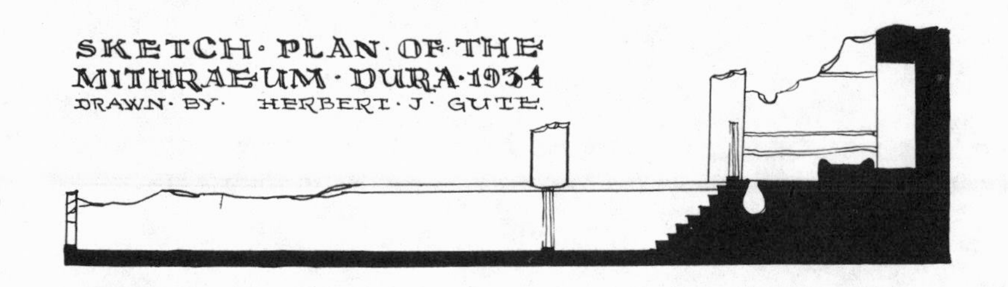

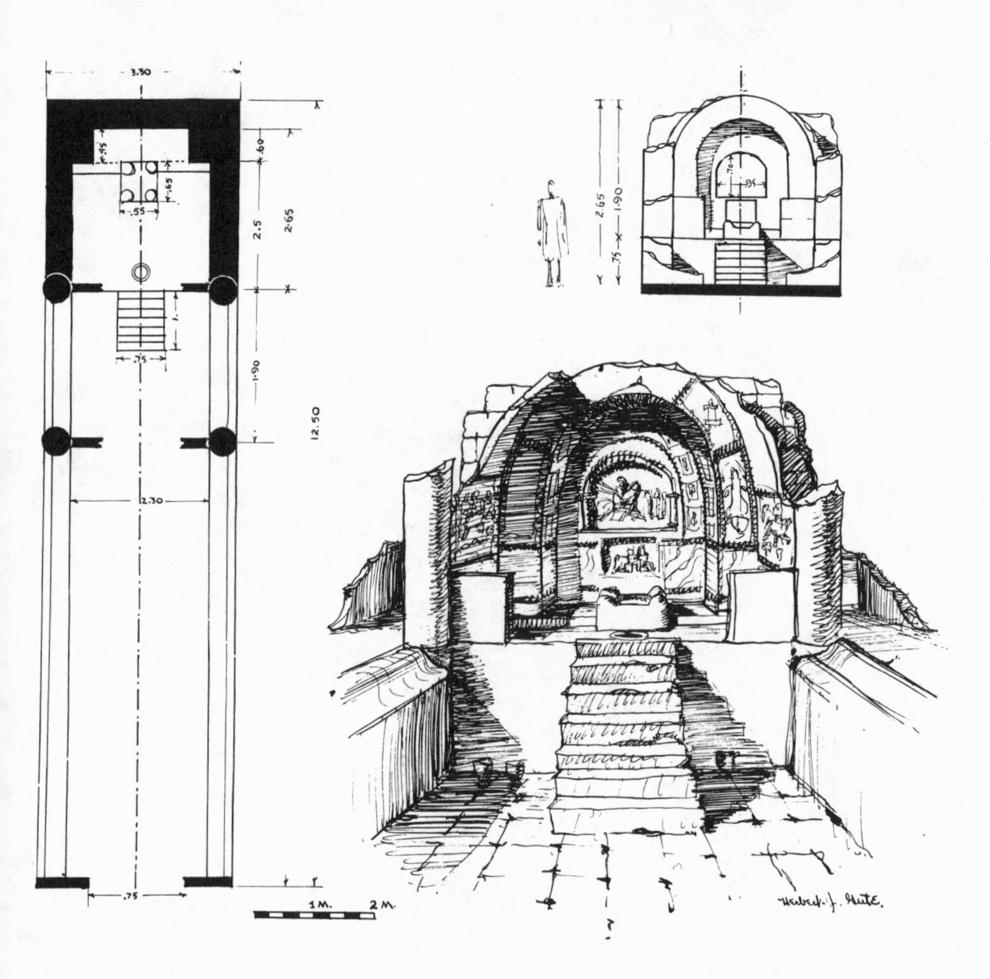

Sketch plan of the Mithraeum done by the staff artist, Herbert Gute.

Top: Removing the Synagogue paintings. The two top registers of paintings have been cut out; the workman is cutting away the brick wall behind the painted plaster preparatory to attaching the wood reinforcement lattice.

Bottom: Removing the Synagogue murals. As the brick walling of the building is removed and loosened from the plaster holding the mural, the plaster is supported on the back on a cradle of wood latticework. This view from the outside of the building shows a portion of the reinforced plaster where the brickwork has been dismantled.

was simple and primitive, but it worked. The task was extremely delicate, for the original plaster, laved onto mud-brick wall that had been scarred with pick or adze so the coating could key onto it, was sometimes nearly three inches thick and thinned off in spots to scarcely more than a quarter inch. Removing the backing wall, securing the new reinforcement, and plastering the thin spots might have tried the patience of Job. Once that portion of the work was completed for a section of the wall, predetermined cuts were made where they would cause the least damage to the face of the painting, and the wall paintings were removed section by section. The upper register had to be removed before the lower ones could be touched. When all the paintings had been taken down, they had to be boxed and sent by trucks two hundred and fifty miles across desert roads to Damascus.

By the time Rostovtzeff arrived, the two top registers had been cut up and taken down; the lower register was now completely uncovered, giving the Rostovtzeffs their first view of that magnificent series of paintings circling the room. As each portion was removed from the wall, it was placed beneath a scaffold and photographed so that the scale remained the same for all pictures. One could then reconstruct the whole wall in a photo-montage and see the walls as individual units long before the reconstruction of the actual panels in the museum was possible. As soon as the Rostovtzeffs had observed the lower register and dado in situ, these also were removed. At last the complete Synagogue, at least the pictured walls, was in boxes and ready for its museum destination.

Our first consideration was to have adequate copies of the paintings in case the originals were damaged or defaced. The work of the copyist was exacting, for it required a careful study of the original colors, a minute scrutiny of detail, and a special search for traces of color and design already fading. One astonishing feature of the Synagogue was the brilliance and clarity of the colors. Just the same, there was the inevitable fading when exposed to light after more than a millennium and a half of protection under the embankment. Some weathering had occurred where the panels were buried close to the surface; on these parts only faint traces of color remained. Blue and white faded faster than red, gray, and black. Du Mesnil and I disagreed

on some details of the painting even while we stood before the wall examining the same panel. The reredos with the overpainting was the most difficult to decipher, while the design on the robe of Aaron in the panel of the purification and sacrifice at the Temple caused our greatest differences of opinion. Hence, the copying required great patience and care in addition to expert brushwork.

Herbert Gute, just graduated from the Yale School of Fine Arts, had been carefully chosen as a competent painter interested in making renderings. How far interest and competence as an artist can carry a new recruit in the type of technical drawing required on a dig is always a vital question. Herb turned out to be a painter of unusual ability; perhaps his German ancestry gave him his flair for meticulous detail. He too was astonished at the extent and richness of the Dura paintings. He recorded them exactly as he saw them, even to the minutest detail on the smallest corner of a panel. Many of the details of the original paintings were brought out in greater clarity in the copies, and where subsequent fading has occurred, the copies have preserved the original lines and colors visible before the fading. Careful study of fragments, where original colors were best preserved in their original shades, was of valuable assistance in making the copies. To study the combination of color, detail, and design today, one will find the copies better than the originals. Of course, one supplements the other, and both are indispensable.

Finally we turned to the walls, the great circuit of stone fortification around the deserted city, now a city largely devoid of ancient remains. As excavations brought their full height and stretch into view, it became apparent that they were built in different periods, and in some parts different methods of construction had been employed. The circuit of a mile and more required study on an extensive scale. The archaeologist observes a good deal more than the amateur; the expert sees far beyond the archaeologist. There is no way to bring city walls into a museum, and there is every need for expert opinion on the site, especially if the fortifications are distant and difficult of access.

So it was that Armin von Gerkan, the well-known expert on walls of the Greco-Hellenistic periods, was invited by Rostovtzeff to come with him to Dura, make a special study of the city walls, and write a report for our seventh season. The Greeks, the Parthians, and the Romans had all made contributions in their respective periods, and special measures had been taken to meet the final siege of the Sasanians. What was the original plan? How and why had changes been

made? It was all written into the walls if one could decipher it, and von Gerkan was available and interested. At Dura he seemed rather silent, partly because his spoken English appeared no better than my oral German, and lengthy conversations were difficult. He did not have time to make an intensive study of the Citadel at Dura nor of the Redoubt. He requested only two or three workmen and probed along the circuit with expert rather than intuitive genius. His results and the way they fit into the overall city plan will be taken up in a later chapter, an example of how the study of one feature will add to the overall history of a site.

March was magnificent that year, the temperature hovering around seventy degrees by day, sunny with little wind or blown sand, and cool nights. The rolling terrain of the desert once more took on the green mantle of grass; desert flowers appeared miraculously, much to the delight of Mary Sue. There were fleas and bites to scratch, however. After the departure of the Rostovtzeffs, von Gerkan, and Du Mesnil, the season was officially at an end. We finished the crating of the last paintings, the final drawings and plans, the notes and reports, and the packing preparatory to closing down the house. At last we had time to enjoy an unhurried outdoor life.

One night a rifle bullet, fired from the top of the wall of the court, smashed through the window of the foreman's house inside the court and through his bed. Luckily he had not yet gone to bed. The military authorities made an investigation and took action against a local sheik. The gossip among the workmen was that the local sheik, enraged because the camp guard was composed of Syrian soldiers rather than of volunteers from his tribe, had tried to take the foreman's life. The underground news cable seemed always to have a reasonable explanation of incidents, and it was not the last time we received useful information from it. Unfortunately, the information always came after the event and never gave us prior warning. The desert remained the desert; the tribal authority was very strong, and life moved in ways mysterious to the foreigner.

At the end of March the camp was closed; the camions with the movable antiquities, including the Synagogue paintings, were loaded. I still had one unanswered question: where to send our treasure of paintings?

Susan wrote home early in March that the great question in camp was what Yale might obtain in the division of finds. The paintings of the Christian Chapel were at last all crated, and the Synagogue paint-

ings composed a monumental series of crates. The Mithraeum pictures were still on the walls and so technically outside the partition of finds for that season, but forming a very real treasure for future division. Seyrig, head of the French archaeological institute in Beirut, and the representative of the Syrian government, recognized the Jewish paintings would not be popular for a time at least in Syria, that reconstructing the Synagogue would be a monumental task for any museum, and that they comprised an exhibit to which few museums could do justice. Tentatively, then, the Synagogue would go to Yale; the paintings of the Chapel and the Mithraeum would go to Syria, along with all the crated finds of the seventh season over and beyond the paintings. President Angell had written to inquire how much extra the shipping of the paintings would require beyond the season's budget.

On March 26 the question of allocation was suddenly thrown open once more when no responsible official of the Syrian government would sign the necessary papers. It was now the end of the second year since the discovery of the Synagogue; reports of the find had spread widely, and rumors of its importance and value had soared. Syrian politics had raised its ugly head, with one party favoring the division agreed upon, another strongly opposed. The underground reported a value of at least a million dollars; and someone thought that retention of the paintings by Syria might bring sizable monetary gains, hinting that the party favoring the suggested division was perhaps bartering away a national treasure.

When the camions departed Dura for Beirut, we were still confident that the agreement reached between Seyrig and Rostovtzeff would stand. We were stunned when we reached Beirut to find the decision had been reversed by the Syrian senate. They would take the Synagogue; Yale would receive the Christian Chapel, the Mithraeum, and the miscellaneous finds of the season.

I could do nothing but turn the camions around, head them for Damascus, and then order the camions now in Damascus with the Chapel paintings and other finds to reload and come on to Beirut. After all, the division I had made of all the antiquities into what I considered two equal groups meant a fair partition no matter which group was chosen by the host country. There could be no appeal. Only our bitter disappointment remained.

On the other side, Seyrig was both chagrined and concerned, not so much, I believe, because his decision had been reversed, but as to what would become of the paintings and the whole Synagogue under

Syrian control. Seyrig was an able scholar as well as competent administrator, and he was devoted to the archaeology and antiquities of Syria. He preferred the life of the roving archaeologist to that of the student and curator in Paris. He was justifiably proud of the work of his department in Palmyra and Baalbek. He was rebuilding the past of Syria into vast monuments that might rival those of Egypt in the ancient Mediterranean world.

With the original paintings to go to Syria, Yale at least received the copies of the paintings. Photographs and copies do not compare with the original work, but these were no ordinary copies. Sometimes an infrared photograph will restore details invisible to the naked eye. A copy made before a picture fades can grasp and retain the original shades; the discerning eye of the artist can sometimes trace what even the camera has lost and the original has left obscured. Photographs taken at Dura reflected our equipment—adequate, but no more. The entire west wall could not be photographed at one time because the top registers had to be taken down before the lower could be uncovered, but a photo mosaic gave us a complete view before the wall was restored in the museum.

Du Mesnil had been in charge of the team excavating the Synagogue; he had supervised the construction of the roof to protect the paintings; and he was the first to identify the nature of the building through his reading of the Aramaic inscription beneath the feet of Moses. Nevertheless, his publication on the Synagogue, *Les Peintures de la synagogue de Doura-Europos,* in 1939, came as a great surprise to us all. The sixth *Preliminary Report,* containing photographs of the walls and a selection of the individual paintings, had been published three years before by Yale, in 1936, but we had to wait twenty years, until 1956, for Yale's *Final Report* by Carl H. Kraeling with contributions by Torrey, Welles, and Geiger. Yet I think Du Mesnil's study served a real purpose as a preliminary comprehensive study between our excavation report and the final, full-scale report. It well deserved receiving the Vatican prize for its contribution to biblical archaeology.

Kraeling was a thorough scholar. His volume examined the discovery, the architecture, the successive periods, the paintings, their interpretation, and their significance. It is a monument to his unflagging zeal and a volume of which Yale may well be proud.

There are few Old Testament scenes of the Classical or even of the Byzantine period with which those of Dura may be compared. If an amazing amount of the walls were preserved, we must remember that

close to forty percent was lost and the interpretations of parts and of the whole must remain open to broad speculation. Perhaps the best compliment of the Kraeling report is to say that all speculation now starts with his painstaking analysis, judgments based on a very sound review, a wide survey of ancient remains, and a thorough knowledge of biblical literature.

Since 1932–33, when the first paintings were uncovered, there have been countless reports, suggestions, reproductions, articles, and whole volumes dealing with the Dura Synagogue. I mention only the contributions of the Dura excavators and the contributions of Yale scholars. To this last group belong the three volumes of Erwin R. Goodenough (which comprise volumes 9–11 in his monumental study of *Jewish Symbols in the Greco-Roman Period*). He made a very wide survey of comparative pictures in the art of various periods, offered a large number of new interpretations, and presented a whole new series of photographs, largely in color. Not completely satisfied with the photographs in Kraeling's volume and the photos at Yale, he invited Frederick Anderegg of the University of Michigan to take new photos in color of the Synagogue paintings in the Damascus museum. In addition to color photos, Anderegg took some infrared exposures, which he found allowed him to penetrate farther into the layers of the paintings, one over the other, above the central Torah shrine. On the day he was to leave Damascus he made one last visit to the museum. He found the morning light just right for one overall picture, and with a wide-angle lens was able to photograph the entire back wall. That unified view, which we never had at Dura, forms the magnificent frontispiece of Goodenough's volume of plates.

11 The Final Campaigns, 1934–37

On the evening before I embarked on the *General von Steuben* of the Bremen Line to begin the eighth season at Dura, my wife and I enjoyed a Gilbert and Sullivan performance in New York. Susan was to stay home this year with our daughter, but the rest of the staff from the previous season was returning. Henry Pearson and Herbert Gute departed with me on September 29, 1934. Francis Comstock and Frank Brown came through Rihaniyeh to meet us in Aleppo, and we all visited the ancient stone church at Qalaat Seman, driving through heavy rains. We had to put chains on the car wheels and get out and push from time to time to make our way through the thick mud.

Joined by Du Mesnil, we officially opened the season on the last day of October, 1934. From the four hundred Arabs who gathered for work, we selected a crew of one hundred men and fifty boys. The camions with our baggage, supplies, and some beds and wicker chairs followed us from Beirut. We unloaded our belongings and then began to lay the tracks for the dump carts we had borrowed from the French. There seemed to be fewer flies this season, which began under a brilliant sun, the air cool and fresh. High-standing fields of maize greeted me when I looked down from the high cliff to the fields below, beside the river. To save valuable time traveling back and forth from Dura to Deir-ez-Zor, I went into town and arranged to have vegetables, meat, and gasoline delivered to us on a regular basis. With that plan in effect we made the trip only once every twenty or so days.

Our initial plans for the eighth campaign called more for continuation, consolidation, and study, rather than new explorations. Brown continued clearing in the Roman quarter, and Pearson returned to the Synagogue to lift the floor and see what lay beneath. Du Mesnil and Comstock worked on the embankment north of the Great Gate, while I excavated the embankment south of the Gate and finished the clearing of the Temple of Zeus Kyrios. Gute set to work making drawings

---

Clark Hopkins had planned but not written the narrative of the eighth, his last, season at Dura. Susan Hopkins preserved his letters, which he wrote almost every third day, sent to her in Wisconsin and then New Haven. She also saved copies sent to her of the three field reports written by Hopkins to President Angell of Yale. Therefore, parts of this chapter are reconstructed from that material, supported by the *Preliminary Reports*.

of the pottery and Comstock inherited from Susan the task of keeping the field catalogue. Toll, who was to arrive later, would continue to excavate the tombs of the Necropolis.

Hardly three days into the season, the Synagogue, as if to remind us of its pre-eminent role in our years at Dura, yielded the first discovery. Pearson found that the older, smaller synagogue continued as far as the south wall of the new, where the wall benches had preserved some of the old wall behind them. Fragments of decorative wall painting, originally forming the dado of the old building, appeared loose under the floor of the new synagogue. Most interesting was a painting of a foot, but unfortunately it remained the only suggestion that the old building may have held figural representations. We found some gilded and red-painted plaster relief fragments that had originally been fixed to the wall, probably with pegs.

A few days later we uncovered the doorway of the Temple of Zeus Kyrios below Tower 16, with its lintel still in place, but lacking any inscriptions. Small pieces of hexagonal blue glass indicated that the naos must once have contained a fairly good sized window. No cult reliefs remained, but the temple altar still held the remains of the last sacrifices; among the ashes were the bones of small birds, probably pigeons.

After two weeks of excavation, Du Mesnil had not made any exciting finds, and he did not conceal his growing sense of frustration. Perhaps Dura's spectacular record of producing treasures of architectural information, paintings, and sculpture had somewhat spoiled us for the uneventful clearing and studying that are the usual fare of fieldwork. Even our constant stream of papyrus seemed to have almost dried up, yielding to Du Mesnil's crew only a single three-quarter-inch fragment. A bronze cuirass complete with its cloth backing provided some momentary excitement, but the small *squamatae*, the metal scales, were so thoroughly oxidized that they crumbled at the touch, while the cloth turned to powder. The weather, however, cool and fine, buoyed us up. We may have been finding practically nothing after four weeks, but we were having a very good time doing it. At least our horseshoe games were keenly played, the competition sharp, and the teams fairly well matched. The arrival of five guards offered us ample protection, although we experienced no threats to our safety.

Max Mallowan, the distinguished British archaeologist who many years later would resume the nineteenth-century excavations of the Assyrian palace complex at Nimrud, stopped in for lunch on Novem-

ber 8. With him were his equally famous wife, Agatha Christie, the creator of Hercule Poirot, along with the architect Robin Macartney, and Hamoudi, the exforeman from the excavations at Ur.

They had left England in the autumn to find a promising site for excavating, one not overlaid with Byzantine and Roman remains. Their prospecting tour was to last three months, visiting various tells, finally settling on Tell Chagar Bazar, where they made arrangements to come back for a season of work. They had been camping in tents at Meyadin when they came down to visit our site. Working in Iraq, they informed us, was impossible. Any group excavating there, said Mallowan, must hire an Arab as a member of the professional staff, paying him in accordance with the pay scale of the other members, and for every year one dug, you had to pay to send an Iraqi to school one year in England or America. I did not notice then, but now discover in Mrs. Mallowan's account of her life as the wife of a field archaeologist that she had "an increasing difficulty in listening to or in taking part in the conversation," discovering on her return to camp that she was running a temperature of 102 degrees.[1]

Kraeling, who would later write the *Final Report* on the Synagogue, arrived from Aleppo for a visit on the fifteenth, along with W. F. Albright and two students studying in Jerusalem. Albright, the dean of American biblical archaeologists, brought us all the gossip from Palestine. They stayed a day and then left the following morning for Palmyra. A few weeks later Donald Wilber, who had been working with the Princeton excavations at Antioch and was to become one of the foremost writers on the early history of Iran and Afghanistan, stopped by on his way to Persia, where he was to do some architectural measuring and drawing for Arthur Upham Pope.

But we also had one unwelcome visitor in the form of a letter from René Dussaud informing us that an antique dealer had for sale a parchment, twenty by six centimeters, covered with Aramaic writing on one side and on half of the other side. The dealer said the parchment had come from a dig between Deir and Aleppo. Dussaud wanted to know if he should buy it for us. I felt sure it had come from Dura. How disconcerting to think that we had to buy back our own finds!

Despite the lack of major finds, the site almost every day produced small discoveries, which, though modest in themselves, continued to add bits and pieces to our monumental jigsaw puzzle. A fine inscription turned up in the excavations, and Frank Brown found two hoards of Roman coins, each containing about one hundred pieces. While my

clearing of the Temple of Aphlad produced no striking results, I did come across the Persian sap nearby. It ran from beside the Sasanian ramp outside the city wall, straight under the wall and into the city, almost three meters wide.

Partly due to the sparsity of finds, no doubt, I eagerly awaited the arrival of Toll, who was to continue the work on the Necropolis, where we had high hopes of finding previously undiscovered, untouched tombs. I picked him up in Deir-ez-Zor the first of December and released my crew to him.

I had decided to begin a careful study of the architecture of the Redoubt and the Citadel preparatory to making isometric drawings. There is no better way to force oneself to make the most careful and detailed observations than to start work on exact drawings. In the end, of course, my drawings had to be more than half conjectural because of the paucity of the evidence. I discovered that the earliest buildings—the early palace on the Citadel, the early Redoubt, and the early Temple of Artemis—were all built with small stones set with mud mortar. The later construction of large stones—as in the second palace on the Citadel, the early market, the Redoubt presently standing, and the circuit walls—appeared to have been built at about the same time, probably in the first century B.C.

Despite several rainy days in December and chilly temperatures that dropped down into the low forties, Toll finished cleaning most of the cemetery before the end of the month. Almost all the tombs, both chambers and loculi, had been robbed, probably in antiquity. He recovered some common pottery and faience ware. Towards the close of the season Toll did find an underground tomb that had been spared by the robbers, but the finds were no more spectacular.

During this time I was in constant correspondence with Henri Seyrig, attempting to obtain some of the Synagogue paintings for Yale. Rostovtzeff strongly urged me to persist, but my best efforts were doomed to failure. Seyrig wrote me in December that the Syrian government had decided they would keep the entire Synagogue in Damascus, where they would reconstruct it in a new building to become the museum. "I am sorry that this solution does not fulfill your wishes," Seyrig continued, "but I hope that the Mithraeum and the whole of the other finds, including the interesting bas-reliefs, will be a welcome addition to the collections of Yale all the same." I made one further attempt, suggesting Syria take the Mithraeum in exchange for the Ezekiel scenes, the Moses scenes, and the painted shield. But the Synagogue paintings were to remain intact and in Damascus.

Seyrig, combining scholarly integrity with practical insight, quickly took steps to disillusion any legislators who might consider the sale of the Synagogue as a new source of revenue. He pointed out that the paintings must be kept together, must be exhibited, and the display of the whole must be worthy of the Syrian state. In these endeavors he was singularly successful. The new museum in Damascus was built around the Dura synagogue, now reconstructed completely with walls containing every painting and the ceiling beams supporting the decorated baked brick ceiling tiles of the original building, with copies to fill the vacant spaces. Even the columns of the court were reconstructed, and the court in part rebuilt. Pearson was engaged to undertake the extraordinary task of reconstruction, and his genius shone as brilliantly in that task as it had at Dura. The museum at Damascus now houses the splendid archaeological finds of the Classical period in Syria, but the Dura Synagogue is its greatest treasure. Worth remembering also is the fact that Syria, in spite of the devastating Druse rebellion in the 1920s and the worldwide depression in the 1930s, still voted money in 1934–35 to build its new museum in the grand style.

The lack of major discoveries in either the city or the cemetery outside the walls, and the unsatisfactory resolution of the Synagogue negotiations were not our only disappointments. Rostovtzeff had written that the decision on funds for Dura by the Rockefeller Foundation should be settled early in December. If I did not receive a cable very soon, I was to assume money was not forthcoming. By payday on December 13, I knew that the season would have to be shortened, probably by a month, and indeed, the possibility of further campaigns at Dura clearly hung in the balance. Rostovtzeff wrote I should be prepared if necessary to close the dig permanently, for by the end of January there was still no money promised.

But Christmas was fast approaching. Because Comstock and Brown were to visit Mosul for the holiday, we had an early Christmas dinner of turkey, mince pie, local tomatoes (about the size of English walnuts), and champagne. We listened to our favorite records from among the group I had received from home: "You're Not the Only Oyster in the Stew" and "O.K., Toots." We invited André Parrot and his mission over from their site for New Year's dinner. They were excavating at Tell Hariri, better known now as ancient Mari, an important capital of the second millennium B.C. in the Middle Euphrates region that had been destroyed by Hammurabi. But Parrot could not come; he was awaiting the arrival of his wife, and he asked that we return the dump carts loaned to us.

Du Mesnil received permission from Seyrig and the Syrian government to excavate with Toll and Comstock the necropolis around the tower tomb of Irzi at Baghouz. I approved their going off for a few weeks. They found during their January excavations a tremendous cemetery about ten kilometers long in which they opened almost eighty tombs dating to the Parthian period.

As winter descended, the thermometer frequently dropped below freezing, and the close of our shortened season loomed near. But now more interesting finds began to appear. Frank Brown found some raw silk cloth with a simple geometric design, and we recovered (in Tower 22) our first parchment of the season, carrying nineteen lines of Greek and dated A.D. 52–53. Some fragments of relief carving and a very fine piece of horn carved with recumbent stags, reminiscent of the couchant animals characteristic of Scytho-Siberian art, were added to Toll's discovery of a Palmyrene inscription dedicating a shrine in the Necropolis at Dura. The structure itself had escaped our notice previously because the mud-brick walls had dissolved to form only a slight rise in the desert landscape. Once identified, however, the shrine's foundations were recoverable just below the surface, about 350 meters from the circuit wall opposite Tower 19.[2] The shrine had been dedicated to Bel and Iarhibol in 33 B.C. At first we thought it might have been a sepulchral shrine associated with the graves, but the early date clearly indicated that the building was constructed before the cemetery had extended that far out from the city walls. The dedicants were two Palmyrenes, Zabdibol and Maliku. Thus, the dedication in Palmyrene, by two Palmyrenes, to two Palmyrene gods may provide the reason for placing the temple outside the city: a foreign enclave in the city may have felt more comfortable erecting its shrine at a distance, removed from the long resident Durene cults and centers. Because the shrine was level with the desert floor, finds were meager. Some fragments of painting indicated that the walls had been muraled; we found the base of a statuette and a very damaged relief of Heracles.

In preparation for the closing, we began to make the large packing crates for shipping, and the reliefs from the Mithraeum were safely stowed in their boxes. Having grown tired of watching the workmen find little if anything, I looked forward to the close of the season. Roman Ghirshman and his wife stopped in for lunch and tea on January 12. Ghirshman, who had worked with George Contenau at the crucial prehistoric sites of Siyalk and Giyan in Iran, was later to assume the leadership of the French Persian mission, where, among other splendid

excavations, he began again the clearing of the great capital at Susa, opened at the beginning of the century by Jacques de Morgan.

As we might have forecast, the most exciting finds were made during the last weeks of the regular season. On January 17 our workmen digging near Tower 24 came upon three shields, one on top of the other.[3] We needed the entire morning to extricate them, and extreme caution was required not only because of the fragility of the wood but also because the painting on the face of one shield had in places adhered to the back of the shield above it. Constructed of planks of poplar wood with a leather edging, the shields were sufficiently well preserved to permit Gute to copy the painting.

One shield displayed scenes from the Trojan War: on the bottom half the Trojans, marked by their Phrygian caps, stood gazing at the wooden horse before their walls; on the top half, the taking of Troy was marked by the victorious Greeks slaughtering their hapless victims.[4] The second shield showed the equally familiar theme of duels between equestrian Amazons and Greeks on foot.[5] The third shield revealed a single large figure of a warrior god, probably the Palmyrene deity Arzu.[6] The shields, which seemed never to have had the central metal umbo affixed, and thus apparently had never been used, may have been display pieces from an armorer's shop, cast into the embankment when it was erected for the final siege of A.D. 256.

I had paid off the workmen and divided the season's finds into two equal parts for the division between Yale and Syria. Seyrig had come and chosen the half that was to go to Damascus. Du Mesnil had left for Egypt. A letter from Rostovtzeff confirmed that he still had no money to support another season of work. And Pearson was completing a new plan of the city as it now lay revealed, when he found a new temple on February 6.

In the process of clearing, he had come across a large doorway in Block H1, located at an important street corner, the first block north of the Temple of Hadad and Atargatis. Inside he found a court giving onto a naos with an arched doorway. The structure, which had begun as a typical well-to-do Durene house built around a court, had developed in four stages.[7] The house had been converted into a naos and pronaos with forecourt. Then a sacristy and meeting room were provided before, in its last reconstruction phase, it occupied the entire building and adjacent house areas.[8] In the last two stages the walls had been covered with paintings, of which only tiny fragments remained, insufficient to determine the scenes they once composed.

Relief of the Gad of Dura. The enthroned lord of the city is represented between a priest, who makes an offering at a small altar, and the military costumed founder of the city, Seleucos Nicanor, who presents the laurel crown.

Four broken bas-reliefs and fragments of another showed the gods Iarhibol and Apollo-Nebo. One relief carried the Palmyrene inscription "The Gad of Dura, made by Hairan bar Maliku bar Nasar in the month of Nisan, the year 470 [A.D. 159]."[9] Another relief, also with a Palmyrene inscription, had been dedicated to the Gad of Palmyra. This temple (of the Gaddé, Semitic title for fortune or guardian god of a city) then could not have been built after A.D. 159, and because of the very strong Western character of the reliefs, we could assume it was erected not too long before that date. The style of the painting fragments of stages three and four confirmed the approximate date, coming from the early first century A.D. and to the last part of the second century A.D., respectively. The warrior in armor shown crowning the seated Gad of Dura on the relief was identified by an inscription as Seleucus Nicanor, founder of the city of Dura.

Here, then, was an important structure that could not be ignored, left undone, particularly if we were pessimistic about the possibility of returning to Dura. Frank Brown offered to stay an additional two weeks with about twenty men to complete the clearing. I sent a man to

Temple of the Gaddé: the Gad of Palmyra. (Courtesy of the Gallery of Fine Arts, Yale University.)

Abou Kemal on February 13 with a cablegram to Rostovtzeff: "Palmyrene Temple and four reliefs discovered. Exceeding budget slightly. Finishing temple unless vetoed."

On February 20, at 5:55 A.M., I drove away from Dura for the last time as field director. (I visited the site again Christmas, 1935.) On my way to Beirut, I visited Tell Hariri (Mari) to see the new palace Parrot

had found, and then followed the pipeline to Palmyra, staying overnight to visit once again the ruins. R. Amy had just recently found some fine reliefs in a tomb there. I continued on to Homs and Baalbek and finally arrived at the Hotel St. Georges in Beirut.

The season's finds had already been divided between Damascus and Yale before we found the reliefs in the Temple of the Gaddé, and, hence, the question remained open as to whom they should be sent. I took the matter up with Seyrig one night at dinner, and he kindly gave the sculptures to Yale. The camions with our part of the division should have met me in Beirut, but a few anxious days of waiting produced no sign of them. But my concern was needless, for they finally arrived on February 25 with a tale of having gotten stuck in the mud after heavy rains. I had then only to see to the loading of the cases onto the boat, followed by a few days in Damascus and Jerusalem, before I sailed on the *Deutschland* for New York.

By now Rostovtzeff had begun to plan to go out on one last campaign the year after this (rather than in the fall of the year) to complete the excavations. As it turned out, funds were obtained, and two more years of work were undertaken with Frank Brown as field director. I moved over to the University of Michigan and undertook the continuation of its excavations at Seleucia-on-the-Tigris.

Frank Brown led the ninth campaign, as well as the tenth and last, for Yale with the experienced Dura staff: du Mesnil as assistant director, Pearson and Henry Detweiler serving as architects, and Toll and Comstock as assistants. These last seasons were given more to completing work begun than to searching for new possibilities. Work was continued on the Agora, while Toll returned to his tasks in the Necropolis. The Dolicheneum (a temple dedicated to a local god Dolichenus associated with Jupiter by the Romans), Temple of Zeus Megistos, and the Temple of the Gaddé were cleared, along with a house belonging to one Lysias, and some private houses.

Du Mesnil and Comstock set about finishing the clearing of the Roman bath (never leave a building only partially excavated lest it deteriorate before it can be adequately examined) and discovered the splendid Palace of the Dux Ripae, or military commander of the river bank. His residence appropriately stood on the edge of the cliff overlooking the banks of the river, the river itself, and the broad sweep of Mesopotamia beyond.

In the course of the fifth season at Dura, my first as director, I explored briefly what appeared to be a fairly deserted part of the city

between the wadi at the northwest end of the Citadel and the north wall of the city.[10] Surface examination near the edge of the cliff where a wall was in evidence had already revealed brick tiles and quantities of ashes, suggesting the presence of a Roman bath. We began a preliminary clearing of the building, which proved to be composed of a long, narrow room with an apse paved with tiles marked underneath with soot and ashes, perhaps from the heating system. Several hollow bricks probably came from the *calidarium* (hot room), but we progressed little further. Obviously this must have been an important location, although I could not see then why this area had been chosen for the bath. Had the section served a military role at Dura for which the bath had been reserved for a special part of the camp? Or had this been a convenient location for drawing water from the river below and, hence, an ideal site for such an installation?

But more propitious sections of the city demanded our attention at the time, and we left this corner with its questions unanswered. Nor did we return to it during my directorship in the subsequent three campaigns. And that was how close we came to discovering the palace of the Roman commander, the Dux Ripae, commander of the riverbank, and the Dolicheneum in its immediate proximity. The little Roman bath had been adjacent to the palace, but the palace walls had been covered without a trace on the surface.

If I had known, or even guessed, that the bath was part of an imperial palace, I certainly would have returned to this section. But in the field one must dig the most promising and the most obviously important places first and in the order of their presumed significance. We had, in the fifth season, discovered the Christian Chapel and marked out the Synagogue for the next season's work, to be followed naturally by continuation of the clearing along the city wall. In the center of the town we were fully occupied with the Agora, which covered an area of many blocks, and the excavating of buildings already identified as temples through their inscriptions, while outside the city wall the potentially rich remains of the Necropolis spread across the desert.

The belated discovery of the palace illustrates the problem every excavator must face. It is not impossible, but it is most difficult, to know when he has finished digging the significant features in his archaeological site. Year after year the sharp eyes of our staff had combed the site for bits of painted plaster and for telltale bits of walls in unusual arrangement, but even with the discovery of the bath re-

mains, none of us had suggested exploring further this rather out-of-the-way section of the city. Primarily responsible for our lack of interest was the paucity of clues to indicate interesting remains beneath the surface. Fragments of plaster, potsherds, green traces of oxidized bronze were far fewer here than in other sections of the city.

In retrospect, I found it most appropriate that the Palace of the Dux and the adjacent temple should grace the work of the last excavations at Dura; the Palace of the Roman Commander was a crown to our excavations as a whole.

The century of Roman occupation was apparently divided into two parts. The altar and little temple dedicated by the Second Ulpian Cohort and the Fourth Scythian Legion lay in the open space in front of the Citadel. The area in front of the north end had been quarried and this open space stretching along the whole face of the wall was taken over by the Roman Camp, now that the Citadel was deserted. One may imagine that this frontier city was held by the soldiers of Verus and Marcus Aurelius while the general marched farther downriver and took Seleucia-on-the-Tigris, just as Trajan had done. When Verus returned to Dura he made it a permanent Roman frontier post, replacing in importance Circesium at the confluence of the Euphrates and the Khabur. Dura then established the frontier line of the Romans on the west bank of the Euphrates, while Circesium on the east bank controlled the valley of the Khabur as well as that of the Euphrates. Under Caracalla (A.D. 211–217) Dura became a Roman colony. If up to then there had been any doubt as to whether Roman occupancy was permanent or transitory, it was now dispelled.

The northwest corner of the town, including the Temple of the Palmyrene Gods and the Temple of Azzanathkona, became the Roman quarter, holding the Roman camp with Praetorium, barracks, bath, and theater. Beside this quarter, on the northeast plateau, was built the Palace of the Dux Ripae (Commander of the Riverbank). A papyrus containing his title and office probably explained a graffito found in the fourth Yale campaign, mentioning the time (A.D. 238) when the "Persians descended upon us." Since our excavations revealed no indication of Sasanian occupancy, and we have only literary accounts of the Persians advancing to Antioch, the phrase was ambiguous. If Dura and her soldiers ruled the riverbank, the graffito probably meant the territory on either side of the Euphrates up to, or partly up to, the Khabur River, and the Sasanians would have overrun much of the territory without taking the city.

The rule of the commander (dux) marked the second change in normal Roman policy, at least as far as European tactics were concerned. It was customary along the Rhine and the Danube for Roman camps to lie outside the settlements, to have their own fortifications, to be more or less sufficient unto themselves. At Dura the city walls served as the camp defenses on two sides, and a strong mud-brick wall divided the camp from the city center. The Palace of the Commander lay outside this compound, a great peristyle court surrounded by rooms with an imposing entryway on one side. A second peristyle court, also built of mud brick laid in mud mortar, stood immediately behind the first. It was the largest single building we found at Dura, a combination of public and private rooms. A dipinto on a sherd found in one of the rooms was dated in the Seleucid year 561 (A.D. 249/50), although construction of the palace probably occurred some time after 211.[11] The Dux Ripae (Domitius Pompeianus, Commander of the Riverbank) is mentioned in a dipinto found in one of the rooms. Several Romans successively bore this title at Dura, the officers in charge of the Euphrates frontier.

The shallow ruins of the palace produced few finds, the most splendid being a gold brooch encrusted with garnet and blue-green glass around a green stone carrying an intaglio design of Narcissus from the first quarter of the third century A.D.[12]

Comstock cleared the temple, that the excavators conveniently named after the god Dolichenus just southwest of the Palace of the Dux. The building had the unusual feature of being composed of twin chapels, standing side by side, to serve the rites of the Roman camp rather than those of the citizens of Dura. The structure goes back to at least the first decade of the third century, while the last phase of the temple construction was dedicated A.D. 251.[13]

One inscription mentions the mountain god Turmasgade, and the other deity of the temple was very probably Dolichenus, a god of the city of Doliche in Commagene, who was popular with the Roman soldiery and identified in the inscription with Jupiter.[14]

Dura was now a Roman colony, but in the war zone, led by a "governor" who controlled the district. The third century was a period of continuing and increasing difficulty for Rome along the great stretch of frontiers both European and Asiatic, and the trend toward separate districts or zones of defense would be natural. This was our first positive indication of such management and it is interesting; it is not merely the army commander who would have his command post in the Prae-

torium, but also a civil leader acting in control of the Roman colony and district but in close cooperation with the army whose camp lay close to his door.

From the point of view of the history of art, the Roman occupation and the fall of the city came at an opportune time. The end of Dura in A.D. 256 occurred half a century before the accession of Constantine (A.D. 312) and three-quarters of a century before Byzantium became capital of the world in A.D. 330 and changed its name to Constantinople. The art of the Byzantines radically changed basic artistic concepts, but the stages from the Greco-Roman to the Byzantine had been largely undocumented. We recognized that the Byzantine represented a shift toward Oriental concepts, but art never changes over night and the gradual alterations, the steps of amalgamation, had to be guessed at rather than documented.

At Dura under the Parthians we see the blending of Oriental and Hellenistic elements in the Temple of the Palmyrene Gods. The Parthians extended and guarded the trade routes along which art elements from South Russia, China, and, most of all, the old Mesopotamian empires and Persia helped form the new artistic medium. The Romans accepted these artistic changes as appropriate to the region and even pictured their cohort lined up in strict frontality, beside the Hellenistic symbols for Dura and Palmyra and beneath the sky gods represented in Hellenistic and Persian costume.

The second astonishing feature, as far as archaeology is concerned, was the coexistence of four religious systems: Greco-Semitic, Mithraic, Judaic, and Christian. The Synagogue represented the age-old faith of the Jews looking to Jerusalem and its Temple, burned in the Roman siege of A.D. 70, but retaining its sanctity. The Dura Synagogue of A.D. 246, with the remains of an earlier chapel beneath, was separated from the city wall by the soldiers' road, suggesting it had not been built until the Roman occupation.

The Christians, not yet asserting the intolerance that became prominent after the first council in A.D. 326, promised rather immediate salvation and immortality to everyone who adopted the new faith and recognized Christ as son of God, God's divine messenger on earth. Their Chapel was built in the second decade of the third century for the faith introduced, not necessarily by the Romans, but by the Western culture which flowed in under Roman suzerainty.

One thinks of Mithra as primarily an Oriental god. More exactly, he was a Near Eastern conception developed from the old Persian

religion by the Hellenistic Greeks and the local inhabitants in southern Asia Minor and Syria. The cult in Dura belonged to the Roman camp. It had two stages of architectural development, as the smaller relief placed immediately below the larger indicated, but its position in the camp pointed to its introduction after Dura had become a colony and established the large occupation force in the northwest corner.

The syncretism of beliefs, East and West, Semitic and Greek, was well represented in the temples of the Palmyrenes, of Artemis, of Atargatis, and of Aphlad and Azzanathkona. Religious tolerance of the East and the West permitted borrowing and adaptation that produced a splendid series of buildings at Dura. Hence, it is not a question whether the new cults and beliefs were Western or Eastern, but rather the definition of the characteristic features that were accepted and promulgated under Roman authority. In Palmyra we see very much the same trends and results in the extensive bas-reliefs and paintings as we do in the mosaics of Antioch, but both places lack the particular religious interpretations that we find at Dura.

We may speculate that the religious emphasis would vitiate the force of artistic tradition because it interprets special interests. The contrary is the case. We find in the Synagogue, in the Christian Chapel, and in the Mithraeum the religious concepts strongly influenced by the local Hellenistic-Oriental trends.

The religious art of Dura was by no means confined to the statement of abstract concepts. In the Synagogue, for example, the artists portrayed persons and events drawn from the widest possible context of the history, story, and myths of the Old Testament. In Byzantium and in the centuries of Byzantine authority, we must not forget, art was largely confined to the exposition of religious concepts and the portrayal of royalty.

At Dura the art of the Jews and Christians was subordinated to the special treatment of subjects. It was not art for art's sake but art to portray vivid scenes and actions. The message, the Bible story, was the important element, and everything else was subjected to it. Of contact with Roman illusionism and Greek grace and beauty, only the remnants remained. There was an entirely new spirit. Conventional lines superseded plastic shapes. The important figure was magnified in size and placed in the center of the scene. Careful balance and static arrangements of figures were reinforced with formalized roles and staring eyes. These were the Oriental traits that helped form Byzantine art, developed in Dura a hundred years before such art became prevalent.

Thus, the careful work of the ninth campaign at Dura gave further definition to the picture of the Roman city before it fell to the invaders from the East. Meanwhile, Toll continued his work on the Necropolis this year and on into the tenth campaign (1936–37), particularly concentrating on the tombs protected under the Roman debris opposite the Great Gate. The west wall of Dura runs for almost half a mile in a straight line from the edge of the southern wadi, angling slightly the last fifty meters in order conveniently to reach the edge of the northern wadi. Beyond the wall lay the Syrian desert of the steppe, which attains a slightly higher level at the city. The two wadis drain the surface water down to the river from the plateau, which is tipped slightly to the west. The desert floor beyond the city, therefore, carries little vegetation in the best of seasons and none at all in the poor ones. An occasional depression will show grass and the tiny little flowers of the desert, but the soil is extremely thin; the stony surface spread over a rock mantel gives little encouragement to vegetation.

Because the land outside the western wall was removed from the river waters by the width of the city, there was no reason to cultivate or to build there and, therefore, as far as one may see, no established private property. But it was an ideal location for the cemetery of the city. The gypsum rock of the cliff and the steppe provided a hard crust with a much softer stratum beneath, which could be quarried fairly conveniently for walls and building foundations. Easily excavated for catacombs, the rock hardens after exposure, reacting to the atmosphere. Quarrying for building materials and excavating for tombs may be conveniently combined: the blocks beneath the top level were cut out and brought into the city for construction work and the resulting caves became the catacombs. It was much simpler to cut a loculus, a space sufficiently large for a coffin in the wall of an underground tomb, than to try to cut a single grave above in the gypsum surface, long hardened by the atmosphere.

The long, low chambers that constituted the tombs might, however, collapse. It was difficult for the ancient contractor to determine the thickness and durability of the upper crust of his catacomb. In many cases, therefore, he left a pillar of the virgin rock in the center of the tomb to support the span of the natural vault.

A variant to these underground chamber tombs was the occasional tower, more or less square, erected on a stepped pedestal in which loculi were left for the introduction of sarcophagi. In Palmyra the inhabitants built surprisingly high towers with floor after floor of rooms

containing the niches for the sarcophagi of the dead, piled vault on vault. Each loculus was sealed with a stone plaque, usually adorned with the bust of the deceased, the name and date, and sometimes the titles. At Dura the vaults were on the outside of the base, and there were no reliefs. The tower on top of this mass contained only a staircase winding around a central pillar and mounting to the roof.

The area along the whole stretch of the wall at Dura, and extending at its widest point almost half a mile into the desert, is marked with the small hillocks and slight depressions that indicate tombs and entryways. The cemetery continues beyond the southern wadi, and the edges of the wadi also have been cut here and there for tombs. Cumont found above the chamber tombs a construction enclosing roughly the extent of the tomb, one wall established directly above the entrance.[15] This enclosure would serve as a meetingplace for the relatives of the dead, for appropriate ceremonies, and to prevent placing undue pressure on, or digging into, the ceiling of the tomb, as well as to assure the sealing of the tomb after the burial. One may assume that the roofs of the towers, apparently flat as at Palmyra, served also for reunions of families, some last rites, and sacrifices for the dead.

Looters had emptied almost all the tombs excavated by the soldiers under Cumont in 1922–23. The single undisturbed tomb contained a skeleton and a single vase of common ware. The largest of three chamber tombs cleared once held twenty-five burials.[16] All Cumont recovered were a brick with glazed surface, one green-and-white glass vessel, and parts of some larger vessels of coarse, heavy clay. Robbers had removed everything else of any value.

Most disappointing, from the excavator's point of view, was the lack of stone reliefs and inscriptions such as might be expected in a mausoleum, and which were otherwise so common at Dura. An occasional little piece of flat bronze was recovered, suggesting that perhaps thin bronze plates with the names of the dead were attached to wooden coffins.

In continuing Cumont's work at the Necropolis, we obviously had to do something with the tombs. The situation was challenging, but discouraging. There were the tombs—hundreds, possibly thousands—but the chance of significant finds was slim. Underground work was hot, the atmosphere close and stuffy, the air permeated with the dust of decayed wood and bones. We had to work very slowly, not only to maintain scientific control but also so as not to stir up the dust that prevented close observation and clogged the lungs. Only one man at a

time could ascend or descend the narrow entrance. The excavated dirt could not be piled on top of the next tomb but had to be carefully removed from the cemetery. In short, each tomb could be excavated by one man, one tomb at a time, with two or three workmen to assist. The cemetery digging proved a long, slow, painstaking task, requiring constant personal supervision.

In the third campaign, Pillet discovered the foundations of the triumphal arch of Trajan and distinguished it from the foundations of the tower tombs Cumont had noticed and one of which he had excavated. In the fourth campaign the foundations of the arch and the inscription were uncovered; then work beyond the walls was discontinued. In the middle of the seventh campaign, just before January, Toll had arrived and begun once again to investigate the Necropolis.

He had begun cautiously with two or three of the smaller chambers on the outskirts of the area. Some of the coffins were in part preserved, and he noted that the deceased were installed in the loculi feet first, with the head close to the opening. Not all had been carefully sealed up, and in some the dead did not seem to be satisfactorily installed. Some, perhaps, had been disturbed by the grave robbers, but it gradually became clear that the families and friends of the dead did not go beyond the entrance to the tomb. The final placement was probably carefully done, but we suspected the ancient undertakers removed whatever they considered of adequate value before the dead reached their final resting place.

All the tombs excavated in that seventh season had been looted, but some silver jewelry and trinkets had been overlooked. Some broken pottery was very valuable, not in itself but to help establish the chronology. In some cases it seemed that the families, anticipating the looters, attempted to forestall them by breaking the expensive glazed pottery, eliminating the temptation to rob the dead. The dump heap we excavated in the eighth campaign, hoping it would provide the layering of strata we lacked inside the city, was a disappointment. There were coins in it, surprisingly enough going all the way from Seleucus I to Gordian III, but the levels were confused. It was clear that dirt thrown out with the ashes came from cuts for foundation walls or other digging in the city, and they did not represent the dates of their deposition in the dump.

But if our rubbish mound proved useless for stratification of the city before the Romans, underneath we came upon those tombs that had been constructed before the Roman fill and were close to the Gate.

Toll shifted his explorations to this new area, and we all waited anxiously to hear whether they might be tombs of an earlier period, whether they had escaped the depredations of the ancient tomb robbers, and whether they were rich in tomb furniture. These tombs, lying closest to the Great Gate, belonged in part to the earliest period of the city. But Toll found that even here those who had last handled the caskets had usually exacted their due of jewelry and trinkets. None of the graves was particularly rich, but two were intact, unmolested until Toll broke into the moldering silence. The pottery, figurines, jewelry, and lamps furnished a sequence of forms, just the stages of development we had hoped to find in the dump heap. There were burials without any of the green glazed ware that came in with the Parthians, and other tombs had earlier forms of glazed pottery, different from any we had hitherto found at Dura.

A difficulty facing Toll was the depth of the dump heap and the extent of the tombs. Numerous soundings through the heap to locate the tombs would have been slow and laborious indeed, but wholesale removal of the dump debris to get down to the level of the plateau would have required enormous amounts of earth moving. The problem of how to proceed was happily solved by the proximity of graves. The area close to the Gate and beside the road was, of course, the most popular for burials. Crowding of tombs in this coveted district occasionally resulted in the loculus of one chamber touching, or even breaking into, the neighboring tomb.

Taking advantage of this congestion of graves, we discovered a dramatic and economical means of locating neighboring tombs by sounding the side walls of the tomb being excavated. Toll, often working ten feet underground with only a very pale light from the entrance, as much as had already seeped down through the ten to twelve feet of dump debris above the entrance, worked his way with lantern or flashlight through the caverns. Patrolling in the close atmosphere—absolutely still with the silence of the dead—one tapped with a little mallet on the solid walls and listened for the resonant tone that would mean a hollow space not far removed. A dull, subdued tone indicated solid rock, the gypsum subsoil of the plateau. And so one inched along in the dark, tapping, hearing the monotonously dull reply. Then suddenly, after a seemingly endless, careful search, one would be rewarded with a sound that, after the discouraging thump-thump of solid rock, seemed to ring with a bell-like clarity.

When Toll found an area that promised a hollow spot nearby, he

very carefully tested where the neighboring tomb came closest. Then a cut was made a few or several inches in the rock wall until the workman's pick broke through into the neighboring chamber. I remember in particular one tomb (found while I was still director) where our exploratory hole broke into the main chamber. The hole was much too small for the passage of a person, but it was large enough for an arm and a flashlight, and the intervening wall was sufficiently thin to permit the light to be flashed to several points. Our eager eyes followed the beam, surveying from our own dark catacomb the tomb that had remained silent and undisturbed for two thousand years, not even touched by the spade of the archaeologist as yet. A single vase, untouched since the ancient mourner had placed it on the ground before he left, stood before the central pillar. The beam of the flashlight, roaming over the walls, probed the mouths of the burial loculi with their dusty remains, a thin shaft of light cutting through the eternal darkness.

So Toll moved underground from one tomb to another without necessarily disturbing the dump heap. Eventually he was able to draw a complete plan of this section of the cemetery with a minimum of surface disturbance. The rewards, beyond those obtained from the study of the tombs themselves, were the sequence of vases, a splendid late Hellenistic figurine of Mercury, god of the dead, some silver earrings, and a jumble of small finds comparable to those uncovered in the private houses. A rather attractive little bronze was of a gazelle, modeled with a suggestion of the South Russian nomadic animal style that grasps the special characteristic grace of the animal. The gazelle, small enough for an earring, stood on a little pedestal to which tiny bells were attached. Suspended from a lady's ear the bells would have rung with an alluring tinkle when she shook her head; in the tomb they served an apotropaic purpose, keeping away the evil spirits.

Our underground tombs at Dura were paralleled by ones of the same type, but much richer, at Palmyra. There the magnificent tower tombs contained great chambers ringed by the loculi, sealed with stone busts, the tombs ornamented with pillars and decorative architectural details. Obviously they were places of pride to the relatives and friends of the deceased. Instead of the elaborate and expensive tombstone of the modern cemetery, the Palmyrene had the tomb chamber and the long sequence of busts of ancestors, family, and friends.

The tower tombs at Dura, with their interior staircases, the loculi in the stone foundation, and the ornamental facade, were closer in form to monuments up and down the river, at Baghouz on the heights oppo-

site Abou Kemal, thirty miles down the river, and at another necropolis fifty miles or so above Deir-ez-Zor. One tower at Dura, in collapsing, had buried its facade in the ruins, and this could in part be recovered to give us a fair picture of the original appearance of the building.

One interesting question concerned the use and purpose of the flat roof. The high Palmyrene towers had flat roofs and stairs leading up. The enclosures on top of the underground tombs at Dura suggested both a reservation of the area and a meeting place for those associated with the funeral, a place for the last rites and sacrifice. I have conjectured the roof was simply used as a meeting place and stage for ceremonies, but did it also relate to the later towers of silence so prominent in the Persian religion of the Parsees? There was no evidence in either Palmyra or Dura for the exposure of the bodies to be devoured by birds of prey before the final disposal of the bones. The tower itself served as a monumental memorial for the dead and such was an appropriate burial structure. Still, the towers belonged to the Parthian period, and the towers of silence followed in the Persian development.

An appealing prayer addressed to the great mother goddess of Asia Minor, a prayer for the dead, reads: "O thou who brings us into being, and creates for us the many forms of plants and animals and who receiveth us all again in the end." Some such thought the Parsees had in dedicating the flesh of the dead to the birds of the air, their bones to the repositories under the tower roofs. On the towers at Dura, the roof lifted the participants of the funerals above the life of the city. There they must have prayed, in the Persian style, to earth and air and sky, to the sun as giver of light, as conqueror of evil and darkness. Then the body of the deceased was consigned to its niche in the tower's foundations.

The tower tombs across from Abou Kemal at Baghouz were better preserved than those at Dura. Four of the five tombs were known by the local names of Abu Zimbel, Abu Gelal, Irzi, and Shaq el Haman.[17] It seemed worthwhile to make a brief sortie to investigate, to see if they belonged to the same period, to secure a record for comparative purposes, and, if possible, to explore one or two underground tombs in the hope of establishing their dates and relationship to those of Dura. Toward the close of the seventh season, as mentioned previously, Toll, Du Mesnil, and Pearson had made the brief safari to Abou Kemal, crossed the river, photographed and measured three of the towers, and were able to dig an underground chamber, where they found the remains of the composite bow.

Many questions still lie unanswered among the bones of the Dura

Excavating the Necropolis under the heap of Roman refuse outside the city walls.

Necropolis in spite of the splendid series of excavations. Did the Christians have a catacomb of their own? And what rites and ceremonies did they perform in those early days before Constantine? Did they repeat in the sepulchral setting, in painting or in sculpture, some of the hopes expressed in the Christian Chapel? The Jews of the Synagogue pictured the resurrection of the bones of the dead with the prophecy of the New Jerusalem on earth. How were their hopes and prayers expressed in the graves? There was the great mass of Durene citizens who subscribed to local and general Oriental religions, and to the religions of Greeks and Romans amalgamated with local cults. Did all of these religions share this one large necropolis? Or was it divided into sections, the cults of which we cannot identify? And in what manner was the cult practice of the Parthians different from that of the local inhabitants? Where does the religion of Mithra, the great cult of the sun and immortality, come in the funerary practices of Dura? Cremation apparently belonged only to the Romans, and to very few of them. Some graves were found within the city itself, and many more built in the wadi walls must have been lost to rain and flooding. Still more lay untouched in the Necropolis after several seasons of work. But, as Toll remarked, "The complete clearing of almost a thousand tombs would have required many years of excavating activities."

The tenth and last campaign at Dura, led by Frank Brown, was largely concerned with cleaning up the previously excavated areas. The Temple of Zeus Megistos in its earliest phase, west of the Redoubt, was cleared, investigations were made under the Temple of Atargatis, and the excavation of the Agora was completed. While Toll completed his work on the Necropolis, final clearing of the Redoubt Palace was accomplished. Detweiler and Brown were to have collaborated on publishing their study of the architecture of the walls, but Detweiler died. Unfortunately, the *Preliminary Report* for the last season has not been published.

12 The Fall of Dura

There stand the stone fortification walls of Dura, stretching across the desert challengingly from one wadi end to the other, running along the edges of the ravines and across the cliff overlooking the river as far as the crumbling ramparts support them. We must turn to these walls not only to learn about the building of Dura, but also to recreate the sudden death of the city in A.D. 256.

The history of the fortifications is most difficult to reconstruct for we have no literary evidence that records the building contracts let, the teams of laborers employed, or commemoration of the date of completion. It is solely an archaeological puzzle, to be fit together on the basis of the physical evidence alone.

As director in the fifth campaign, I had carried on where my predecessor, Pillet, had left off. In the sixth campaign, von Gerkan had been invited by Rostovtzeff to make a careful examination of the walls. Von Gerkan was an authority in Greek wall construction and would be able to provide the added insight of the expert to that of the general field archaeologist. Breasted, of course, had had no time in a single day to examine the fortifications, but Cumont, in spite of the fact that his two campaigns comprised in all less than a month's time, had not only traced the grid pattern of blocks in the city, but also had made a careful though hurried study of the entire circuit of walls. Our primary interest in the walls had been in what their towers contained, the information that could be gleaned from the contained debris, and in the buildings concealed and preserved in the embankment against the protective curtain of stone.

Two basic facts needed to be determined: the date of the construction of the walls and whether the walls represented one or more building phases. Cumont noted that the paintings of the Temple of the Palmyrene Gods dated from the first century, and, therefore, the walls on which the paintings were fixed—an intimate part of the fortifications—should antedate that period. The saw-toothed pattern of walls along the south wadi was of Hellenistic tradition. Thus, Cumont suggested that Seleucus Nicanor, founder of the city, had constructed the fortifications at the very beginning of the city's history, about 300 B.C.[1]

He also pointed to a graffito found in the Temple of the Palmyrene Gods, representing high walls, an arched gate, and the crenelations of the ramparts on top, as showing the fortifications of the city complete.[2] I think we may follow Cumont's reasoning so far as to say that the city grid plan of blocks was laid out at the very foundation of the city, and that this grid pattern suggests a line of fortification walls along the edge of the river cliff, the wadis to the south and north, and a stretch straight across the desert from the south to the north wadi. Although there is question as to the date of the paintings in the Temple of the Palmyrene Gods, inscriptions beginning in the middle of the first century leave no doubt that the temple existed by that time. Since the temple occupied the corner of the walls, the probability is strong that the high fortification walls also had been constructed by then at least.

Since Cumont's time at Dura we had uncovered the Temple of Zeus Kyrios, built against the walls south of the Great Gate. The relief work of the temple was set into the city wall, which confirmed that temple and wall were constructed at that time (second decade of the first century A.D.). In the Tower of the Archers, a tower built next to the Temple of the Palmyrene Gods, Cumont found a parchment of 195 B.C. buried in the sand, indicating the tower must have been in existence when the parchment was lost in it.[3]

But it was the tower of the Temple of the Palmyrene Gods that told us that a single date for the building of the walls was not to be considered. The front of the lower part of this long tower was constructed of big, carefully cut blocks fitted so closely together that no mortar was employed. The rear of the tower, on the other hand, was built of blocks only roughly finished and loosely set, joined together with thick plaster. The striking feature was that just above door level, at a height of six feet, the carefully joined work of the lower stone courses on the front side ended, and the upper courses bonded with the rear blocks of the tower.

It was no surprise, therefore, to find that the lower courses of well-cut blocks in the lower part of the tower did not bond with the rough-cut blocks of the lower rear part. Digging beneath the floor, we uncovered the foundations of a cross wall of carefully cut blocks and saw the jagged cut through the original back wall of the tower. Clearly, then, the tower had first been made of the well-cut blocks; one wall was then torn down to provide for a larger tower, now partially restored in the rougher work. If the inhabitants of Dura had wished to spell out, in architectural terms, the fact that the circuit walls had undergone two

distinct building periods, they could not have done so more clearly than in this tower.

For the earliest possible dating of the towers there was mixed archaeological evidence. No clear signs of a very early date, in the way of coins or pottery, were found in either the Tower of the Temple of the Palmyrene Gods or in the Great Gate. The earliest date reported by Rostovtzeff from the inscriptions of the Great Gate was 66/65 B.C.[4] Johnson listed 101 inscriptions from the Great Gate, of which relatively few were dated. The best we could say was that the series of inscriptions went from close to the end of the first quarter of the first century A.D. back into the B.C. period with less continuity.

One of the most interesting results of von Gerkan's study was the discovery that the Great Gate did not belong to the original plan.[5] Von Gerkan placed the Citadel wall in the early part of the Parthian occupation (120–65 B.C.) and the rebuilding of the towers in stone and the stone curtain walls over the entire circuit (except for some parts between Towers 22 and 24, and Towers 24 and 1) in the period 65–19 B.C. This conclusion fits very well into the history of the Near East in the last century of the pre-Christian era.

Thus, the history of the walls may be briefly summarized. The founders, the Macedonian Greeks, first erected for defense a wall of mud brick, but on the desert side, the most vulnerable side of the young settlement, they strengthened the wall with a masonry socle. This was the defense system the Parthians found at Dura when they took over the city in the last quarter of the second century B.C., perhaps in 113 B.C., but at least probably not before the capital city of Seleucia became finally and firmly held by the Parthians in 129/128 B.C.[6] The first Parthian occupation was military, and the interior Citadel was built by them as a fortified acropolis. When Pompey took over Syria and Palestine for Rome, the Romans threatened Greeks and Parthians alike in Dura, and the Parthians responded by undertaking a massive strengthening of the walls. They completely rebuilt the Great Gate and enlarged the Tower of the Palmyrene Gods. The mud-brick curtain walls along the ravines and facing the river were turned to stone, and high towers were erected along the desert, while stone masonry replaced the mud brick above the socle.

But the Roman threat seemed to evaporate in 20/19 B.C., when peace was made between Romans and Parthians, and the Roman standards captured by the Parthians in their defeat of Crassus were returned. The rebuilding of the desert wall was not quite completed, and

now the need for this monumental stone protective shield had disappeared. So the work, which had proceeded from south to north and included the Great Gate, was given up, and the mud brick above the northern end of the wall (the curtain between Towers 22 and 24 and between 24 and 1) was left standing.

To this composite analysis, I would suggest one or two possible changes. I am bothered by the suggestion that the Parthians would undertake the construction of a monumental entrance in the desert wall as a reaction to a grave Roman threat. This and certain other construction features are left unresolved by von Gerkan's brilliant study. A different solution to the early history of the walls may be suggested. I would propose the alternate: that when the Parthians reached Dura, they found walls and towers of mud brick except in the northern part of the desert curtain, where stone towers were replacing the mud brick and a stone socle was being constructed between the towers. In the early period of the Parthian occupation, the Greek citizens of the city had sympathizers in Hellenistic Syria, and the Parthians built a fortified Citadel Palace on the highest land, the acropolis, as a chief part of their military occupation.

After the Roman conquest of Syria by Pompey, the Parthians were allied with the Greeks against the "barbarian" Romans, and they then replaced the mud brick in the south half of the desert curtain, and parts incomplete in the northern half, with stone masonry, building the southern half according to their own specifications, but retaining the Hellenistic character in the northern half.

When the Roman peace was concluded in 20/19 B.C., Dura became not only the Parthian frontier city against the Romans, but the district center, and seat of the Parthian provincial governor, the *arcapat*. Then the west, desert curtain wall was altered to allow for a monumental gate. The Pax Romana, continuing unbroken in the first century of our era, obviated the need for excessive defences, and the mud brick above the stone socle was allowed to stand in places, between the towers at the northern end mentioned above.[7]

If one accepts the Parthian origin of the Great Gate, the sequence of graffiti found there falls logically into place, from the earliest dates to A.D. 32/33. I question the dates ascribed to the so-called earliest graffiti from the Great Gate, those that go back into the second century B.C. The continuity of the series of graffiti and their frequent use in the first two and a half centuries of our era casts chief doubt on the dating of the "early" ones. There are but three inscriptions ascribed to the two centuries before our era. Especially are the dates suspicious because

those three inscriptions are separated one from the other and then from the third by one hundred and fifty and more years.

The building of the Great Gate just before and in the first years of our era would coincide with the great period of prosperity at Dura, evidenced by the great number of new temples: those of Artemis-Nanaia, the Palmyrene Gods, Azzanathkona, and Zeus Kyrios. This century of prosperity that began with Roman peace also witnessed the growth of Palmyra and Hatra, a reflection of the establishment of the great trade routes up and down the rivers of Mesopotamia and along east-west lines stretching from the coast to India and China beyond. These were the physical rewards of a century of peace between the two great empires of the Classical world.

If we must be somewhat uncertain about the growth of the walls during the Hellenistic and Parthian periods at Dura, we are even more unsure about the exact time when the walls were heightened. Von Gerkan postulated the Roman period. Later investigation dated the additions definitely to the years following A.D. 216. The mud-brick embankment (the *agger*) that sloped against the walls, the great blanket of debris that covered and preserved the marvelous murals, belonged to the desperate efforts of the inhabitants of Dura in the face of the Sasanian attack.

Pillet had noticed the ramp against the walls on the outside in his early excavations, and the clearing of the tower in the fourth campaign had shown how far the ramp had risen above the top of the stone wall. Undoubtedly the mud brick encasing the crenellations along the top of the wall and rising above the door lintel into the tower showed the desperate attempts of the defenders to heighten the walls hurriedly against the rising ramp, with its threat of charging phalanxes of Persians and their rams. But the Sasanians had another major weapon against the walls, the sap.

For the discovery of the Sasanian mines and the development of the story of the siege operations undertaken against the city, I give Du Mesnil full credit. The French journal *L'Illustration* published an article of his in 1933 on "Underground War, 1700 Years Ago." I had already published in the *Illustrated London News* short accounts of the discovery of the Christian Chapel and of the Synagogue at Dura. The *News* also wanted me to write an account of the siege operations. My short article in the *Illustrated London News* of September 1934 was based largely on the work of Du Mesnil, which raises an interesting question: Who has the responsibility for making excavation results available? And who has the right to make them available? Usually the field director of an

The effects of the Sasanian mine on the city wall.

expedition gives the first report, and the sponsoring university or museum follows as promptly as practicable with a preliminary account, which allows other scholars to use the new evidence even before the final publication is made. The final publication, then, often has the advantage of suggestions made by scholars around the world on the preliminary report.

In the French-American Dura excavations, it seemed very appropriate for the French representative, Du Mesnil, to give the results of his own investigations at Dura to the French magazine *L'Illustration*, and for me to report on the Dura finds to the English-language *Illustrated London News*. However, Du Mesnil's publication in 1939 of a volume on the paintings of the Synagogue under the auspices of the Biblical Pontifical Institute in Rome was felt to be rather an intrusion on the primary right of Yale University to make the first full report. The preliminary report on the Synagogue had been published in 1936, but Du Mesnil used much material that had not been included in the preliminary report and was, therefore, presumably still the property of

Du Mesnil inspects the displacement of the Southwest Tower that resulted from the Sasanian mining of the foundations.

Yale. Or was it? The excavations were financed by Yale, but the French Academy shared in organization and staff.

I was very annoyed when von Gerkan, who was at Dura at Yale's request to examine the fortifications, published in a German periodical a plan and analysis of the Christian Chapel.[8] He thought our plan and conclusion on the date of the building were incorrect and said so. I thought some of his suggestions were good, but that others were erroneous. It appeared to me that his publication should have been delayed until our preliminary report appeared in 1934. The differing points of view should have been discussed on the site, but he had made

no mention while at Dura of being interested in the Christian building or of making a study of it. I felt his publication without asking permission, or even taking up the different interpretation with us before publishing his own views, was a breach of confidence, a confidence not binding but taken for granted. Von Gerkan never acknowledged exceeding the bounds of propriety. Certainly Du Mesnil's publication on the Synagogue was different from that of von Gerkan's action, while Du Mesnil's first account of the "Underground War" was most appropriate.

As we now know, in the middle of the third century A.D., when Dura was in Roman hands, the Persian Sasanians arrived before the walls of the city. Anticipating a siege, which would no doubt entail an attempt to breach the desert wall, the military commander took countermeasures. A broad glacis was piled up against the inside of the curtain wall, filling the buildings along Wall Street with dirt and gravel. A much sharper slope of debris was thrown up against the outside face of the wall. These tactics were to counteract direct attacks against the face of the wall as well as the possibility of collapsing the wall by undermining. The foe would have to drive an excessively long shaft underground to reach the footings of the wall and on through into the city itself. Or if the enemy sappers did undermine the wall, it might collapse in place but would not tumble, for the double slopes would keep the masonry structure upright. This is precisely what the Sasanians did, with just these results. The story was pieced together slowly as we worked on the towers and the embankment.

I had excavated the southwest tower during the fifth campaign and remarked with some astonishment the cave beneath its walls and the charred beams, not to speak of the corner of the tower which had sunk, though yet remaining upright, leaving the upper wall in a perilous position. Pillet, it is to be remembered, had interpreted this dislocation as the result of the earthquake in A.D. 160. I had accepted this reasonable verdict, being much more concerned with clearing the tower before any further collapse occurred, than I was with scientifically examining underground passages.

The following season Du Mesnil, after working on the Synagogue, continued work along the wall and excavated Tower 19, which also showed evidence of having been shaken by the earthquake. At the base of the tower, he found planking for an underground tunnel still in place and, in the embankment, an extraordinary collection of human remains and martial equipment. The ribs of one skeleton were covered with a

Two views of a soldier who died in the Persian mine under the city wall. The remains of his mail shirt cover his chest and raised right arm.

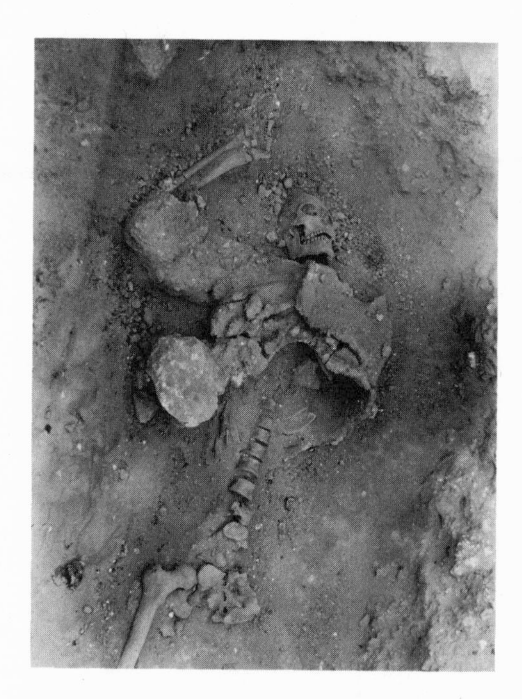

chain-armor corselet; the swords the soldiers had carried lay beside the bones. There is question as to which side—Sasanian or Roman—the unfortunate party belonged. Under their armor, three of the soldiers still retained their handfuls of Roman coins belonging to the very last period of the city's existence. Yet because of the type of armor and swords they carried, the possibility does exist that they were Persians carrying (looted?) Roman money. Perhaps we had here a group of defending soldiers engaged in digging and shoring up a countermine in the embankment on the inside of the curtain wall against the Sasanians cutting underneath the wall from the outside to undermine the tower. A sudden shift of earth, and the Romans were caught in the collapse of their tunnel. We well knew from our own experience how the earth could move without a sound: in our fourth campaign two workmen were lost tragically in the sudden collapse of a narrow trench we were running through the embankment; there had been no warning and no escape.

At Tower 19 a clear demonstration was made once and for all that no earthquake was responsible for the partial collapse of the towers. Du Mesnil traced the cavern beneath the southwest tower, through a series of tombs, out to the entranceway of a ramp cleverly hidden in

the south wadi just beneath the hard crust of surface rock. While the southwest tower was being undermined by the Sasanian sappers, the ramp was erected just beside the tower next to the corner; its height would preclude effective defense with bow and arrow from the tower to the southwest angle. When the timbers supporting the roof of the mine under the tower were fired, the mine would collapse, carrying with it the tower above. Simultaneously the attackers would rush over the ramp and the fallen tower, throwing the whole southwestern corner of the city into enemy hands.

It must have been planned that the mine under Tower 19 would not only displace its defenders and allow the attackers to climb up over the steep outside embankment, but also that an extension of the tunnel under the tower, pushing through the interior embankment, would serve to introduce the Sasanians into the city to take the defenders of the tower from the rear. The coins found with the skeletons in the collapsed mine gave conclusive proof that the attack had taken place in the decade of A.D. 250–260.

At the beginning of our seventh season it seemed wise as we continued our operations along the embankment inside the wall to explore a little farther along the outside also. Von Gerkan's work both inside and outside the walls had shown there was little hope of recovering anything of value from the fairly steep glacis heaped up against the outside face of the circuit wall. The embankment inside the city had not only preserved the buildings it buried, but it also contained all sorts of materials that were refuse for the citizens of Dura, but most valuable for the archaeologist. The embankment outside was free of debris, except for an occasional stone.

Our discovery and interpretation of the mines had now given us the reason for the construction of the embankments inside and outside the walls by the defenders of the city: attempts to undermine and collapse the walls would be frustrated by the double embankments, which would continue to hold the walls upright even though their footings might be displaced. Clearly, the outer embankment was made steep to impede a rush of troops, while the interior glacis was broad and sloping to prevent easy penetration by an attacker's tunnel.

If the outside embankment promised little, the hasty collection of material used by the Sasanians to build the great ramp against the walls might hold richer returns. We began trenching through the ramp mound not too far from the wall, where the height of the ramp might have preserved materials from rain and damp, and where original de-

posits would have remained in place. Entryways to numerous tombs were still visible in the ground adjacent to the ramp; perhaps the buried tombs would still hold tombstones or inscriptions.

Our expectations along these lines were disappointed, but we did discover the hood trench of a tunnel reaching beneath the outer embankment toward the wall. Du Mesnil took charge at once with great enthusiasm and followed the course of the tunnel, almost but not entirely filled with earth. The tunnel ran beneath the surface crust, straight under the wall, and up into the interior glacis within the city. Only a lamp or two and a few bits of metal were our reward. We could not tell if the mine had been carried to completion, for there were no boards or charred beams to reveal how far it had extended.

But its purpose was clear. The siege troops and machines—battering ram and ballista—on top of the ramp would push the attack and draw the defenders to the obvious point of attack. The firing of the mine under the adjacent southwest tower would collapse the walls sufficiently to allow the attackers to breach them at that point. At the very least, the collapse of the tower would cause a moment of great confusion. At that moment, the mine beneath the ramp, broad enough for four men abreast and high enough for the crouching figures to move forward quickly, would be pushed through the last meter or so, and the emerging troops would take the defenders from the rear. The beleaguered Romans at the southwest tower would be too occupied to reinforce their comrades at the mine under the ramp. So the Sasanian plan of attack slowly became clear to us.

Now we knew that the two towers adjudged by Pillet to have collapsed through the earthquake had fallen instead at the end of the city's Roman period with the successful completion of the Sasanian mines. The only other collapsed or sunken tower was that of the northwest corner of the Great Gate. In digging this section in the second campaign, Pillet made a great point of claiming that the debris at the bottom must belong to the period before A.D. 160. Johnson had shown, however, that this thesis would not hold and that the tower had been occupied until the end of the city, or at the least had not been tumbled in A.D. 160.

Was the siege successful, the well-planned assault on the walls crowned with victory for the Sasanian Persians? Since the latest date we have for activity in the city, close to the year A.D. 256, was the same as that of the coins in the pockets of the defenders caught in the mine, and since the city seems to have been completely deserted suddenly,

we must assume the Persians stormed the walls, carried off men, women, and animals, and left the desert stronghold a desolate ruin.

How long did such an attack last? Signs of danger, and a very real premonition of it, must have reached the city well in advance of the actual assault troops, sufficient to allow time for the building of the great embankments. The Synagogue was dated, and so completed, in A.D. 246. Obviously there had been no thought of throwing up the embankment then, and hence no sense of impending danger.

The Persian documents speak of a prior taking of Dura in 253 by Shapur. Dura had been a valuable frontier city for three hundred and fifty years under the Parthians, and it seems unlikely that it would have been so utterly destroyed without cause. If, however, the Persian documents are correct and the city was taken and occupied by the Sasanians in 253, then it seems likely that if they had been subsequently ousted by the Romans with the help of the Palmyrenes, there might well be talk of reprisals beyond merely retaking the city. The threatened return of the Sasanians may have brought the frantic efforts to defend the city at all costs. It seems much more logical, also, that the city, knowing it had earned the wrath of the Persians, could anticipate the attack and have time to prepare its elaborate defense of the double embankment, a time-consuming task.

On the other side, the Sasanians must have come well prepared for siege operations, for mines cannot be dug with swords, or ramps fashioned with spears. No matter what the size of the Dura Roman garrison and its supporting citizens' militia, the storming of Dura's massive walls against a determined force would be most difficult. The Romans required two years to take Jerusalem with the devastating allies of exhaustion and starvation. At hilltop Masada a tiny force held out against the Romans who, preparing for the long siege, built their *castra* in the desert below and slowly erected the great earth ramp up to the walls that can still be climbed today. I doubt that the Sasanian siege lasted over the winter season, however, and it therefore could not have extended more than several months. The desert turns to mud with the winter rains, and provisioning a large attacking force by pack animal and wagon over the soggy remains of desert roads would have been enormously difficult. The countryside was probably cultivated, but the Durenes, anticipating the siege, would have harvested crops against famine as well as to deny the Persians the possibility of living off the land. It is true that some winters at Dura may be dry, but no commander would risk a sizable army on the lucky chance that the season would be clement.

And so the picture of Dura's last days became increasingly clear to us. The city had spared no effort to prepare itself for the vengeance of the Persians. Enormous effort was spent to raise the embankments inside and out; buildings public and private were sacrificed for the common weal. The Sasanian commander, meanwhile, picked A.D. 256 as a favorable year for the assault. He arrived before Dura with his siege equipment and plan for taking the city well in hand, having projected success in a single season of campaigning. The hoards of coins with dates coming up to A.D. 256 and the buried jewelry suggest ample time for action and recognition by at least some members of the community that capture was not only possible but close at hand. The Persian sappers went to work on their mines as soon as the city gates were sealed. We found the saps under the southwest tower and close to the Synagogue. I think another sap was driven under the Great Gate, where a corner of one of the towers had caved in. We did not find that particular mine, but that is because we did not look for it. That Persian mine still lies waiting discovery and exploring at Dura.

The building of the siege ramp and the tunneling must have occupied several weeks if not months. There must have been ample time for the defenders of Dura to have sent out messengers, slipping through the circle of Persian troops, to summon help from Palmyra and Antioch. Yet there is no evidence that aid was forthcoming from the outside. Very probably both Palmyra and Antioch feared for their own safety and hesitated to divide their forces. Dura, with its back to the Euphrates, must have stood alone, beyond help, to face its final agony.

Dura was the Roman frontier and the first obstacle the Sasanian Persians would meet on a march up the Euphrates. There was, however, the route up the Tigris to Sinegra, Edessa, and Nisibis, and then the return down the Euphrates. If these cities had been taken or surrendered, the chances of assistance coming to Dura from these quarters would have been remote indeed. Dura is placed by the Sasanians at the end of the list of cities taken in A.D. 253. Perhaps a token force of Persians was left to hold the city as the main armies moved on, but as soon as the main body of troops of Shapur were out of sight the Durenes may have ejected the prize troops, breaking their oath of fealty. They knew that eventually a retributive enemy would return, and hence the precaution of the embankments.

The amount of earth, brick, and debris thrown up to form the agger was staggering, as we learned in our slow excavation. Wall Street was filled to the top of the house walls, thirty feet. It takes a stout wall to withstand pressure from earth shoveled thirty feet high. In the

Christian building the west walls were strengthened by mud bricks in front. Apparently in the Chapel the aedicule of the baptistery was felt to be strong enough to withstand the pressure alone. The Synagogue was filled with dirt to the rafters of the back wall, but no mud brick was installed. The roof remained, as evidenced by the roof tiles we found in the upper part of the debris.

Time is required to make mud brick; a great deal of water is needed; and more time is consumed in the slow baking of the brick in the sun. Certainly most of the merchants had stores at their disposal, but still, preparations are seldom made in time for, or adequate against, an attack. There was haste in the measures taken at Dura, although the provident, for example, had added the mud brick to the walls of their buildings facing the great wall so that their structures could be easily restored once the attack was successfully repulsed. If the walls held, the city would be safe; if the fortifications were breached, any and all protection of walls, houses, shrines, and temples was in vain. When the news came to Dura that the Sasanians indeed were on the march to the city, the harvest must have been brought in and the military prepared for its desperate struggle.

The sappers completed their work and the ramp was built. The Romans had worked their countermine. But eventually the Persian preparations were complete. We can imagine a dark, fall night in October or early November after several months of siege in A.D. 256. The inhabitants of Dura call on the gods for heavy rains, which would force the retreat of the Persian army. The citizens well know they can expect no quarter, particularly after their behavior in A.D. 253. Along the desert wall at several points, the enemy sappers have been hard at work on twenty-four-hour shifts. (In tunneling there is no day or night.) The ramp has been steadily growing and now has almost topped the crenellated great wall despite the emergency measures of mud-brick walling on top of the masonry. Which Persian task force will strike the first blow, and from what quarter? How will the signal be given? Perhaps a false alarm is deliberately triggered, a feint made against the northeast bastions; a screen of men may be sent up against the River Gate to divert attention.

Then the major effort begins: the mines are fired; the underground supporting timbers give way; the towers slip down, although the embankments hold them erect. The crash of masonry and the shouts of the Persian infantry turn all to confusion compounded by terror. Covered by vollies of arrows, a fresh force races up the ramp with

scaling ladders for the last few feet of wall, and while the defenders fight to hold the wall, the wide embrasure of the sap under the ramp is broken through into the city, spilling Persians into the streets to take the defenders from the rear. Once a foothold is gained inside, the Main Gate can be rushed, forced, and the Persians pour through.

The numbers of Roman troops in the city must have been small compared to the size of the attacking army, and the civilian militia would have been no match for regulars. The city must have fallen with the taking of the walls, for beyond the evidence found in the tunnels, we discovered no signs of fighting in the town proper and not enough evidence of fire to suggest the city had been put to the torch. We must suppose that the inhabitants were marched off in a body, perhaps to some distant frontier to be held for ransom, or sold as slaves in the market of Ctesiphon.

The taking of Dura is little more than a moment in the long history of frontier wars. It was scarcely mentioned in Persian records. We have a notice that Dura was taken in A.D. 253. Before that, in 238, one of our graffiti noted, the Persians had descended, and Antioch fell in the same year. Our sources for eastern history are fragmentary, and the history of the middle of the third century A.D. is confused and vague. After the siege and victory by the Persians in A.D. 256, the record is blank. The mute testimony that remained was of a site desolate and forlorn, where the lonely and level sands covered the bones of the city and stretched away across the desert.

View from the Citadel looking down toward the expedition house and the Euphrates beyond.

13 The History of Dura

Alexander crossed into Syria fresh from his successful conquest of Asia Minor in 333 B.C. and before he died ten years later he had conquered Egypt and the Near East, taking his Macedonian troops as far as the Oxus and the Indus rivers. His successors, the Diadochi, the generals of his staff, took over the vast holdings of the Macedonians. Ptolemy established himself and founded a dynasty in Egypt. Seleucus became master of Mesopotamia, erecting his new eastern capital, Seleucia-on-the-Tigris, almost due east from Babylon. Lysimachus held Thrace, while Antigonus and his followers were destined to take over Greece and Macedonia. The lands from Greece to India contained the Macedonian kingdoms, separate, independent, not very friendly, and subject to partition under the pressures of ambitious subordinates and political realignments.

Seleucia was no sooner established as the eastern capital of the Seleucid kingdom (Antioch became the western), than Dura (the "Fortress") was founded by Nicanor, probably the general of Seleucus I, and settled by soldiers who were given land as cleruchs. We do not know whether the new center was called Europos by the Greeks in remembrance of the birthplace of Seleucus or as a name for the district, the "Broadlands," or "Broad-faced." The cleruchs were given arable land along the Euphrates and as far away as the Khabur, fifty miles to the north, as we learn from the papyri. But the city itself was located on the flat ground of the rocky plateau, a hundred feet above the river plain, barren and waterless except when winter rains fell in the desert. We found no spring of water and no well dug down to water level. The Euphrates beneath the cliff furnished an unlimited supply of water in all seasons, but the carry from river to fortress was long and difficult.

Dura, as the name suggests, was a fortress, and as the only one along this stretch of the river, it provided shelter for merchant caravans. As the fortress grew, it could control more and more of the district in its immediate vicinity—along the riverbank and over the western desert—and send its patrols over the trade routes that extended north and south following the Euphrates, and east and west across the desert tracts.

We would do well to ask why Dura was not founded fifty miles

upstream at the confluence of the Euphrates and the Khabur, where richer land was abundant and where the future Circesium would be established because the route from the east, along the Khabur, took the caravans to the Euphrates. The answer lies in the fact that the first objective of a military stronghold such as Dura was defensive. The cleruchs were usually established along the frontiers where the veteran soldiers, and then their sons, might hold the land against encroachments from beyond the border. The nomad tribes along the edge of the desert presented the same menace to the Hellenistic settlers as they had in earlier days to the Twelve Tribes of Israel east of the Jordan. The great caravan route between the western capital of Antioch and the eastern one of Seleucia ran down the Euphrates for the greater part of the way, and Dura was located close to its center. This early highway followed the east bank of the Euphrates as far south as the Khabur, then crossed the river and continued along the west bank. The east bank was flat and fertile, but broken by irrigation ditches and the occasional flooding of the river that resulted in deep, almost impassable mud. The west bank had the desert at its back, and ascent to the plateau was inconvenient wherever the serpentine river washed up to the foot of the cliffs, but it was passable for most of the year.

If that be the purpose of Dura, to act as guardian, then we must first look at the city for its defenses, when its first walls were erected by the Macedonian Greeks to protect the caravan route at this juncture. Main Street, which ran from the desert gate that faced Palmyra toward the river, was twice the width of the side streets and a quarter larger than the secondary arteries. This road must have been the caravan route, and it descended toward the river by a wadi cutting through the center of town. There must have been an eastern gate on the river side, now lost with the crumbling of the cliff face. The turn of the road from desert to riverbank is accounted for by the long wadi running back into the desert for miles just south of Dura, which offers serious obstacles to dromedary traffic.

On either side of the road leading to Dura and the river were foundations of early Macedonian palaces. Plutarch, in his account of Eumenes, the secretary of Alexander, told of Alexander's conquest of Asia Minor and his payment of officers in conquered territories by giving them soldiers to help capture fortified estates which they could keep as their rewards. At Dura, both foundations and outer walls are of ashlar masonry, so closely laid as to allow for almost no plaster mortar, but interior walls are of rubble and mud brick. The Redoubt fronted on

the wadi with a bold, embossed facade of a type reminiscent of early Hellenistic Greece. Only the northwest corner of the palace on the Citadel remained, but fragments found there showed the good lime plaster of characteristic Hellenistic type rather than the later gypsum plaster. Potsherds found in both structures dated to the early third century B.C., if not earlier, to the end of the fourth; and coins of Antiochus I, while not conclusive evidence, added to the impression that the buildings were early Hellenistic. The erection of great embossed walls, as we find on the Redoubt, was fashionable around the time of Alexander and down through the turn of the century. It is hard to choose between these two buildings—the Citadel Palace and the Redoubt—but I think the Redoubt was built first, with its site overlooking the caravan route and the River Gate. The second palace built on the Citadel, commanding a broad view of the river and the Mesopotamian plain beyond, largely cut off the view from the Redoubt.

Plutarch's mention of fortified estates in the east leads one to question whether the Macedonians erected the Citadel wall before the city wall to serve as a defense post for the soldiers and a refuge for the local inhabitants in case of attack. At the least, we can say that the plan of the city was laid on generous lines from the start, for the grid of streets not only covers the whole later city area, but the streets are carefully proportioned in order of importance. And it is clear that the line of the projected circuit wall was determined at the same time as the city plan, for all the streets end at the wall.

Alexander's city planners carried to the East this grid plan, along with the location of the Agora on the main street, a characteristic rectangular marketplace open to the flow of traffic, devoted to the business life of city and merchants, and, in contrast to Roman forums, adorned with shrines and temples. The earliest foundations of the Agora (composed of Blocks G1, 2, 3, 4) disclose the carefully cut large blocks laid almost without any mortar, which speak of Greek workmanship. It is surprising that while on the other side of Main Street two great temples—of Artemis-Nanaia, and of Atargatis and Hadad—were early dedicated, none seems to have been built in the vicinity of the Redoubt. However, if the city was planned as a whole, as a single unit, and the center of city life was to be the Agora and the "farmers" market as well as business center for the new town, and the city was to run along democratic lines, then it is reasonable to expect the chief temples to be built in the heart of the city.

The choice of deities worshiped by the Durenes seemed curious at first. Why Artemis and Atargatis? The old Babylonian armed goddess Ishtar had been an important figure in this region, and her characteristics would have been transformed into the huntress Artemis or into the Greek armed goddess Athena, with Nanaia the local name. The strong, uninterrupted religious feeling of the region would continue to find expression in the worship of Atargatis, great goddess of the desert, close relative, with her lions, to Cybele and Magna Mater in Asia Minor. Associated with Atargatis is Hadad, powerful god of the storm, a fruitful sky and earth deity, but dwarfed nevertheless in Asia Minor and Syria by Atargatis, deity of fertility and generation, the animals of the earth and the ranks of men. So at Dura she sits triumphant between her lions, and Hadad is crowded so far off to one side in their sculptured portraits that there is no room for his flanking bulls. He looks harried and worried as he clutches his symbols of power.

The temples of these gods are in the eastern form: square court with double room, pronaos and naos, the broad-room type (a rectangular room with the entrance set into the long side wall). Clearly, the Macedonians left to the local priests the choice of gods and the style of their shrines. Perhaps we have here the influence of wives chosen from the local population. Alexander had encouraged his troops to take native wives and set the example himself with his marriage to Roxane. How many of the soldiers brought, or bought with dowries, wives from Greece and Macedonia? We found the names of many Greek men in the inscriptions, but of few Greek-Macedonian women. It is difficult to estimate how much time passed before local influence was strong enough to determine the religious focus of the community. Possibly simple shrines were first erected in the Greek style opposite the Agora, dedicated to Artemis and Zeus rather than to Nanaia and Hadad, but this remains a hypothesis without any archaeological evidence. That which we found suggests the predominant influence of local religious elements: local gods honored in temples of Oriental plan who took on Greek names as cloaks that only partially concealed their native identity.

In the city blocks surrounding the Agora and its temples there is little that goes back to the Hellenistic period, to the time before 100 B.C. In the center of the city, the depth of soil above the plateau rock was shallow, despite the accumulated remains of six centuries of occupation. Although the relative absence of pottery, coins, and figurines was surprising, we must remember that the Macedonian veteran did

not buy expensive imported Macedonian pottery, and the taste of his native wife would have run to the local, common pottery to which she had been accustomed. In some Eastern settlements, such as Seleucia-on-the-Tigris, figurines were numerous, but comparatively rare in other Greek settlements. But it is puzzling that Hellenistic coins were a rarity except in the vicinity of the Redoubt and the Citadel. One may surmise that when the Redoubt and Citadel were built, the first settlement clustered near the River Gate and below on the river plain. But the question remains open.

The walls of Dura opened far more questions than we could answer. Dura was not primarily a caravan city but the stronghold controlling the district. The high plateau afforded a certain amount of natural protection, with the cliff face fronting the river on the east and the wadis with quite sheer sides running into the desert on the north and south sides. But even these ravines required walls to be built above them for added safety from native incursions, while the entire western flank of the city, from wadi to wadi, lay completely open to the desert beyond. Guard towers were built and a curtain wall of mudbrick was strung between some of them. The walls, with the towers placed almost a block apart, went up about the beginning of our era, as the inscription on the Great Gate suggests. At the north corner of the desert wall was placed the Temple of the Palmyrene Gods, dated to the first century A.D., and at the other, southern, end was built the Temple of Aphlad, also of the first century A.D.

However, at least one of the masonry towers probably predated the Parthian wall. The tower against which the Palmyrene Temple was built was made in two distinct parts. Originally it was a small square tower of heavy blocks of masonry, closely fitted. Then, when the wall was built, it was elongated and raised higher. We see this sequence in the lower courses of the tower, which do not bond with the elongation, but above the fifth course of masonry the courses of stone are a continuous run above and beside the old work.

The kingdom of the Seleucids was relatively quiet until about 185 B.C. when the Parthians were seriously challenging the Seleucid satrapies. The succession had been fairly regular and without special incident. Though Bactria and Parthia had broken away, these were small and distant provinces; the main body of the empire held together very well. There should have been no need for a wall of great strength girdling Dura at this time. Then, suddenly, civil war erupted, a war lasting for about forty years, from 162 to 121 B.C., and the Parthians

invaded Mesopotamia, taking Seleucia in 141 B.C. and establishing a new frontier along the Euphrates against the Hellenistic Greeks. The Seleucid dynasty, pushed out of Palestine by the Maccabees, held only the Syrian coast and as much of the desert and the middle Euphrates as they could.

In 20 B.C., Augustus and the Parthians signed their first treaty, acclaimed on both sides as a lasting pact. Dura fell into Parthian hands, and Circesium to the north went to the Romans. Now there was every need for frontier fortifications. Now the Parthians erected the circuit wall and incorporated the small north tower into the Temple of the Palmyrene Gods complex.

It was the custom of the Parthians to have a walled palace area within the strong outer walls, as we see in the Parthian cities at Hatra and Nippur. The walled citadel of Dura carries the Parthian stamp. The quarry inside the town, beneath the face of the wall, was used to hamper any attempt to scale the wall or approach it with a battering ram. Probably the Parthians utilized the foundations of an earlier Greek palace for their fortified high point. With the dawn of the new, Parthian era at Dura came not only the great palace with triple *liwans* (high arched halls completely open on one side) on the Citadel but also the renovation of the Redoubt.

The arched Great Gate of the city, which received the plodding camel trains from Palmyra, probably was constructed during the Parthian period, for its masonry of rough blocks bound with gypsum mortar contrasts sharply with careful Hellenistic stone cutting and lime mortar. The earliest date for the Gate is 17–16 B.C. with the construction of the curtain wall, while temple construction also occupied the builders of Dura in the beginning of the present era. It is then that the temples along the wall—Aphlad, the Palmyrene Gods, Zeus Kyrios, and Azzanathkona—were raised. Here I disagree with von Gerkan, who suggested the circuit wall belonged to a late Hellenistic rebuilding program just before the present era. I am inclined to believe the building of the heavy Parthian wall was a reflex of the establishment of the frontier just above Dura.

We cannot be sure when the Parthians marched into Dura, perhaps as early as 120 B.C. or as late as 96. The evidence of the coins, admittedly slender, favors 113. The peaceful occupation of the city—for there is no indication of a bitter struggle for, or taking of, the city by the Persians—apparently was immediately followed by the Parthian fortification of the Citadel. It would be natural for a new group,

different in language, religion, and custom from the old inhabitants, to build themselves a strong refuge within the city before they looked forward to erecting a city wall to protect all against an outside enemy.

The Parthians, after taking Seleucia in 141 B.C., were driven out, and then returned. Their war with the Macedonians continued, and the Seleucid armies retired west to Antioch. The frontier remained fluid with the status of Dura not clearly established before 96 B.C., although the Parthians were probably there before then. Roman victories in the western states of Asia eventually brought the two powers face to face. A treaty in 20/19 B.C. fixed the frontiers of the countries on the Euphrates, at the Khabur River, forty miles upstream from Dura. In spite of internal difficulties and external Roman pressures, the Parthians maintained their frontiers intact until the eastern campaign of Trajan in A.D. 115. The half century before the beginning of the present era and the first century A.D. encompass the great period of Parthian glory and influence that ended at Dura with the campaign of Verus, who arrived from Rome in A.D. 164.

At first glance one might assume that Roman culture entering Dura would be greeted with a cheer, but this is not the case. The granting of the title of colony to Dura and citizenship to the Durenes would be taken as token of Roman regard and hailed with acclaim. However, the Roman tax collector followed fast on the heels of privileges given, and Roman farming of tax gathering resulted in unfavorable treatment of the new colonies. There is the strong probability that Caracalla's wide distribution of citizenship was not so much intended to unify the empire as to bring in new taxes.

Dura had an uncommon wealth of formal inscriptions, and most useful they were in providing the names of deities, the dates of buildings, and the identification of significant events. Perhaps I overestimate Dura's contribution in relation to other excavations, but in Olynthos in Macedonia, Seleucia-on-the-Tigris, and Apollonia in Libya, where I had firsthand experience, inscriptions were almost nonexistent. Robinson, I remember, brought with him to Olynthos a great supply of squeeze paper in the expectation of finding records in stone. The paper was almost unused. Seleucia, though a Hellenistic and Parthian capital, yielded almost no records in stone. Stone was plentiful in the Hellenistic port of Apollonia, but inscriptions were very seldom preserved.

In Dura almost every temple presented the name of the god, often the date of the relief dedicated to him, and the names of the men and women who occupied the seats or erected the monuments. The long-

hand record began with the names of men scratched in the walls, often just with the hope that they would be remembered. Frequently the date was affixed, a great help to us in allocating the names to the appropriate period. It seemed as if every one in this desert crossroad was anxious to leave his personal mark, his name behind, to show that he had been there and to be remembered by the deity he worshipped. When the dedication is not specifically addressed to the goddess of good fortune, the pious wish is that the writer "be remembered."

As a student in Wisconsin I had the privilege of working on half a dozen papyri: one concerned the army, another the sale of a house, others the apprenticeship of a young boy as a weaver, a private letter, the sale of a young pregnant burro. They gave an entirely different view of life in an ancient town from that provided by the mute stones and walls of excavations. In the House of Nebuchelus at Dura the note of commercial transactions was scribbled on the walls, permitting a reconstruction of the economic history of the tradesman. We found the records and parts of records of the detachment of the army from both the Greek period of the cleruchs and the Parthian regime. Intimate glimpses were provided by the account of a break-in of a house, of two marriages and two divorces, and of the sale of property. How close to contemporary military practice was the notice concerning the Parthian envoy coming up river who was to be met, entertained, and speeded on his way—an account of expenses to be submitted to headquarters! The envoy was to be given xenia, hospitality, a broad term, and how much does that allow in expenses?

The Roman calendar vividly illustrated the tie with Rome that even this distant outpost retained, while the find of the Diatessaron allowed us a hitherto unsuspected insight into the Christian cult of the Syrian desert. Almost of equal importance was the shield of the Roman soldier found by Cumont, with the list of cities, his itinerary from the Black Sea to Dura, the first and one of the few geographical records. Only by the merest chance had the soldier recorded his march on the leather of the shield that happened to be preserved and then fortuitously discovered.

Dura lay halfway, in the third century, between the birth of Christ and the advent of Mohammed, and the thoughts and beliefs of that distant desert post reflect the religious turmoil of this time when the many old cults and their gods were dying and new ones were being born. Just before the Roman occupation of Dura, citizens friendly to Palmyra set up new and elaborate reliefs in the Temple of the Gaddé,

signaling the kinship of the two cities. When the Romans came, they erected their own military temple below the Citadel and, in the time of Caracalla, the temple near the Great Gate. I believe the Temple of Mithra, with its decoration executed in the Eastern style, was an expression of opposition to, not cooperation with, the Roman military forces. Almost a hundred years later, the Synagogue and its decoration express the Mesopotamian tradition in art as triumphant over both Roman and orthodox Palestinian.

The Parthian imprint on Dura is greater than that of Rome. Of course, the Parthians had controlled Dura for 277 years and were much more amenable to Mesopotamian culture and society than were the Romans. The Parthians accepted easily the different cults and subscribed to the gods of earth and sky, adding only their own emphasis to the sun god and the supreme god of heaven, beyond the sun (Mithra) and the moon.

Our lack of precise knowledge of Parthian art and culture presents a serious difficulty in estimating their importance and extent in Dura. In art we recognize the use of frontality and the Mesopotamian multiple view. In the military sphere there is the use of the composite reflex bow adopted by the mounted archer. The scale armor, the cataphract, is not necessarily Parthian, but certainly Eastern. In religious representation the great sky god wears Parthian-Iranian trousers, and the close association with sun and moon suggests a Parthian trinity on one side and a gradual rise of the Iranian Mithra on the other. At Dura the Parthians introduced the massive work in stone with gypsum mortar, the vaults, and the stone arches.

For the Roman period we have a superabundance of information with which to interpret the Dura materials, but Roman influence here is not to be simply stated. The Eastern Mediterranean remained Greek-speaking, a legacy of Hellenic influence spread through the conquests of Alexander. When the Romans debarked on the eastern seaboard, they found Greek culture, Greek customs, and Greek religion firmly entrenched. Except for very brief interruptions, Roman control of Syria and Palestine was unbroken from 63 B.C. (the time of Pompey) and the consolidation under Antony in 49 B.C., until the Arab conquest seven hundred years later. The Roman army played a conservative role in the East, and, as we see it at Dura, army organization remained intact, the Latin language was retained, the Roman calendar observed, and cult worship preserved. But Roman civilization was diffused in the centers of Antioch, Baalbek, and Palmyra. The Baals and

Bels and Bols (Aglibol and Iarhibol) were syncretized with their Roman equivalents.

Dura, founded by Greeks, was from the start dominated by Greek language and culture, although, as I have noted, Oriental habits and religious customs soon made their appearance, perhaps in part due to the taking of native wives. The Parthians introduced Persian and the Pahlavi alphabet, but apparently never insisted on their general use and were content to translate their offices and customs into Greek terms. The Christianity introduced in the Roman period of Dura is not of the Latin Gospel, but the Greek, and more particularly the composite Gospel of Tatian, composed in Syria, written in Greek, and addressed to the Greek-speaking community, not the Roman. The Synagogue, also built in Roman times, was a product of the Greek-Jewish community with leanings toward the Persian and, more specifically, the Mesopotamian. As far as I can see, the decorations of the Synagogue contained almost no Roman elements, except perhaps for the costumes of enemy soldiers.

When the Romans came to Dura, they walled off their section of town and placed the Palace of the Governor away from the center of the city, close to the Roman camp. They introduced the Roman bath and the little Roman military and political temple, borrowing only the transverse arches of the Parthians above the square pillars to support the roof. One's first thought is that the Temple of Mithra situated within the confines of the camp would express Roman religious influence. However, it is clear that when the temple was built, its location was in the then Parthian and local military zone rather than the Roman. The Mithraeum also shows the Oriental side—Eastern horsemen and Iranian magi—rather than Roman, Western elements. Once the Roman camp was shifted to the northwest corner of the city, the Mithraeum was accepted as part of the camp's religious establishment, but it must have been strange to most of the soldiers. The Roman detachments felt more akin to the Palmyrene culture than the Parthian. The Twentieth Palmyrene Mounted Cohort, as part of the military establishment, linked Palmyra and Dura closely together, and the common worship of a sky god with sun and moon belonged to both cultures.

The language of the Roman Camp was Latin, and Latin remained an intrusive foreign tongue in Dura. The official documents were all in Latin; with recruits from far-flung outposts of the empire, the Latin language was the essential medium of interchange. But it helped to make the Romans a foreign force, a force worshiping the Roman em-

peror and saluting the standards of Rome with her eagle at the top. Greek remained the official language, the speech of the elite, while the native Semites spoke Aramaic and Palmyrene, the local dialect, and wrote in a script of their own. Although Hebrew prayers were recited in the Synagogue, some graffiti in the paintings are in Pahlavi, indicating that a number of distinguished citizens retained Parthian speech and letters from the days before Verus.

A polyglot town was Dura, as were so many along the desert. Like a sea, the desert nourished border towns, the meeting places of caravans and peoples. Mesopotamia was the broad field of empire after empire—Assyrian, Babylonian, Persian, Greek, Parthian, Roman—all impressing characteristic symbols on the towns, but all gradually being absorbed in the great ocean of Semitic-Arabic development.

For the first fifty years of Roman occupation, Roman arms were supreme in the East and so remained until the rejuvenated Persians under able Sasanian kings replaced the fragmented Parthian leadership in A.D. 224 and looked westward over the troubled Roman Empire. After Verus, Septimius Severus had marched down the Euphrates and up the Tigris. His triumphal arch could display the signs of his victory over Seleucia and its adjacent Parthian capital of Ctesiphon, but Parthian Hatra in the desert, with its rich temple of the sun god, checked him. The Romans never fully mastered desert warfare, a warfare that demanded first of all an intimate knowledge of the terrain and, second, a mobility greater even than that of the swift-moving, but marching Roman legions. Caracalla was preparing for an Eastern campaign, and as the third century developed, Roman military problems in both the East and the West deepened. With increasing emphasis on military leaders, imperial power became more and more the reward for military success, and the empire was crippled by wars of succession. While the Goths and Visigoths poured over the Danube in the West, on the Eastern front the Sasanians, claiming inheritance of the old Persian empire, menaced the Roman frontiers in Syria and in Asia Minor. And Dura, for good or evil, lay on the main route to Western conquest by the Sasanian kings Ardashir (A.D. 22?–241) and Shapur I (A.D. 241–272).

The Sasanian Persians who replaced the Parthians in Mesopotamia thirty years before Dura fell were always the enemy. The chances of new, distinctly Sasanian elements playing a part at Dura were very small, although one drawing has been called Sasanid. Artistic changes come particularly slowly in the Orient, and the extensive remains of artworks found in Dura seem all of a piece.

The Sasanian Ardashir I, who overthrew the Parthian Artabanus V

in A.D. 224 or 226 (the Persians dated their regime from October of 226), was a religious zealot as well as national reformer. He desired as much to restore the rigid precepts and lofty aspirations of Zoroaster as he did to reoccupy all the territories formerly claimed and occupied by the Achaemenian Persians. His son and successor, Shapur I, was a military leader of ability even greater than that of his father; although more tolerant of foreign religions and a patron of Greek learning, he was equally ambitious in the political and military fields.

Reconquest of the old Persian empire, which included Asia Minor as well as the Syrian coast and Egypt, may have seemed a dream, but restoration of the territory held by the Parthians in accordance with the treaty made with Augustus seemed very practical indeed. Armenia in Asia Minor was always a satellite of the Parthians when a strong king was on the throne, but then would become tributary to Rome when her emperor was strong. So Nisibis and Carrhae were Sasanian targets in northern Syria, while Antioch, the great Hellenistic center on the Orontes and capital of north Syria, had early figured in the Persian plans, with Shapur I attacking in 238, while Ardashir was still alive. The two great Sasanian leaders ruled for almost half a century (A.D. 226/7–272), elevating Persia to a position of power, a dominant force in the political strategies of the East.

If the New Persian Empire of the Sasanians were to expand westward, the first logical step was Dura. By the Parthian treaty with Augustus, Dura had been given to the Parthians and the frontier fixed at the Khabur fifty miles up the river. Thus, the Sasanians, in their defeat of the Parthians, could make a legal claim on Dura. There must have been claims and counterclaims, exchanges of embassies and crisscrossing of emissaries. The Sasanians undoubtedly demanded the evacuation of Roman troops from Dura and no doubt issued a threat to the citizens if they did not accede to Persian authority. Dura was not just a frontier fortress in the eyes of the Sasanians: it was a walled city held by a Roman force in disregard of the Augustan agreement. More important, it was also a closed door blocking Sasanian western expansion.

The mortally precarious position of Dura in the struggle between Rome and Persia must have been apparent to the men and women of the desert caravan cities. By A.D. 256, which marked thirty years of rule by, first, Ardashir and now Shapur I, the tides of battle had ebbed and flowed. The Persian attack of 238 had reached Antioch, almost on the shore of the Mediterranean. Alexander Severus had threatened a massive attack along the Tigris and Euphrates and across the desert.

Gordian III had advanced down the Euphrates, to be halted just above Dura when he had been murdered by Philip the Arab in A.D. 244. Palmyra was rapidly rising to a position of independent strength.

Shapur's projected attack on Dura in 256 was not just a threat to the Roman force there, but to the life of the city. The vital question was whether the Romans would insist on holding Dura and retain their garrison within the walls, or withdraw their force, leaving an open city that could hope to live for many more years under a new authority. Perhaps the thickness of the city's walls and the size of the resident military forced the fatal decision. The Romans had a fairly good-sized force, even if not equal to that of an invading army: part of the Third Cyrenaica Legion, men recruited in the Greek-speaking half of the North African shore; the detachment of the Fourth Scythian Legion, stemming from the Black Sea region of southern Russia and northern Asia Minor; the Second Ulpian, which had been represented in the early occupation, although it is not clear if these troops still remained; the Twentieth Palmyrene Archers, camel troops, roaming the desert as the present-day Meharists do, preserving peace and keeping a watchful eye on frontiers. Local boys interested in military service might have enlisted in the Palmyrene force as the cohorts attached to legions, local units acquainted with the district and trained in the desert. Among the citizenry were riders and hunters, caravan men accustomed to guard their charges from both wild animals and freebooters. They could help in defense, but against trained soldiers they would be at a distinct disadvantage. However, the walls of Dura, though somewhat vulnerable because of their great length, were high and strong and supplemented by the slopes of the wadis to the north and south and the sheer cliff on the Euphrates side.

If by some trick of time we could stand before Shapur I, we would much like to ask him why he destroyed Dura in A.D. 256? Why, once taken, was Dura not turned into a Sasanian stronghold rather than evacuated and abandoned to the desert? I think Rostovtzeff was right in accepting the accuracy of the Sasanian records, which noted the capture of Dura by Sasanian troops in A.D. 253. Perhaps, then, Shapur would answer that when he first took the city he had warned the Durenes that revolt would mean destruction. But they did not heed his warning: they revolted as soon as the main Sasanian task force was well out of sight. Thus, he must return in full strength when conveniently possible and carry out his promised revenge. And so the city was given back to the desert sands.

The year 253 was ideal for the first Sasanian attack. The Roman

emperor had been killed in the barbarian invasions along the Danube in 251. As Roman military strength in the East faltered, Palmyra assumed greater independence. Shapur I had ruled at Ctesiphon for twelve years, and his boast and intention were to regain the broad frontiers of the old Persian empire.

Valerian, who succeeded Decius after a short reign of the former general Aemilianus, had held important civil and military posts under Decius and Gallus, but he was incompetent and everyone knew it. Northern Italy, Greece, and Asia Minor were threatened by barbarian inroads, and the best Valerian could do was proclaim a general persecution of Christians in A.D. 257. With no hope or help in sight, Dura wisely capitulated in 253, and unwisely revoked her submission immediately afterwards. The lack of almost any Sasanian coins at Dura indicates that Dura cast off the Sasanian yoke as soon as it was safe. We have the dipinto of the Palmyrene general coming in triumph and the painting which I believe to be of the Sasanian defeat to mark the liberation of the city. The wonder is that Dura revolted from the Sasanians when Roman support was so weak. I suspect the Durenes counted on Palmyrene support, which was not forthcoming.

A second question may then be asked: Why did the Sasanians destroy a flourishing city which, before serving an important role protecting an important commercial highway and border for one hundred years under the Romans, had stood as a western bulwark of the Oriental world against Greeks and Romans for two hundred years? I believe there are two important reasons for Shapur's action. His ambition no longer saw the old Parthian western border as the limit of his domain; hence, Dura would not be required as a frontier fortress, for that frontier would be moved. Second, a strong city left intact would always be a risk and obstacle to movements up and down the river. Better to eliminate the fortress once and for all than to live with an ever-present threat at one's back. The policy of Shapur I was liberal and progressive. He rebuilt Persian economy and promoted a broad program of public works. He also wanted to concentrate the power in the south and to eliminate the powerful, and thus possibly rival, cities in the north. So Hatra, the influential Parthian-Arab center near the ancient Assyrian capital of Nimrud (Calah) was destroyed, sharing Dura's fate.

Less than two decades later, in 272, the Romans destroyed Palmyra, reducing the proud city of Zenobia to little more than a crossroad oasis. The gap left by the fall of these two cities in the desert would be filled only by the camel caravan and herdsmen of the nomadic

tribes. The Byzantines eventually moved into the Syrian coastal regions, and Byzantium's towns spread into the desert along the Hauran, which enjoyed the runoff from the Anti-Lebanon mountains. In Palestine, Gadara held the riverbank west of the Jordan and kept open the north-south trade route along the edge of the desert.

But the great stretch of desert between the Jordan and the Euphrates remained open and void. In the seventh century A.D. the lands north of Arabia resounded to the marching armies of Mohammed, and the Moslem cavalry and Arab youths trained in the desert swept away both Byzantines and Persians. The destruction of the Syrian desert cities that began with Dura and Palmyra was now complete.

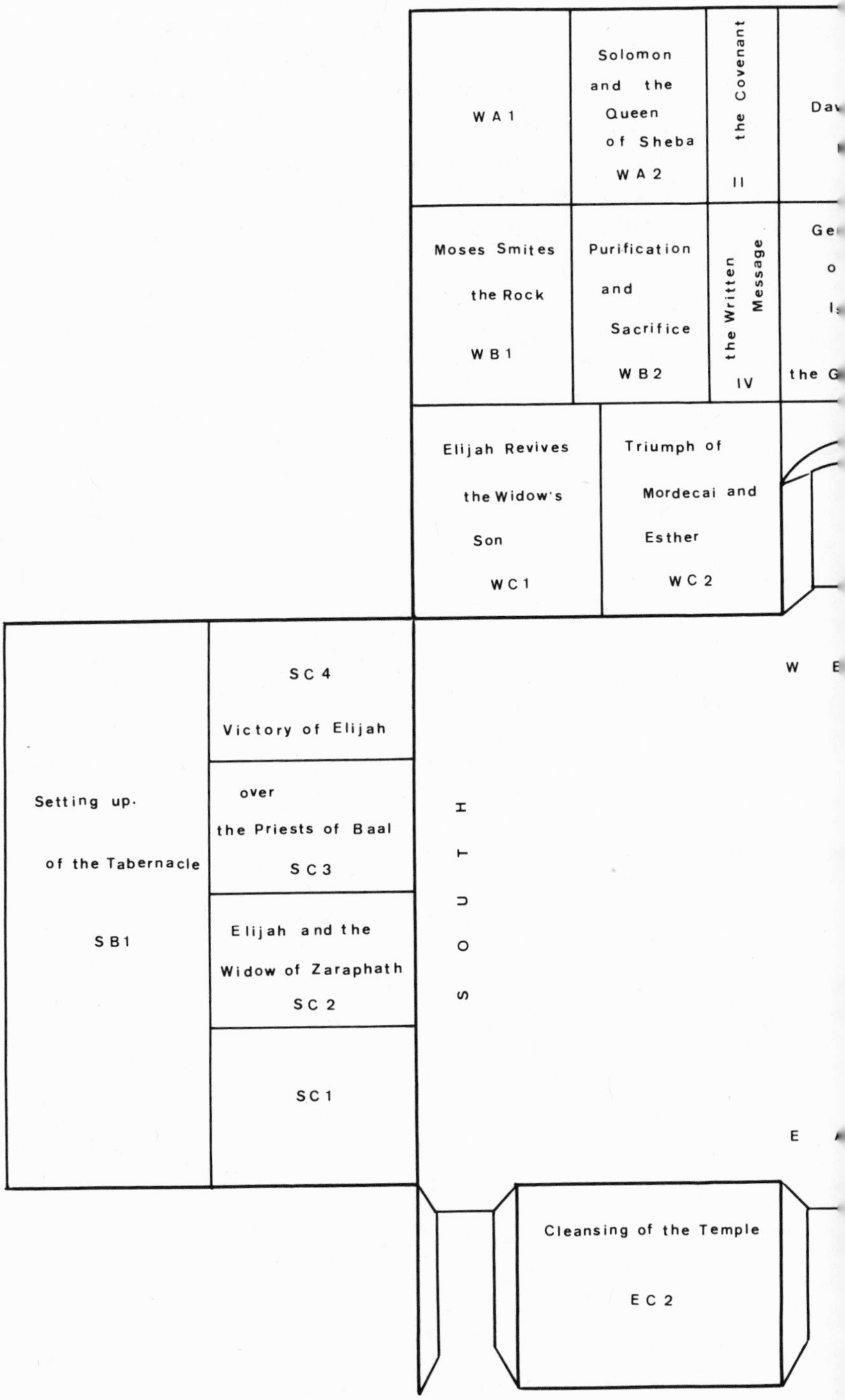

WA 1

Solomon and the Queen of Sheba
WA 2

the Covenant
II

Dav

Moses Smites the Rock
WB 1

Purification and Sacrifice
WB 2

the Written Message
IV

Ge
o
I
the G

Elijah Revives the Widow's Son
WC 1

Triumph of Mordecai and Esther
WC 2

W E

SC 4
Victory of Elijah over the Priests of Baal
SC 3

Setting up of the Tabernacle
SB1

Elijah and the Widow of Zaraphath
SC 2

SC 1

SOUTH

E

Cleansing of the Temple
EC 2

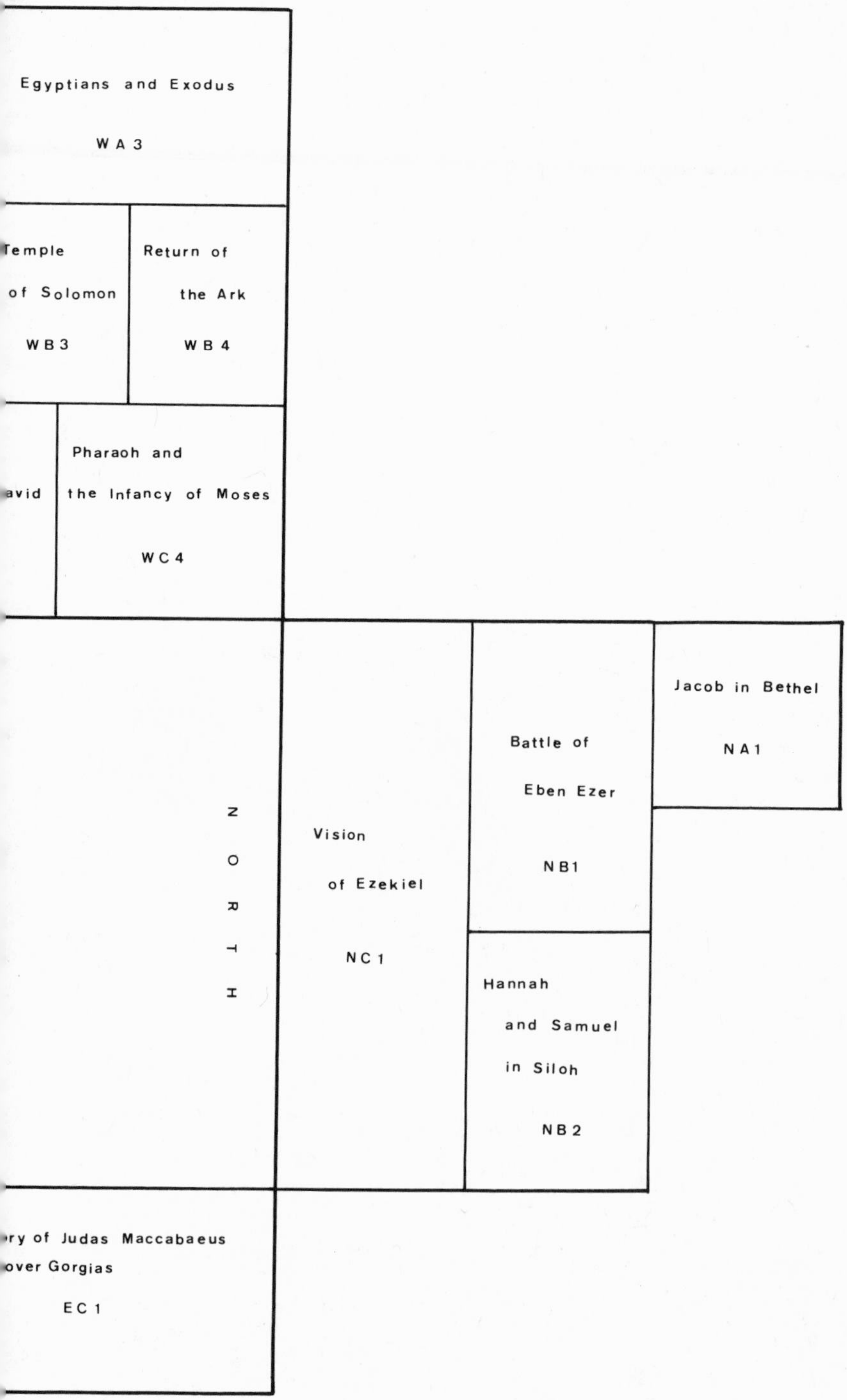

Egyptians and Exodus

WA 3

[T]emple of Solomon

WB 3

Return of the Ark

WB 4

[D]avid

Pharaoh and the Infancy of Moses

WC 4

NORTH

Vision of Ezekiel

NC 1

Battle of Eben Ezer

NB1

Jacob in Bethel

NA 1

Hannah and Samuel in Siloh

NB 2

[Victo]ry of Judas Maccabaeus over Gorgias

EC 1

Chronology

Monarchs: Hellenistic (H)
Roman (R)
Parthian (P)
Sasanian (S)

	Events at Dura	General Events Affecting Dura
B.C.		Seleucid-Macedonian calendar begins: 312/11
		Seleucus I (H): 312–281
		Seleucia-on-the-Tigris founded: ca. 306/05
300	Dura-Europos founded ca. 300: grid plan of city established; line of city walls set; Redoubt built; first Citadel Palace; Agora begun	Antiochus I (H): 281–261
250		
		Beginning of Parthian dynasty with Arsaces: 247
		Antiochus IV (P): 175–164/63
150		Mithridates I (P): ca. 171–138
		Parthians take Seleucia-on-the-Tigris: 141
	Citadel wall: 120–65	Rome makes Asia a province: 130
		Mithridates II (P): ca. 124/23–87
	Parthians take Dura ca. 113	
100		
	Begin to enlarge Agora as Oriental bazaar	Sulla makes first Roman contact with Parthians: 92
	Construction of masonry city wall and some of the towers: 65–19	Rome in control of Syria: 63
		Parthian lord Suren defeats Crassus at Carrhae: 53
50	Second Citadel Palace	
	Temple of Bel and Iarhibol dedicated in necropolis outside the walls: 33	Phraates IV (P): ca. 38/37–2 B.C.
	Dura becomes seat of Parthian provincial governor	Treaty between Rome and Parthia establishing boundaries; return of Roman standards: 20
B.C.	Great Gate begun: 17–16	

	Events at Dura	General Events Affecting Dura
A.D.	Temple of Zeus Kyrios: 29 Temple of Atargatis and Hadad: 31 Temple of Artemis-Nanaia begun: 31 Relief of Aphlad: 54	Civil war in Parthia: 39–47
50	Andron dedicated to Aphlad: 54	Vologases I (P): 51/52–76/77 Nero (R): 54–68 Parthians defeat Roman general Paetus: 62 Corbulo sent by Nero to check Parthians: 63 Vespasian (R): 69–79 Jerusalem falls and is destroyed: 70 Titus (R): 79–81 Domitian (R): 81–96 Pacorus II (P): 76/77–115
100	Foundation of Temple of Zeus-Theos: 114 Commemoration of Zeus-Bel, Temple of the Palmyrene Gods: 115 Romans briefly take Dura Triumphal Arch of Trajan: 116 Parthians again control Dura: ca. 121	Trajan begins Parthian campaign: 114 Trajan takes Seleucia-on-the-Tigris: 115/16 Mesopotamia and Syria become Roman provinces: 116 Death of Trajan: 117 Hadrian (R) 117–38 Jewish revolt suppressed by Rome: 132–35
150	Last stage of Temple of Gaddé, just prior to 159 Earthquake: 160 Romans under Verus take Dura: 164 First Mithraeum: 168–71 House converted into Synagogue: ca. 165–200	Antoninus Pius (R): 138–61 Vologoses IV (P): ca. 148–92 Rome undertakes war against Parthians: 162 Verus retakes Seleucia-on-the-Tigris: 166 Tatian's Diatessaron: ca. 172 Commodus (R): 180–92
200	Roman garrison enlarged: ca. 210 Dura made a Roman colony: ca. 211 Palace of the Dux: post-211 First phase of Dolicheneum: 210/11 Praetorium of Roman camp: 211–17 Heightening of city walls: post-216 House of the Merchant Nebuchelus: 218 Conversion of house into Christian Chapel; wall paintings: 232–56	Septimius Severus (R): 193–211 Jewish and Christian persecutions by Septimius Severus: 202 Caracalla (R): 211–17 Marcus Antoninus (R): 218–22 Alexander Severus (R): 222–35 End of Parthian rule: defeat of Artabanus V (P) by Ardashir (S): 224 or 226

	Events at Dura	General Events Affecting Dura
		Beginning of Sasanian dynasty: Ardashir 224(?)–240/41
	Grafitto: "Persians descended on us": 238	Sasanian raid into Syria: 238
		Sasanians repulsed at Antioch: 243
	Final rebuilding of Mithraeum: ca. 240	Shapur I (S): 241–72?
	New roof on enlarged Synagogue: 243/44	
	Synagogue paintings: 243/44–253/54	Decius (R): 249–51
250		Organized persecution of Christians: 249–51
		Hatra destroyed by Shapur I (S): ca. 250
	Last building phase of Dolicheneum: 251	
	First Sasanian attack and possible occupation of Dura: 253	
	First building of embankment: post-254	
	Embankment inside city wall enlarged: 256	
	Sasanian attack and fall of Dura: 256	
		Richest period of Palmyra: 261–71
		Palmyra falls to Rome: 272

Glossary

Abou Kemal Located on the road between Baghdad and Aleppo, approximately five and one half miles north of the Syria-Iraq border, on the right bank of the Euphrates. Approximately 2,200 inhabitants in the early 1930s.

Achaemenian The Persian royal family that ruled the Near East and Egypt after the fall of the Assyrian Empire and until its destruction by the Macedonians under Alexander.

acropolis Upper, usually fortified administration or religious center of a Greek city.

acroterion (pl. *acroteria*) Free-standing decoration on the corners of pediments on classical buildings.

aedicule Niche or shrine to hold the cult image.

agger Sloping embankment against a wall.

Aglibol Early Palmyrene moon god, usually associated with Bel (lord, master) and Iarhibol.

Ahura Mazda Supreme deity of Persians, god of light and creation as opposed to Angra Mainyu (later Ahriman), lord of evil power. Also called Ohrmazd later.

akinakes The Persian short sword.

Albright, William Foxwell (1891–1971) Professor of Semitics at Johns Hopkins University; director of American School of Oriental Research in Jerusalem (1919–36); noted biblical archaeologist, author of *The Archaeology of Palestine, Archaeology and Religion of Israel, From Stone Age to Christianity.*

Alexander Severus (Marcus Aurelius Severus Alexander) Roman emperor A.D. 222–35; favored Christians and accepted holiness, but not divinity, of Christ.

andron Room for men in Greek home.

Angell, James Rowland (1869–1949) Psychologist, professor, University of Chicago (1894–1919); president, Yale University (1921–37); educational counselor to the National Broadcasting Company.

Antioch-on-the-Orontes On the left bank of the Orontes River, approximately fifteen miles from the Mediterranean, founded 300 B.C. by Seleucus I as the capital of Seleucid Syria. Annexed by Pompey (A.D. 64), it became the capital of the Roman province of Syria.

Antiochus I, Soter Son of Seleucus I; came to the throne 280 B.C.; died 262/61.

Aphrodite Greek goddess of love, beauty, generation, known as Aphrodite-Urania in her cults in Cyprus, Cythera, and Corinth.

Aphlad Semitic god, probably the son of the Semitic storm god Hadad; the Baal, "great god," of the middle Euphrates town of Anath.

Aramaic Semitic language of the Aramaeans that gradually became the lingua franca of Asia until displaced by Arabic.

arcapat A rank or office, such as for a vassal lord or citadel commander.

archivolt Molded banding on a round arch over door or window.

arcosolium Arched niche, particularly to hold a coffin in a catacomb.

Ardashir I (ca. A.D. 224–41) First of the Sasanid dynasty, under whose reign the Parthians were conquered.

Artabanus V (ca. A.D. 213–24) Last of the Parthian kings, defeated by Ardashir I.

Artemis Daughter of Zeus, sister of Apollo, goddess of the hunt and birth; in Seleucid times, identified with Nanaia, old Mesopotamian lunar deity.

Arzu (or Azizou) Arab god known from Palmyra, associated with astral deities; represented mounted on a camel.

ashlar Dressed stonework laid in regular courses.

Assur (Ashur) Modern Tell Sharqat, Iraq; oldest Assyrian capital on Tigris; still important during Parthian times, when it was renamed Libanae, but disappears from history after Sasanian period until excavated 1903–14.

Atargatis Syrian goddess, consort of Hadad; in her role as fertility deity, she was associated by the Greeks with Aphrodite.

Avroman Site in Kurdistan that yielded Greek legal documents from the Parthian period, demonstrating the persistence of Greek law and language in the Parthian state.

Azzanathkona Major Semitic goddess, associated by the Greeks with Artemis; sometimes a major divine force in fertility and war.

Baalbek City with Greco-Roman ruins, about fifty miles northeast of Beirut; contains best examples of Hellenistic architecture of first to third centuries A.D.

Baal-Shamin Semitic "Lord of the Heavens"; associated at Palmyra with Aglibol, Iarhibol, and Malakbel.

Baghouz Site across the Euphrates from Abou Kemal which yielded, in

addition to its tower tombs, prehistoric painted pottery during explorations of 1935–36.

baptistery Section of church holding the font for baptism services.

Bell, Gertrude M. L. (1868–1926) Took first in history at Oxford, began study and exploration of Syria and Palestine in 1899; served in Egypt during World War I, administrator under Sir Percy Cox and Sir Arnold T. Wilson in Iraq between Armistice and rebellion of 1920; founded archaeological museum in Baghdad, where she died.

Boghaz-Keui (Boghaz-Köy) The modern Turkish village near the sixteenth-century-B.C. Hittite capital of Hattusas.

boule (bouleuterion) Greek senate house or council chamber.

Breasted, James Henry (1865–1935) Professor of Egyptology and director of the Oriental Institute, University of Chicago (1919–35); coined phrase "Fertile Crescent," which he explored as ancillary to his major work of gathering a complete corpus of Egyptian inscriptions, published as *Ancient Records of Egypt* (1906–07).

broad room Rectangular room with entrance in one of the long walls.

Byblos Biblical Gebal, modern Jebeil, called Byblos in Greco-Roman times. Phoenician port and trading depot, located about twenty miles north of Beirut.

calidarium Hot room of a Roman bath.

campus Field or open space for exercises, assemblies, etc.

Caracalla (Marcus Aurelius Antoninus) Roman emperor A.D. 211–17. In imitation of Alexander he sought to incorporate all Asia into a single Roman land; invaded Parthian Media in A.D. 216.

caravan route Any one of the major trade roads across the Syrian desert for shipment of goods from the Mediterranean to Iran, Turkmenistan, and China, or south to the Persian Gulf and India, or to the Gulf of Aqaba and Egypt.

castra Military encampment.

Carrhae Modern Harran, Turkey. Site of Parthians' defeat, under Surenas, of Crassus, the Roman proconsul of Syria and commander of the Roman army, in 53 B.C.

cataphract Suit of armor that covered the entire body. The *cataphractii* were armored cavalry, usually archers or lancers (*clibanarii*), of the Parthians.

centurion Roman officer commanding a "century," 100 men.

Charax Spasinu Charax, located at the head of the Persian Gulf,

thrived as southern terminal for trade on caravan route running down the Tigris-Euphrates valley destined for India.

chiton Basic Greek tunic, worn long or short, sometimes belted, by men and women.

Circesium (Qarqisiya) Roman site on the left bank of Euphrates near confluence with the Khabur River.

citadel Fortress or stronghold that commands a city, usually located, as at Dura, on the highest ground (acropolis).

clavi Vertical stripes of colored material at each shoulder on shirts or chitons.

cleruch Citizen with military obligations who was settled in Greek colonies abroad but still retained Greek citizenship.

cohort One of ten divisions of a Roman legion, composed of 300–500 soldiers.

Commodus (Lucius Aelius Aurelius) Eldest son of Marcus Aurelius and Roman emperor A.D. 180–92.

consignatio The anointing with oil at the closing of the baptismal rite that ends with the words of good will.

Constantine (Flavius Valerius Constantinus) As coemperor with his brother-in-law, Lucinius, he granted privileges to the christian Church, issued regulations in favor of Christianity (Edict of Milan, A.D. 313), and was baptized on his death bed.

Contenau, Georges (1877–1964) Curator in chief of the Oriental Antiquities Department of the Louvre, professor at the École du Louvre, the University of Brussels, and Catholic Institute of Paris; director-general of the French Archaeological Mission in Iran.

Corbulo (Gnaeus Domitius) Roman general and proconsul; became governor of Syria (A.D. 58) and worked out peace treaty with the Parthians (A.D. 63).

Crassus Dives (Marcus Licinius) One of the "first triumvirate" with Caesar and Pompey; defeated at the Battle of Carrhae (A.D. 53) by the Parthians. He died in the fight and lost the Roman standards ("eagles," symbols of Rome's power) to the enemy.

Ctesiphon Founded by the Parthians, approximately twelve miles southeast of Baghdad across the Tigris from Seleucia, ca. A.D. 221.

Cumont, Franz V. M. (1868–1947) Professor at the University of Ghent, curator of the Brussels Royal Museum, foreign member of the French Academy; particularly known for his work on Mithraism, Manichaeism, and Roman paganism.

curia The assembly hall for the representatives of the several tribes of Rome.

Cyprian Christian saint (ca. 200–58), bishop of Carthage, martyred in the persecution of Valerian.

Cyrus II, the Great (559–530 B.C.) Son of Cambyses, founder of the Persian Achaemenian empire.

Darius I (522–486 B.C.) Achaemenian king, builder of the imperial palace complex of Persepolis; considered the most brilliant of Persian kings after Cyrus the Great, but best known to the Western world for his failure in the invasion of Greece.

Decius (Gaius Messius Quintus) Roman emperor, A.D. 249–51; persecuted the Christians, whom he held responsible for dividing the empire.

Deir-ez-Zor Located on the right bank of the Euphrates in Syria, served as the capital of the district of Zor; population of 6,659 (in 1949); lies at the juncture of the river route between Aleppo and Baghdad and the road between Damascus and Mosul.

Diadochi The successors of Alexander the Great who ruled the Hellenistic world.

Diatessaron Literally, "out of the four (Gospels)"; a synthesis or harmony of the Gospels into a single narrative. See *Tatian.*

dipinto (pl. *dipinti*) Painted grafitto or drawing.

Dog River Near Beirut, on the ancient invasion route of the Mediterranean littoral, its cliffs carry carvings and inscriptions left by the several armies passing through; Assyrian, Babylonian, Egyptian, and so forth.

Druse A people and religion of Syria and Lebanon with different customs from their Moslem and Christian neighbors, with whom they were in conflict in the nineteenth century. Despite the regional autonomy granted them by the French in 1921, they led an uprising, the Druse Revolt, in 1925.

Dunand, Maurice Excavator of coastal city of Byblos, begun in 1926 and continued through the 1940s.

Dussaud, René (1868–1958) Curator of Oriental Antiquities, Louvre, in the 1930s; wrote extensively on Near Eastern religions and Phoenician art.

exorcisterium Ritual chamber for driving out evil spirits during baptismal rites.

frigidarium Hall of the Roman bath furnished with a cold-water bath.

frontality That form of artistic presentation where the human figures in a composition are oriented toward the observor and not related to each other. All major aspects of the body face to the front, while the profile is rarely used.

flying gallop Running animal represented with forelegs extended almost parallel with the body, but the hoofs of the hind legs touching the ground.

Gad (pl. *Gaddé*) Semitic generic name for the guardian spirit, the genius of a city.

genearchos The head or chief of a tribe or clan.

glacis Earthwork slope forming a rampart against a wall.

Gotarzes II Parthian king, ca. A.D. 38–51; succeeded Artabanus I and divided the empire between himself and his brother Vardanes. Victory portrait carved at Bisitun (ca. A.D. 50).

graffito (pl. *graffiti*) Figure or inscription scratched on a wall.

Hadad (Adad) Semitic god of storm and thunder.

Hadrian (Publius Aelius Hadrianus) Governor of Syria A.D. 114, and Roman emperor A.D. 117–38; he discontinued the imperialism of his protector, Trajan, in the Near East.

Han Dynasty Imperial dynasty of China 206 B.C.–A.D. 220; Parthian and Chinese envoys and travelers inaugurated the Silk Route between the Near and Far East.

Hatra (el-Hadhr) Located about fifty miles south-southwest of Mosul, capital of small northern kingdom, an important Parthian caravan city in the first to third centuries A.D.

Hauran Desert plateau south of Damascus in Syria, east of the Jordan River.

headers and stretchers Laying building bricks or blocks with the blocks alternately placed short side (header) and long side (stretcher).

Hellenistic period The age of Macedonian Greek hegemony in the ancient world when the dynastic families descended from the generals of Alexander—Seleucids, Ptolemies, Attalids, Antigonids—ruled. The period is dated from the death of Alexander, 323 B.C., to (usually) the Battle of Actium, 31 B.C., when Rome became a world power.

Heracles Most popular of Greek heroes, but not a god; usually represented with lion skin.

Herzfeld, Ernst (1879–1948) Trained as an architect, he became the foremost authority on Iranian art and architecture, particularly early

Islamic. Surveyed the monuments of the Tigris and Euphrates with Friedrich Sarre; excavated Persepolis.

himation Greek long mantle worn draped over one shoulder.

Hippodamean plan Regular laying out of city blocks in a grid pattern; ascribed by Aristotle to Hippodamus of Miletus, although its use predates him.

Hormizd II Sasanian king A.D. 302–09.

Iarhibol Palmyrene solar god, usually part of a trinity with Bel and Aglibol.

illusionism The use of optical illusion in painting, particularly popular in Rome, to give the effect of open space on the flat wall surface.

Ingholt, Harald (1896–) Danish-American archaeologist; assistant curator at the Ny Carlsberg Glyptothek, Copenhagen, lecturer at the American University in Beirut, associate professor of Hebrew and Old Testament at Aarhus University, Denmark, retired as professor of archaeology, Yale University; excavated Palmyra, Hama, and Shimshāra; editor of the journal *Berytus*; authority on Palmyrene, Parthian, and Gandharan sculpture.

Irenaeus Christian saint of the second century A.D., probably born in Asia Minor.

Irzi Local name for a tower tomb in the necropolis at Baghouz, opposite Abou Kemal.

Isidorus of Charax Asian Greek who wrote (ca. A.D. 25) a "geography" of Parthia, which describes the way stations from Zeugma on the Euphrates, eastward through Iran to Alexandria (modern Kandahar, Afghanistan).

Khabur Tributary river of the Euphrates in northwest Mesopotamia.

limes Boundaries (of the Roman provinces).

liwan (iwan) Large, roofed hall with one side completely open, giving onto a courtyard.

loculus (pl. *loculi*) Small chamber or recess to hold a body or funeral urn.

Lucius (Verus Lucius) Jointly held tribunal powers with his "adopted" brother, Marcus Aurelius; one of the weak Roman consuls (A.D. 130–69).

Marcus Aurelius Antoninus Roman emperor A.D. 161–80; Stoic; reintroduced persecution of Christians, whom he saw as the cause of Rome's ills.

Mari (modern Tell Hariri) City founded at the end of the fourth mil-

lennium B.C. on the middle Euphrates. A major center until it was taken by Hammurabi, the city continued to exist on a more modest scale in Assyrian times to the sixth century B.C. This ancient capital was identified and excavated by Parrot from December 1934 through 1939.

Martin, Richard A. Anthropologist with the archaeological expeditions of the Oriental Institute in Syria and Anatolia.

martyrium Building or hall early Christians used as a burial place and memorial for the relics of martyrs.

megalographia Mode of drawing where figures are presented large, dominating the pictorial space, usually covering the full height of the panel in which they are drawn.

Megarian ware Hellenistic red ceramic pottery characterized by modeled figural and floral designs around the outside.

Meharists The Syrian desert police, mounted on *mehair,* a breed of swift dromedaries.

Mesopotamia Literally "between the rivers" of the Tigris and the Euphrates, roughly corresponding today to Iraq.

Meyadin Town located on the west bank of the Euphrates, midway between Dura and Deir-ez-Zor, on the highway between Deir and Abou Kemal.

Middle Persian Written language of the Parthians and Sasanians, called Pahlavi.

Midrash Collection of Jewish expositions of and elaborations on the Old Testament.

Mishrife-Qatna (Tell el-Mishrifeh) Ancient Akkadian settlement in Syria, south of Hamath, later under influence of Hittites, Hurrians, and Amorites.

Mithridates I Parthian king, ca. 171–138/37 B.C., who separated Iran from the Seleucid empire and held as Parthian the lands east of the Euphrates.

Mithridates II During his reign, ca. 124/23–88/7 B.C., Parthia became a world power and came into first contact with the ambitious plans of Sulla's Rome.

Mithridates III His short reign, ca. 58/7–55 B.C., as Parthian king began with the slaying of his father, Phraates III, and ended with the surrender of the kingdom to his brother.

Morgan, Jacques de (1857–1924) Mining engineer and archaeologist; as chief of the French Archaeological Delegation in Persia, he began

the excavations of Susa, which were continued after his death until halted by World War II.

Mouterde, Father René Professor at St. Joseph University, Beirut, writer on Hellenistic antiquities of Lebanon.

Nabateans An Arab tribe settled in northern Arabia that grew prosperous through control of the caravan route between Syria-Mesopotamia and the south. From their capital at Petra, they extended their power as the Seleucid dynasty crumbled, became allies of Rome, and were finally annexed by Trajan A.D. 105/06. Particularly strong Hellenistic influence in their architecture.

Nanaia Fertility goddess of Parthian times; related to Semitic Ishtar and assimilated with Artemis (Artemis-Nanaia) at Dura.

naos The inner chamber or shrine with the cult image of the temple.

Naqsh-i Rustam Site, approximately four miles north of Persepolis in southwest Iran, of rock-cut tombs of Achaemenid kings, with series of Sasanian reliefs depicting the victories of the Persian kings.

Nemesis Greek personification of retribution.

Nero Claudius Caesar Ruled Rome from A.D. 54 until his suicide A.D. 68.

Nicanor See Seleucus Nicanor, founder of Seleucia.

Nimrud Dagh Mountain of Commagene in Northern Syria holding the monumental tomb and giant sculpture of the Seleucid king Antiochus, 66–34 B.C.

Olynthus Greek city north of Potidaea; became the capital of the Chalcidian League in Classical times.

Opis Ancient Babylonian city mentioned as located where the Tigris and the Euphrates come closest together and, hence, thought to be at or near Seleucia-on-the-Tigris.

Oriental Institute Founded by James Henry Breasted at the University of Chicago as a research center for the recovery of the monuments and history of the ancient Near East, supported by John D. Rockefeller, Jr., in 1919. The present center and its museum are housed in a building designed and built for the Institute in 1930–31.

Origen (Origenes Adamantius) Alexandrian Christian, ca. A.D. 185/6–254/5, teacher, writer on Christian dogma, the commentaries on Holy Scripture; tortured during the Decian persecution.

Ormazd Late Iranian form of the name of Ahura Mazda.

Paetus (Lucius Caesennius) Roman consul and general, sent to Armenia and appointed governor of Syria by Vespasian in A.D. 70.

Pahlavi (Pehlevi) Middle Persian, the written language of the Parthians and the Sasanians.

palaestra Greek or Roman gymnasium or training ground.

Palmyra (Tadmor) Oasis in the Syrian desert that became rich and virtually independent in the time of the Roman Empire because of its location on the major east-west trade route. Made a Roman colony by Caracalla, it rose to its full height under Odaenathus and Zenobia as Roman power declined in the East, but it was finally neutralized by Aurelian in A.D. 273, a blow from which it never recovered.

Palmyrene language Aramaic written in a local variety of alphabet related to Hebrew and Nabatean.

Parrot, André (1901–) Curator of Asiatic Antiquities of the Louvre, professor at École du Louvre, authority on biblical remains and architecture, director of excavations at Mari (Tell Hariri) on the middle Euphrates north of Abou Kemal.

Parthians Originally a seminomadic Iranian people from east of the Caspian Sea who, under the Arsacid dynasty, detached the Iranian highlands from the Seleucids. Under the great Mithridates I, their capital was established at Ctesiphon, the Euphrates forming the frontier between Roman and Parthian spheres of power. The last Parthian king, Artabanus V, was killed A.D. 224 by a local prince of southwest Iran (province of Fars), marking the fall of the Parthian state and the beginning of Sasanian rule.

Parthian gallop The representation of horses with all four legs extended, almost parallel with body; the far back leg is drawn just beneath the near leg, the rear hoofs turned back.

Parthian Stations See *Isidorus of Charax.*

Persepolis One of the royal seats of the Achaemenian Empire; planned by Darius I, it was under almost continuous construction until its sack and the collapse of the Persian Empire in 330 B.C. The extensive remains are located in the modern Iranian province of Fars before the plain of Marv Dasht; east of Shiraz.

Petra Capital of the Nabateans; gained commercial importance as a caravan stop on the trade route from Damascus to the Gulf of Aqaba.

Phrygians Peoples of a kingdom founded by Indo-European-language-speaking invaders on the central plateau of Asia Minor; became subject to Persians, Seleucids, Attalids, and, in 116 B.C., were absorbed into the Roman province of Asia.

Pompey (Gnaeus Pompeius) Roman general and consul, 106–48 B.C.;

defeated the Parthian king Mithridates and annexed Syria and Judaea.

Pope, Arthur Upham (1881–1969) Distinguished modern scholar of Persian art, editor of the monumental *A Survey of Persian Art*, and one-time director of the Asia Institute.

praepositus Commander of an auxiliary military unit or, possibly, of a legionary detachment.

praetorium Official residence of a Roman governor.

pronaos Antechamber which leads into the shrine, or naos.

Qumran Khirbet Qumran, located near the Dead Sea, contained a community in the early centuries A.D. whose parchments of the Old Testament, Apocrypha, and secular works were written in Hebrew, Aramaic, and Greek.

Ras Shamra The site of ancient Ugarit, about ten miles north of Latakia on the Syrian coast; produced documents written in an alphabetic cuneiform script of twenty-nine characters in a language related to Hebrew.

Reredos The decorated wall behind an altar or shrine.

Ramadan The sacred month of the Muslim year when fasting from sunup to sunset is mandated.

Rawlinson, George (1812–1902) Orientalist. Camden professor of ancient history at Oxford, canon of Canterbury. His brother, Sir Henry Creswicke Rawlinson (1810–95), served in various governmental positions in the Near East, continued Layard's excavations for the British Museum, deciphered the Old Persian inscription of Darius I at Bisitun.

Rostovtzeff, Michael Ivanovitch (1870–1952) Born near Kiev, educated at the universities of Kiev and St. Petersberg; on the faculty of the University of Wisconsin, 1920–25, and of Yale University, 1925–44, when he retired. Died in New Haven, 1952. Among his voluminous writings are *Ancient Decorative Painting in South Russia* (in Russian), 1913, on painting of the Greek colonies north of the Black Sea; *Iranians and Greeks in South Russia,* 1922, dealing with the relations between Greeks and Scythians in that region; *A History of the Ancient World,* 1926–28, an historical survey; *The Animal Style in South Russia and China,* 1929, a study of the art of the nomads; and the two multivolumed, monumental studies; *The Social and Economic History of the Roman Empire,* 1926; and *The Social and Economic History of the Hellenistic World,* 1941.

Rowell, Henry Thompson (1904–74) Ph.D. from Yale University, oc-

cupied in fieldwork in the late 1920s and early 1930s; chairman of the Department of Classics of Johns Hopkins, and director of the American Academy in Rome.

San Remo Treaty Lloyd George (Great Britain), Alexandre Millerand (France), Francesco Saverio Nitti (Italy), and K. Matsui (Japan) convened in San Remo on the Italian Riviera, April 19–26, 1920, to agree on the disposition of the Turkish territories as part of the Turkish treaty. Syria-Lebanon was mandated to France, Iraq and Palestine to Great Britain, until they were ready for full independence.

San Vitale (Ravenna, Italy) Byzantine church begun ca. A.D. 525 containing mosaic portraits of Justinian and Theodora with their retinues, displayed in formal, frontal poses that Breasted recognized as originating in the style of the paintings of Dura-Europos.

Sarmatians Nomadic tribes, related to the Scythians, who moved slowly westward across South Russia after ca. 250 B.C. and eventually settled in Roman European territories.

Sarre, Friedrich (1865–1945) German archaeologist; founded the Islamic art section of the Berlin Museum; served there as curator and director.

Sasanians Iranian people from southwestern area of Persia (Fars) who rose under Ardashir Papakan in revolt against their Parthian overlords, ca. A.D. 220. The dynasty established by Ardashir, A.D. 226, who, with his son Shapur, came into conflict with Rome. The Sasanians considered themselves the inheritors of the old Persian Achaemenian empire, and sought with some success to reestablish it. The dynasty was finally destroyed by the invading Arabs in A.D. 651.

Schaeffer, Claude F. A. (1898–   ) French archaeologist, excavator in Cyprus, Syria, and Turkey, best known for his discovery of the Ugaritic tablets at Ras Shamra on the Phoenician coast.

Schlumberger, Daniel (1904–1972) French Orientalist and archaeologist; explored Syrian sites around Palmyra and excavated in Afghanistan after World War II.

Scythians Indo-European nomads who composed various tribes described by Herodotus as occupying the broad band of steppe land from South Russia to the Altai Mountains. In the late seventh century B.C. they invaded the Near East, but soon retired and eventually formed settled communities on the north shore of the Black Sea and traded extensively with the Greeks. The Ashkuzai of the Old Testament.

Seleucia-on-the-Tigris Located eighteen miles south of Baghdad, it be-

came one of the major Hellenistic cities and the Seleucid capital; later occupied by Parthians. Located on the caravan route Palmyra-Dura-Selucia-Nippur-Charax, it was also astride one of the main east-west trade routes. After excavating at Dura-Europos, Clark Hopkins continued the excavations of Seleucia in 1936–37.

Seleucids Kings of the Macedonian dynasty founded by Seleucus Nicanor, one of the Diadochi, or successors of Alexander. At its height, the Seleucid Empire stretched from Syria to Afghanistan.

Seleucus I One of Alexander's generals, satrap (governor) of Babylonia; on Alexander's death ruled over the eastern provinces, and following the Battle of Ipsus (301 B.C.), inherited most of the old Persian lands. The Selucid era and calendar begin 312 B.C., the date Seleucus entered Babylon.

Septimius Severus (Lucius) Roman emperor A.D. 193–211; campaigned in the East, annexed Mesopotamia (A.D. 199), and divided Syria into two provinces, Coele and Phoenice.

Seyrig, Henri (1895–1973) Director of the French Archaeological Institute at Beirut, director of the Archaeological Service of Syria; published annual studies in the journal *Syria* on all aspects of Syrian archaeology.

shadow line The drawing on the ground of a figure's cast shadow.

Shapur I Sasanian king, ca. A.D. 241–72; excellent military strategist, builder of the Sasanian Empire. He was successful against the Romans, defeating the Emperor Valerian. Established friendly relations with the Jews after their Roman "captivity." Destroyed the city of Dura-Europos.

socle The projecting footing of a wall, sometimes high enough to be used for the dado.

soffit The underside of an arch, cornice, or architectural overhang.

squamatae The overlapping metal scales sewn on cloth to make defensive armor.

squinch An architectural support carried across a corner under a superimposed mass, usually a dome.

Stein, Sir Marc Aurel (1862–1943) Born in Budapest; served in India at Lahore Oriental College, and as inspector-general of education for the northwest frontier province; made three geographical and archaeological explorations into Central Asia and two expeditions to Baluchistan and south Persia. His most valuable published works on these travels are the eleven-volume *Khotan* (1907), *Serindia* (1911), and *Innermost Asia* (1928).

strategus Leader of an army; chief officer of the boule.

Surena Parthian noble who, with his masterful use of cavalry and cataphractii, defeated the Roman army of Crassus at Carrhae, 53 B.C.

Targum Aramaic translation and paraphrasing of portions of the Old Testament.

Tarsus Early Hellenized capital of Cilicia (in Turkey), renamed Antioch-on-the-Cydnus by the Seleucids, and annexed by Rome in 66 B.C.

Tatian Christian apologist and editor of the Diatessaron; born of Syrian parents in Mesopotamia, and converted to Christianity. The Diatessaron is a synthesis of the Gospels into a single narrative, ca. A.D. 170. The Diatessaron was widely used in Syria as the standard Gospel text.

tell An artificial mound or hill; the remains of an ancient site.

thymiaterion Incense burner.

Tiridates Brother of the Parthian king Vologases I; king of Armenia, deposed by Rome and then renamed to the throne by Nero A.D. 66.

Titus Flavius Vespasianus Son of Vespasian; prosecuted the Jewish Wars, captured Jerusalem in A.D. 70; named Roman emperor in A.D. 79.

Tombs of Gordion Princely burials covered over with huge tumuli of earth and stone at the site of Gordion, Turkey.

Torah The "Law," the first five books of the Old Testament, the Books of Moses.

Trajan (Marcus Ulpius Traianus) Roman emperor, A.D. 98–117; conquered Nabataea and later Armenia and Mesopotamia; took the Parthian capital of Ctesiphon.

tribune Roman commander of troops; elected from the common people (the tribes) as their representative.

triclinium Dining room of Roman house.

Tyche Personification of good fortune, identified with Roman Fortuna; particularly popular in Hellenistic and Roman times as patroness of cities. As Chance, she is the unforeseen in human affairs.

Uruk Ancient Sumerian city, the biblical Erech, modern Warka.

Valerian (Publius Licinius Valerianus) Roman emperor A.D. 253–60; issued edicts against the Christians resulting in general persecutions of A.D. 257–58, confiscation of Christian property, and reduction of Christian communities to second-class citizenship.

Vardanes Parthian ruler who shared throne with his brother Gotarzes, ca. A.D. 39–47; captured Seleucia-on-the-Tigris.

Virolleaud, Charles J. G. (1879–1968) French advisor for archaeology and the arts in the 1920s, student of Mesopotamian mythology, reporter on work of the French School of Jerusalem.

Vologases I King of Parthia, ca. A.D. 51–80; spent most of his years at war with Rome, but later established amicable relations with Vespasian.

voussoir Wedge-shaped block used to form an arch.

wadi Sharp declivity, cut, or depression in arid region usually made by the erosion of land from the runoff of heavy rains.

Wilber, Donald Newton (1907–) Graduate of Princeton University, member of archaeological expeditions in Egypt, Greece, France, Syria, and Iran; served as U.S. government official; has written extensively on the past and present of Iran, Afghanistan, Pakistan, and Ceylon.

Wilson, Sir Arnold Talbot (1884–1940) Served in various posts in India and Persia before World War I and explored in Luristan and Fars; deputy chief political officer of the Indian Expeditionary Force "D" in 1915; deputy and then acting civil commissioner in Iraq until he resigned government service in 1920. Shot down and killed behind enemy lines in World War II.

Xerxes Persian Achaemenian king, 486–465 B.C.; son of Darius the Great; invaded Greece, but his forces were defeated in the naval battle of Salamis.

Zenobia Aramaic Bat Zabbai, wife of Odaenathus. She and her husband were particularly powerful as rulers of Palmyra and surrounding areas in Syria. Aurelian's armies took Palmyra and transported Zenobia to Rome.

Zeus Greek sky god, father of gods and men. Epithets added to his name describe his attributes: thus Zeus-Kyrios (Zeus the Lord), Zeus-Megistos (Zeus the Greatest), Zeus-Theos (Zeus God). In Hellenistic Syria he was syncretized with and recognized in the local gods: hence he was called Zeus-Baal, Zeus-Bel, or Zeus-Ahura-Mazda.

Notes

Abbreviations

Breasted, *Forerunners*	Breasted, J. H. *Oriental Forerunners of Byzantine Painting.* Chicago, 1924.
Cumont, *Fouilles*	Cumont, F. *Fouilles de Doura-Europos.* Paris, 1926. 2 vols.
Du Mesnil, *Peintures*	Comte Du Mesnil du Buisson. *Les Peintures de la synagogue de Doura-Europos, 245–56 après J.-C.* Rome, 1939.
Goodenough, *Syn.*	Goodenough, E. R. *Jewish Symbols in the Greco-Roman Period.* New York, 1964. Vols. 9–11.
Kraeling, *Chr. Bldg.*	Welles, C. B., ed. *The Excavations at Dura-Europos. Final Report* VIII, pt. ii. C. H. Kraeling. *The Christian Building.* New Haven, 1967.
Kraeling, *Syn.*	Bellinger, A. R. et al., eds. *The Excavations at Dura-Europos. Final Report* VIII, pt. i. C. H. Kraeling. *The Synagogue.* New Haven, 1956.
Perkins, *Dura*	Perkins, A. *The Art of Dura-Europos.* Oxford, 1973.
Rep.	Baur, P. V. C. et al., eds. *The Excavations at Dura-Europos. Preliminary Report of the —— Season of Work ——.* New Haven, 1929–52. 8 vols. in 10 parts.
Wischnitzer, *Messianic*	Wischnitzer, R. *The Messianic Theme of the Dura Synagogue.* Chicago, 1948.

Chapter 1

1. Breasted, *Forerunners,* p. 53.
2. Ibid.
3. Ibid., p. 52.
4. Bell, G., *Amurath to Amurath* (London, 1911), p. 80.
5. Breasted, C., *Pioneer to the Past* (New York, 1945), p. 261.
6. Breasted, *Forerunners,* p. 54.
7. Ibid., p. 50.
8. Ibid., p. 54.
9. Breasted, *Forerunners,* p. 98.
10. Cernik, J., *Technische Studien-Expedition durch die Gebiete des Euphrat und Tigris. Petermann's Geographischen Mittheilungen,* Ergängungsheft 44 (1875), pp. 39ff, Taf. 2.

11. Bell, pp. 111–13.
12. Sarre, F., and E. Herzfeld, *Archäologische Reise im Euphrat- und Tigris-Gebiet* (Berlin, 1911–20), II, pp. 386–95; III, Tafn. 81–83.
13. Breasted, *Pioneer,* p. 287.

Chapter 2

1. Cumont, F., "Rapport sur une mission a Sâlihïyeh sur l'Euphrate," *Comptes rendus, Académie des inscriptions et belles lettres* (1923), pp. 12–13.
2. Cumont, F., "Les Fouilles de Sâliḥîyeh sur l'Euphrate," *Syria* 4 (1923): 39–41.
3. Breasted, *Forerunners,* pp. 3–4.
4. Cumont, pp. 14–15.
5. Breasted, *Forerunners,* p. 86.
6. Cumont, *Fouilles,* parchment IIA, line 10.
7. Perkins, *Dura,* p. 41.
8. *Rep.* VII–VIII, p. 201, fig. 50.
9. Cumont, *Fouilles,* pp. v–vi.
10. Cumont, *Fouilles,* pp. 278–79.
11. Cumont, *Fouilles,* pl. lv.
12. The chariot variously used as bringer of light, e.g. Daniélou, J., *Primitive Christian Symbols* (Baltimore, 1961), p. 78; Barnett, R. D., *A Catalogue of the Nimrud Ivories* (London, 1957), p. 212.

Chapter 3

1. Cumont, *Fouilles,* pp. ix–x.
2. Rostovtzeff, M. I., *Caravan Cities* (Oxford, 1932), p. 158.
3. *Rep.* I, p. 2.
4. *Rep.* I, p. 2.
5. Rostovtzeff, p. 159.
6. *Rep.* I, p. 19.
7. *Rep.* I, pp. 59, 65, 67.
8. *Rep.* I, pp. 6–7.
9. *Rep.* I, pp. 50–51.
10. *Rep.* I, p. 23.

Chapter 4

1. *Rep.* II, pp. 1–2.
2. *Rep.* II, p. 3.
3. *Rep.* II, pl. xxxix.
4. *Rep.* II, pp. 163–64.

5. *Rep.* II, pl. xli, 2.
6. *Rep.* II, p. 201.
7. *Rep.* II, p. 33.
8. *Rep.* II, pp. 117, 132.
9. Mirsky, J., *Sir Aurel Stein* (Chicago, 1977), p. 459.

Chapter 5

1. *Rep.* III, p. 2.
2. *Rep.* III, pl. xiv.
3. *Rep.* III, pp. 35–36.
4. Cumont, *Fouilles,* p. 183.
5. *Rep.* III, p. 31.
6. *Rep.* III, pl. vii.
7. Andrae, W., and H. Lenzen, *Die Partherstadt Assur. WVDOG* 57 (1933), Taf. 10.
8. *Rep.* III, p. 39.
9. *Rep.* III, p. 89.
10. Downey, S., *The Heracles Sculpture* (Yale, 1969), p. 57.
11. *Rep.* IV, pl. xxii, 2.
12. *Rep.* IV, p. 27, pl. iv.
13. *Rep.* IV, p. 28.
14. *Rep.* IV, pp. 73–74.
15. *Rep.* IV, pp. 182 and 199.
16. *Rep.* IV, pp. 183, 187, pls. xvii, xviii.
17. Lloyd, S., *Art of the Ancient Near East* (London, 1961), fig. 160.
18. *Rep.* IV, p. 284.

Chapter 6

1. *Rep.* V, p. 24.
2. Goldman, B., *The Sacred Portal* (Detroit, 1966), p. 123.
3. *Rep.* V, p. 241.
4. Kraeling, *Chr. Bldg.,* p. 96.
5. *Rep.*V, pp. 246–47.
6. Kraeling, *Chr. Bldg.,* pp. 140–41.
7. von Gerkan, A., "Die frühchristliche Kirchenanlage von Dura," *Römische Quartalschrift* 42 (1934): 219–32.
8. Kraeling, *Chr. Bldg.*, p. 9.
9. Kraeling, *Chr. Bldg.*, p. 138.
10. Graham, J. W., "The Greek and the Roman House," *Phoenix* 20 (1966): 3–31.
11. *Rep.* VII–VIII, pp. 50, 58.
12. Kraeling, *Chr. Bldg.*, p. 38.

13. Kraeling, *Chr. Bldg.*, pp. 18–19.
14. Kraeling, *Chr. Bldg.*, p. 132.
15. See Kraeling's masterful article "The Sator Acrostic," *The Crozier Quarterly* 12 (1945): 28–38.

Chapter 7

1. C. H. Kraeling, "A Greek Fragment of Tatian's Diatessaron from Dura," in K. Lake and S. Lake, eds., *Studies and Documents* 3 (1935): 3–35.
2. Ibid., p. 19.
3. Kraeling, *Chr. Bldg.,* pp. 57, 202.
4. Perkins, *Dura,* p. 53.
5. See the excellent account of Kurt Weitzmann, "Narration in Early Christendom," *American Journal of Archaeology* 61 (1957): 83–91.
6. Kraeling, *Chr. Bldg.*, p. 61.
7. Weitzmann, "Narration," p. 85.
8. Kraeling, *Chr. Bldg.*, p. 82.
9. Kraeling, *Chr. Bldg.*, pp. 84–85.
10. Kraeling, *Chr. Bldg.*, p. 85.
11. Kraeling, *Chr. Bldg.,* p. 151.
12. Kraeling, *Chr. Bldg.,* p. 146.
13. Waterman, L., *The Religion of Jesus* (New York, 1952), p. 72.

Chapter 8

1. *Rep.* VI, pp. 31–32.
2. *Rep.* V, p. 219.
3. *Rep.* VI, p. 19.
4. *Rep.* VI, p. 152.
5. *Rep.* VI, pp. 51, 63.
6. *Rep.* VI, p. 66.
7. *Rep.* VI, frontispiece, pls. xxv, xxv-A.
8. *Rep.* VI, pl. xxii.

Chapter 9

1. *Rep.* VI, p. 335.
2. Kraeling, *Syn.,* p. 263, tiles A and B; Aramaic text by Torrey; Goodenough, *Syn.* XI, pp. 46–47.
3. Kraeling, *Syn.,* pp. 38–39.

4. Kraeling, *Syn.,* p. 338.
5. Wischnitzer, *Messianic,* p. 5.
6. Kraeling, *Syn.,* p. 62.
7. Kraeling, *Syn.,* p. 63.
8. Frankfort, H., *Cylinder Seals* (London, 1939), pp. 210–11.
9. Goldman, B., *The Sacred Portal* (Detroit, 1966), p. 90.
10. Kraeling, *Syn.,* p. 363.
11. Josephus, *Antiquities* XX, 2–4.
12. Kraeling, *Syn.,* pp. 230–31; Wischnitzer, *Messianic,* pp. 79–80.
13. Grabar, A., "Le Thème religieux des fresques de la synagogue de Doura (245–56 après J.-C.)," *Revue de l'histoire des religions* 123 (1941): 178–79.
14. Du Mesnil, *Peintures,* p. 115; Kraeling, *Syn.,* p. 232; Goodenough, *Syn.,* IX, p. 113; Grabar, pp. 178–80; Wischnitzer, *Messianic,* pp. 83–85; Sonne, I., "The Paintings of the Dura Synagogue," *Hebrew Union College Annual* 20 (1947): 300.
15. Sonne, p. 300.
16. Kraeling, *Syn.,* p. 234, that the covered receptacle was the scroll chest from which the reader has taken the sacred scriptures.
17. Gutmann, J., ed., *The Dura-Europos Synagogue* (Missoula, 1973), p. 149.
18. On the winged angels in Jewish and Christian art, see F. Landsberger, "The Origin of the Winged Angel in Jewish Art," *Hebrew Union College Annual* 20 (1947): 227–56.
19. Kraeling, *Syn.,* p. 90.
20. Kraeling, *Syn.,* pl. xxviii.
21. Kraeling, *Syn.,* p. 94, pl. xxix; Goodenough, *Syn.* XI, fig. 346.
22. Kraeling, *Syn.,* pls. liv, lv; Goodenough, *Syn.* XI, fig. 347.
23. Kraeling, *Syn.,* pp. 107, 110.
24. Kraeling, *Syn.,* p. 125.
25. Kraeling, *Syn.,* pp. 130–131; Goodenough, *Syn.* X, p. 20.
26. Kraeling, *Syn.,* p. 113; Goodenough, *Syn.* X, p. 144.
27. Kraeling, *Syn.,* p. 177, note 680.
28. Kraeling, *Syn.,* pp. 136–37; Goodenough, *Syn.* XI, fig. 340.
29. Gutmann, "Dura-Europos Synagogue," p. 149.

Chapter 10

1. *Rep.* V, p. 291.
2. *Rep.* VII-VIII, pl. xxxi-1.
3. *Rep.* VII-VIII, pl. xxxi-2.
4. Cumont, *Fouilles,* pp. 274–77.
5. *Rep.* VII-VIII, pp. 85–87; Cumont, F., "The Dura Mithraeum, in J. R. Hinnells, ed., *Mithraic Studies* I (Manchester, 1975), p. 152.
6. To Cumont fell the task of writing the final report on the Mithraeum, which was finally edited and published by E. D. Francis in *Mithraic Studies* I, pp. 151–214.
7. *Rep.* VII-VIII, pp. 110–11.

Chapter 11

1. Mallowan, A. C., *Come, Tell Me How You Live* (London, 1946), pp. 40–41.
2. *Rep.* VII-VIII, p. 310.
3. *Rep.* VII-VIII, p. 326.
4. *Rep.* VII-VIII, pls. xli–xlii.
5. *Rep.* VII-VIII, pls. xliv–xlv.
6. *Rep.* VII-VIII, p. 367, pl. xlvi.
7. *Rep.* VII-VIII, p. 218.
8. *Rep.* VII-VIII, p. 234, fig. 67.
9. *Rep.* VII-VIII, p. 244.
10. *Rep.* V, p. 289.
11. *Rep.* IX-3, pp. 55–56, 94–95.
12. *Rep.* IX-3, pp. 58–59, pl. xii.
13. *Rep.* IX-3, p. 130, fig. 9.
14. *Rep.* IX-3, pp. 96–97.
15. Cumont, *Fouilles,* p. 275.
16. Cumont, *Fouilles,* p. 276.
17. *Rep.* IX-2, pp. 146–47.

Chapter 12

1. Cumont, *Fouilles,* p. 22.
2. Cumont, *Fouilles,* p. 24, fig. 7.
3. Cumont, *Fouilles,* p. 286.
4. *Rep.* I, p. 35, Inscription R5b.
5. *Rep.* VII-VIII, p. 26.
6. Bellinger, A. R., *The Coins* (New Haven, 1949), p. 201.
7. *Rep.* VII-VIII, p. 20.
8. von Gerkan, A., "Die Frühchristliche Kirchenanlage von Dura," *Römische Quartalschrift* 42 (1934): 219–32.

Bibliography

The Excavations at Dura-Europos. Preliminary Reports. New Haven.
First Season, Spring 1928. Baur, P. V. C., and M. I. Rostovtzeff, eds. 1929.
Second Season, 1928–1929. Baur, P. V. C., and M. I. Rostovtzeff, eds. 1931.
Third Season, 1929–1930. Baur, P. V. C., M. I. Rostovtzeff, and A. R. Bellinger, eds. 1932.
Fourth Season, 1930–1931. Baur, P. V. C., M. I. Rostovtzeff, and A. R. Bellinger, eds. 1933.
Fifth Season, 1931–1932. Rostovtzeff, M. I., ed. 1934.
Sixth Season, 1932–1933. Rostovtzeff, M. I., A. R. Bellinger, C. Hopkins, and C. B. Welles, eds. 1936.
Seventh and Eighth Seasons, 1933–1934 and 1934–1935. Rostovtzeff, M. I., F. E. Brown, and C. B. Welles, eds. 1939.
Ninth Season, 1935–1936. Rostovtzeff, M. I., A. R. Bellinger, F. E. Brown, and C. B. Welles, eds.
Part I: *The Agora and Bazaar.* 1944.
Part II: *The Necropolis.* 1946.
Part III: *The Palace of the Dux Ripae and the Dolicheneum.* 1952.
The Excavations at Dura-Europos. Final Reports. New Haven.
III, Part I, Fascicle 1: Downey, S. *The Heracles Sculpture.* 1969.
Part II, Fascicle 2: Downey S. *The Stone and Plaster Sculpture.* 1977 (Los Angeles).
IV, Part I, Fascicle 1: Toll, N. *The Green Glazed Pottery.* 1943.
Fascicle 2: Cox, D. H. *The Greek and Roman Pottery.* 1949.
Part II, Pfister, R., and L. Bellinger. *The Textiles.* 1945.
Part III, Baur, P. V. C. *The Lamps.* 1947.
Part IV, Fascicle 1: Frisch, T. G., and N. Toll. *The Bronze Objects.* 1949.
Part V, Clairmont, C. *The Glass Vessels.* 1963.
Fascicle 3: Dyson, S. L. *The Commonware Pottery, the Brittle Ware.* 1968.
V, Part I, Welles, C. B., R. O. Fink, and J. F. Gilliam. *The Parchments and Papyri.* 1959.
VI, Bellinger, A. R. *The Coins.* 1949.
VIII, Part I, Kraeling, C. H. *The Synagogue.* 1956.
Part II, Kraeling, C. H. *The Christian Building.* 1967.

Alföldi, A. "Die Hauptereignisse der Jahre 253–61 n. Chr. im Orient im Spiegel der Münzprägung." *Berytus* 4 (1937): 41–68.
Altheim, F., and R. Stiehl. *Asien und Rom, Neue Urkunden aus sasanischer Frühzeit.* Tübingen, 1952.
———. "Inscriptions of the Synagogue of Dura-Europos." *East and West* 9 (1958): 7–28.

Aubert, M. "Les Fouilles de Doura-Europos." *Bulletin monumental* 92 (1934): 397–407.
———. "La Peinture de la synagogue de Doura." *Gazette des Beaux-Arts* 20 (1938): 1–24.
———. "Les Peintures de la chapelle chrétienne de la synagogue de Doura-Europos." *Bulletin monumental* 97 (1939): 121–24.
Baur, P. V. C. "The Christian Chapel at Dura." *American Journal of Archaeology* 37 (1933): 377–80.
———. "Les Peintures de la chapelle chrétienne de Doura." *Gazette des Beaux-Arts* 14 (1933): 65–78.
———. "Megarian Bowls in the Rebecca Darlington Stoddard Collection of Greek and Italian Vases in Yale University." *American Journal of Archaeology* 45 (1941): 229–48.
———. "The Cock and Scorpion in the Orthonobazos Relief at Dura-Europos." in *Studies Presented to D. M. Robinson,* vol I. Ed. by G. E. Mylonas. St. Louis, 1951.
Bellinger, A. R. *Two Roman Hoards from Dura-Europos. Numismatic Notes* 49. New York, 1931.
———. *The Third and Fourth Dura Hoards. Numismatic Notes* 55. New York, 1932.
———. *The Sixth, Seventh, and Tenth Dura Hoards. Numismatic Notes* 69. New York, 1935.
———. *The Eighth and Ninth Dura Hoards. Numismatic Notes* 85. New York, 1939.
———. "The Numismatic Evidence from Dura." *Berytus* 8 (1943): 61–71.
———. "Seleucid Dura; The Evidence of the Coins." *Berytus* 9 (1948): 51–67.
Ben-Shammai, M. H. "The Legends of the Destruction of the Temple among the Paintings of the Dura Synagogue," (in Hebrew). *Bulletin of the Jewish Palestine Exploration Society* 9 (1942): 93–97.
Bickerman, E. J. "Symbolism in the Dura Synogogue." *Harvard Theological Review* 58 (1965): 127–52.
Breasted, J. H. "Peinture d'époque romaine dans le désert de Syrie." *Syria* 3 (1922): 177–213.
———. *Oriental Forerunners of Byzantine Painting.* Chicago, 1924.
———. *The Oriental Institute.* Chicago, 1933.
Casel, O. "Älteste christliche Kunst und Christusmysterium." *Jahrbuch für Liturgiewissenschaft* 12 (1934): 74.
Contenau, G. "La Scène pastorale de Doura." *Revue Assyrologique* 37 (1940): 123–26.
Cumont, F. "Rapport sur une mission a Sâlihīyeh sur l'Euphrate." *Comptes rendus, Académie des inscriptions et belles-lettres* (1923): 12–41.
———. "Le Temple aux gradins découvert à Salihiyeh et ses inscriptions." *Syria* 4 (1923): 203–23.
———. "Les Fortifications de Doura-Europos." *Syria* 5 (1924): 24–43.
———. *Fouilles de Doura-Europos* (1922–1923). 2 vols. Paris, 1926.
———. "Une Campagne de fouilles à Doura." *Revue archéologique* 4 (1934): 173–79.
———. "The Dura Mithraeum." trans. and ed. by E. D. Francis, in *Mithraic Studies* I. Ed. J. R. Hinnells. Manchester, 1975.
Dinkler, E. "Dura-Europos III, Bedeutung für die christliche Kunst." *Religion in Geschichte und Gegenwart* 2 (1958), cols. 290–92.
Dussaud, R. "La Résurrection des ossements desséchés." *Syria* 20 (1939): 91–92.
Ehrenstein, T. *Über die Fresken der Synagoge von Dura Europos, eine Studie.* Vienna, 1937.

Eissfeldt, R. "Dura-Europos." *Reallexikon für Antike und Christentum* IV (1958), cols. 362–67.
Ferrua, A. "Dura-Europo cristiana." *Civiltà cattolica* 4 (1939): 334–37.
———. "Dura-Europo e la sua sinegoga." *Civiltà cattolica* 4 (1939): 75–85.
Fink, R. O., A. S. Hoey, and W. F. Snyder. "The Feriale Duranum." *Yale Classical Studies* VII. New Haven, 1940.
Francis, E. D. "Mithraic Graffiti from Dura-Europos." In J. R. Hinnells, ed., *Mithraic Studies* II. Manchester, 1975.
Frey, J.-B. "La Question des images chez les juifs à la lumière des récentes découvertes." *Biblica* 15 (1934): 265–300.
Frye, R. N., J. F. Gilliam, H. Ingholt, and C. B. Welles. "Inscriptions from Dura-Europos." *Yale Classical Studies* XIV. New Haven, 1955.
von Gerkan, A. "Die frühchristliche Kirchenanlage von Dura." *Römische Quartalschrift* 42 (1934): 219–32.
———. "Zur Hauskirche von Dura-Europos." *Mullus, Festschrift Theodor Klausner. Reallexikon zur byzantinischen Kunst, Ergänzungsband* I (1964): 143–49.
Gilliam, J. F. "The Dux Ripae at Dura." *Transactions and Proceedings of the American Philological Association* 72 (1941): 157–75.
Goodenough, E. R. *By Light, Light.* New Haven, 1935.
———. "The Crown of Victory in Judaism." *Art Bulletin* 28 (1946): 139–59.
———. "The Evaluation of Symbols Recurrent in Time as Illustrated in Judaism." *Eranos Jahrbuch* 20 (1951): 285–319.
———. "The Paintings of the Dura-Europos Synagogue: Method and an Application." *Israel Exploration Journal* 8 (1958): 69–79.
———. *Jewish Symbols in the Greco-Roman Period,* vols. IX–XI. New York, 1964.
———. "The Greek Garments on Jewish Heroes in the Dura Synagogue." In A. Altman, ed., *Biblical Motifs, Origins and Transformations.* Cambridge, 1966.
Grabar, A. "Les Fresques de la synagogue de Doura-Europos." *Comptes rendus, Académie des inscriptions et belles-lettres* (1941), pp. 77–90.
———. "Le Thème religieux des fresques de la synagogue de Doura (245–56 après J.-C.)." *Revue de l'histoire des religions* 123 (1941): 143–92; 124 (1941): 5–35.
———. "Le Fresque des saintes femmes au tombeau à Doura." *Cahiers archéologiques* 8 (1956): 9–26.
Grégoire, H. "Les Baptistères de Cuicul et de Doura." *Byzantion* 13 (1938): 589–93.
———. "Encore les baptistères de Cuicul et de Doura." *Byzantion* 14 (1939): 317.
Gutmann, J. "The Dura Frescoes." *The Reconstructionist* 31 (1964): 20–25.
———. "Die Synagoge von Dura-Europos." K. Wessel, ed., *Reallexikon zur byzantinischen Kunst* I (1966), cols. 1230–40.
Gutmann, J., ed. *The Dura-Europos Synagogue.* Missoula, 1973.
Harmatta, J. "The Parthian Parchment from Dura-Europos." *Acta Antiqua Academiae Scientiarum Hungaricae* 5 (1957): 261–308.
Hempel, H. L. "Zum Problem der Anfänge der AT-Illustration." *Zeitschrift für die Alttestamentliche Wissenschaft* 69 (1957): 103–31.
Hill, E. "Roman Elements in the Settings of the Synagogue Frescoes at Dura." *Marsyas* 1 (1941): 1–15.
Hopkins, C. "The Palmyrene Gods at Dura-Europos." *Journal of the American Oriental Society* 51 (1931): 119–37.

———. "Dura-Europos Discoveries: the Unexpected in Archaeology." *Illustrated London News,* 13 August 1932, 239–41.
———. "Horse-Armour and a Unique Painted Shield." *Illustrated London News,* 2 September 1933, 362–63.
———. "Jewish Prototypes of Early Christian Art." *Illustrated London News,* 29 July 1933, 188–91.
———. "The Sixth Campaign at Dura-Europos, 1932–33." *American Journal of Archaeology* 37 (1933): 471–74.
———. "The Tragedy of a Buried City Told by Its Ruins." *Illustrated London News,* 22 September 1934, 421–23.
———. "A Flood of New Light on Mithraism." *Illustrated London News,* 8 December 1934, 963–65.
———. "The Christian Chapel at Dura-Europos." *Atti del III congresso internazionale di archeologia cristiana.* (Rome, 1934), pp. 483–92.
———. "Roman Painted Shields and Temple Sculptures from Dura-Europos." *Illustrated London News,* 31 August 1935, 350–51.
———. "The Season 1933–34 at Dura." *American Journal of Archaeology* 39 (1935): 293–99.
———. "New Monuments from Dura." *Bulletin of the Associates in Fine Arts at Yale University* 9 (1935): 55–58.
———. "Yale Excavations at Dura-Europos, 1934–35." *American Journal of Archaeology* 40 (1936): 192–93.
———. "A Note on Frontality in Near Eastern Art." *Ars Islamica* 3 (1936): 187–200.
———. "Aspects of Parthian Art." *Berytus* 3 (1936): 1–30.
———. "Architectural Background in the Paintings at Dura-Europos." *American Journal of Archaeology* 45 (1941): 18–29.
———. "The Christian Chapel at Dura-Europos." *St. Joseph's Lilies* 36 (1947): 127–33.
———. "The Siege of Dura." *Classical Journal* 42 (1947): 251–59.
Hopkins, C., and M. I. Rostovtzeff. "La Dernière campagne de fouilles de Doura-Europos," *Comptes rendus, Académie des inscriptions et belles-lettres* (1932), pp. 314–28.
Hopkins, C., and Comte Du Mesnil du Buisson. "La Synagogue de Doura-Europos." *Comptes rendus, Académie des inscriptions et belles-lettres* (1933), pp. 243–54.
de Jerphanion, G. "Bulletin d'archéologie chrétienne." *Orientalia christiana* 28 (1932): 296.
Johnson, J. *Dura Studies.* Philadelphia, 1932.
Kirsch, J. P. "Die Entdeckung eines christlichen Gotteshauses und einer jüdischen Synagoge mit Malereien aus der ersten Hälfte des 3. Jahrhunderts in Dura-Europos in Mesopotamien." *Oriens christianus,* ser. 3, vol. 8 (1933): 201–04.
———. "Die vorkonstantinischen Kultusgebäude im Lichte der neuesten Entdeckungen im Orient." *Römische Quartalschrift* 41 (1933): 15–28.
———. "La *Domus Ecclesiae* cristiana del III. secolo a Doura-Europos in Mesopotamia." *Studi dedicati alla memoria di Paolo Ubaldi.* Milan, 1937, pp. 73–82.
———. "La Basilica cristiana nell-antichità." *Atti de IV. congresso internazionale di archeologia cristiana* I. Rome, 1940, pp. 113–26.
Kittel, G. "Die ältesten jüdischen Bilder: eine Aufgabe für die wissenschaftliche Gemeinschaftsarbeit." *Forschungen zur Judenfrage* 4 (1940): 237–49.
Kollwitz, J. "Beziehungen zur christlichen Archäologie." *Jahrbuch für Liturgiewissenschaft* 13 (1935): 310–12.

Kraeling, C. H. "The Earliest Synagogue Architecture." *Bulletin of the American Schools for Oriental Research* 54 (1934): 18–20.
———. "A Greek Fragment of Tatian's Diatessaron from Dura." in K. Lake and S. Lake, eds., *Studies and Documents* 3 (London, 1935): 3–37.
———. "The Sator Acrostic." *The Crozier Quarterly* 12 (1945): 28–38.
———. "The Meaning of the Ezekiel Panel in the Synagogue at Dura." *Bulletin of the American Schools of Oriental Research* 78 (1940): 12–18.
Kümmel, W. G. "Die älteste religiöse Kunst der Juden." *Judaica* 2 (1946): 1–56.
Lambert, E. "La Synagogue de Doura-Europos et les origines de la mosquée." *Semitica* 3 (1950): 67–72.
Leveen, J. *The Hebrew Bible in Art.* London, 1944.
Louis, R. "La Visite des saintes femmes au tombeau dans le plus ancien art chrètien." *Recueil publié à l'occasion du cent-cinquantenaire de la société nationale des antiquaires de France* (Paris, 1954), pp. 109–22.
Mesnil du Buisson, Comte R. du. "La Guerre de sape il y a dix-sept siècles." *L'Illustration,* 5 August 1933, pp. 481–83.
———. "Les Fouilles de Doura-Europos, 1932–33." *L'Illustration,* 29 July 1933, pp. 454–57.
———. "Les Fouilles de Doura-Europos en 1932–33." *Revue archéologique* 4 (1934): 69–75.
———. "Les Nouvelles découvertes de la synagogue de Doura-Europos." *Revue biblique* 43 (1934): 1–18; 105–19; 546–63.
———. "Une Peinture de la synagogue de Doura-Europos." *Gazette des beaux-arts* 14 (1935): 193–203.
———. "Les Miracles de l'eau dans le désert d'après les peintures de la synagogue de Doura-Europos." *Revue de l'histoire des religions* 111 (1935): 110–17.
———. "Un Temple du soleil dans la synagogue de Doura-Europos." *Gazette des beaux-arts* 16 (1936): 83–94.
———. "Nouvelles observations sur une peinture de Doura-Europos." *Gazette des beaux-arts* 16 (1936): 305–06.
———. "Les deux synagogues successives à Doura-Europos." *Revue biblique* 45 (1936): 72–90.
———. "L'Étendard d'Atargatis et Hadad à Doura-Europos ou la déesse Sèmia." *Revue des arts asiatique* 11 (1937): 75–87.
———. *Les Peintures de la synagogue de Doura-Europos.* Rome, 1939.
———. "Inscriptions sur jarres de Doura-Europos." *Mélanges de l'Université Saint-Joseph* 36 (1959): 3–49.
Meyer, R. "Betrachtungen zu drei Fresken der Synagoge von Dura-Europos." *Theologische Literaturzeitung* 74 (1949): 29–38.
Millet, G. "La Scène pastorale de Doura." *Syria* 7 (1926): 142–51.
———. "Dura et Bagawat," *Cahiers archéologiques* 8 (1956): 1–8.
Newell, E. T. *The Fifth Dura Hoard. Numismatic Notes* 58. New York, 1932.
Nock, A. D. "The Synagogue Murals of Dura-Europos." *Harry A. Wolfson Jubilee,* II. Jerusalem, 1965.
Nordström, C.-O. "The Water Miracles of Moses in Jewish Legend and Byzantine Art." *Orientalia Suecana* 7 (1958): 78–109.
Noth, M. "Dura-Europos und seine Synagoge." *Zeitschrift des deutschen Palästina-Vereins* 75 (1959): 164–81.

Obermann, J. "Inscribed Tiles from the Synagogue of Dura." *Berytus* 7 (1942): 89–138.
Orlandos, A. "The New Mithraeum at Dura." *American Journal of Archaeology* 39 (1935): 4–5.
Pagliaro, A. "Date e pittori nella sinagoga di Dura-Europo." *Rivista degli studi orientalni* 28 (1953): 170–73.
Pearson, H. F. *A Guide to the Synagogue of Doura-Europos.* Beirut, 1939. (French edition, 1940.)
Pelekanides, J. "The Early Christian Baptisterium" (in Greek). *Neas Sion* (1936): 52–57.
Perkins, A. *The Art of Dura-Europos.* Oxford, 1973.
Pijoan, J. "The Parable of the Virgins from Dura-Europos." *Art Bulletin* 19 (1937): 592–95.
Quasten, J. "Das Bild des Guten Hirten in den altchristlichen Baptisterien und in den Taufliturgien des Ostens und Westens." *Pisciculi. Antike und Christentum Ergänzungsband* I (Münster, 1939): 220–44.
———. "The Painting of the Good Shepherd at Dura-Europos," *Mediaeval Studies* 9 (1947): 1–18.
Renov, I. "A View of Herod's Temple from Nicanor's Gate in a Mural Panel of the Dura-Europos Synagogue." *Israel Exploration Journal* 20 (1970): 67–74; 21 (1971): 220–21.
Riesenfeld, H. "The Resurrection in Ezekiel XXXVII and in the Dura-Europos Paintings." *Uppsala University Arsskrift* 11 (1948): 27–38.
Rostovtzeff, M. I. "Un Contrat de prêt de l'an 121 ap. J.-C. trouvé à Doura." *Comptes rendus, Académie des inscriptions et belles-lettres* (1930): 158–89.
———. "Yale's Work at Doura." *Bulletin of the Associates in Fine Arts at Yale University* 4 (1930): 74–85.
———. "L'Art gréco-iranien," *Revue des arts asiatique* 7 (1931–32): 202–22.
———. *Caravan Cities.* Oxford, 1932.
———. "Das Militärchiv von Dura." *Münchener Beiträge zur Papyrusforschung und antiken Rechtsgeschichte* 19 (1934): 351–78.
———. "Il Rebus Sator." *Annali della R. Scuola Normale Superiore di Pisa* 3 (1934): 103–05.
———. "Das Mithraeum von Dura." *Mitteilungen des deutschen archäologischen Instituts, Römische Abteilung* 49 (1934): 180–207.
———. "Die Synagogue von Dura." *Römische Quartalschrift* 42 (1934): 203–18.
———. "Dura and the Problem of Parthian Art." *Yale Classical Studies* V. New Haven, 1935.
———. "Progonoi." *Journal of Hellenic Studies* 55 (1935): 56–66.
———. "The Squatting Gods in Babylonia and Dura." *Iraq* 4 (1937): 19–21.
———. *Dura-Europos and Its Art.* Oxford, 1938.
———. "Le Gad de Doura et Seleucus Nicator." *Mélanges Syriens, R. Dussaud* I. Paris, 1939.
——— "Res Gestae Divi Saporis and Dura." *Berytus* 8 (1943): 17–60.
Rostovtzeff, M. I., and C. B. Welles. "A Parchment Contract of Loan from Dura-Europos on the Euphrates." *Yale Classical Studies* II. New Haven, 1931.
Rowet, J. "De picturis sacris in D.-E. detectis." *Verbum Domini* 1936 19 (1939): 366–70.

Schneid, N. *The Paintings of the Synagogue at Dura-Europos* (in Hebrew). Tel Aviv, 1946.
Seston, W. "L'Église et le baptistère de Doura-Europos." *Annales de l'école des hautes-études de Gand,* 1937, pp. 161–77.
Seyrig, H. "Travaux archéologiques en Syrie, 1930–31." *Jahrbuch des deutschen archäologischen Instituts* 46 (1931), cols. 575–96.
Shoe, L. "Architectural Mouldings of Dura-Europos." *Berytus* 9 (1948): 1–40.
Simon, M. "Remarques sur les synagogues à images de Doura et de Palestine." *Recherches d'histoire judéo-chrétienne* (Paris, 1962), pp. 188–208.
Sonne, I. "The Paintings of the Dura Synagogue." *Hebrew Union College Annual* 20 (1947): 255–362.
Stechow, W. "Jacob Blessing the Sons of Joseph." *Gazette des beaux-arts* 23 (1943): 193–208.
Stephens, F. J. "A Cuneiform Tablet from Dura-Europos." *Revue d'assyriologie* 34 (1937): 183–90.
Stern, H. "The Orpheus in the Synagogue of Dura-Europos." *Journal of the Warburg and Courtauld Institutes* 21 (1958): 1–6.
———. "Quelques problèmes d'iconographie paléochrétienne et juive," *Cahiers archéologiques* 12 (1962): 99–113.
Sukenik, E. "The Ezekiel Panel in the Wall Decoration of the Synagogue of Dura-Europos." *Journal of the Jewish Palestine Oriental Society* 18 (1938): 57–62.
———. "Discovery of an Ancient Synagogue," *Art and Archaeology* 33 (1932): 206–12.
———. *The Synagogue of Dura-Europos and Its Paintings* (in Hebrew). Jerusalem, 1947.
de Vaux, R. "Un Détail de la synagogue de Doura." *Revue biblique* 47 (1938): 383–87.
Villette, J. "Que représente la grande fresque de la maison chrétienne de Doura?" *Revue biblique* 60 (1953): 398–413.
Watzinger, C. "Die Ausgrabungen von Dura-Europos." *Die Welt als Geschichte* 2 (1936): 397–410.
———. "Die Christen Duras." *Theologische Blätter* 18 (1938): 117–19.
Welles, C. B. "Die zivilen Archiv in Dura." *Münchener Beiträge zur Papyrologie und antiken Rechtsgeschichte* 18 (1934): 379–99.
———. "Inscriptions from Dura-Europos." *Yale Classical Studies* 14. New Haven, 1955.
———. "The Population of Roman Dura" in P. R. Coleman-Norton, ed., *Studies in Roman Economic and Social History in Honor of Allan Chester Johnson.* Princeton, 1951.
———. "Dura Parchment I." *Archiv für Papyrusforschung* 16 (1956): 1–12.
Wessel, K. "Dura-Europos." In *Reallexikon zur byzantinischen Kunst.* Stuttgart, 1966, cols. 1217–30.
Wischnitzer-Bernstein, R. "The Conception of the Resurrection in the Ezekiel Panel of the Dura Synagogue." *Journal of Biblical Literature* 60 (1941): 43–55.
———. "The Samuel Cycle in the Wall Decoration of the Synagogue at Dura-Europos." *Proceedings of the American Academy for Jewish Research* 11 (1941): 85–103.
———. "Studies in Jewish Art." *Jewish Quarterly Review* 36 (1945): 47–59.
———. *The Messianic Theme in the Paintings of the Dura Synagogue.* Chicago, 1948.
———. "The 'Closed Temple' Panel in the Synagogue of Dura-Europos." *Journal of the American Oriental Society* 91 (1971): 367–78.
Wodtke, G. "Malereien der Synagoge in Dura und ihre Parallelen in der christlichen Kunst." *Zeitschrift für die neutestamentliche Wissenschaft* 34 (1935): 51–62.

Index

(A listing of biblical references appears at the end of the index.)

Biblical References

MACCONNAL-MASO
MERCEDES G-SERIES
181

PREFACE

I have always been part of a team when training horses at Ivyleaze, my mother's home near Badminton, where I have the best back-up team in the world. Our combined knowledge has been pooled for this book, which is another team effort. It could not have been written without the help of my mother, Heather Holgate, and our friend Dorothy Willis, who has been based at Ivyleaze since 1980; they have both acquired a vast reservoir of knowledge. I am also indebted to our vet, Don Attenburrow, whose painstaking research has done so much to improve the treatment and welfare of horses.

There have been many others who have given me valuable advice over the years, thereby widening my own knowledge of a wonderful sport in which you never stop learning. With apologies to all those I am leaving out, I will mention three trainers who have had a tremendous influence. Pat Manning has been my guiding light in dressage; Lady Hugh Russell was largely responsible for teaching me how to ride cross-country fences and Pat Burgess has given me an immense amount of help in show jumping.

Heartfelt thanks are also due to my wonderful sponsors at Citibank Savings. Without their continuing encouragement and support, it would have been impossible to keep a yardful of horses.

Finally, I must thank the girls – both past and present – who have looked after the horses at Ivyleaze with such dedication. They have kept our eventers well and happy, thus enabling them to give of their best in competitions.

Opposite: *It takes time for the horse to jump into water with complete confidence. Beneficial is leaping in boldly and obviously enjoying himself*

1

UNDERSTANDING THE HORSE

I was once asked to pinpoint the most important key to the successful training of a three-day event horse. Was it the lungeing, the ridden flat work, the feeding, the fitness programme or the cross-country schooling? I had no need to rack my brains to answer, 'It's a combination of them all.'

At Ivyleaze, we all have our separate tasks which contribute towards the same goal. Together we are aiming to train each and every horse in the yard to the stage where it can realise its full potential as an eventer. No specific part is more important than the others; they are all essential to the one shared aim. We endeavour to make everything that happens to the horse, from the time it is first handled right through to advanced three-day events, seem like a natural progression from what went before.

There are no short cuts to training an event horse. Each animal has to be treated as an individual and given as much time as it requires for the various stages of its education. This means that each step has to be accomplished satisfactorily before the next step is introduced. Problems, based on misunderstanding, invariably arise when you try to speed up the process. Training has to be slow and painstaking, because that is the only way we can establish the co-operation and confidence which lie at the heart of all successful partnerships between horse and rider.

At Ivyleaze we firmly believe in the principle that prevention is better than cure. Those words are likely to be repeated many times in this book, because they influence our whole outlook on training. We avoid problems like the plague, because we know how difficult they are to overcome.

Horses have a sharp awareness of what is going on around them and a very long memory. If a problem arises, it may be impossible to cure completely because it is stored away in the horse's memory box, ready to re-emerge in any situation which is similar to the one which sparked it off in the first place. Refusals are just one example of something that is much easier to prevent than to cure.

Our other golden rule is never to grind on. This maxim is applied to every stage of training, at whatever level. You serve only to accentuate problems by grinding on, because the horse then becomes bored, confused or resentful. He will therefore become unco-operative.

Because each horse is an individual, it is impossible to say how long each stage of his training will take. I have attempted to give some rough guidelines simply because the reader might not know whether a particular stage is likely to take three minutes or three months – but there can be no hard and fast rule. The trainer has to listen to the horse and proceed with the utmost caution.

If the horse is not progressing, you obviously need to check that you are doing nothing wrong yourself. But, having done that, you should take stock of the fact that some horses do take longer than others. You may have expected results in two days and are therefore tempted to move on to something new when they are not achieved. If you were to persevere for two weeks – never for more than a short time per training session – the problem would probably disappear.

The rider must learn to listen to the horse. A few people are lucky enough to have an inborn instinct which enables them to do this quite easily, but most have to teach themselves to be more receptive. The partnership has to be based on mutual trust and respect, which can be achieved only through the rider's sensitivity. Abusive measures are a sign of failure, indicating that the rider's understanding and knowledge have been exhausted. The same applies to the use of gadgets, to which people tend to resort when there are shortcomings in their own training methods.

Before administering punishment, you need to be sure that the problem is not of your own making; in nine cases out of ten opposition stems from the rider's mistakes, not the horse's recalcitrance. In other words, it is not the rider who has a problem with the horse, but vice versa! For clear and careful aids to be given, the rider needs to be like a tree, with arms and legs (the branches and roots) able to move independently of each other, while the trunk remains still. Aids come from the legs, weight of the body and, finally, the hands. The use of the seat is widely misinterpreted and therefore wrongly applied; such misunderstandings are less likely to occur if you think of your weight (rather than your seat) as an aid.

Everything is dependent on the quality of the horse's flatwork. If he is not straight over fences, there is only one reason: it is because he is not straight on the flat. Dressage is therefore the basis of all successful training. The horse will not be required to learn anything that he cannot do already. Out on his own in the field he walks, trots, canters, jumps; he performs flying changes and piaffe. Dressage simply teaches him to do these things when asked, with the weight of a rider on his back.

It is that additional weight which interferes with his natural balance. He is built in such a way that his own body is heavier in front and, with incorrect training, the rider's weight is liable to push him further on to his forehand. When that happens, the forelegs bear the burden that should be carried by the hindlegs. This in turn produces strain and stress. Horses left alone in the wild rarely sustain injuries to their tendons or ligaments; these normally occur because the horse has not learnt to carry his own weight, plus that of his rider, in a way that causes least stress.

People often refer to riders having good or bad luck according to whether their horses stay sound or go lame. Obviously there is an element of truth in this; but it is also true to say that we can help to make our own good luck by training the horse to carry himself in balance – *with* the weight of a rider on top. This will not only help to conserve the horse, it will also enable him to jump with accuracy and rhythm and to reach the level of fitness required for the endurance aspect of a three-day event.

The horse can give of his best only if worked in a calm and loving atmosphere. The actual rate of progress will depend on his temperament, size, physique, maturity, fitness and soundness. Training the horse is a slow but fascinating process and, when we succeed, it is wonderfully rewarding.

2

FINDING THE RIGHT HORSE

Since I have been lucky enough to compete on British teams, I am always looking for horses that are capable of reaching international level. It has to be admitted that I am not the easiest person in the world to please in this respect. When my mother and Dot believe they have found the ideal horse, I am still not necessarily satisfied. Possibly I have been spoilt by choice!

I like the horse to go in a certain way, so that I feel we can both hit it off together. It worries me when I seem unable to find the key to a particular horse; I begin wondering whether it would have a better chance of achieving its full potential with another rider. Master Craftsman was one such horse. I worried about him all through his novice and intermediate stages, and if he hadn't been my mother's pride and joy, he would almost certainly have been sold on. I came to realise what a dreadful mistake that would have been when I rode Crafty at the 1988 Olympic Games in Seoul. I have therefore learnt that perseverance can be rewarded!

Some of the horses we buy are unbroken youngsters whose jumping ability has to be taken on trust. We never pay vast sums of money for a horse, nor do we buy them ready-made at advanced level; we prefer to make them ourselves. The only one of our horses that was already advanced on arrival at Ivyleaze is Griffin, whom we swapped with Ian Stark for Murphy Himself. Griffin was a little small for Ian and Murphy's boisterous temperament made him too big and strong for me; he needed a man on his back.

My goals are not necessarily shared by all other riders. Not everyone wants a horse with international potential – which is just as well, because they do not exactly grow on trees. If I were buying a horse to ride only in novice events, I would obviously be ready to overlook certain points that might be seen as shortcomings in an international eventer. But there would be some priorities that remain the same; whatever our ultimate goal in eventing, we need a horse that is honest, bold and sound.

Because there are certain factors in common, I will begin by describing the type of horse we like to have at Ivyleaze. My mother's success in finding the right material has prompted me to put her thoughts first. She has inherited an instinctive sixth sense from her father, who was regarded as one of the best judges of a horse in the West Country. Someone else's instincts are not particularly easy for the rest of us to follow, nor are they simple to explain, but my mother has done her best.

HEATHER HOLGATE'S INSTINCT

'The first impression the horse gives me is terribly important, rather like meeting a person,' says my mother. 'It's the general feeling I get when I walk into the box that creates this first impression. It's not necessarily whether the horse is pleased to see me or not; the Irish horses are usually displeased to see me, but I have now looked at so many of them that I am aware of this in advance.

'Though the head can tell you an enormous amount, the initial feeling I get comes from the whole horse. It is so instinctive, it's virtually impossible to explain. Later on, of course, I do look for specific things. I like the head to be sensible, but I've never worried about eyes. If I had been influenced by them, we would never have bought Priceless; you could hardly describe his eyes as big and generous.

'Everybody needs to realise and remember that no horse in the world is one hundred per cent perfect. I often work on the theory that the more defects you can see the better; if everything on the horse appears absolutely fantastic, there is very often a hidden problem. Bearing this in mind, I think one should be prepared to overlook things like false curbs. The same applies to splints, as long as they're not so big that they're likely to be knocked and are fully formed. Small things like this should not sway the decision as to whether you buy the horse or not, though it is obviously essential to get it very well vetted.

'Though it is always nice to have a horse that moves well, I am not personally bothered too much about whether it moves perfectly straight. Some people would be put off if they saw the horse dishing but, if most other things seemed to be going for it, I wouldn't let that bother me. I would be more concerned about its rhythm and elevation and the way it uses its hocks. Movement can be improved, assuming that the horse has good natural paces, so I am always thinking of what can be done and how long it will take.

'Unless the horse is too young, I obviously look at the way it jumps as well. I want to see whether it uses its shoulder, how high it raises its forearm, whether it tucks its forelegs up and so on. These are technical things, which I am much more aware of now than I was ten years ago when they wouldn't have worried me. I still look carefully (as I did in the early days) at the horse's face and ears as it comes into a fence. They will tell me how clever it is and whether it enjoys jumping or is slightly apprehensive.

'A good temperament is absolutely vital, whether you're looking for an international horse or a novice. In most cases those that are half bred have better temperaments than Thoroughbreds, but there are always some exceptions to the rule; I have known one or two Thoroughbreds with superb temperaments. I am always looking, first and foremost, for a really kind and generous horse; from what I've seen of eventing, I know you can't get anywhere without these qualities. In my case, it also helps if I can find a horse that my daughter likes as well!'

Inevitably, since I am the one who rides the horses in competitions, those that are sold are usually the ones that suit me least. They are certainly not discards. Among successful horses that were once at Ivyleaze was one that went to Canada, where his new owner rode him to win the national Young Riders' Championship; another was shortlisted for the 1988 Canadian Olympic team.

It is only because we need to sell some horses each year that we usually buy geldings. Mares are far more difficult to sell, so we do not go out of our way to look at them. If the right one turned up on the doorstep, however, we would not turn her down because of her sex; I happen to think that a good mare is hard to beat.

A TECHNICAL ASSESSMENT

Dot and I concentrate on the technical aspects as well as listening to our instincts; neither of us is blessed with my mother's sixth sense. I always look at the horse's legs and feet first. It seems daft to fall in love with the animal's head until you know whether it has the sound limbs and feet that are needed to carry you across country.

Feet

I would not wish to buy a horse with either boxy or flat feet. I am also put off by feet that are not part of a matching pair, because it could mean that the normal development of one foot had been retarded by previous lameness. Soft or sensitive soles are also best avoided in our sport, because the ground on cross-country courses can be rough. You can test the sole by tapping it and seeing whether the horse flinches. The horse's feet have to bear his own weight of around 500 kg (nearly 10 cwt) plus that of his rider. In trot this heavy burden rests on two feet at a time; in canter there is a moment when one foot alone bears the entire weight. It is therefore obvious that the horse needs sound and healthy feet to be an eventer.

Night Cap's head shows boldness and intelligence

Good feet for an eventer

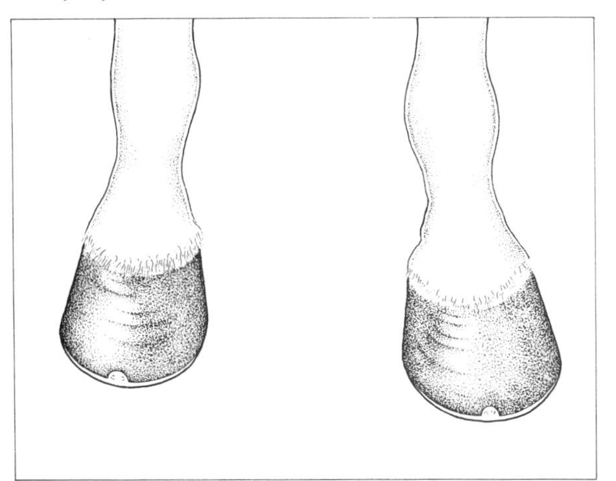

Boxy and flat feet are to be avoided

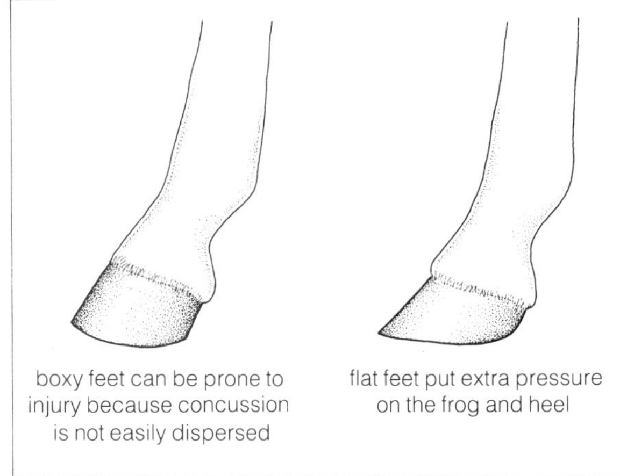

boxy feet can be prone to injury because concussion is not easily dispersed
flat feet put extra pressure on the frog and heel

Legs

I am concerned that the legs should be a good shape, with a strong second thigh. Minor blemishes would not necessarily worry me as long as there was no scar tissue in an area that might hit a fence and could not be protected by boots or bandages. Priceless had a scar on his cannon, but it was always covered by boots; it might have been a problem had it been on a knee or the front of a fetlock.

Like my mother, I would disregard a false curb, which is merely a fault in conformation. It would obviously be considered as a defect in the show-ring, but that does not concern me because I am looking for an athlete rather than a show horse. Priceless had a false curb, but we still went ahead and bought him. However, a true curb, which could be an unsoundness, would probably send me straight home without waiting to see how the horse moved or jumped.

You can recognise a true curb by standing at right angles to the horse's hindleg; from this position it can be seen as an obvious swelling at the bottom of the hock joint. The false curb is less obvious from this angle, but will be seen more clearly if you move a little nearer the horse's head. You can also try picking up the leg and bending the hock joint; in some cases, if it is a false curb, the swelling will seem to disappear.

I would prefer the horse to be physically built so that his natural way of standing brings his hindlegs underneath him. It has to be said that this is not true of Master Craftsman; his hindlegs tend to be further back than most of our other horses, but it does not seem to have been too much of a handicap!

The legs should have good bone and plenty of room for the tendons. I dislike cowhocks but would not automatically turn down the horse because of them. I would obviously turn down one with damaged tendons, because they indicate weakness and possible trouble in the future.

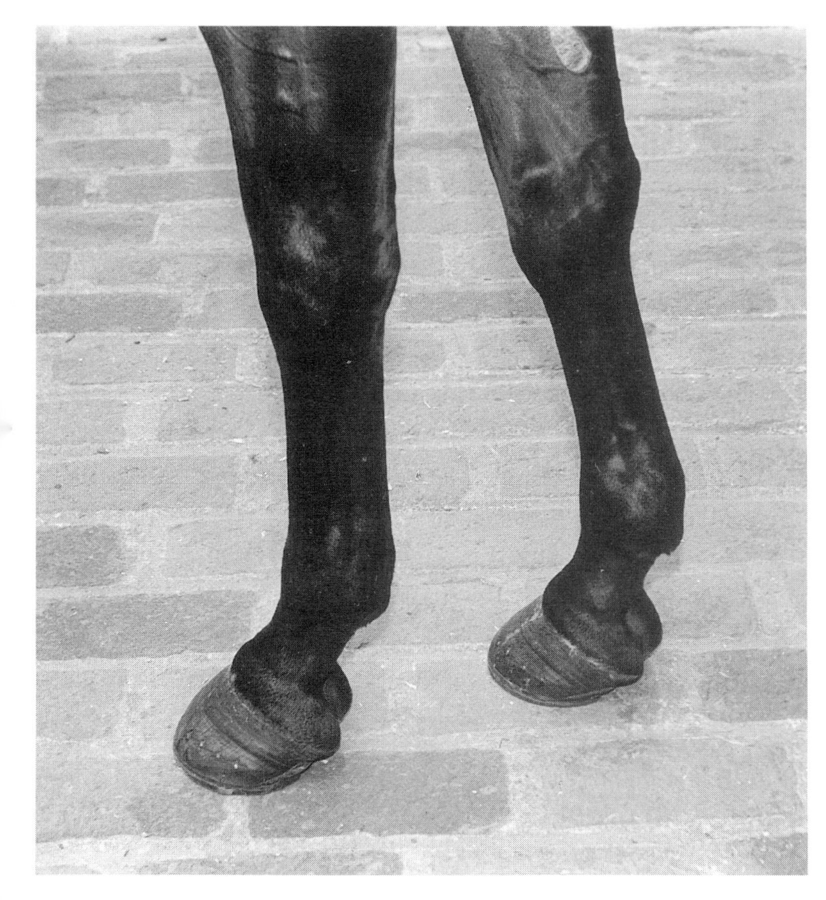

An example of good feet and limbs. Note that this horse is short between knee and fetlock, which suggests that he has strong tendons

Head

I like a good honest head and am definitely averse to pretty faces. Night Cap probably comes closest to my ideal. He has a lovely head, with a kind and honest look. It's the overall picture that pleases me; it would be hopeless to try to break it down into different parts.

Priceless has a real workman's head. We like to think he is similar to Arkle in this respect, even though he does have piggy eyes. However, we feel he has done quite enough to prove that a horse with small eyes can have a big heart and a sharp brain. It is the expression in the eyes, rather than their size, that can tell you a fair bit about the horse's temperament. It can indicate such things as bad temper, a suspicious nature, intelligence, or a friendly attitude towards people.

Welton Chit Chat: the ideal type of event horse, who naturally stands with a round frame

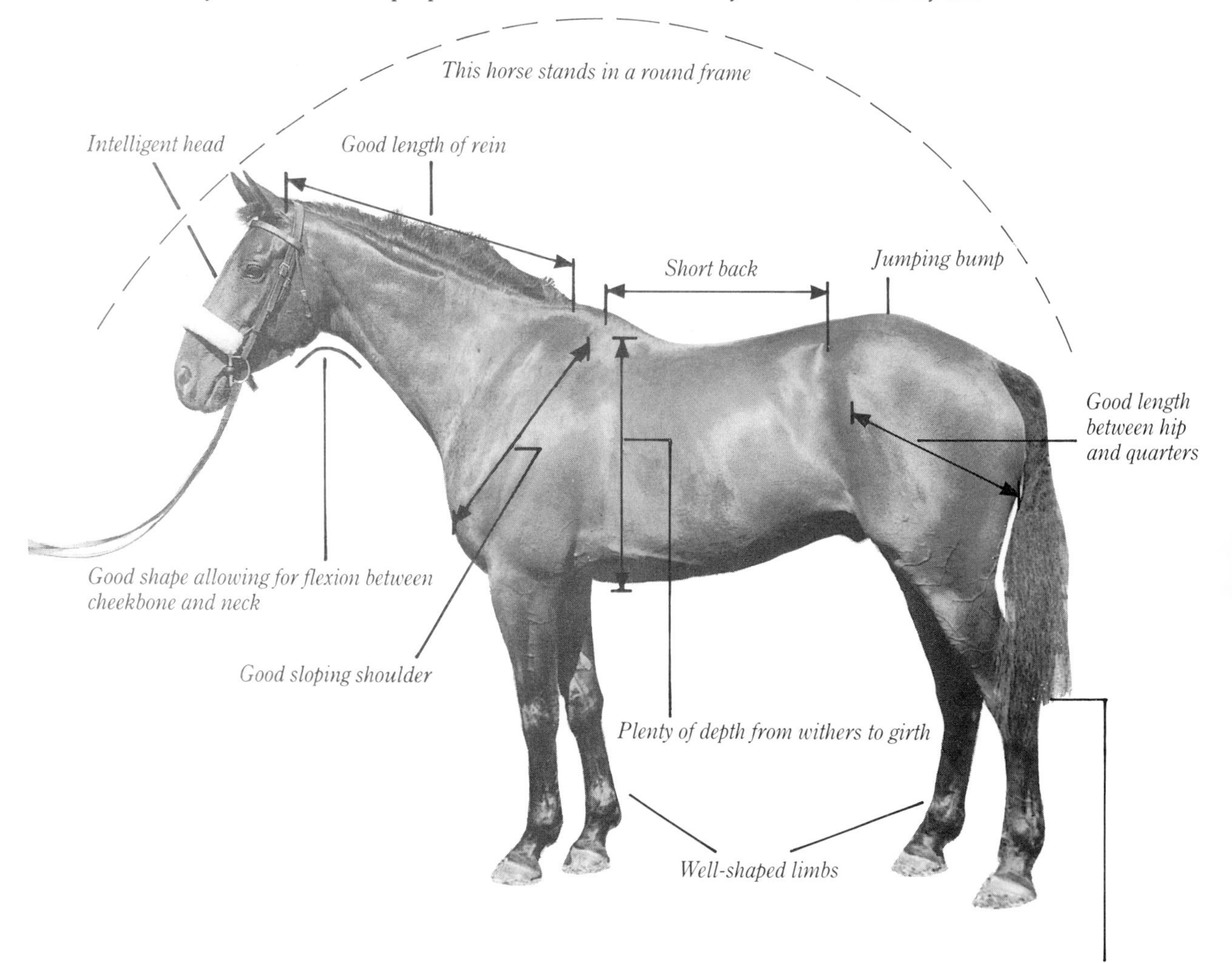

Body

My potential three-day eventer would need to have a good front, with a sloping shoulder and lean withers. Much can be done to improve the quarters, so they don't receive quite such close scrutiny. I like the horse to be in proportion, so that its overall shape looks right and all the parts appear to fit together and make a pleasing picture. I would prefer him not to be either too long or too short through his body, from nose to tail.

Griffin: another good type of event horse, with some small defects in conformation. His heavy head makes it more difficult for him to balance himself and engage his hindlegs but, because of good conformation, he is not on his forehand

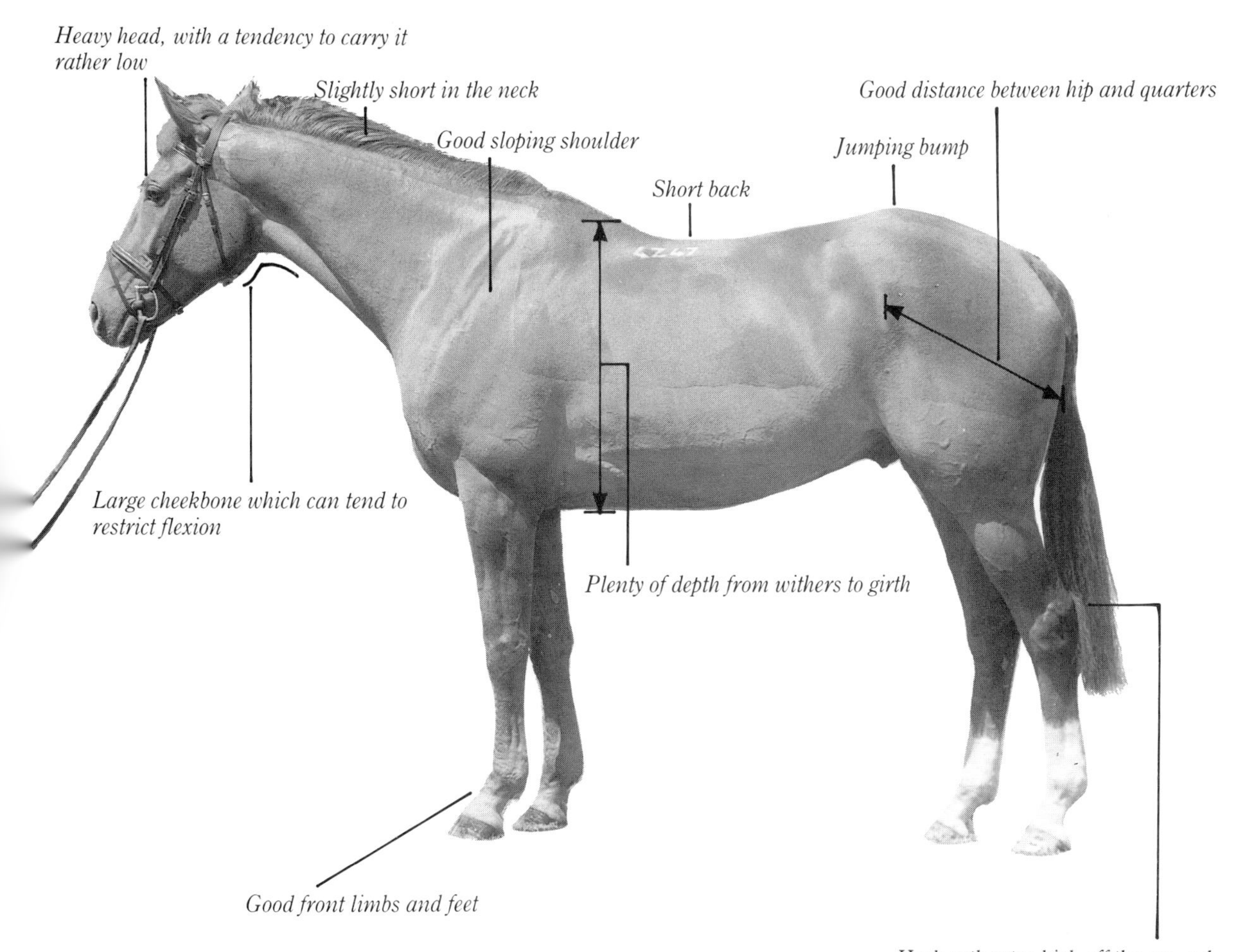

Movement and jumping

Next I would want to see the horse walked and trotted in hand, so that I could look at it from all angles – coming towards me, going away and moving past to give me a sideways view. I prefer the horse to move straight, but a slight dish would not be the end of the world. It must, however, have a natural spring to its trot and I would like it to give the impression that it enjoys moving.

My next request would be to see the horse ridden, by someone other than myself, so that I can assess its way of going and style of jumping. It would not worry me if the horse were not on the bit or if it failed to bend in the right direction; I would be far more interested in its natural balance and cadence. Other things, like any obvious stiffness, would be noted for future reference

Welton Houdini: another good type, also with a few faults. His natural way of standing gives him a hollow frame with shoulders down and head up, so you would expect him to find the same problem when ridden. However, thanks to his co-operative temperament, he has learnt to overcome these defects

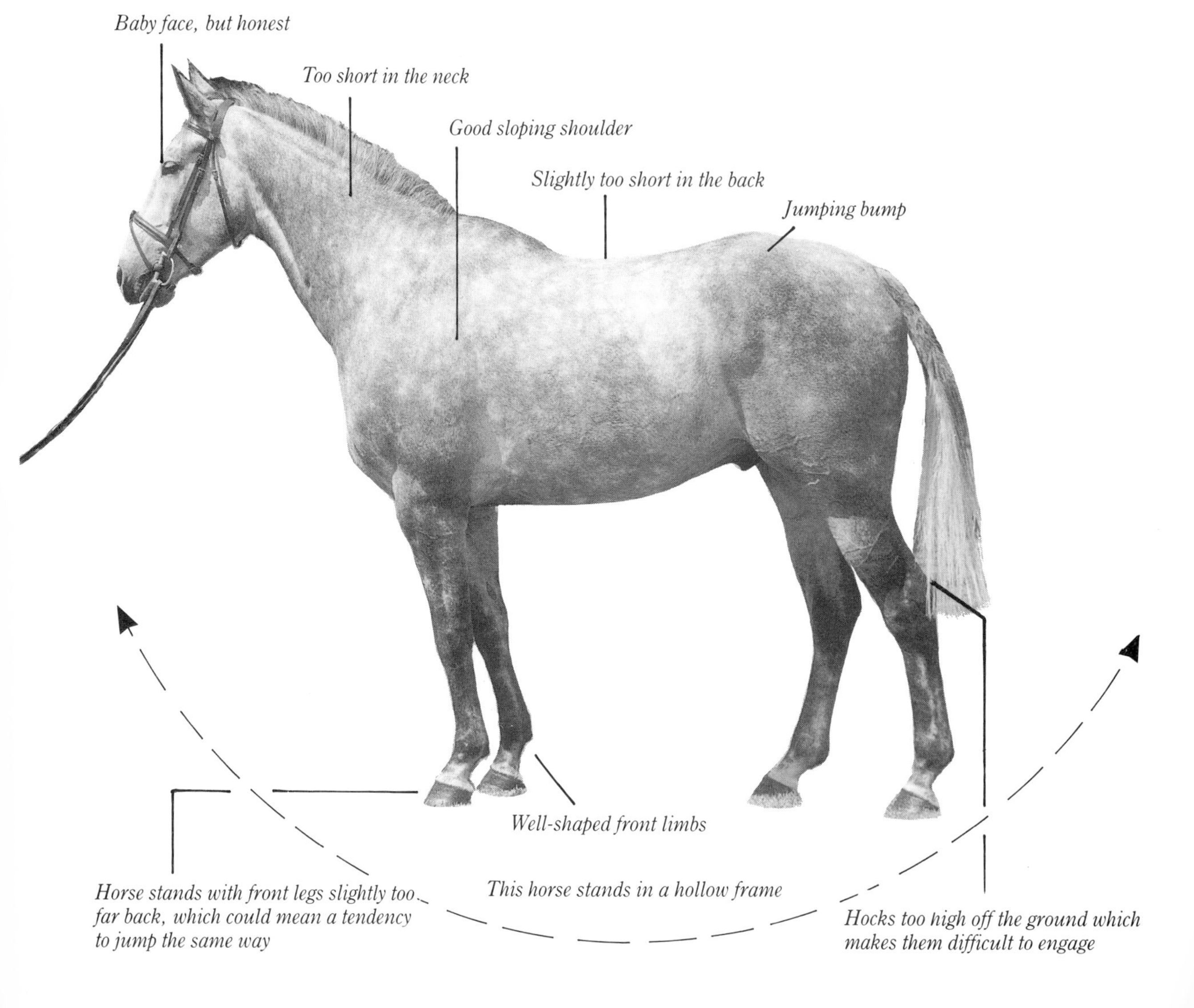

when I came to ride the horse myself.

Before that happens I would want to see the horse jump with someone else on board. If he already has some jumping experience, I would hope to see him build up from a small fence (such as cross-rails) to a decent-sized obstacle of about 3 feet 6 inches. Virtually any horse jumps with a rounded bascule when it has never been ridden over fences. The problems arise through incorrect schooling and, once bad habits have been learnt, they are hard to eradicate. At the risk of sounding ultra-fussy, I have to admit that I would not buy a horse that jumped with a hollow back or one that dangled. The latter describes a horse that fails to raise its forearm sufficiently when jumping. This would not be much of a handicap at novice level; indeed, I have known some danglers who went on to become successful advanced horses. I simply feel that they are less capable of extracting themselves from the sort of trouble one can easily meet in a championship three-day event, because they are virtually incapable of getting close to a fair-sized fence without hitting it.

If the horse refused, I would not automatically assume that it was chicken-hearted. It could, of course, be just that – but it is equally possible that the problem occurred because the rider was nervous or the horse simply lacked confidence. I would have to get on the horse myself to try to discover the root of this type of problem.

I would do flatwork first and then jumping, with my mind focused on any shortcomings I had seen while watching from the ground. Usually I can tell, more or less, if the horse is green or careless, if it lacks confidence or is just plain chicken. I will keep on asking myself whether I think we can make improvements, using our own methods. In our experience, very few horses are careless by nature. Mistakes that look like carelessness are usually the result of insecurity – which, apart from the odd exception, is always man-made.

Temperament

Psychoanalysing the horse on first acquaintance is never easy. Temperament is the number one priority, yet often there are only small clues to tell us whether the odd little quirk will work for us – with the help of confidence and training – or against us. Usually a bad temperament is man-made, but if you suspect that the horse has a mean or cowardly streak, it is better to find another mount.

There are all sorts of small indications that will tell you whether a horse is brave and clever. When he is led out of his stable you might ask to see him go further than his own familiar surroundings, so that you can see how he copes with different sights and sounds. Maybe you could ask for him to be led past a car – either with dogs inside barking at him or with one or two doors left open on purpose – so that you can see how he reacts. If the horse walks on confidently, you will know that he is brave.

Ears are the best indicators of intelligence. When you see a horse cock an ear at a car going up the road, you know that he is aware of what is going on. When you are aware of something, you are capable of dealing with it, so it is almost the same as being clever.

Breeding and colour

My ideas have changed over the years. Since riding Thoroughbreds, like Master Craftsman, I would be more inclined to buy one than when I rode Priceless and Night Cap, who both had some non-Thoroughbred blood. They are not, however, suitable for inexperienced riders and many of them would not be sufficiently tough for three-day eventing. Having been bred to race, Thoroughbreds are notoriously lazy when it comes to schooling and you have to take care not to overdo the flatwork. Asking them to walk, trot and canter in the correct shape for dressage goes against their breeding, whereas the warm-blood horses have been specifically bred for that purpose.

I am obviously interested to know something about the horse's sire and dam, though it is far less relevant than in racing. Some sires have undesirable reputations, perhaps for bad temperament or for tendon trouble in their progeny. Others are highly thought of because they have produced a number of good offspring. Ben Fairie (the sire of Priceless and Night Cap) and Master Spiritus (who sired Master Craftsman) come into the latter category. We have horses by both these stallions, but we did not buy them just because of the sire's reputation; we happened to like them as well.

Colour is immaterial. I would personally love to ride a piebald or skewbald; they are often regarded as a bit plebeian, but I think they look really smart when well turned out. My favourite colours are grey and dark brown. I am not that keen on chestnuts, but I would certainly not be put off buying one if I liked everything else about it.

Vetting

We very rarely buy a horse without first getting our vet, Don Attenburrow, to look at it. There are few absolutely positive things we can be told about the animal but, thanks to Don's radio stethoscope, we can have the condition of its wind in black and white.

The radio stethoscope is a more effective way of testing lung function than the alternative endoscope, which involves pushing a tube down the horse's throat. The endoscope allows examination of the upper airway but since it can be used only on a horse that is standing still, its findings are inconclusive. Don's radio stethoscope, on the other hand, can test the lungs while the horse is galloping – and it is, after all, during exercise that we are concerned about his wind. It involves applying a microphone to the skin over the horse's windpipe; this transmits sounds to a tape recorder and produces a graph, similar to the one seen when a human heart is tested. Don is therefore able to see and hear the condition of the horse's lungs.

Our indispensable vet also tests the sight of our potential purchases and listens to their hearts, using a conventional stethoscope. At the same time he makes sure that there are no old injuries that might prove troublesome in the future. Unfortunately, the veterinary certificate can do no more than state that the vet has found no defect at the actual time of the inspection; it can offer no reassuring guarantee for the future. The vet is not a miracle man; he cannot promise you that the horse will be sound for the next ten years. It could slip ten minutes later and go lame.

A horse for the novice rider

The more horses you look at, the more you will learn – so try not to be in too much of a hurry to buy. You will need to enlist the help of someone experienced, whether you are answering advertisements (which do not necessarily tell the whole truth) or go to see a horse that you have heard about by word of mouth. Such help is absolutely vital if you are thinking of going to sales, where it is all too easy to be swayed by a pair of trusting eyes. Never, ever buy without getting the horse properly vetted.

If you are allowed to take the horse on trial (which obviously won't be the case if you buy from sales) you will have the chance to find out how well you suit each other. Once you have bought the horse it will become part of your life, so compatibility is important. Taking the animal on trial is, however, a big responsibility and you need to be adequately insured.

Each rider's individual circumstances, skills and ambitions are likely to be different – and all three have to be taken into account. In making the following suggestions, I am considering an inexperienced rider with limited funds, who is looking for a horse that would be fun to compete with in novice events.

The funds must not, of course, be too restricted. Buying the horse is only the first part of the outlay. It will be followed by payments for saddlery, forage, bedding, shoeing, veterinary treatment, transport and entry fees. There should then be some money left in reserve for training. Every rider, at whatever level, needs some help to iron out the faults that tend to creep in unnoticed, otherwise the horse's performance tends to get worse rather than better.

As already mentioned, there are priorities at all levels of the sport. Boldness and honesty are obviously essential – there is no fun to be had riding a horse that stops or is in any way ungenuine. I would also look for good feet and legs, because there will be a certain amount of stress and strain on them, even in novice events. You cannot guarantee that the horse will stay sound, but he is less likely to go lame if he has the right limbs and feet for the job. His conformation should be basically correct as well, because faults in this department can have a detrimental effect on the way he moves and jumps.

Breeding is not important for novice events; it does not matter in the least if the chosen animal is related to a cart-horse. Indeed, if funds are limited this could be an advantage because such a horse could live outside most (if not all) of the year and would therefore be easier and cheaper to keep than one with a large percentage of Thoroughbred blood. The cart-horse relative would also be likely to have a more equable temperament than the animal with classier breeding; very often he will move well and possess both stamina and a good jump.

There is no sustained galloping in a one-day event, but there can be plenty of twists and turns; so you need a horse that is handy rather than fast. It doesn't matter a hoot if he is a bit common and plain; it should still be possible to get round the cross-country within the optimum time once he is ready to attempt it.

Age can cover a fairly wide range. I would recommend a minimum age of six for a novice rider, because the horse should then be mature, but an older horse of up to twelve years old would be equally acceptable and would have the advantage of costing less. He may also be an excellent school-master.

Experience would be an advantage for the novice rider. Although it is not impossible for horse and rider to learn together, I would certainly advise any newcomer to the sport to start with a horse that has already done some jumping. It should also enjoy jumping; novice riders should never consider buying an animal that they have seen refusing. Horses can be pretty cute, regardless of age and breeding; they know when they have someone inexperienced upstairs and they try them out, much as a naughty child might do with a new babysitter.

Short-striding horses may have their limitations, but they also have distinct advantages for the novice rider. A horse with a long stride can be much more difficult to train over jumps than the rabbits of this world, who are far more manoeuvrable and usually meet each fence just about right. It can seem like the difference between driving an articulated lorry and a mini.

Whistling is not necessarily a drawback if the animal is to be aimed at novice and intermediate one-day events with, perhaps, a novice three-day event as the ultimate goal. This would become a handicap only if you were to attempt an intermediate three-day test, because the whistler is unlikely to be able to cope with the faster time required, although there are exceptions. Don has passed horses that were whistlers; with the help of his radio stethoscope he was confident that the condition would not have a significant effect on their performance.

Crib-biting or *weaving* would not necessarily matter either, as long as the horse was not losing weight as a result and there were no others in the yard to copy him. If you take your crib-biter or weaver to stay in another yard, you do have to warn people beforehand. Otherwise you might lose some friends through their horses picking up one of these habits.

Size is important for novice riders. Unless you have plenty of experience, you should avoid buying a horse that is too big for you. The same applies to a little horse that is a very big ride, because of his extravagant movement. He may seem perfectly amenable when you first ride him, but he could become too strong for you once he is really fit. You can often get away with a horse that is slightly small, but being over-horsed can be frightening, if not downright dangerous. Let us not forget that the sport is supposed to be fun!

SUMMARY

Basic requirements for horse with international potential

Feet:	Should be part of a matching pair. Should not be boxy or flat. Avoid horse with soft or sensitive soles.
Legs:	Good shape required and should be in proportion to the horse's body. Avoid horse with true curb; ignore false curb. Should have good bone and room for the tendons. Leave well alone if there is any sign of damaged tendons.
Head:	Look for good honest head and kind eye.
Body:	Sloping shoulder and lean withers required. Also correct conformation, with all parts in proportion.
Movement:	Look for natural spring at trot, plus balance and enjoyment of moving.
Jumping:	Look for rounded back, raising of forearm and folding of fetlocks. Should enjoy jumping.
Temperament:	Must be brave and honest. Watch how horse reacts to strange sights and sounds. His ears will tell you how aware he is of surroundings and whether he has a good brain.
Breeding:	At least 50 per cent Thoroughbred.
Colour:	Immaterial.
Vetting:	Essential.
Conclusion:	Nothing is perfect, and handsome is as handsome does.

Basic requirements for horse to be ridden by inexperienced rider in novice events

As above:	Feet, legs, head, temperament, colour, vetting.
Age:	Six to twelve years.
Breeding:	Immaterial, but undiluted Thoroughbred blood not recommended.
Movement:	Short-striding horse easier to ride, but may lack paces for good dressage marks.
Body:	Conformation should be basically correct.
Jumping:	Avoid any horse that refuses. Look for enjoyment in jumping.
Size:	Must not be too big for the rider.

3

LESSONS ON THE LUNGE

Dorothy, who does most of the lungeing at Ivyleaze with some help from my mother and myself, says that it involves being 'part of a triangle which is formed by the lunge line, the horse and the whip'. She can either widen or close up the triangle by positioning the whip closer to the horse's hocks or further away. It is much the same as in ridden work; Dot comes closer with the whip when she wants to produce more energy, just as the rider would increase pressure with the legs. If the horse were fairly busy and showing plenty of energy, the whip is kept further away on the lunge, just as the rider's legs would be passive under saddle.

We all regard these lungeing lessons as crucial, especially with young horses who are being asked to work for the first time. They are used to being handled, but they have not yet been expected to concentrate and make an effort. It is therefore the beginning of the horse's working relationship with a person and it will colour his future attitude towards both work and people.

We start to lunge the young horses at three years of age, normally in the autumn when we are approaching the end of the eventing season and will have more time to devote to the babies. They are not clipped, because they spend most of the time at grass and therefore

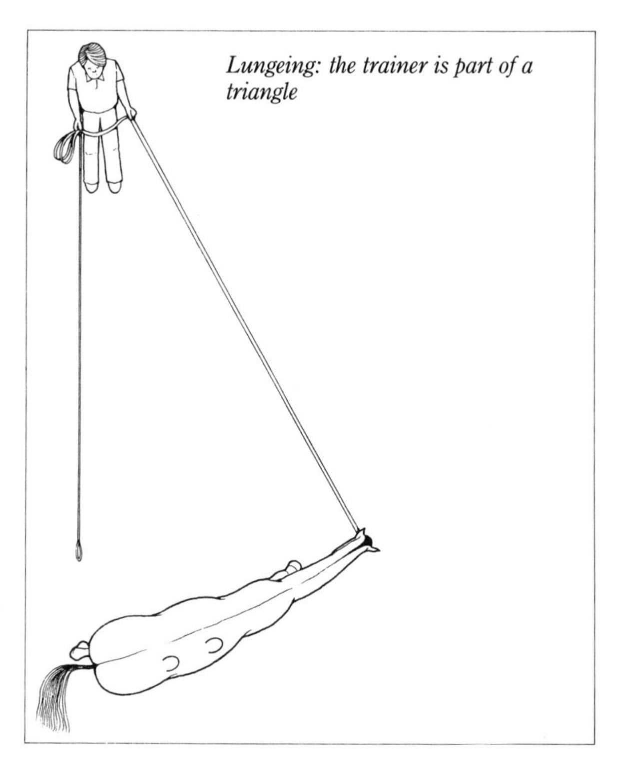

Lungeing: the trainer is part of a triangle

need their winter coats. Nor are they shod unless the ground is very hard, in which case they would wear front shoes. The hindfeet remain unshod for reasons of safety; it reduces the amount of damage they can do if they kick each other or their handler.

Those that are older when they arrive at Ivyleaze also begin their schooling with us by taking lungeing lessons. This will form an integral part of their training for the rest of their competitive lives; we therefore want to teach the horse to be obedient on the lunge. We rarely use it as a means of letting off an excess of high spirits. The youngsters can buck and play as much as they like when turned out in the field, but they are expected to listen and be obedient when they are wearing tack. If they were allowed to perform rodeo acts on the lunge, they might see no reason why they shouldn't do the same when they have a rider on top.

Every effort is made to avoid introducing too much at once. When the youngster is led out in a lungeing cavesson for the first time, we start by walking him around the vicinity of his stable where everything is reassuringly familiar. We do not want him to be distracted by strange surroundings until he has been given a little time to become accustomed to each new item of equipment. Later he will be led around the school, in circles and large rectangles of about 20 by 40 metres – in other words approximately the same size as the small dressage arena used at one-day events.

The horse is led from both sides during these lessons in hand and he is frequently made to halt, always using the voice command as well as an indication on the lunge line – plus, if necessary, a restraining hand in front of his shoulder. The handler is aiming to make him stop without turning his head or swinging his quarters, so that he remains on the same straight line that he was following at walk.

BEFORE THE FIRST LUNGEING LESSON

Our youngster will have been led in hand since he was a yearling (assuming he has been with us that long) so he would have become used to the human voice and to some of the basic commands such as 'walk on', 'slow down' and 'whoa'. The voice, accompanied by pats on the neck or shoulder, is also used to give encouragement to the youngster, thus helping him to form a happy association with a person.

If the horse's stable manners leave something to be desired – maybe because he is arrogant or mistrustful – he will be given part of his lesson in the stable until he learns to behave. Even with a well-mannered youngster, we make sure that someone (not necessarily the person responsible for leading him on the lunge) spends time with him in the stable to get him used to a human presence.

Before proper lungeing lessons begin, the youngster is taught to walk and halt in a straight line during short five- to ten-minute lessons given three or four times a day. We use a lightweight lungeing cavesson with a drop noseband for these lessons, with a lunge line attached to the front ring. The line is looped and held by the handler who walks close to the horse's shoulder so that he can be prodded with an elbow if he falls in (by leaning towards the handler) or barges.

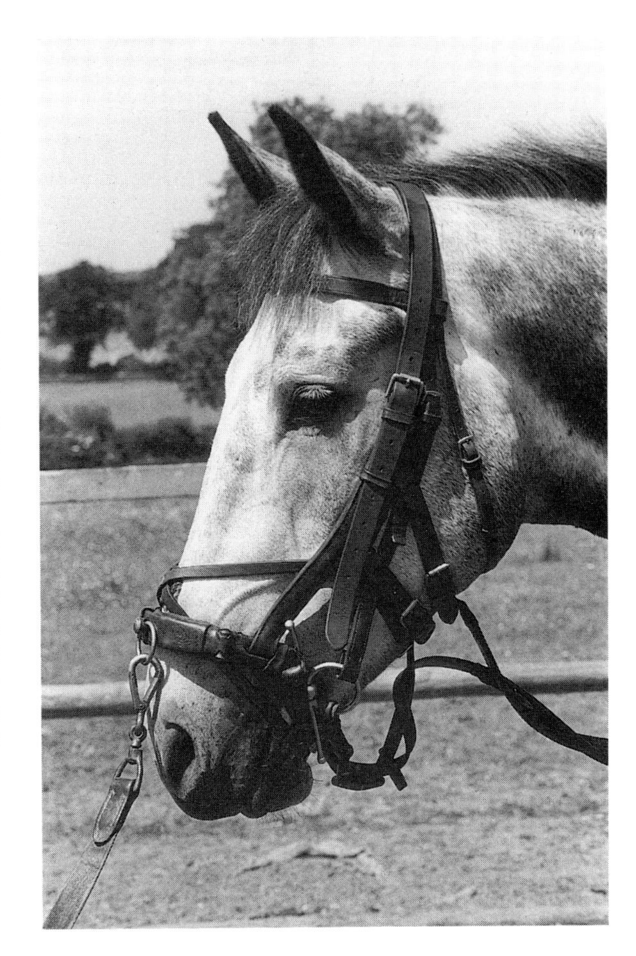

The cavesson correctly fitted

During this pre-lungeing stage, the horse will become used to wearing lightweight brushing boots on all four legs. We use the type with Velcro fastenings rather than buckles; young horses can be fidgety and these are much easier and quicker to put on. Another new experience is encountered in the stable, which is where we let him get used to the feeling of having a lightweight canvas roller tightened around his middle. The horse will accept a saddle and girths more readily if he has been accustomed to wearing a roller in his stable, whether it is securing a rug or worn on its own.

All these experiences will make the next stage easier. By the time the horse is ready to move away from his handler on the lunge line, he will have learnt the voice commands, become familiar with the area to be used for lungeing and he will be used to going there in boots and cavesson, with the lunge line attached.

LUNGEING BEGINS

The first proper lungeing lesson begins in exactly the same way as those that preceded it. The youngster is led out as usual; he is asked to walk and halt in a straight line and he is led around the school. Having done this without any fuss, the lunge line is fed out to the horse by about 3.5 metres so that he moves on to a circle at walk without anyone alongside him.

Once this is achieved, the lesson should be ended more or less immediately, on a happy note. At all stages of training, the handler (and later the rider) must resist the temptation to keep going a little longer because everything is going sweetly. That is precisely the time to stop – before problems arise.

The horse should be rewarded after his work, perhaps by being allowed to pick at some grass on his way back to the stable or by the offer of nuts from the handler's pocket. We never give nuts before the lesson begins; they would encourage the horse to start looking for titbits instead of concentrating on his lessons.

The lesson is repeated the following day, with the horse spending a little longer circling round on the lunge at walk. At the same time a little more of the lunge line is fed out to him in order to enlarge the circle. The trainer has an assistant on hand during these early lessons on the lunge but, unless called upon to help with a particular problem, the second person remains a spectator. The assistant may have an active role to play later, if the horse is constantly falling in or fails to move at the required pace. In these instances, the second person would walk in a small circle half-way between the trainer and the horse. If the animal is lazy and tends to hang back or keep stopping, the assistant carries the stick. This will not necessarily be used on the horse; it is usually enough for him to know that the stick is now much closer. If he always wants to go too fast, the assistant holds the lunge line so that he feels some extra restraining pressure.

Size of circle

The horse will eventually be on a circle of between 18 and 20 metres in diameter, which means that he will more or less be using the width of the small dressage arena. If the horse is big as well as inexperienced, this may seem too small for him. In such cases both control and balance can be achieved more easily by enlarging the area, which is done by the handler walking a smaller circle of perhaps 6 to 10 metres. This should be seen only as a short-term measure which might last for about two weeks; eventually the handler (after going round in ever-decreasing circles) will aim to pivot on one spot.

It is only by remaining in the same place when lungeing that you can find out whether the horse is describing a proper circle. If the lunge line goes slack, you will know that he has fallen in; when the pull becomes stronger, you will be aware that he is leaning out. You cannot be aware of such things while you are moving on a circle yourself, because you (rather than the horse) are maintaining contact on the lunge line. That process will be reversed when you stay in one spot; the horse will then take contact himself, which is a necessary part of his education. We are, after all, going to want him to take contact on the reins when he is ridden.

Length of lessons

Lessons must be kept short. Lungeing can be boring; it also puts tremendous stress on the hocks of an immature three-year-old. If the horse becomes sore, he is going to associate work with discomfort which is something we are always striving to avoid. We would therefore limit each lesson to between twenty and thirty minutes, including the initial walking in hand that would probably take up the first ten minutes. It is therefore essential for the trainer to make a note of the time before each lesson begins.

Our youngster might have two or three lessons per day, but none of them would be longer than half an hour and they could be as short as fifteen minutes. We never believe in ploughing on when a lungeing session (or any other form of schooling for that matter) is not going well. Perhaps the horse is resistant because he is tired or upset; if so, it is much better to put him away and bring him out again later the same day.

Once our youngsters have started work on the lunge, they do not have a day off for the first two or three weeks. Any break in the daily lessons could cause a setback, with the horse being a little too fresh (and therefore less obedient) after his day of rest. We believe in waiting until the horse has absorbed those early lessons on the lunge and something positive has been established.

Most horses have a natural bend to the right. Because of this, they tend to carry more weight on the left shoulder and have the left hip lower, while the quarters move out to the right. They therefore find it much easier to circle right and are noticeably stiffer when you work them on the left rein. It is important to keep changing the rein; we would never ask the horse to continue circling in one direction for more than ten minutes.

In the early stages, we are not looking for a perfect outline; we are more concerned to establish a nice rhythm and even length of stride. The horse should be both obedient and happy, as well as listening.

The voice

Horses are normally much better listeners than people and the human voice plays a major part in their training. The inflexion can be used as a rebuke or reward – and they know exactly which one is meant without the need for endorsement with stick or carrot. They also learn to recognise certain words, which makes it important for all of us who are working with the horses at Ivyleaze to use exactly the same words of command.

All our eventers are particularly receptive to the words 'Good boy'. When they hear them spoken while on the lunge, it usually signifies that the lesson will soon be ending. Some of them stop dead in their tracks as soon as they hear those two words, so we have to be careful not to say them too early.

PROGRESSING STEP BY STEP

The young horse will have worn lungeing cavesson and boots for his first lessons. If all goes well, a roller is added on the third day – fairly loosely fastened at first and then tightened if he seems quite happy. At this point he would probably be led over a single trotting pole at walk to get him used to going over something on the ground.

Trotting is added to the lesson when the trainer has decided that the time is ripe. It should not be introduced until the horse has learnt to pay attention; if he won't listen to you at walk and halt, you have not the slightest hope of getting him to concentrate at trot. It is not possible to predict how long this will take with each individual, because you can never foretell which youngster will respond quickly on the lunge and which one will be a complete featherbrain in the initial stages.

When the horse is walking and trotting satisfactorily, the roller is replaced by saddle and girths, without the stirrup irons and leathers. Depending on his reaction, this stage lasts a minimum of two days and possibly a week or more. Once he has accepted the saddle as part and parcel of everyday life, we add the stirrups. They are put up very short at jockey's length (with the leather looped an extra time around the iron if necessary) so that they are lying against the saddle flap.

Each lesson still begins with the horse being led in hand. By moving the stirrups – first when standing still and then while walking – the handler can help the youngster to get used to the mobility of the irons. Again it will depend on his reaction to this new experience whether he trots straightaway on the lunge, or goes back to walk and halt for a while.

Once he is going well at walk and trot, the stirrups can be let down gradually, over a period of perhaps a week or more, until they are hanging below the flap. They must not, however, be long enough to touch his elbow; this could cause him pain and would almost certainly create anxiety, which we are keen to avoid at all costs. It may take a week for the horse to accept the stirrups below the flap; they are bound to fly about while he is trotting on the lunge, and even if he stays relaxed, he will certainly be aware of them.

When this stage is accomplished to our satisfaction, we introduce the bridle which is fitted with a rubber bit, but with the reins removed. When the horse is ready for the reins to be added, they are tied up and attached to the throatlash so that they flap slightly on the horse's neck. Although the reins are slack rather than taut, it is at this stage that we discover whether the horse is anxious or fussy about the bit in his mouth. In many ways the whole of the training process is a voyage of discovery – for the trainer as well as the horse.

After a minimum of two days with the reins attached (or as long as is needed for the horse to be going happily), leather side-reins are introduced. They are not long enough for the horse to eat grass or, should he feel so inclined, to chew his boots. Neither are they short enough to put any pressure on the bit when the animal's head is carried normally.

The side-reins are shortened gradually over a period of several weeks, until the horse feels a slight contact and is prevented from turning his head too far to either right or left. He would also be unable to raise his head too high or to reach below his knee; though the side-reins are not being used to pull the head in towards the chest, there would be enough contact to encourage a little flexion at the poll.

The young horse requires a different length of rein in walk and trot. Whereas an advanced horse has the same frame at both paces, the novice is longer and less collected at walk and the side-reins have to be adjusted accordingly.

Having reached this stage in walk and trot only (cantering on the lunge will not be introduced until later), the young horse is ready to be backed, as described in the next chapter. We have known youngsters who could be backed within a few weeks of their first lungeing lesson, but they were exceptions to the norm. Usually you encounter some problems on the way which slow you down. The average time would be seven weeks and we have spent as long as three months on certain horses, whom we felt needed the extra time to secure the foundations of their training. The better this is established, the fewer problems you run into at a later stage.

A young horse tacked up for a lungeing lesson under saddle

Opposite: *Side reins – correct fitting for the young horse* (above), *and incorrect fitting* (below), *which leads to the horse being overbent*

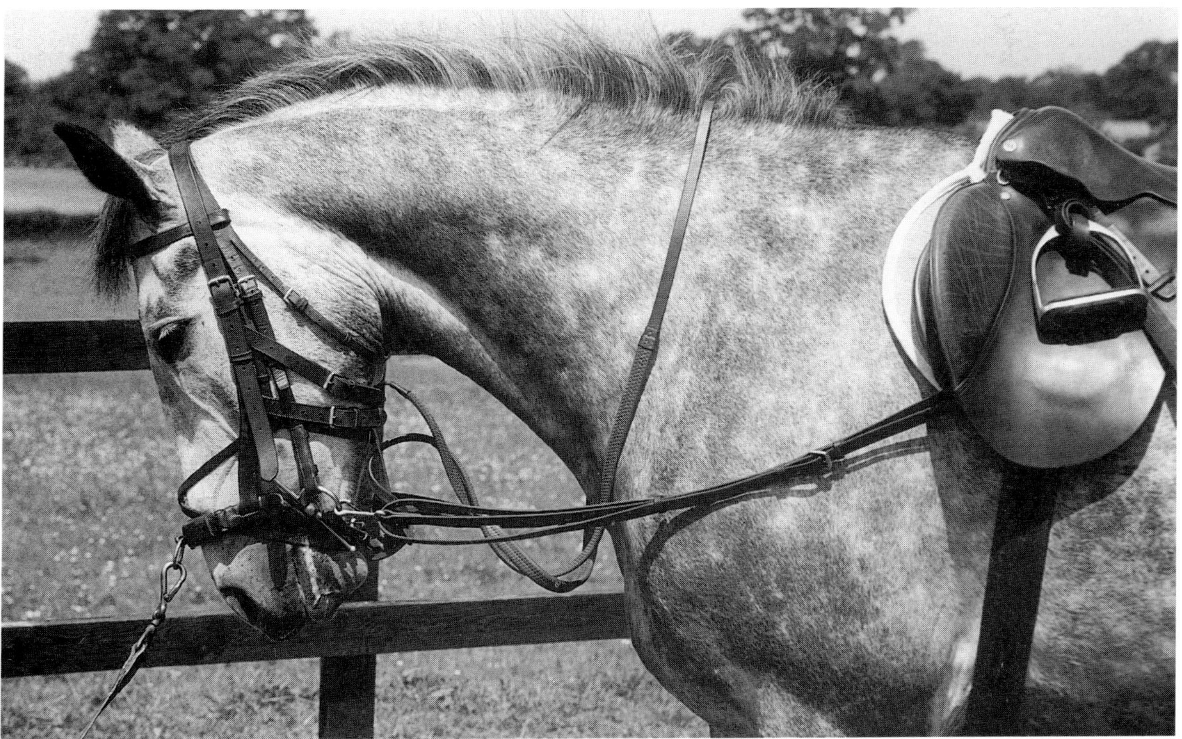

Lungeing an older horse

The mature horse that has already been backed will obviously bypass the earlier stages of lungeing. He starts with cavesson, boots, saddle and bridle, with the side-reins added a couple of days later. Very little leading is normally required before he is lunged for a maximum of ten minutes on each rein; he is used to work, so more of the lesson can be done in trot. At this stage we are trying to achieve a secure rhythm before he progresses to any serious ridden work.

As already mentioned, we will continue to use lungeing as part of the horse's training through every stage of his eventing career. These lessons supplement the ridden work and they have the same goal; whether on the lunge or under saddle we are aiming to get the horse to move in a balanced and rhythmic way, with a nice length of stride and correct bend.

Most horses favour the right rein, just as most people are more comfortable using their right hand. Because we want the horse to be ambidextrous, so to speak, it is lunged a little more on the rein it favours less. This helps to eradicate stiffness, but it should not be overdone. You can concentrate on the less-favoured rein to the extent where it becomes the one that the horse prefers; the opposite rein then becomes the problem. Sometimes the horse is so much stiffer on one side that he becomes resentful if worked for too long on the difficult rein, so that must also be taken into account. You can't expect co-operation from a horse that has been driven to resentment.

Lungeing should be regarded as an extension of ridden work, not as a separate entity. A horse on the lunge will often want to turn in and face the trainer when he halts, but it should be remembered that the rider will want him to halt without turning. So we train the horses to maintain the line of the circle they are describing on the lunge and to keep facing that way when they stop.

Sometimes we will be aiming to increase the length of the horse's stride – or, conversely, to introduce shorter strides if the horse is naturally very long and is likely to find collection difficult under saddle. The stride can be lengthened by having the horse on a slightly bigger circle (so that you are again walking some of the way with him) and by using the voice and whip to engage the hindquarters and produce increased energy. For a shorter stride, you would use a smaller circle. In this case energy still needs to be created, but the forward momentum has to be contained so that the horse becomes more springlike and less stretched out. The restraining influence comes through leverage on the lunge line and use of the voice, with words such as 'steady' and 'slow down' with which the horse is already familiar.

Dealing with problems

Some horses become stick-shy, invariably because they have been abused at some stage in their lives, and this can be a real drawback in lungeing. You have to assess whether the horse is over-reacting to the whip because he is frightened, in which case you will have to be ultra-careful. Normally the stick would be fairly close to his hindlegs and he might be given a touch on the hocks with it, if he needed a reminder to move forward. If it is pointed at the horse's shoulder, the stick has the same function as the rider's inside leg and can therefore be used to lighten a weighted shoulder (normally the left). A horse that is genuinely stick-shy will tend to run away from the whip and it has to be used with the utmost finesse.

On the other hand, he may be reacting because he is confused about what the trainer is asking him to do – or because he doesn't want to do it. It is essential to assess the situation rather than jump to a premature conclusion.

The aim is always to encourage the horse to co-operate, never to frighten him into submission. There are, however, occasions when it can be useful to give him a little assistance in frightening himself. If he stops and moves backwards on the lunge, for instance, you can use his disobedience effectively by telling him to continue going back until he bumps into something. Similarly, if he drags you across the school on the end of the lunge rein, keep him moving in that direction. As long as he is not likely to hurt himself, a slight bump can work wonders for discipline.

The training of horses requires a great deal of basic common sense, together with prudent regard for the old adage that prevention is better than cure. To take another example of a common misdemeanour on the lunge, you may have a horse that keeps swinging round at one particular spot. The answer could be to have someone strategically placed, holding a lunge whip. The assistant should let the horse see the whip, without frightening him. We don't mind him frightening himself on occasions, but we do not want him to be afraid of us. If he is to be any good at eventing, he must retain a certain independence of character; we are aiming to mould his personality, not to crush it.

Trotting poles

Once the horse has begun to establish a rhythm on the complete circle, he can be introduced to trotting over poles on the ground. They will help to loosen him and give him something new to look at and think about, as well as creating more elevation. Most horses enjoy going over things; those that don't like it will never make good eventers.

For the first lesson, we would have the pole on a straight line rather than a circle. This means that the handler has to walk or run (depending on whether the horse has been asked to walk or trot) on a parallel line about 10 to 12 feet away from the animal. If he accepts this single pole happily, more can be added, one by one, with a distance of 4-4½ feet between them.

Three poles are then placed in a fan shape on part of a circle. The average distance will be 4-4½ feet between the poles, but this will depend on the length of the individual horse's stride and it is up to the handler to acquire an eye for what is right, normally by trial and error. If you watch the horse's footfalls, you should be able to see whether the poles are too far apart or too close together and adjust them accordingly. The number of poles can be increased one by one over a period of about two weeks, until you have six on a fan-shaped curve.

Our horses do this exercise three or four times during the first fortnight, depending on their fitness and maturity. They are rarely clumsy without a rider and will normally look after themselves by lengthening their necks to look down and see where their feet are going. At this stage we are looking for the quality of stride, which includes rhythm, balance and elevation. It may be necessary to encourage the horse to lengthen its stride, which can be done by placing the poles a little further apart.

The trainer stands within the triangle and maintains contact with the horse's head

The poles can then be gradually raised (by 2 or 3 inches and later, perhaps, by 6 inches which is the highest we would have them), assuming that the horse looks as though he will be able to cope with this slightly more difficult exercise. We use our discretion and do not raise the poles for every horse, because some would find it a problem. Those that can cope with the raised poles are encouraged to loosen their backs and shoulders and improve their elevation and rhythm by this exercise.

Trotting over poles, whether lying on the ground or raised, can be arduous work for the horse. We would never do more than five minutes on each rein, always at trot, and after the first fortnight the horse would be asked to do this exercise only every three or four days. We never fail to make a careful note of the time the lesson starts, so that we can avoid the all too easy trap of creating problems by carrying on for too long. This is an obvious temptation, because you want the horse to advance as quickly as possible; the handler therefore has to learn to be satisfied with achieving a little progress at a time.

Jumping

The Ivyleaze horses usually do some jumping on the lunge at the age of three. They start over small cross-rails, with a placing pole 9 feet in front. This helps the youngster to take off at the right point, thereby giving him confidence over fences. We would then introduce him to an upright, again with a placing pole in front, which is gradually raised during successive lessons until it is about 3 feet high.

By now the horse will have been backed; he will be working with a rider as well as taking lungeing lessons. The ridden work rarely includes jumping until he has reached the age of four.

SUMMARY

Prior to first lungeing lesson
Horse led in hand, learns to walk and halt in straight line; is also led on large circles and rectangles around the school. Gets used to wearing cavesson and boots outside, and roller in his stable. Learns basic voice commands.

Sequence of lungeing lessons
(All preceded by approximately ten minutes' walking in hand.)
Horse wears cavesson (with lunge rein attached to front ring) and boots on all four legs. The lunge line is fed out so that he moves away from trainer and on to a circle at walk.
A roller is fitted (about two days later, if no problems).
Trotting is introduced when horse is behaving at walk.
Roller is replaced with saddle and girths (no stirrups).
Add stirrups, with leathers very short so that irons lie against saddle flaps.
Lengthen stirrups gradually until they are below saddle flaps, making sure that irons cannot touch horse's elbows.
Add bridle (no reins).
Add reins, tied up and attached to throatlash.
Introduce side-reins.
Shorten side-reins.

Trotting poles
Horse is led over single pole on the ground, usually at about the stage when he is being lunged at walk in cavesson, boots and roller.
After he is wearing side-reins at the correct length for trot, the horse is taken over trotting poles placed on a straight line.
Trotting poles are introduced on the curve of a circle.
If horse is able to cope, the poles on this circle are raised.

Golden rules
Keep lessons short (no more than thirty minutes, including ten minutes walking in hand).
Do not introduce anything new until each step has been accomplished satisfactorily.
Be strict with the horse – explosions of high spirits should not be allowed on the lunge.
Keep the horse happy and interested in his work (i.e. do not grind on with anything he finds difficult).
Keep to walk and trot only – cantering will be introduced at a later stage.

ungeing over poles of varying heights nd distances will improve the horse's alance and cadence

4

BACKING AND EARLY RIDDEN WORK

BACKING

It is normally one of the girls who gets on a young horse when we are backing it for the first time, which is good experience for working pupils. This is always done in the security and familiar surroundings of the horse's own stable, with someone holding his head.

We will have prepared the three-year-old for this moment by pulling downwards on the stirrup leathers in his stable so that the saddle feels heavier on his back. If the youngster tends to be difficult when faced with a new experience, the rider will start by lying across its saddle a few times. More often than not, however, the rider mounts from a solid, non-wobbly stool or is legged straight up to take a normal seat in the saddle. The horse is patted quietly for a minute or two, then the rider dismounts.

This would be done about three times, and if the horse behaved perfectly, we would leave it at that for the day. If he were difficult in any way, the rider might get on four or five times – always with the intention of finishing on a good note, with the horse behaving reasonably well. Should the animal find all this rather alarming, we might repeat the process later the same day.

Once the horse seems quite happy to have someone getting on and off his back, he will be led round the stable with the rider on top. Being in such an enclosed space restricts his movements, which obviously makes it much safer than being in the outdoor arena. Our horses are usually ridden in the stable for about a week before they venture outside.

We always use a mounting block for getting on the horse once it has left the confines of its stable – and for ever after when it is possible. This means there is less chance of the saddle slipping; it also reduces the risk of pulling the horse's back or digging a toe into his ribs. Obviously the animal has to be taught to accept a rider mounting from the ground, because I might need to get on him in some place where there is no mounting block available, but otherwise we try to avoid it. At an event, or when we travel by horsebox to go cantering, I either stand on the ramp of the vehicle to get mounted or have a leg-up.

The mounting block at home has an additional advantage with young horses. It is always hard to assess how much they are blowing themselves out, but if you are getting on from the ground, you have to make sure that

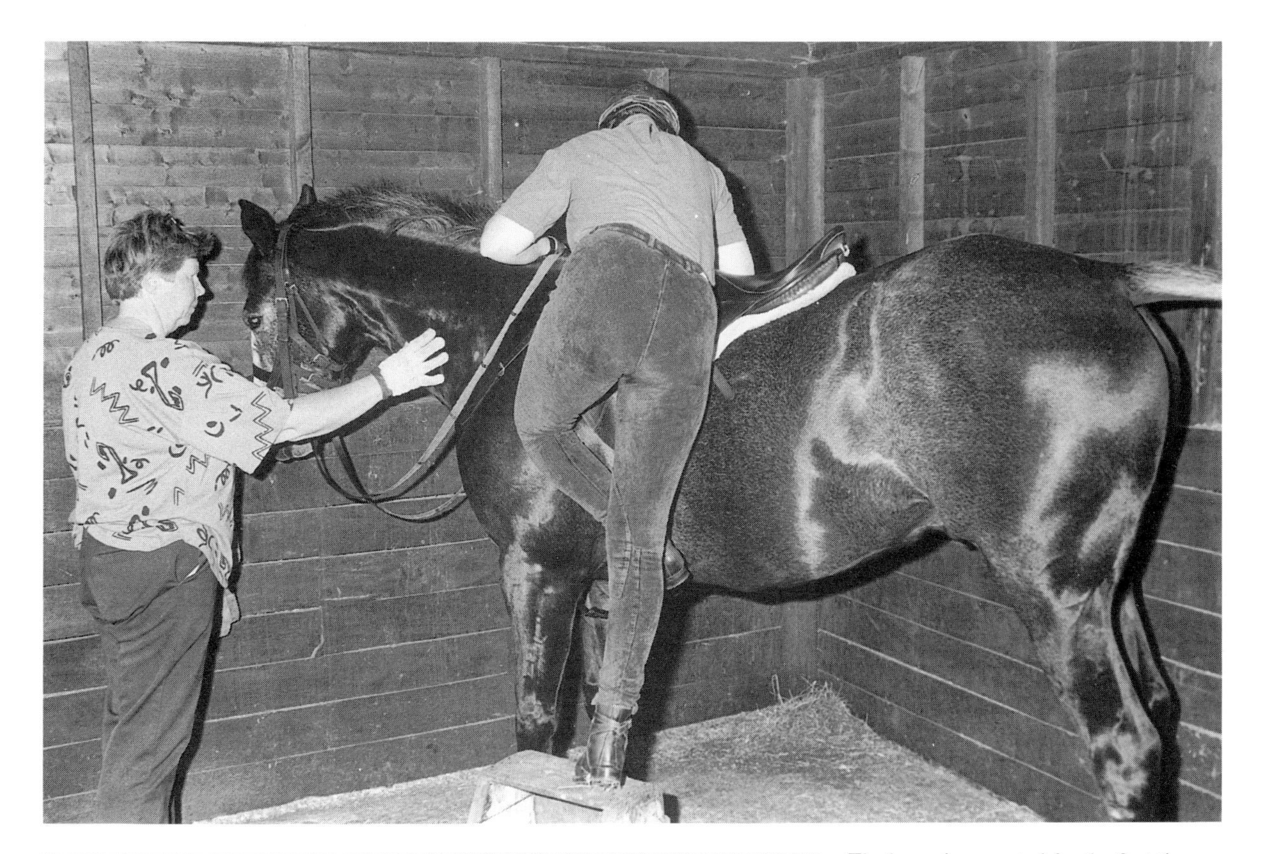

The horse is mounted for the first time in his stable

Mounting outside in the arena for the first time, with a handler to hold the young horse

the girths are tight. We do not want the horse to associate the saddle with discomfort, which would be the case if we over-estimated how much he was blowing himself out and tightened the girths like a vice. They can afford to be a little less tight when you are getting on from the mounting block, with an assistant holding the horse's head.

ACCEPTING THE RIDER

To begin with our young horse wears a cavesson over its bridle, with a lunge-rein attached. He would normally be bitted with a cheek snaffle that had a mild rubber mouthpiece. Steering can be difficult with a young horse and the shafts at either side of this bit are more suitable than rings, because they prevent it from sliding through the mouth when the rider gives the aids to turn. The equipment would also include a breastplate or neck-strap.

During his first lesson with a rider on board, the horse is led round at walk on both reins. We prefer him to be led, rather than lunged, with a rider because it enables him to walk on a larger area. At this stage, he should pay attention to the person on the ground rather than the rider on top. Until now all his instructions have come from someone on foot; he needs time to get used to the idea of transferring his attention to someone who is out of sight and whose voice is coming from higher up and further back.

The ideal rider for our immature three-year-old would obviously need to be light. He (or she, as is usually the case at Ivyleaze) also needs to be totally laid back. These early lessons have to be learnt with someone who is relaxed and fairly passive; any hint of tension and anxiety will make it more difficult to gain the horse's calm acceptance of carrying a person on his back.

Quite often, as he walks forward with his handler at his side, the horse is unaware that there is someone sitting quietly on top. At first the handler will be giving the voice commands; then the person on the ground stays quiet and the rider does the talking. Some horses turn their heads to look up and back in order to find out what is going on. After that, they normally accept our totally laid-back rider quite calmly.

Assuming that the horse is behaving, we would expect to remove the lunge line after about a week – though the youngster would continue to wear the cavesson for two or three more lessons, just in case it was necessary to attach the line again. Although control is now being transferred to the rider, the handler still walks alongside. More often than not, the horse is unaware that he is no longer being led. When the handler moves away, transferring total responsibility to the rider, the youngster often follows. He has to be pushed back on his own separate line with a hand held out towards his shoulder.

Reassurance is all-important!

Opposite:
First lunge lesson with the rider (above)

Responsibility is transferred to the rider (below)

Lungeing over poles will help the horse to use his shoulders correctly and engage his hindlegs

A youngster takes his first jump from trot. Dangling forelegs show that he is unsure about his technique

Later the same day. Increased confidence enables the horse to jump with greatly improved style

It is important to walk the disiance correctly. The horse's experience, physique, natural ability and length of stride then have to be taken into account

A difficult open intermediate coffin: rail, one stride, ditch, bounce, ditch, one stride, rail

The approach requires a short, balanced and bouncy canter to the base of the first rail

The horse needs encouragement and freedom of head and neck over the ditch, with the rider sitting up

The horse gains confidence while the rider still encourages him to go forward

Freedom of head and neck is still required, with the rider staying in the centre of balance

An open intermediate double bounce of three white verticals, with both distances at 15 feet (4.57 metres). The correct pace on approach depends on the horse's length of stride as well as the distance between elements and can only be learnt through practice at home. There is no room to make adjustments if the approach pace is wrong. The rider should stay well within the centre of balance, allowing the horse freedom of his head and neck without losing contact. Note that the horse lands equidistant between the vertical rails

LEARNING THE AIDS

Throughout these early stages the horse is learning, by constant repetition, to associate the rider's leg aids with the familiar voice commands of 'walk on' and 'whoa'. Normally he is given a moment's advance warning by using the word 'and' before the instruction. If we want to prepare him to move forward, for instance, the command would be 'and . . . walk on'.

Assuming that he has been correctly handled, the young horse is remarkably quick to learn that 'and' means some instruction is about to follow; we will continue to use it at a later stage as a preparation for transitions. It makes no difference who says the words, as long as they are always exactly the same for each command. These words form a reassuring connection throughout the early stages of training. The young horse first learns them when he is led in hand; they are repeated (not necessarily by the same person) when he is being lunged; they are used by someone else when he is ridden. The importance of using the same words during each stage of training cannot be overemphasised. They are the great link that helps us to continue getting our message through to the horse without confusing him.

The voice is obviously an essential part of the early ridden work, when we are teaching the horse to respond to leg aids. The instruction to 'walk on', given at first by the handler, is accompanied by a light squeeze from the rider's legs. Unless the horse is whip-shy, it may be helpful for the rider to carry a short stick, which can be used to reinforce the legs by giving a light tap behind the girth as a forward aid. This message can also be underlined by the handler tapping the horse by hand behind the girth. When used on the shoulder, the stick helps to correct the horse when he has fallen in or out.

The horse again has to learn to walk and halt in a straight line, this time with a rider on top. Before the handler moves away, he must be taught to stop when the rider tells him to do so.

Though we want the horse to accept a light contact on the reins, the rider has to be ultra-careful to avoid spoiling his mouth. We would use the voice command together with a light feel on the reins – reinforced, if necessary, by a tug on the neck-strap. The handler can also help with light pressure on the lunge-rein and by placing a restraining hand in front of the horse's shoulder. If the horse were to acquire a bad mouth through the rider pulling on the reins at this stage, it would be with him for the rest of his life.

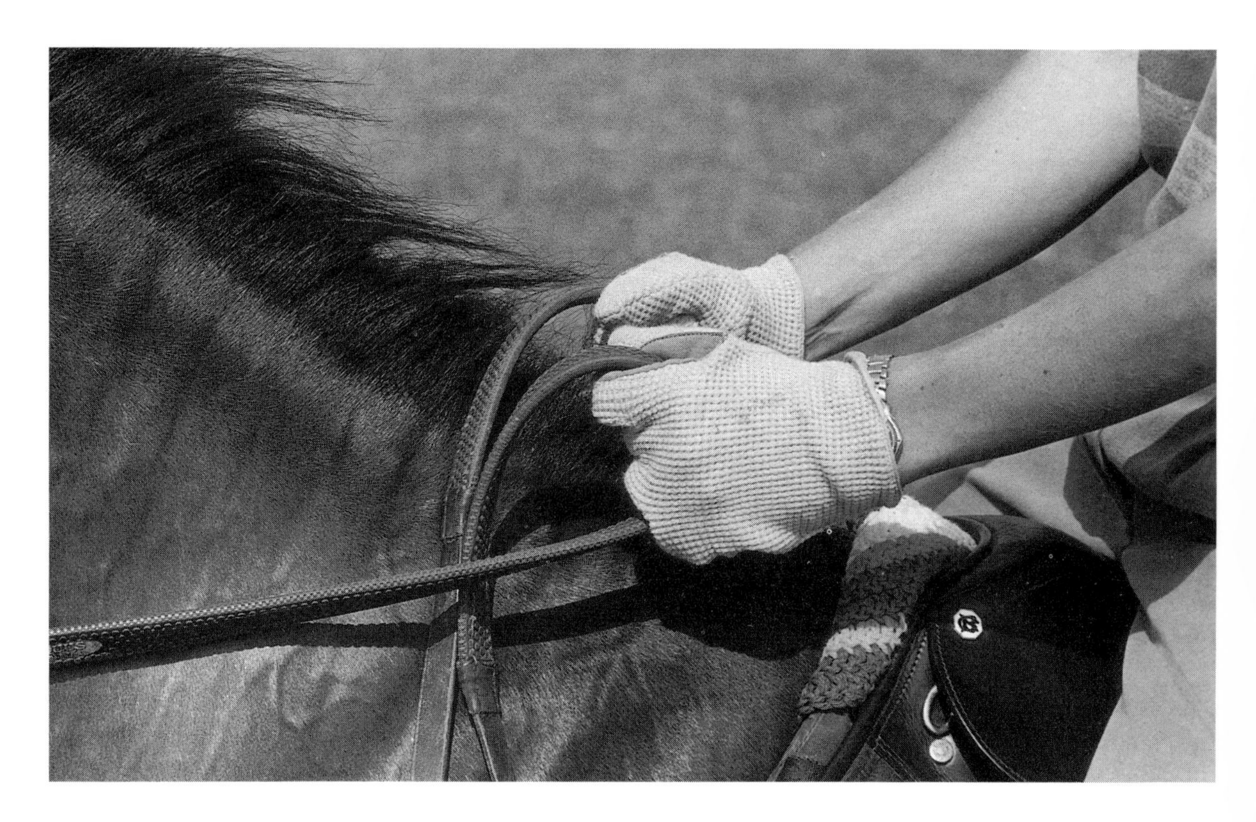

Through constant repetition, praise and discipline (plus some help from his remarkable memory), the horse will slowly learn to respond to the signals given by the rider's legs, weight of body and hands. Aids for the basic movements he will learn are described at the end of this chapter.

THE RIDER

You need to check your own position before you begin to think about what the horse is doing. This is best achieved by riding on a loose rein round the school while you are sorting yourself out.

Only a light contact, which allows a forward feel, should be attempted in the initial stages. You must be careful not to pull back on the reins but to relax your arms and let your hands move forward instead; they should maintain contact as though they were a piece of elastic extending to the horse's mouth. To do this, you need to be completely relaxed in body, arms, wrists, hands and fingers. You should hold the reins as you would a bird – gently enough to avoid squashing it, but not too lightly because it would fly away.

The amount of contact, which will vary according to the individual preference of each horse, needs to be maintained at a constant level. We sometimes help to explain this to our working pupils in the tack room, with someone holding the bit end of the reins and simulating the movements a horse might make. The pupil holds the reins at the rider's end and tries to follow these movements, with arms moving forward and back to maintain a level contact.

If you are tense on a horse, your shoulders, elbows and wrists will be tight, making it difficult for you to use your hands in a sympathetic and responsive way or to distribute your weight correctly. The effort of concentration makes it all too easy to become tense without being aware of it, so you need constantly to question yourself to make sure that every part of you is relaxed. If you are aware of some tension, take deep breaths or sing a song to help yourself relax!

Hands – correct position (left), *and incorrect position* (right), *which could lead to 'blocking' the horse*

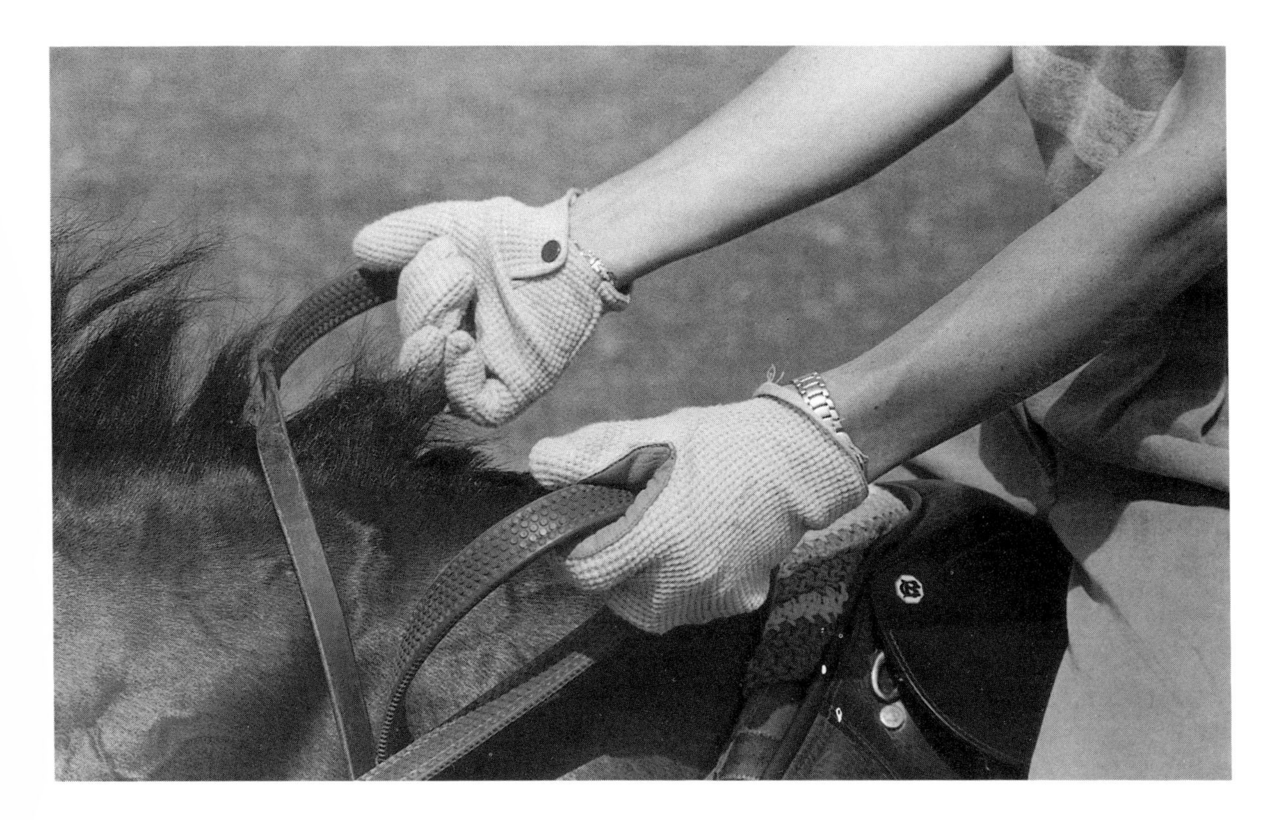

There should never be more contact through the hands than through the legs. Novice riders often try to grab the unfortunate horse between the upper leg and crutch, which gives their thighs the same action as a clothes peg and tends to move them upwards rather than down into the saddle. Instead your legs should be resting quietly against the horse's side, with thighs and knees relaxed. If you think of your ankle being on the horse, you will probably get the right part of your leg on him. The horse can be either helped or hindered by the way you use your legs or distribute your weight.

Some riders have a natural feel, which helps them to listen to the horse. Others have to teach themselves to be aware of everything the horse is doing. This awareness will be a key part of successful training because it tells you what you want to achieve and when you should start trying to achieve it. You cannot plan a horse's training on a computer; you have to listen to him, feel your way forward and be flexible.

SCHOOLWORK

You will need an enclosure of some description, whether it be a corner of a flat field or a permanent arena. The young horse requires 40 by 40 metres or (as we have at Ivyleaze) 30 by 60 metres for his schoolwork; anything smaller is too restrictive.

To begin with, lungeing will still occupy the larger part of the youngster's lessons after he has been backed. He is given two short sessions of work each day for the first week, which start with him being led or lunged for about ten minutes at walk. This is followed by five minutes trotting on the lunge. He then has a maximum of fifteen minutes' ridden work at walk, using the half of our school that is further away from the exit. If he is going well, the rider will finish the lesson early to ensure that it ends on a happy note. After about a week, he is given just one daily lesson of combined lungeing and ridden work lasting a total forty-five to fifty minutes.

During the course of the next few months the lessons will slowly change, with a decreasing amount of time given to lungeing and more of each session devoted to riding. At the start of his ridden work, he is walked on 25-metre circles and squares at the far end of the school. After that we would aim to get the youngster used to going all the way round the school, to crossing it on the diagonal and to being ridden on large loops, circles and oblongs. He is kept away from the post-and-rails at the side of the school, probably by no more than 5 feet, because young horses have a tendency to lean towards any type of fencing.

Ten to fifteen minutes' ridden work in the school is quite long enough. We do not want him to regard this as a place of tedious work, any more than we encourage him to see other areas where he is ridden as places to let off steam. Whether in the school, being ridden in the fields or later, when the horse is considered safe to take out on a hack, he is expected to listen to his rider. There will, of course, be occasions when the horse wants to do nothing of the kind. It is our job to try to combine strictness with consideration, so that he is happy in his work and wants to co-operate. We must also be on our guard to foresee situations in which arguments might arise, in order to avoid them.

If, for instance, the horse has been known to nap, you could have someone standing quietly at the exit with a lunge whip as you are riding past during the early work under saddle. If you are any good at listening to the horse, you should know when you can dispense with your assistant, always bearing in mind that you must do everything you can to avoid a battle of wills.

If our youngster is going really well at trot on the lunge and is behaving sensibly, we will include trotting at an early stage of the ridden work. This is done without making too many demands on the horse. As my mother says, 'You would obviously trot to the best of your ability, but this is not the time to have the vapours if the horse falls in coming round corners at trot, or if he loses some of his rhythm and balance.' Rome was not built in a day and we should not expect to get results from a young horse overnight.

To begin with, the rider will simply do a little trotting to add variety to the lesson in between trying to get a good walk. By the time we begin to ask for something more at trot, we will want the horse to be walking quite well, with good rhythm and balance.

Learning the basics

The horse has to learn to go forward on a light contact, which is best done by keeping him on the move and stopping as little as possible. If he is reluctant to go forward, the rider's voice and leg aids can be reinforced by using the stick behind the girth. It may also help to exploit his herding instinct by having another horse some distance in front, but within sight, for him to follow.

He must also learn to be straight, with the line of his body from ears to tail following the line he is taking. Thus on a circle, his body has to be curved to be 'straight'. Each hindfoot should follow in the same track as the corresponding forefoot, but do not be alarmed if your youngster tends to wander off that line. It is quite normal for the young horse to be crooked, as you will probably have seen for yourself when he was on the lunge.

It is only through work on the flat that he becomes straight. To do so, he has to learn how to balance himself with the weight of a rider on his back, and take an even contact on both reins.

Falling in on a turn or circle is also quite normal. It happens when the horse moves his balance to the inside shoulder and his feet usually follow by drifting inwards. His head invariably bends to the outside, which leaves his inside shoulder leading. If you watch a horse out in the field, you will see that this comes naturally to him. He automatically moves round corners with his inside shoulder leading and his head bent in the opposite direction. We therefore have to exercise great patience when we ask him to stay upright and bend his body from head to tail on the same curve as the turn or circle he is following. This is something that is not natural to him and he will need time to learn how to do it.

We are looking for a light contact as soon as the horse is backed, but we will not be asking for him to carry his head in any particular way. If you try to improve his shape before he is happy to have contact with the rider's hand, he will learn to evade the bit. To begin with, he will move his head in all directions and your hands have to follow wherever it goes. Once he is happy to accept a light feel on the reins the horse's head will remain still.

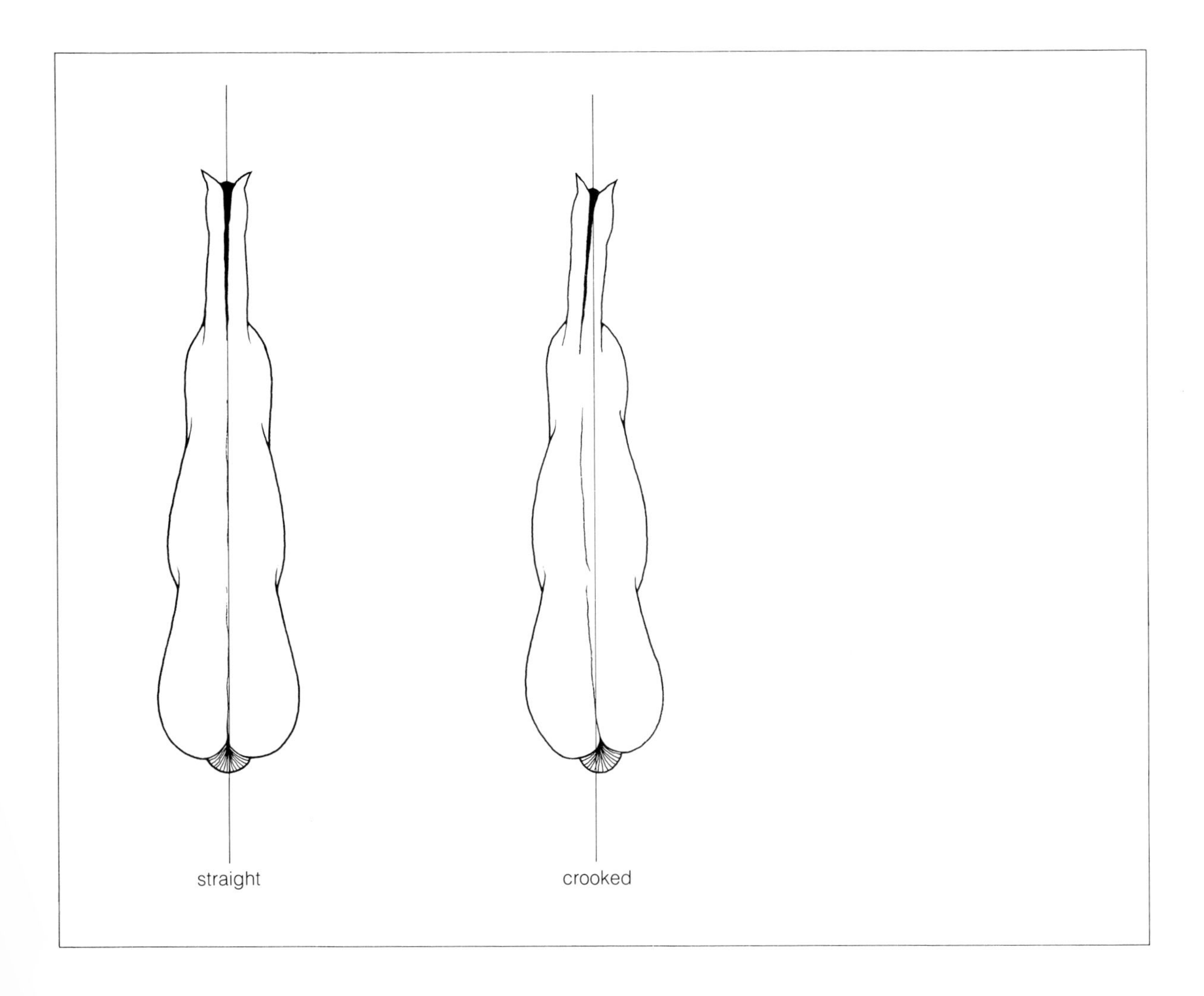

Rhythm and balance

Work on the flat helps the youngster to carry himself in a way that makes it easier to support the weight of a rider. He has to learn to carry the extra load on his hindlegs and he can do this only if the rider's body is controlled, in balance and still, moving only when there is a specific reason for doing so. The horse is capable of balancing himself on his own; we must therefore avoid interfering with this natural balance through our own lack of suppleness or an inability to sit still.

Young horses are often in a hurry; they tend to go faster than the speed at which they can best achieve rhythm and balance. This is often a throwback to nature, which has given the horse an instinct to use flight as a means of escaping from anything he dislikes or distrusts. Even when happy and trusting, he may like to remind you that he was born to gallop!

It could be that the youngster's exuberance is caused by over-feeding, which is a common mistake. As will be discussed in a later chapter, we believe in giving as little hard feed as possible to get the horse fit enough for the job in hand.

When he goes too fast, the horse tends to shorten his stride. This encourages him to fall in on turns and circles, simply because that is the only way he can balance himself at the speed he is going. The trainer has to decide on the right speed for each individual horse, otherwise he will not be properly balanced with the weight of the rider. This usually means slowing down a little in trot, but not necessarily in walk.

If you do not have a metronome in your head, it may help to play some music in the school so that you can recognise whether or not the horse is in rhythm. The choice of music would obviously have to be different for walk (which is in four-time, with no moment of suspension) and trot (which has a distinctive one–two beat, with suspension between steps).

Turns and circles

Riding a correct turn is exceptionally difficult. There is only one way of doing it properly and, according to a renowned dressage expert, 3982 ways of getting it wrong – which could easily lead one to despair! The rider must be able to move each hand and leg independently, because each has a different function. When turning to the left, your inside (left) leg is used on the girth to stop the horse falling in and to keep him moving forward. The outside leg rests behind the girth, ready to be used if he swings his quarters towards that side. The inside rein asks the horse to make the turn and bend in that direction. The outside rein allows for this flexion, while maintaining contact and (if necessary) controlling the speed or correcting any excessive bending. Your shoulders need to turn with the horse and there should be slightly more weight on your inside seat-bone.

In order to get your weight in the right place, you have to overcome a natural tendency to sit on the outside seat-bone. If you try turning your shoulders to the left while you are sitting in a chair, you will probably discover this for yourself. It is natural for the right hip to collapse and for the weight to sink to that side. Your position will be crooked if you do the same thing on a horse; he will then have to compensate by going crooked himself, thus losing his balance, impulsion and rhythm.

If the horse encounters problems at a later stage, it will almost certainly be because he never learnt to do basic turns and circles correctly. They will not be achieved in one session. As already mentioned, most horses tend to be stiff on the left side. This means that they carry their quarters slightly to the right, which results in them leaning on the left shoulder. Turning to the right will seem much easier, but will nevertheless be incorrect because the horse's weight will be on his left shoulder and he will therefore have a tendency to fall out. Turning to the left will reveal more obvious faults – and plenty of them! The horse will tend to fall in with his left shoulder, tip his quarters to the left, cross his left hindleg in front of the right hindleg, lean on the inside rein and lose contact with the outer one. All these things have to be corrected.

Circles are a continuation of turns, so the same principles apply. If the horse consistently falls in when circling to the left, we use an exercise called counter-balancing. This is done by using your inside (left) leg on the girth to encourage him to lighten his left shoulder and keep moving forward, while feeling on the right rein. As a result, the horse bends to the right for one or two strides, which helps to take the weight off his left shoulder. He is then asked for the correct bend to the left, which he is in a better position to achieve because his weighted left shoulder has been released. Once he has learnt to go on a circle with the correct bend, the horse will discover that it is far more comfortable than doing it the wrong way and will therefore be happy to co-operate.

The counter-balance exercise can also be used on a straight line to encourage the horse to balance himself correctly and take an even contact on both reins. Another method to help overcome his natural stiffness is to give and take with the left hand while maintaining your leg aids; this will stop him leaning on the left rein and encourage him to take contact on the right.

Introducing canter

The horse is usually ready to canter after five to six weeks of ridden work but his physical development, as well as his state of training, has to be taken into account before it is introduced. If the horse is still physically immature, we would not attempt canter until later in the year – or we might wait until the following year. We invariably use a minuscule cross-rail, about 6 inches off the floor at the centre, when asking for the first canter strides. This is placed three-quarters of the way down the long side of the school, which will encourage the youngster to strike off with the inside leg leading.

The cross-rail saves the necessity for strong aids – and avoids the possibility of the horse going into a fast and unbalanced trot – since he will invariably land in canter over this tiny fence. Once started we would probably include canter in the lesson two or three times a week, asking for no more than one large circle to begin with.

Within the next fortnight, the horse will probably be asked to strike off in canter using normal aids (see end of chapter) instead of the cross-rails. If he is sufficiently strong and balanced, he might do two or three circuits at canter – preferably using a slightly larger area than 20 by 40 metres which is rather confined for a young horse. The trot needs to be re-established for a period of four to five minutes immediately after the horse has cantered. He may have become excited and will need this time to return to the calm, relaxed and rhythmic trot that should have been established by now.

Hacking

The length of time required before the horse can be hacked out in safety obviously depends on the individual youngster and the local environment. Usually our country lanes are reasonably quiet and we would normally reckon to take a horse out on them between eight and ten weeks after he had been backed. First he

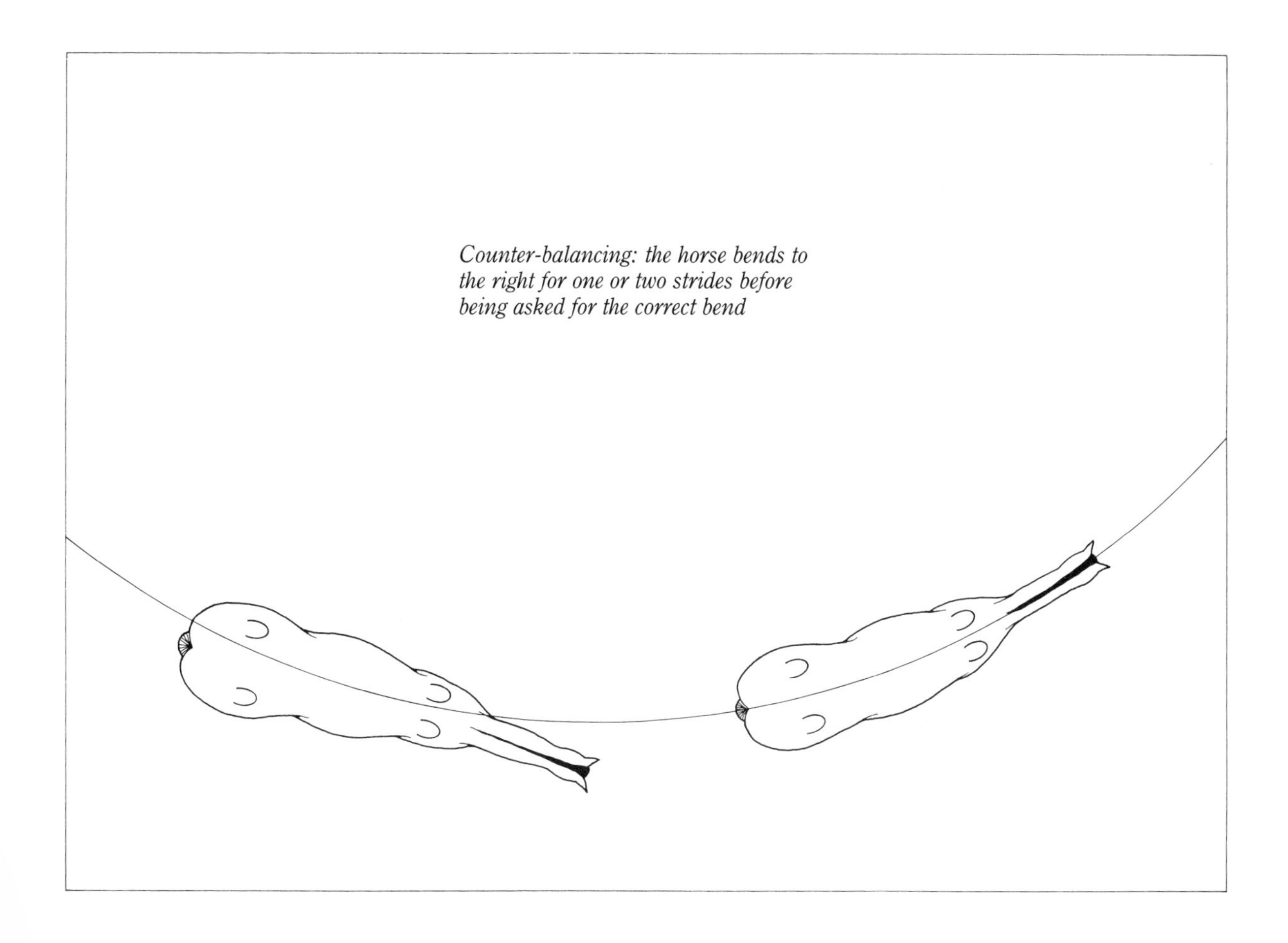

Counter-balancing: the horse bends to the right for one or two strides before being asked for the correct bend

will be taken down the drive, then a little way along the road, followed by a short trip around the village block. Assuming he is behaving sensibly, he will then venture further afield. We rarely hack a youngster out alone, so the horse will normally have one or more companions with him. These early outings last between twenty minutes and half an hour, using different routes but avoiding anything in the locality that might cause alarm.

By now the horse will be working for one hour six days a week. He will have a combination of lungeing, ridden work in the school and hacking – though not necessarily all three on the same day. The art of successful training lies in keeping the horse interested by giving him a variety of things to do; you can never hope to hold his attention if he is bored stiff with his lessons. Hacking should not, however, be seen as something entirely different. We use these outings as an extension of the school work, not as occasions for meandering along country lanes.

At this stage the youngster should be able to stay straight at walk and trot on a 20-metre circle and a large 20 by 40 metre rectangle, without falling in or leaning out on either rein. He should also be able to walk and trot on a large figure of eight. Being a baby, he will still be slightly wobbly, so little more is expected of him until the following year.

After three to four months' work, the youngster is turned away. Thanks to his long memory, we can be confident that his lessons will not have been forgotten when he comes back into work about four months later.

SUMMARY

Horse is first backed and ridden in stable.
Rider gets on outside from mounting block.
Horse is led round at walk on both reins, with handler on ground giving voice commands.
Rider gives voice commands.
Horse learns aids by rider applying them at the same time as the voice command. The stick is used only as a last resort.
Lunge line is removed from cavesson, but handler continues to walk beside the horse.
Handler moves away, transferring control to rider.
More lungeing than ridden work initially, slowly changing until horse is ridden more than lunged.
Horse ridden on large circles and squares at one end of school.
The whole school is used for oblongs, loops and riding across it on the diagonal.
If horse is sensible and trotting well on lunge, he starts trotting with rider.
Canter is introduced after five to six weeks, if horse has made sufficient progress in training and physical development.

The rider

Check position before thinking about the horse.
Aim for light contact with horse's mouth.
Relax arms and let hands follow horse's head movements.

Be completely relaxed and in balance.
Never use more contact through hands than legs.
Learn to listen to the horse.

Basic aids

Upward transitions (halt to walk, walk to trot):
Slight sink of body, with weight moved back a little.
Squeeze evenly with both legs.
Ease the reins as horse moves forward, but do not lose contact.
Trot to canter (left leg leading):
Weight on left seat-bone, slightly angling left hip forward.
Ask for a little bend with left (direction) rein.
Allow for bend by giving with right (supporting) rein.
Left leg on the girth.
Right leg behind the girth.
Allow a little with the hand as horse strikes off, but do not lose contact.
Downward transitions (canter to trot, trot to walk, walk to halt):
Prepare with one or two half-halts three strides before transition (i.e. squeeze momentarily with legs, while using hands to restrain faster movement and increase collection).
Weight up and slightly back.
Squeeze with both legs.
Retard with give and take of reins, followed by a longer retarding feel until the transition is executed.
(N.B. Your hands do not allow horse to continue at same pace, but he should not freeze up against you.)
Turn or circle to left:
Weight a little more on left seat-bone.
Ask for bend with left rein.
Right rein allows for bend and controls speed, while maintaining contact.
Left leg on girth to stop horse falling in and keep him moving forward into both reins.
Right leg behind girth to control quarters.
Turn shoulders with movement, taking care to keep weight off right seat-bone.

Aims for the trainer

At this stage we want the horse to learn to:

- go forward
- stay in rhythm
- accept contact on both reins
- balance himself with the weight of the rider
- be straight
- maintain the correct bend.

If he leans on left shoulder, use counter-balance when circling on left rein: using left leg and right rein, ask him to bend right for one or two strides. Immediately ask for correct bend to left, while weighted left shoulder is released. Counter-balance can also be used on a straight line to improve balance and encourage even contact on both reins.

Golden rule: Keep lessons short!

5

PROGRESS ON THE FLAT

The young horse is more mature, both physically and mentally, when he continues his education at the age of four. We believe in giving him plenty of time at this stage, so that he can learn without being hurried along. The better these foundations are established, the easier it will be the following year when he is preparing for his first competition.

Though he needs enough food to help him grow, his education will suffer if he is fed too much and is over-fresh as a result. He will then be too excited to listen to his trainer and the work he does will be less concerned with learning than with letting off some of his excess energy.

When he comes back into work, the four-year-old will need to get fit with a minimum of three weeks' walking. During that time, his daily work starts with ten to fifteen minutes on the lunge. This helps to make him fitter and more obedient, which also makes him safer to be taken out for a hack immediately afterwards. Work is limited to fifty minutes for the first week and then builds up to between one and one and a half hours. After the first fortnight, he still goes out for his daily hack and is either ridden or lunged in the school. Hacking is also part of his schoolwork; at Ivyleaze we do not believe that dressage training has to be confined to an enclosed area and that you can forget all about it when you leave the arena. The horse probably finds it more enjoyable to be out on the roads or in the park, where there is so much more to see, but he is nevertheless asked to move to the best of his ability and to stay on a straight line. Once he is fit, he can do more strenuous schoolwork and hacking. He will also start the jumping lessons which are described in the next chapter.

IN THE SCHOOL

The shapes that the horse follows under saddle can now become more demanding, but the emphasis on forward movement remains. He is asked to walk smaller circles of 10 metres in diameter and to trot 20-metre circles. As well as oblongs and large 20-metre squares, he walks a smaller square of 15 metres. Diamonds, also of 15 metres, are introduced. These help to make the horse's shoulders more mobile and teach him to keep his quarters in the right place as he turns. As we have already discovered, making a correct turn is by no means easy.

Schooling exercises
10 and 20 metre circles
10 × 20 metre rectangle
diamond
3 and 10 metre loops
figure of eight
3 loop serpentine

Next we would introduce 3- and 5-metre loops, followed by figures of eight within the small dressage arena size of 20 by 40 metres. The figure of eight incorporates two diagonals and constant changes of rein, which helps us to discover how well the horse balances himself, with the weight of his rider, in a small area.

More canter work is then introduced, both out hacking and in the school. The horse is asked to strike off, using a corner or circle to encourage him to lead on the inside leg. If he fails to do so, he will not be reprimanded. He has been asked to canter and has produced the right pace, albeit on the wrong lead; it would only confuse him to be told that was not what was wanted.

Usually he will come back to trot of his own accord if he is leading on the outside leg, and he should be

Turns and circles are difficult to perform with the correct bend, as shown here

allowed to do so. He is then trotted round the school and asked to canter at exactly the same place as before. This exercise is repeated, if necessary, until he strikes off on the required leg. He would then build up the number of canter circuits on each rein, depending on how long he can keep going without any sense of struggling.

Three loop serpentines at walk and trot are now added to the other shapes the horse was following in the school. We also like to introduce our four-year-olds to leg yielding, which is always the first lateral work they do. This exercise, in which the horse yields to the rider's leg by moving away from it, encourages more response to the aids. It also improves the horse's suppleness and elevation, at the same time helping to produce a rounder outline. At this stage, we would attempt it only at walk.

Horse and rider enjoying their work at canter

For his first lesson in leg yielding, the horse is ridden along the three-quarter line of the school and then asked to move diagonally for a few steps, crossing one fore and hindleg in front of the other. We are looking for only a little flexion from head to tail, with the horse bent slightly away from the direction in which he is going. It is always the inside of the bend (rather than the direction of the movement) which is meant when referring to the inside rein or leg – which, in this case, is on the opposite side to the direction the horse is taking.

For leg yielding to the left, he should be slightly bent around the rider's inside (right) leg, which is applied behind the girth to keep the horse moving diagonally away from it on four tracks. The outside leg, used lightly near the girth, maintains impulsion, controls the quarters and prevents falling out. The inside rein helps to keep the horse's shoulder on the correct line and asks for a slight bend, while the outside rein controls the speed.

Leg yielding from three-quarter line of school

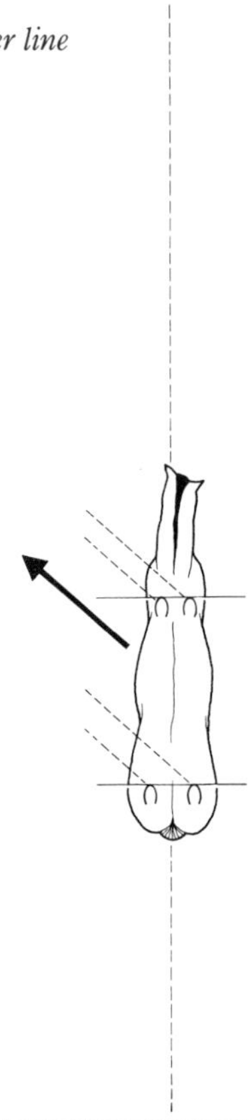

At this stage of the horse's ridden work, we also use trotting poles – at first one on its own to be ridden on a straight line, gradually building up to about four on the curve of a 20-metre circle. These would be spaced at the same distances that were used for lungeing. Walking and trotting over these poles on the ground encourages the horse to look and see where he is putting his feet. They also help to improve his cadence and elevation, with the weight of a rider on his back. If the horse is supple and relaxed, we might add more poles or raise them off the ground by no more than 6 inches.

By now the horse's trot should have improved. We want him to be calm and relaxed, in rhythm and listening. Each horse has a distinctive one–two beat in trot and the trainer needs to know that beat in order to achieve good rhythm.

It is at this stage – assuming that the horse is performing the above movements satisfactorily and has accepted an even contact with the rider's hands – that we can also start asking for some flexion from the poll and so introduce the correct outline for dressage.

Lengthening and shortening

As soon as the horse has rhythm and co-ordination at working trot or canter, I would think about asking him to lengthen and shorten within that pace. We tend to start with canter, which is normally the easiest because it comes naturally to the horse. The walk is the most difficult, because it is harder to create energy at a slower pace; for this reason, we normally leave lengthening and shortening at walk until last.

By now the horse's jumping lessons (described in the next chapter) will have started and I will want him to be able to shorten and lengthen in canter on the flat before I think about cantering into a fence. His own length of stride dictates which of the two we work on first. If short, I would begin by asking him to lengthen for five strides and then come back to working canter. This is done by creating the necessary energy with your legs and then giving with your hands, thus allowing him to stretch his head further forward without losing contact. The horse is unable to lengthen without this extra freedom of his head and neck, because his feet cannot hit the ground any further forward than his nose. You will be looking for lengthened strides, not a faster speed; if you are failing to achieve it, you will have to come back to working canter and try again.

Opposite:

Good lengthening at trot, except for horse's head being slightly behind the vertical (above) *and a little overbent* (below)

Since the horse is quick to anticipate his rider, you should avoid asking for lengthened strides in the same place. Once he gets the idea that the long side of the school is the place where you want him to perform this new exercise, he is likely to start doing it without being asked. So use the short side as well and, maybe, part of a large circle.

Some horses have a naturally long stride and they need to work at shortening. In this case the rider's legs again need to create energy, but instead of encouraging the horse to stretch his head and neck forward, the hands ask for that extra impulsion to be contained in shorter and more active strides, almost like a series of half-halts. Though the immediate aim may be to help him with his jumping, shortening will also improve his working canter and be of benefit when he comes to perform serpentine loops and other more difficult movements at canter. There will be similar benefits for the short-striding horse when he learns to lengthen; these lessons will eventually produce a better medium and extended canter, as well as helping with his jumping and galloping.

Having said all this, we should not be too greedy. Flatwork can help to improve the horse's natural paces, but only up to the natural limit imposed by his conformation and action.

The rider

The rein must always be used with a give and take of the hand, never with a continuous pull. Sometimes the hand will be passive and it can be slightly restraining at the same time, while waiting for the horse to submit to the aids. The passive hand, when incorporating that element of restraint, can be an effective means of correction if the horse is leaning on the bit or unwilling to give to the rider's hand. It must always be used with sensitivity and backed up by leg aids. The restraining element is omitted if the horse is evading the bit. In this case the hand is entirely passive, waiting for the horse to come to the bit and accept an even contact. When this is achieved, the hand should yield (but without losing the contact), which is a way of conveying thanks and praise.

The level of contact you have with the horse's mouth will depend on his balance, size, cadence and individual preference. Some horses like a fairly strong contact while others prefer a much lighter one, so you have to assess your own partner's specific requirements and go along with him.

If you suddenly drop the reins and lose contact during a movement or transition, he will wonder what has happened to you and lose his balance. This is not because you are supporting him; the connection between you and the horse, established through the reins, is more like holding a child by the hand while you are crossing the road: you are holding (rather than supporting) the youngster – but if you suddenly let go of that hand, the child will be thrown off balance and will probably stop still.

You will also affect the horse's balance if you are out of synchronisation when riding a circle at rising trot. When circling left, you should sink into the saddle as the horse's right (outside) forefoot and left (inside) hindfoot touch down. Since you should frequently change the diagonal – and therefore the direction – this means sitting for one stride as you change the rein. It is always the horse's inside hindleg which produces the required energy and impulsion.

Whenever you are sitting into the saddle, you must be fluid – moving your weight with the motion of the horse's back. If you go against his natural movement, he will stiffen against you. This can lead to a tense back – and trouble.

During these lessons you should continue to concentrate on keeping the horse straight, with his hindfeet following the same track as the forefeet. Since the engine is located in the rear, it might be easier to think of the hindlegs propelling the forehand ahead, rather than following in its wake. To stay straight, you need to look forward and see your line; if you close your eyes, you are bound to go crooked.

Lungeing

We continue to use lungeing as part of the horse's schooling throughout his eventing career. It replaces one, or maybe two, dressage lessons each week and is regarded as schoolwork, not simply as a type of relaxation. While on the lunge, we can work at getting the horse comfortable in the rounded shape that we want him to adopt, which is not natural to him; when left to his own devices, he moves around with his head in the air. We can also encourage him to stay loose in his body while he goes over trotting poles, which help his cadence, elevation and footwork. It is obviously easier for him to do this on the lunge, without the extra effort and change of balance required when he is carrying a rider on his back.

Most of our lungeing is done at trot. We can work on the rhythm of this pace, with and without trotting poles. Our horses rarely canter on the lunge and they do very little walking. Without a rider on board to reinforce the aids, there is always the danger that the horse will shorten in front and not behind when he is lunged at walk which leads to uneven footfalls.

HACKING

Schoolwork is not forgotten while out hacking, which occupies most of the four-year-old's ridden work each day. The one great difference outside is that he is not constantly required to turn, but can keep on a fairly straight line for most of the ride, which is better in this stage of his training.

My mother, who enjoys riding our youngsters both in the school and out hacking, says that there are always three questions in the forefront of her mind: Is the horse in rhythm? Is he going forward? Is he straight? She is also aiming to get him on the bit and going evenly in both hands, without putting his shoulders or quarters in the wrong place. Much as she enjoys these outings, she knows that it is not the time for gazing at the countryside or planning the menu for lunch!

The rider's awareness while out on the roads has to take account of possible hazards ahead, usually in the shape of large vehicles. It might be necessary to pull into a gateway if you spot a large lorry coming towards you. We prefer to keep the horse moving because then it has less time to get itself into trouble, but you have to use your intelligence and know when it is more prudent to stop and let the vehicle go past. It is also important to choose the route out hacking in a positive way, because a horse will always know if the rider is uncertain about which way to go.

There is no time for gazing at the countryside while out hacking

Improving the paces

The walk can be the most difficult pace of all. We have had plenty of horses at Ivyleaze who proved that point, usually because their walk was short and choppy. There is no set solution to this problem; we would try several different things in the hope of finding one that worked. Equally important, we would discard any exercise that was clearly failing in its objective. It's a great mistake to grind on when you are obviously getting nowhere fast.

The horse might lengthen his stride if the rider pushes him on – or he may, on the other hand, shorten even more. Sometimes it helps to slow him to a crawl or to ride him on a long rein for three or four minutes, using the legs quietly to ask him to move on. A horse will normally walk with its head and neck down when given more rein, much as it would in its wild and natural state. The walk produced under these circumstances is usually good and sometimes excellent.

Though he will probably revert to his short stride once the reins are taken up again, the horse has shown the rider – even if only for a few strides – that he is capable of something much better. Having felt the improvement, the rider now has something at which to aim. You must make a fuss of the horse to tell him that those few strides, whichever way they were achieved, were just what you wanted. You must also resist the temptation to keep slogging on with that particular exercise in the hope of attaining a perfect walk in one lesson. Those are the very occasions when you are likely to lose the little you had achieved by asking for too much.

The horse needs plenty of variety, rather than repetition, and it is up to the rider to provide it. You could, for instance, do a circle at trot before making another attempt to establish a longer stride at walk, which is hard work for a young horse. It might help him to relax if you give him a short trot or canter afterwards. There is no easy rule of thumb; you have to listen to the horse and try to understand what is going through his mind in order to find out what is right for each individual.

Asking for an extra effort, such as a longer stride at walk, should not be restricted to work in the school. The next time, it can be attempted while you are out hacking, which is another way of breaking up the work and keeping it varied.

Sometimes, especially when dealing with a Thoroughbred horse that has been bred for racing, we encounter the opposite problem at walk. This time the stride is very long and it needs to be shortened slightly so that the feet are quicker off the ground. In this case, you need to take up the reins with a give-and-take movement and ask, with your legs, for a shorter and quicker stride.

You do not want the horse to lose his natural long walk (which you will need when extended walk is required in a dressage test), nor do you want him to start waddling, which sometimes happens when the balance changes. You also need to take care that he does not shorten too much, which you will know has happened if he either stumbles or begins to jog, because then you have to go back to square one. You therefore need to be constantly aware of what the horse is doing, always remembering that he must shorten from the back rather than the front end.

Some horses are reluctant to walk at all out hacking; instead they jog and always want to be out in front. The answer may be to trot on, until you are quite a long distance ahead of the ride. The horse will then start wondering what has happened to its companions and, perhaps, be a little less eager to go off in front next time. If this is tried a few times and it fails to have any effect, you will have to think of something else.

It could be that the horse jogs and generally behaves in a silly way because he is highly strung, which is usually the result of him having been asked to do something too soon. He easily gets in a panic, and when he does, he is prompted to have an argument. In this case it can be a good idea for the rider to make no demands, but to sit quietly and pat the horse reassuringly. Since it takes two to make an argument, the horse will usually become aware that he is wasting his time.

No horse is one hundred per cent perfect, so you are bound to encounter problems with at least one pace. If he has a wonderful swinging trot, for instance, you will probably have to work hard at improving his canter. Having done that, you should not be too surprised to find that you have lost some of the excellence at trot. Improvement in one department is often gained at the cost of a step backwards somewhere else, but you invariably regain what was lost if you are patient.

You can sometimes use the canter to achieve a better trot. If, for instance, the horse is naturally correct and balanced at canter, it can help to use that pace before trying to improve and lengthen a trot that tends to be short and choppy. The key is to use the horse's best pace, because that is the one in which he feels most comfortable.

Napping

We can never hope to conquer a horse physically, but we can usually get the upper hand mentally if we are prepared to use our imagination and common sense. Arguments are best avoided wherever possible, which is why you need to be cautious about using a stick on a horse that naps. One good smack could be the answer. Should that treatment fail I would rather use patience as a weapon and, if the horse refuses to move in the required direction, be prepared to sit on him all day.

This form of resistance has to be nipped in the bud or it will become a major – and possibly insurmountable – problem. A typical case is when a horse stops in its tracks and makes it abundantly clear that he has absolutely no intention of taking another step in that direction. Instead he tries to whip round and retrace his steps. We would not allow him to turn or to go backwards, but would make him stand facing the direction in which he was supposed to go. He is not asked to move forward, because that would almost certainly prompt him to fight back. He has to stand there until he is so bored that he eventually decides to yield to the rider's will.

There are some horses that rear when they are prevented from turning or going backwards. They don't normally rear very high, but the rider must be able to stay on without hanging on to the reins. Dismounting should be regarded only as a last resort, when you feel it would be impossible to stay in the saddle. If you do dismount you will need to run the stirrups up and take the reins over the horse's head. Should he still refuse to move forward, you have to stand and wait. Every so often you can say, 'Come on, let's go,' in a positive way, but without attempting to drag the horse forward. If he still refuses to budge, keep him standing there. We have never yet come across a horse at Ivyleaze that didn't eventually succumb to boredom and submit to moving forward.

You have not entirely lost the battle by dismounting, but it is always won more effectively from the saddle. It will be important to assess the individual horse, taking its past history into account, to ascertain whether the problem is likely to recur. If another confrontation seems probable, you will need to enlist the help of a stronger and more experienced rider who can sort out the problem without dismounting. Meanwhile, avoid having a confrontation with the horse yourself. You can always lunge him or do some other form of exercise that has not given any problems in the past.

We have enlisted outside assistance at Ivyleaze on a number of occasions. One horse that persistently napped was cured by a friend, who employed our favoured method of sitting still and doing absolutely nothing until the horse decided that it was all a big bore and submitted. We knew the rider had the temperament and discipline to deal with the problem, and that he was a strong enough horseman to sort out our wayward youngster from the saddle.

Though patience is normally far more effective than punishment, there was one occasion when the stick worked wonders for us. We had a horse that constantly swung round while out hacking and then did its level best to go whizzing back the way it had come. Having tried without success to put an end to this infuriating habit, my mother asked another friend if he could instil some discipline. She had a feeling that a few sharp whacks from someone stronger than herself might do the trick. At the first wallop, the horse unleashed a spectacular series of leaps and jumped a fence unbidden. He must have been disconcerted to find that he still had a rider on top, and at the second wallop he capitulated. Our friend rode him a few more times and when my mother got back on the horse she found he was a reformed character. We might have been able to crack the problem by other means, but the stick proved far quicker and more suitable for this particular horse's temperament.

The stick would have created a further problem instead of effecting a cure, had we used it on another of our horses, who was suspicious of people – as a result, we felt sure, of being badly treated in the past. This horse, whose brain is razor sharp, used to make the excuse of having seen something unusual while out on a hack and then whip round to the left. It would probably have been quite easy to turn him a full circle and then continue the way he was going, but we never allow our horses to do that; if they swing round to the left, they have to go back to the right in order to face the required direction. This form of resistance can otherwise lead to running out at a show jump, because the horse knows it can swing left or right and do a complete circle.

My mother dealt with this particular horse herself. She never raised her voice or got angry with him when he whipped round. Instead she sat quietly; he was not allowed to keep turning left, but there were no heavy aids from legs or stick to try to get him back the way he came. It took about five minutes before he yielded and she then rewarded him with much patting and repetition of the words 'good boy', which all our horses recognise as a sign that we are pleased with them.

SCHOOLING ON GRASS

Horses can all too easily pick up the idea that grass has only two purposes: to be eaten when they are loose in the field or to be galloped on when ridden. This is not a helpful attitude when you get to a one-day event and have to ride a dressage test on grass!

We had one horse, who was already backed when we bought him, that used to go loopy every time he felt grass under his feet. He was to test our patience and consume much of our time when we set about curing him. Three days a week, one of us would ride him at walk on grass. He soon learned to behave reasonably well when kept on the move, but if you stopped and stood still, he would immediately tense up and start plunging and rearing. There was no hint of a problem if you stopped him on the road; it was only grass that produced these antics.

We used to take this horse to the field when some of the older eventers were cantering, and he would have to walk while they went sailing past. Eventually our efforts paid dividends; you could stand him in the field while the other horses galloped past and he was as good as gold. Had we not been ready to devote as much time to him at that stage, we could have had a big problem on our hands.

Young horses, especially those that are normally schooled on an all-weather surface, need to be educated on grass. If I had no field available, I would try to persuade a local farmer to let me use a corner of his land for occasional schooling. There is no need to spend a long time there; ten minutes is enough to remind the horse that fields are sometimes used for dressage.

The next day I might take the same horse for a happy ride in the park, with a little trotting and some quiet cantering. The day after that I might do dressage in the park, which all helps to tell him that he is expected to behave whatever we happen to be doing on the day. If he always does the same thing at the same place, he will begin to anticipate instead of listening.

SUMMARY

Horse gets fit with three weeks' walking (ten to fifteen minutes on lunge, followed by roadwork).
Exercises in school include 10-metre circles at walk, 20-metre circles at trot, 20-metre squares and diamonds at walk. Also 15-metre squares and diamonds.
Three- and five-metre loops and figures of eight are introduced.
Canter work increased, with horse asked to strike off on inside leg.
Three-loop serpentines at walk and trot are added to the other shapes.
Lengthening and shortening is introduced, using horse's best pace.
Leg yielding is taught at walk.
The horse is asked for flexion from his poll.
Trotting poles are included in ridden work.
Lungeing and hacking continue.
(Jumping lessons have begun – see next chapter.)

Aims for the trainer

The priorities remain the same: forward movement, rhythm, balance, even contact on both reins, straightness.
We are looking for improved paces, and some lengthening and shortening of strides within horse's easiest pace(s).
Horse must be educated to behave on grass.
If he naps, do not let him turn a full circle.
Leave him with only one option: to go the way you want.

Aids for new movement

Leg yielding to left from three-quarter line of school, with horse bent slightly around the rider's inside (right) leg in the opposite direction to the way he is going:
Weight even on both seat-bones.
Left (outside) leg – used near girth – maintains impulsion, controls the quarters and prevents falling out.
Right leg, used just behind the girth, keeps horse moving diagonally and forward.
Left rein prevents horse falling out or going too fast.
Right rein helps to keep the shoulder on correct line and asks for slight bend.
(N.B. Inside rein and leg are on the opposite side to the direction in which the horse is going.)

6

LEARNING TO JUMP

The horse must be given confidence through every stage of his education. This is particularly vital when you are teaching him to jump; anything that undermines his self-assurance will be a serious setback, requiring a frustrating return to square one.

Our youngsters begin their jumping lessons when they are four years of age, which is when I start to play a bigger part in their lives. By the time I get on to give them their first jumping lessons, they will have learnt about co-operating with their human minders, without losing their individual personalities. The Ivyleaze method has never involved bludgeoning horses into submission.

FIRST LESSONS

The horse is already familiar with trotting poles, both on the lunge and during his early ridden work as a three-year-old. I use them again before jumping the four-year-old for the first time, though I would probably trot across them only a couple of times. They help to make the horse a little bit sharper and get him used to the weight and balance of the rider.

I normally have the trotting poles on one side of the arena and tiny cross-rails, about 1 foot high, on the opposite side. This small fence has a placing pole 9 feet in front of it, which will help the horse to take off from a point that will make the jump seem easy. Having ridden him over the trotting poles, I ask him to take the cross-rails from trot.

As with everything else connected with the horse's training, this should seem like a natural progression from the lessons that went before. If the rider gets worked up about it, the horse is likely to lose his cool and become over-excited whenever he is presented with a fence to be jumped. If he does tend to rush, I put some trotting poles in front of the placing pole; these help him to concentrate and stay within the rhythm.

I want the horse to be happy jumping the cross-rails from trot. Having popped over them a few times, I might finish jumping for that day – or I might take him over a small vertical of 1 foot high. I would not, however, start with a vertical. We always use cross-rails to begin with because their shape encourages the horse to aim for the lowest point in the centre, which helps to keep him straight.

Teaching the horse to stay straight in his early jumping lessons will be of enormous benefit later, when I want to take the corner of an angled fence. Short-coupled horses usually find staying straight much easier than the longer animal with a big stride like Master Craftsman, who tended to drift to the left whenever he thought he was getting uncomfortably close to a fence. This gave him some extra space and was easier than shortening up like a concertina, which he finds particularly hard work because of his conformation. Drifting to the left or right might not be a problem over single, straightforward fences, but it could land you in all sorts of trouble when jumping corners and angles.

After two or three weeks, I would expect to be jumping three or four fences of about 2 feet high from trot, sometimes with place poles and at other times without them. These would either be single fences or built in line to form a grid. Progress obviously depends on the individual horse, who has to be given as much time as he needs to gain confidence.

I jump the youngsters twice a week until their balance and confidence are established. This must be done without interfering with the natural bascule that all horses have until they are required to carry a rider over fences. After that, I jump only once a week. I might pop the horse over the odd fence to give him some fun and loosen him up on non-jumping days, but he would have only the one proper lesson.

Jumping will occupy only part of a twenty-minute session in the school, which would also include flat work. I want the horse to enjoy jumping and look forward to it, which would not be the case if I were to make this part of the lesson any longer. Since I never do the same thing twice running, the horse has the added interest of waiting to see what fences he will be jumping next. He might have a quick jump through a grid and then do a round of little fences. He may just concentrate on grids, which can be built in many different ways to allow for plenty of variety. Constant repetition is as boring for horses as it is for people, so I am always thinking of ways in which to make the exercise more interesting.

SINGLE FENCES

I use a placing pole in front of single fences only until the horse has learnt to trot into the obstacle and pick up confidently. After that, I want to encourage him to see his own stride and learn to look after himself. Without the placing pole to put him right at take-off, the youngster invariably starts by standing much too far back from these single fences. Then he will probably come in too close until he learns to make his own adjustments of stride.

Apart from telling the horse where to go, and keeping him in rhythm and balance, I am a passive passenger during these lessons. I need to give the youngster plenty of freedom in his head and neck, which encourages him to think for himself, maintain his balance and produce a nice rounded jump. At this stage he will be jumping little spread fences as well as verticals in trot.

On the familiar 'prevention is better than cure' theme, I refuse to allow the possibility of the young horse stopping at a fence and having to be turned back for a second attempt. For that reason, the four-year-old always starts jumping from trot, over fences that are small enough to be taken from a standstill should he grind to a halt. If you have to turn back, the horse will have learnt the last thing you wish to put into his head: he will know that stopping can be used as an alternative to jumping. I would not consider any obstacle that was too large to take from a standstill unless I was as sure as one ever can be that the horse would jump it.

The same is true whether I am jumping small fences at home or popping over little logs and ditches in Badminton Park, where we are lucky enough to be able to ride. If the horse does stop, he is not allowed to turn away but has to keep facing the obstacle until he jumps it from a standstill. In such cases the rider has to be unyielding.

BUILDING A GRID

Gridwork makes the young horse more athletic and helps him to be quicker with his feet – or to pat the ground, as we call it at Ivyleaze. With the use of a placing pole and correct distances between fences, the grid will also help to build his confidence because he will reach each of these small jumps at exactly the right point of take-off. When he becomes more self-assured, we can introduce different distances which encourage the short-striding horse to lengthen, or the long-strider to shorten.

The first grid would consist of two small cross-rails, with 18–19 feet (5.5–5.8 metres) between them, which would give room for one non-jumping stride. I would probably add a third and then a fourth cross-rail, with the same space between them, before introducing other types of fences and varying distances.

We might then change to a cross-rail followed by two

A place pole brings the young horse to the correct point of take-off and he jumps with confidence. He is allowed to use himself by being given freedom of his head and neck

verticals. The second vertical can later be converted to an oxer, after which I might add another vertical to give four fences on a line. Once the horse is happy to land and take off again one stride later, the fences can be gradually raised. The actual height and distance will be at the rider's or trainer's discretion, because you have to take account of the horse's temperament, capacity and natural balance. I have known horses that had so much confidence they were ready to jump 3 feet in the first week or so, but these were exceptions. Others might take as long as two months to reach that height.

By this time bounces will have been introduced. I always use small cross-rails when first asking the horse to bounce – in other words to jump two fences without taking a non-jumping stride between them. As already mentioned, the cross-rails help to keep the horse straight, which is particularly useful with youngsters who invariably have a tendency to wander off the line. My first bounce is always in a grid over cross-rails, set at no less than 10 feet (3.05 metres) apart. Depending on the height of the cross-rails (and remembering the bigger the fences the greater the distance required between them), I might extend this to 12 feet (4.57 metres).

The grid for this first bounce consists of three cross-rails, with one non-jumping stride to the second element and a bounce to the third. If you are feeling ambitious, you can then add a fourth cross-rail to give a second bounce. This type of exercise helps to get the horse on his hocks and makes him sharp and quick on his feet.

Once he is happy with this exercise, the bounce can be introduced into a grid that includes different types of fences, such as parallel bars and uprights. We continually alter our grids to avoid making the lessons repetitive; we also set them in a different place each time. I might, for instance, start the horse with a grid on the left-hand side of the school going away from home. If it stayed in that place, the youngster would soon begin leaning towards the post-and-rail fencing on his left and he would anticipate the right-hand turn at the end of the grid. The next time we would therefore make sure that he had to turn left at the end.

Placing poles

The grid can be made more adventurous by using placing poles before or after the fences. A pole strategically placed before a jump can encourage the horse to get close, so that he rounds his back and bascules over the fence instead of taking off in a flatter shape, more like an aeroplane. Often the youngster tends to panic slightly and take off from too far back, so this exercise helps to build confidence by teaching him that he can get closer and round his back. It also encourages him to use his shoulders in order to tuck up his forearms.

A pole placed 9 or 10 feet away from the fence on the landing side will also improve the bascule by encouraging the horse to land a little steeper. This is not recommended for an inexperienced youngster, who might well find the sight of this pole lying close to the spot where he was expecting to land such a shock to his system that he freezes in mid-air.

With a green horse, I prefer to put the pole equidistant from two fences in the grid. It can then be gradually moved in until it is 9–10 feet (2.74–3.05 metres) away from the landing side of one fence and further away from the obstacle which follows. The bigger the fence, the more room he needs to land, so you must exercise caution. The pole will be as close as 9 feet only after a small fence of up to 3 feet (0.9 metre); a distance of at least 10 feet will be used for anything bigger. Placing poles can also be used to encourage the youngster – or any horse that has a tendency to wander off line – to remain straight over his fences. This is done by laying the poles on the ground at right angles to one or more fences. Normally I use poles on the take-off side only for a green horse, if he has a tendency to drift on his approach to the jump. He has to be more experienced before I put them on the landing side as well.

The poles will probably be at the same width as the fence itself when first introduced; then, slowly but surely, they are moved in so that any deviation from a straight line is restricted. I have had the poles as close as 3 feet on an advanced horse, but I would not pull them in by more than 1 foot (0.3 metre) on either side of the stands for a baby. You can be sure that the horse will not land on one of these poles more than once; instead, he will learn to look down and see where he is putting his feet! He will therefore be encouraged to jump straight at all times.

JUMPING AT CANTER

Approaching a fence at canter is a big step forward for a young horse. I never attempt it until I am satisfied that the youngster can canter in balance on the flat and has learnt to shorten and lengthen his stride. You will know when your own horse is ready simply by asking yourself, while cantering around the school, whether you would feel comfortable jumping at that pace. The type of canter the horse is producing on the flat will give you the answer, as long as you are honest with yourself. If

The horse is dangling with his forearm, leading to the inevitable fence down (above). *Fortunately most horses are apt pupils. This youngster has learnt by his mistakes and now produces a good jump* (below)

it doesn't feel too great, you will know that the time has not yet come. Some of our horses do not canter into a fence until they are five; others have such a natural canter that they are ready quite early in the year they reach the age of four.

We have to use our natural instincts to feel our way forward with each individual horse. There is no programme that can be planned in advance, with set lessons on given days. I may appear to be ultra-cautious, but that is because I care for my neck too much to risk taking short-cuts with the horse's education over fences!

In jumping, as with everything else connected with horses, we have to take the animal's physical assets and defects into account. This means, for instance, accepting that the horse with a huge, long-striding canter is going to find this stage more difficult than the one with a short, choppy stride. Some, like Murphy Himself, can overcome the problems of a long stride through natural balance and athleticism. Others, like Master Craftsman, take much longer because the athleticism does not come naturally. He was bred to race and lengthen, but not to shorten.

Like most horses, Crafty was slightly on his forehand when he was younger and that made it difficult for him to shorten his stride. Beneficial, on the other hand, had a short, choppy stride which was by no means ideal for dressage but did make jumping easier, because he was never far away from the best point of take-off. Ben was one of those rare horses who was ready to pop over a 2-foot fence at canter after his first grid work. We built a 3-foot fence within the grid for Crafty, but his canter work on the flat had not progressed sufficiently to jump him over single fences at canter in those early stages.

When a young horse has had a lesson which includes jumping at canter, I invariably bring him back to trot the following week. Otherwise he might begin to find it all so exhilarating that he lights up as soon as he sees a fence. Sometimes the rider gets over-excited; the horse then gets the impression that he is supposed to screech round and go flat out into his fences. Instead, he should be learning to land in a relaxed way and maintain a rhythmic canter between the jumps.

If the horse is jumping with plenty of confidence at canter, I would consider clear-round show jumping at this stage. One of our horses, To Be Sure, did this as a four-year-old and loved it. I could feel him growing in conceit after trotting into the first two fences and clearing them. He landed over the second in canter and, because he felt happy and balanced, we stayed at that pace for the rest of the course. Had he lost his balance and rhythm, or tried to rush his fences, I would have brought him back to trot.

I might also consider taking the horse cub-hunting in the autumn, with a view to taking him fox-hunting when he is a five-year-old. This is a marvellous form of education for young horses. It teaches them how to look after themselves across country and to find a 'fifth leg' when required to keep upright. They therefore gain enormous confidence and, because they love jumping in company with other horses, they find it great fun. Priceless had a full season's hunting as a youngster and it did him the world of good.

If the horse does hunt, he needs to do so at least ten times – if not for the whole season. Taking him out three or four times is worse than useless; he simply learns to associate grass with galloping. Given a longer period, he would discover that the sport is not just about charging across country and he would learn to settle down.

CROSS-COUNTRY FENCES

I always jump a few coloured poles at home before going out to jump natural obstacles. It is much easier for the horse to learn how to leave the ground and land back on to it without such additional problems as unlevel ground or an awkward approach. He is also less likely to hurt himself over movable poles than over solid natural fences, and as I want him to associate jumping with pleasure, any pain he experienced would be an obvious setback.

The first natural fences our horses jump are usually little logs and ditches in and around Badminton Park. These are taken at trot and are small enough to be jumped from a standstill if the horse happens to stop. I would try to find some water for him to walk through as a four-year-old, and I might jump on and off a small bank. I might also take him to a schooling area that had a selection of trotting fences, but I would not attempt anything more ambitious.

The horse would need to be confident over show jumps at home, including small doubles and combinations, and he would have to be happy to take them in canter, before I would consider attempting a decent-sized novice cross-country fence.

THE RIDER

If you and the horse can establish a good style over small fences, you will have fewer problems when the time comes to jump larger obstacles. During the approach, you should be slightly in front of the vertical with your shoulders on about the same line as your knees. It is important to remain still; if your weight moves, the horse will be thrown off balance. As a result, he loses impulsion and may not have enough power left to clear the fence. At this stage your seat should be lightened within the rhythm of the canter; if you sit deep in the saddle, the horse's back will stiffen. It is your job to control the pace and direction and to keep him in rhythm and balance, so you must not lean to either side.

When he reaches the fence, the horse uses his head and neck to balance himself before he thrusts off with his forelegs. There is a natural instinct for the rider to lean further forward at this point, with the mistaken idea that it will help the horse. In fact it hinders him, putting more weight on to his shoulders at the precise moment when he is trying to use them to thrust forwards and upwards. Instead of flopping forward on that last stride, you should slightly open the angle between your body and the horse by becoming a little more upright, thus reducing the weight on his shoulders.

Almost as soon as the horse's forefeet leave the ground, his hindfeet touch down at roughly the same spot. His weight is then shifted back to his hindlegs, which contain the energy of a coiled spring for the final push at take-off. The rider's body should therefore move forward when the forelegs are in the air and before the hindlegs have left the ground. This does not mean flinging yourself forward; you simply need to go with the movement and stay in balance.

Meanwhile, your hands should also have followed the movement. They will need to go forward on the final stride, when the horse's head and neck are lowered and stretched out towards the fence. His head will then be raised and his neck arched as his forelegs leave the ground, so your hands have to come back to maintain contact. They move forward again as his hindlegs touch down in preparation for take-off, which is when his head and neck stretch out for the start of his bascule.

Your application of the aids should depend on the horse. He needs to bring his hindlegs well under his body to gain enough impulsion for take-off, but this does not necessarily mean using strong aids. Some horses prefer to have no leg aids until the last stride before take-off; others jump better if you use your legs for the last three strides. The same individual preferences pertain to the amount of contact on the reins. I have one horse who dislikes anything more than the lightest of contact on the final stride when he is lowering and stretching his head and neck. Others prefer a fairly firm contact throughout the approach. It is up to the rider to assess each horse's preference and learn to adapt accordingly.

There is nothing you can do to help the horse once he is airborne, but you can be a positive hindrance if you fail to stay in balance – or if you lean to either side and, in the process, encourage him to be crooked. You will need to remain a passive passenger until he has taken one stride away from the fence after landing. He will touch down on one foreleg and put all his weight on to the fetlock, which is pushed down until the pastern is roughly horizontal. He needs to be given time to regain his balance after landing, so you should wait until he takes his second stride away from the fence before bringing him back under control.

It will help the horse if you are able to see a stride at canter. Some people can do this naturally; others have to work at improving their eye. One useful exercise is to have a small cavalletto four to five non-jumping strides from a fence of about 2 feet 6 inches (0.8 metre). The average horse has a canter stride of 12 feet (3.7 metres) at canter, so you can measure the approximate distance by taking the relevant number of 1-yard strides yourself and allowing a total of 7–8 feet (2.1–2.4 metres) for landing over the cavalletto and taking off at the fence. The distance should obviously be adjusted if your horse has a longer or shorter than average stride.

The horse should approach the cavalletto in a rhythmic and balanced canter; you should not make the mistake (as some riders do) of hooking up and then riding like crazy. When he jumps this tiny rail – which he will do by elevating and shortening his canter stride – the rider counts down the number of strides to take-off. You can also learn to see a stride by cantering towards a thistle or past a tree and counting down the strides as though you would be jumping at that particular spot.

SUMMARY

Horse trots over poles on ground and then jumps small cross-rails (with place pole 9 feet in front) from trot.
Horse jumps small single fences from trot.
Grids are introduced.
Horse learns to see his own stride over small single fences (without place pole) from trot.
Bounces are introduced between small cross-rails in grid.
Cross-rails, uprights and spreads are included in grids, with different distances (bounce, one and two non-jumping strides).
Placing poles are used as necessary – to encourage horse to stay straight, take off later, land steeper, etc.
Small natural obstacles (logs, ditches, etc.) are jumped at trot.
Fences are approached at canter when the horse can maintain his balance while lengthening and shortening in canter on the flat.

Aims for the trainer

Horse must stay straight (no veering to left or right on approach).
He should enjoy jumping, so keep lessons short and varied.
He must not learn to refuse (start with fences that are small enough to jump from standstill).
Retain horse's natural bascule.
Improve athleticism through gridwork.

The rider

Should try to develop an eye for a stride.

On approach: Body slightly in front of the vertical. Seat lightened within rhythm of canter. Remain still and in balance.

Last stride: Body slightly more upright until forelegs thrust off. Lean forward with the movement when weight is on horse's hindlegs. Hands follow movement of horse's head and neck.

Over the fence: Stay in balance – do not lean to either side.

Landing: Do not interfere until horse has taken one stride away from fence.

7

FEEDING AND STABLE ROUTINE

The horse needs to be fit, contented and sound if he is to produce his best performance, hence the true saying that competitions are won in the yard as much as in the saddle. Constant attention, much loving care and consideration for each horse's individual needs are among the keys to success.

We have long followed the advice of our vet, Don Attenburrow, who believes that each horse should be given the minimum amount of hard feed to get him fit for the job in hand. He must have plenty of bulk as well (unless he is overweight and needs to be rationed), but the oats and event nuts have to be measured out with great care for each individual. Too little will mean that the horse has insufficient energy; too much will make him too fresh and run the risk of problems that are associated with a build-up of protein, like azoturia and lymphangitis.

You need to take the look of the horse into account – and the feel he gives you while ridden – when trying to ascertain the right quantity of hard feed. We always keep a record of the amount each horse is fed, so that we can refer back to it the next year and follow the same pattern, if it had been successful. If not, we would note that adjustments were needed. Horses that are new to the yard are always more difficult to assess because we have no such guideline.

Because we buy only the best, my mother visited the Spillers mill before deciding to use their event nuts; she had to be satisfied with the way they were made. The hay, which comes from a local farmer, has to be top quality, otherwise it will have a detrimental effect on the horses' wind and digestion. When buying the best you are paying the highest price, so we make sure that nothing is wasted.

Before the horse competes, you will need to check with your feed merchant to make sure that any prepared nuts or mix do not contain forbidden substances. Ointments and fly repellents also have to be checked for the same reason.

HARD FEED

For Ivyleaze horses, the hard feed consists of Spillers event nuts and crushed oats. We also use an event mix from which the molassine has been omitted; we prefer to have this left out because the horse can become so accustomed to having something sweet that he is reluctant to eat a feed without it.

NAME
CRAFT
TOBY
GRIFF
LUKE
FRANK
HOVIS

As already mentioned, the quantity varies for each horse and the work he is doing; we have to rely on sight and sense and feel, especially with any animal that is new to the yard. The old-fashioned guide, which recommends one pound of hard feed per day for every mile the horse is covering, would probably not be a bad starting point – but it has to be adjusted to individual needs.

Weather conditions must also be taken into account. Our horses may have more hard feed while they are wintering at grass than when they first come back into work. They need energy-giving food as well as bulk to keep warm and retain some of their condition. We like them to have plenty of weight on them when they start the long process of getting fit again.

Amounts will also vary according to the season. There is hardly any nutritional value in the grass when we are preparing for the early spring events. Most horses therefore need more hard feed at that time of year than they will in the run-up to summer and autumn events, when the grass is at its best. Hard feed is reduced on their day off and also before and during any long journeys, which will be discussed in a later chapter.

We have a blackboard in the feed room which lists the amount each horse is fed. Even if you look after the animal on your own, it is far better to have the quantities written down rather than filed away in your head. If you are suddenly taken ill, it will then be possible for someone else to take over the feeding. At Ivyleaze, the amounts vary considerably from one horse to the next. We have one Thoroughbred, for example, who had more hard feed than the advanced horses when he was a six-year-old; he was big, gangly and still growing at that age and he needed the extra nourishment to develop. At the same age, Ben Hovis tended to be far too fat and full of himself, so he was given very little.

If the horse is uppity by nature, it is better to cut the hard feed right back or omit it altogether. It is possible to get him one-day-event fit on grass, nuts and other bulk food without giving anything that might make him too fresh.

BULK FOOD

Needless to say, the staple food is grass – eaten in its natural state out in the fields or fed as hay or grass meal. We also use HorseHage, which has the great advantage of being 'dust-free'. The word 'dust' is used in a general sense to cover all the micro-organisms to be found in the atmosphere and, to a greater extent, in hay and straw – both of which contain fungal spores. For this reason all our hay is soaked for twenty-four hours, during which time the spores swell up. Some drop away into the water, while those that remain become harmless because they are too large to reach the lungs.

Though hay-soaking is a tiresome chore, we believe it is well worth the bother because of the enormous benefit to the horse's wind. Without those twenty-four

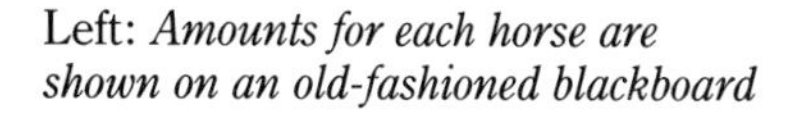

Left: *Amounts for each horse are shown on an old-fashioned blackboard*

Right: *HorseHage provides a useful alternative to hay*

The hay is weighed for each individual horse (left) *and then soaked to eliminate dangerous spores*

hours immersed in water, the hay spores can cause a hypersensitivity or allergy which closes the small airways and so reduces the amount of oxygen he can inhale. This, in turn, affects his performance.

Our hay is weighed for each individual horse and then soaked in a porous sack, which has his name written on it. These sacks are heavy to drag out of the water to be drained. It can be particularly difficult in the winter, when you may have the added problem of snow or ice. During the summer, the sacks of soaked hay have to be kept in the shade, otherwise the contents would either ferment or dry out and produce more spores. Since these spores can be windborne, hay has to be soaked for every horse in the yard in order to create a dust-free environment.

HorseHage is the trade name for grass that has been cut and wilted, but not dried, before being sealed in a bag. It therefore keeps its moisture content, which eliminates the problem of spores. We find it extremely useful but, because it is richer than hay, we feed it sparingly.

Assuming that the horse has no weight problem, he has as much hay as he will eat, within reason. It is fed in handfuls throughout the day, before the night ration is given at nine o'clock in the evening. During the daytime our horses are never left for more than three-quarters of an hour without something to eat; food helps to keep them occupied and stops them getting bored. By feeding little and often we are also following the normal eating habits of the horse in its natural environment. Handfuls of hay or HorseHage are given about fifteen minutes before each of the four main meals of the day; this helps to stimulate the gastric juices and take the edge off the horse's appetite, so he is encouraged to chew his hard feed rather than bolt it.

We provide bulk in the manger (as distinct from the hay-rack) with a mixture that we always refer to as 'slops'. It contains chaff, sugar-beet and grass meal. The proportions vary according to the horse's likes and dislikes; we want each one to enjoy his food and will therefore cut down on anything that he finds unpalatable enough to leave behind in the manger. The sugar-beet and grass meal, which are both used in nut form, are good for putting on weight but need to be reduced if the horse is getting too fat. Both need to be soaked before feeding, otherwise they are likely to cause colic. The beet needs to be soaked for at least twelve hours, which would be the maximum in a hot climate because it would be likely to ferment if left any longer. You should make sure it is ready by inserting your hand and checking that there are no hard lumps – otherwise the sugar-beet can cause a serious blockage of the gullet.

All the horses have chaff, which is made from oat straw and hay. It is freshly cut in an electric chaff cutter once a week and it, too, is soaked before feeding. We never feed bran, because it has little nutritional value and it upsets the balance of minerals – principally by absorbing the calcium content in the food instead of letting it be absorbed by the horse. Calcium is an important aid to bone growth, which is the main reason why Irish horses (brought up on limestone that is rich in calcium) develop such good density of bone at an early age.

During the winter we feed boiled or flaked barley, which helps to put on weight. If the horse looks too thin, he will be given smaller amounts of flaked barley during the summer as well.

ADDITIVES

We have vitamins made up for use mainly in the winter months. Fewer additives are required during the summer; at that stage the grass contains most of the vitamins the horse needs. More salt, however, is required during hot weather to replace the amount lost in sweating. We give two ounces of salt per day during the summer and only one ounce in the winter months. We also give half an eggspoon of cod liver oil and four ounces of limestone flour each day throughout the year.

THE STABLE YARD

If we were designing a stable yard from scratch at Ivyleaze, we would put it in a sheltered spot and try to find a nice view for the horses, so they were not constantly looking at each other and getting bored. A window in the back of the stable is enormously helpful in this respect, because the horse has another scene to view. We have back windows on a couple of our stables and the horses that occupy them seem to be less bored than those who can only look out over the front door.

The yard would be close to the house – or connected by the type of intercom system that is widely used in the United States – so that the people living there could hear if anything went wrong during the night. The dung heap, the place for storing hay and the chaff cutter would have to be at least 80 metres away from the stables in order to keep the horses in a dust-free environment.

Ventilation, which is tremendously important, would preferably come from air ducts in a central line along the middle of the roof. Drainage would have to be carefully planned, with the individual drains in each stable leading to a central one. Drainage lines, shaped in a herring-bone pattern, have the added advantage of stopping the horse from slipping. We like the mangers to be approximately chest high and detachable, so that they can be taken out for cleaning. Hay-racks would also be chest high.

Right: *We use chaff in preference to bran. The feed is well mixed before being given to the horse*

Left: *All horses enjoy a really good roll*

BEDDING

There are four popular types of bedding – peat, straw, shavings and paper. Straw is bad, because it contains many spores. When the other three were tested, they reproduced only a very small quantity of spores and paper was the best of them. Equibed, a product made from sterilised waste material, was the only bedding which did not reproduce any spores. This is unobtainable at present, but it may be back on the market again in the future. When we used it, the horses were reluctant to get out of bed in the morning and the dogs wanted to rush into the stables for a lie-down.

We tried paper for a while; it is better than shavings from the point of view of spores, but a real pain when it comes to disposal. The only way to get rid of the mucked-out paper bedding is by burning. It would be a terrible eyesore if farmers were to spread it on their land, whereas they don't mind using manure in shavings as a fertiliser.

Straw has far too many damaging spores, so shavings it has to be for the time being. We do not use them as deep litter. All the wet patches are removed, together with the droppings, from the stables each morning. During the remainder of the day, the stables are skipped out several times. By regularly picking up the droppings, you avoid any unnecessary waste of shavings as well as making the beds look much better.

THE DAILY ROUTINE

Horses are creatures of habit, especially where mealtimes are concerned. Priceless always knew, as accurately as if he'd been wearing a watch, exactly when he was due to be fed. If his meal was not served promptly, he liked to remind everyone by banging loudly on his stable door. We therefore keep to a fairly strict timetable in the yard at Ivyleaze.

The day's routine is always written out the night before, giving the work each horse will be doing – plus any other relevant details, such as a visit from the farrier which will obviously affect the programme for any horses being shod. The normal routine changes on the days that some of the horses canter, gallop, go for a cross-country school or to the hills for fitness work. Most of these require a journey by horsebox, so they take up more time than hacking or working in the school at home.

All the horses have a day off on Mondays, when they go out in the fields while the hard-working girls who look after them go shopping. The horses always wear protective Velcro boots when turned out; if the weather is cold, they also wear hoods and New Zealand rugs. For the rest of the week, a rota is organised which gives each horse two or three more sessions outside. Being turned out gives them the chance to enjoy a sense of freedom in their natural environment, where they can let off any pent-up high spirits and have a really good roll. If it is not their turn to be let loose, they are led out for ten minutes to pick at some grass, which helps their appetite and stops them from getting bored.

Although we like them to go out, we sometimes have to restrict sessions in the field during the summer for any horse that is inclined to put on weight. It may, however, be enough to put those with a weight problem in the smallest paddock, where they have access to less grass, while their leaner stable-mates graze on a larger area. Some of the novices go out at night and, because of our small acreage, all droppings are removed from the fields the following morning to stop the ground from becoming horse-sick.

The horses that are being aimed at three-day events wear leg bandages day and night – for warmth as well as protection, but not for support. In cold weather, this encourages circulation while the horse is standing in his box, before being taken out for exercise. Those for which no three-day event has been planned are normally bandaged only at night.

In addition to picking up droppings, all the following are done on a daily basis.

The horse's legs are checked to make sure there is no heat or swelling.
Shoes are checked for wear, plus any sign of looseness or risen clenches.
The yard is swept and tidied several times a day.
Stables are skipped out, also several times a day.
In cold weather, the horse is regularly checked for warmth, by putting a hand under the rugs to feel his shoulder and loins. More rugs will be put on if required.
Hay, chaff, grass meal and sugar-beet nuts are soaked.
Water is checked regularly and topped up as necessary – in addition to buckets being emptied and refilled twice a day.
Feed buckets are always scrubbed clean after use.
Mangers are washed once a day.
Tack is cleaned.
Numnahs, boots and bandages are washed.

We can have as many as eight horses competing at different levels, so we have an exceptionally busy yard at the height of the season and we have to start working the horses early. At such times, the rest of the day's routine goes something like this.

6.45 a.m.	Each horse is given a handful of hay or HorseHage. Day-time beds are prepared for horses that slept out. Horses that were out overnight are brought in for breakfast.
7.00 a.m.	Breakfast feeds are given. Water buckets are removed from stables during mucking out. After mucking out, water buckets are emptied, re-filled and returned to stables.
7.15 a.m.	Girls have breakfast.
8.30-9 a.m.	Horses go out for hack or work in school. Hay, HorseHage or grass is given to any horse left behind.
10.30-11 a.m.	Horses that were working in the school and/or any left behind on earlier ride go for hack. One or more horses returning from hack may be schooled. Horses that have finished work are strapped off.
12 noon	Handfuls of hay are given.
12.15-12.30 p.m.	Lunch-time feeds are given.
from 2.00 p.m.	Horses to be turned out (according to rota) go into field for half an hour to one hour. Other horses led out for ten minutes to pick at grass. Strapping for horses that were not groomed in morning.
4.45 p.m.	Handfuls of hay are given. Put on night-time rugs and bandages.
5.00 p.m.	Tea-time feeds are given. Water buckets are emptied and re-filled.
6.30 p.m.	More handfuls of hay.
9.00 p.m.	Night-time hay is taken into stables. Night-time feed is given.

OTHER ESSENTIALS

All our horses are shod approximately every six weeks. If the ground seems likely to become hard, our farrier fits pads under the shoes. Made of chrome leather and soaked in neat's foot oil, they have a cushioning effect which reduces the risk of injury on firm going. The leather is cut so that the frog remains exposed, which means that we have to be careful to prevent any grit or small stones getting through to the area between the pad and sole. This is normally achieved by stuffing a line of cotton wool around the frog.

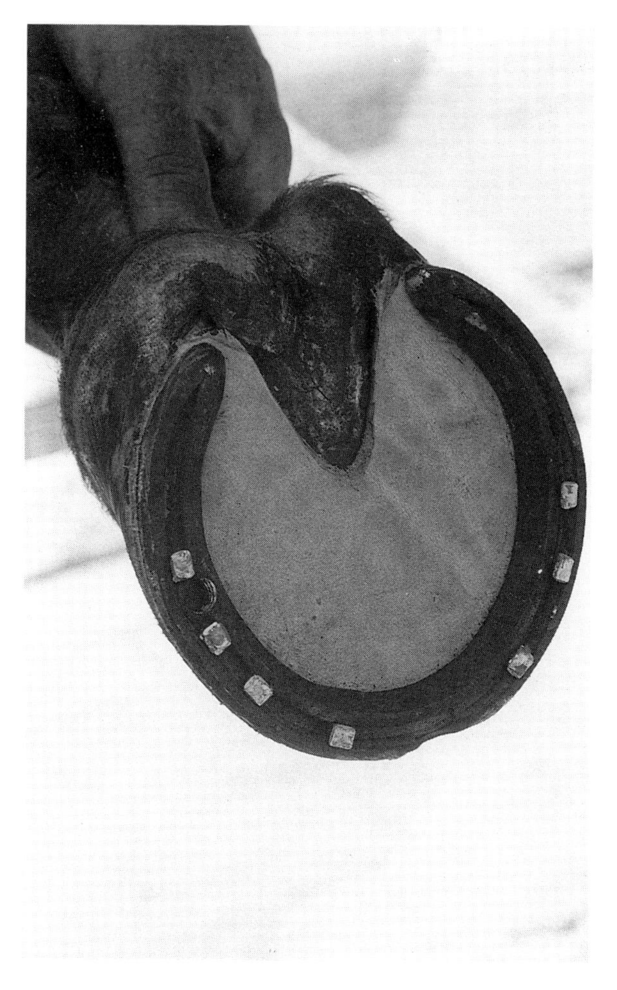

Leather pads act as shock absorbers on hard ground

The horses are also wormed regularly, according to the instructions given with the product used. Each horse is given a worm count when he first arrives at Ivyleaze and another one six months later. After that, he is worm-counted once a year, or more often if we feel it is necessary. Teeth are regularly checked so that any sharp edges can be removed by rasping.

Vaccinations are given early in the New Year, a few days before the horse begins his three weeks of walking on the roads. He therefore has plenty of time to recover before he is asked to do any strenuous exercise. Unless the weather is exceptionally cold, the advanced horses are clipped out at about the same time, while the novices are trace or blanket-clipped. If we happened to be in the grip of a big freeze, the clipping would wait until after they had finished their three weeks of walking.

A trace clip

SUMMARY

Give minimum amount of hard feed (i.e. oats and event nuts or mix) to get horse fit.
Judge quantities of hard feed by horse's appearance and feel he gives when ridden.
If guideline needed, start with 1 lb of hard feed for each mile being covered – and adjust as necessary.
Before competition, check with feed merchant to make sure event nuts or mix do not contain any forbidden substance.
Reduce hard feed on rest day.
Hard feed is given four times a day, except on rest day.
Hay is soaked for 24 hours to eliminate problems caused by fungal spores.
HorseHage can replace part of the hay ration. It contains no spores, but is richer than hay so should be fed sparingly.
A handful of hay or HorseHage is given 15 minutes before each hard feed.
Chaff is soaked to eliminate risk from fungal spores.
Sugar beet nuts and grass nuts are soaked.
Rations of sugar beet nuts and grass nuts are reduced if horse is getting too fat. Grazing is also limited if he is overweight.
Boiled or flaked barley is given to put on weight, especially in winter.
Additives: vitamins (more required in winter), salt (more in summer), cod liver oil and limestone flour.
Horses are creatures of habit, so keep to timetable.
Horses should be shod approximately every six weeks.
Pads are fitted under the shoes when ground is hard.
Horses are wormed regularly and worm-counted once a year.
Vaccinations are given in early January.
Unless weather is particularly cold, horses are clipped in early January.

8

CLOTHING AND EQUIPMENT

You do not need to spend a fortune on your own or your horse's wardrobe in order to compete in your first one-day event. Obviously, you should check an up-to-date rulebook to make sure of the exact requirements – and, equally important, to ascertain the items which are not allowed. Having once been eliminated for inadvertently using the wrong tack, I make a point of studying each year's new rulebook with the utmost care.

THE RIDER

The informal rat-catcher dress (hunting cap or bowler, tweed riding coat, shirt with collar and tie, buff breeches and black or brown boots) is currently acceptable for novice one-day events. If you want to start with clothes that you can continue wearing in intermediate classes, you will require black boots (not brown), a black jacket (not tweed), alternatively a red jacket for a man or blue for a woman, and a white hunting tie (not collar and tie). If you do not possess them already, all the items mentioned are usually available second-hand.

Except for a crash helmet (with an approved BSI number and black or navy cover) the same clothes are worn for show jumping. For cross-country you also need the crash helmet, plus a sweater or shirt and a back protector. A hunting tie, which can be home-made quite easily, is recommended for extra protection and a bib for displaying your number is normally required. This is the one phase in which you can use exactly the same clothes from novice to championship events.

Spurs are optional in novice events, but I always wear them for competitions. They will be compulsory at advanced level, so I feel the horse might as well get used to them at an early stage. Gloves are also optional, but I wear them too – normally leather and string for dressage and string only for show jumping and cross-country.

BITTING

The bit is the most crucial piece of equipment and, unlike most other items, it cannot be adjusted if it happens to be the wrong size. So much is affected by this small piece of metal or rubber that goes inside the horse's mouth (the confidence of both rider and horse for a

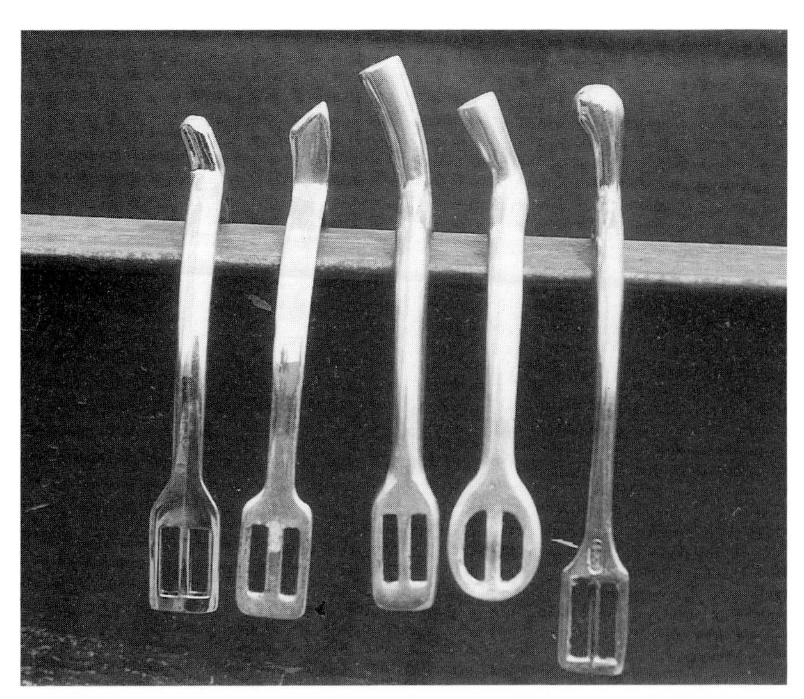

Various types of spurs

A collection of snaffle bits which are suitable for young horses. Left (top to bottom): *French bridoon, rubber snaffle with cheek bars, plain snaffle with cheek bars.* Right (top to bottom): *small bridoon bit, rubber snaffle with D-rings, loose ring snaffle*

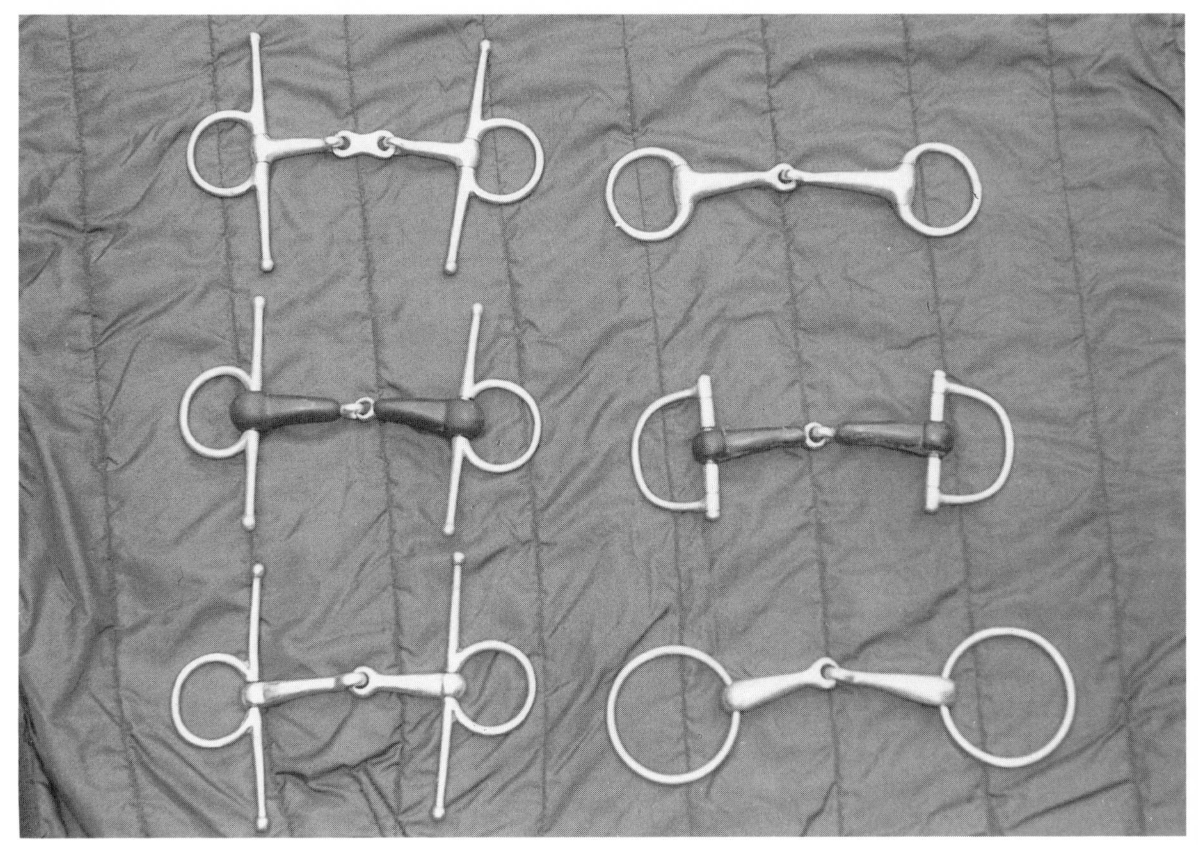

start) that it can prove more difficult to get right than all the other contents of the tack room put together.

Horses' mouths come in all shapes and sizes. One animal may have teeth that are more prominent than usual, another may be particularly sensitive on the corners or the bars of his mouth. The art of bitting has to take such things into account, so that the main action of the mouthpiece avoids any sensitive or delicate area.

If the horse has a problem with his mouth, it is up to the rider to try to pinpoint the cause before automatically changing the ironmongery. Perhaps the horse's teeth are causing the problem and need to be rasped; maybe there is a sore in his mouth. It could be that the bit is the wrong size and that he needs a wider or narrower version of the same mouthpiece rather than a different type. Or it may be the rider's hands that are creating the problem.

The snaffle should be a mild bit, but it can be severe if it fits badly. When pupils come to Dot for the first time, she frequently finds that the bit they are using is the wrong size for the horse. She is loath to put anyone to the expense of buying a new one, but correct bitting is so fundamental that she would be doing her pupil a disservice if she failed to point it out.

Sometimes the problem is in the horse's mind rather than his mouth. He may have had badly fitting tack, too strong a bit or a rider with heavy hands in the past. As a result he has formed a mental association which connects the bit with pain; he therefore becomes agitated as soon as anyone takes up the reins. His new rider may have the most perfect hands in the world, but it will still take a long time before the bad memory fades and he feels confident to take contact on the reins.

The snaffle can do as much, if not more, damage in novice hands than bits that are more severe. If, for instance, the horse is a little too strong across country and the rider is constantly pulling against him, a sore can easily be caused. It could be that a stronger bit, used with a lighter aid, would be more effective. Then, instead of pulling on the reins non-stop, the rider can give an effective aid as and when necessary. Constant pulling is like endless nagging: it is disliked but not respected. As one wise person put it, you would do better with a bucketful of hands than a bucketful of bits!

Finding the right bit

We like to keep the tack simple and the bit as mild as possible, especially in the early stages of the youngster's education when he should be learning to respond through correct training. I will be using an ordinary snaffle for the dressage, but I might feel the need for something a little stronger on the cross-country.

All our horses start with an ordinary unjointed rubber snaffle, which is the mildest bit of all. If I were looking for something stronger, I would then try a jointed rubber snaffle. After that, if I still wasn't happy, I would try the same bit in metal, if necessary followed by a loose-ring snaffle and a French bridoon. Though I have used stronger bits than these fairly mild versions of the snaffle on intermediate and advanced horses, I have never put anything more severe in a novice mouth.

We have naturally acquired a large collection of bits over the years, including different sizes as well as various types, so I can usually try out different mouthpieces without having to buy them. If you do not have your own collection, it would be better to borrow a bit or try to have one on approval. Otherwise you can incur the unnecessary expense of buying one that proves unsuitable.

It should be stressed that we prefer to stay with the mildest bit we can find, but that is not always possible. If we have to resort to a slightly stronger one for the cross-country, we will be hoping that we can switch back to a kinder bit at some later stage. This does sometimes happen, so you should not make the mistake of thinking that the mouthpiece has to become progressively stronger.

BRIDLE, BREASTPLATE AND MARTINGALE

All our horses have one exercise bridle and another for competitions. This helps to reduce the risk of a strap breaking through stress, or of any part disintegrating because the stitching has rotted. Because these things can happen, we take the precaution of checking all straps and stitching on a regular basis. Obviously two bridles are not essential, but you do need to be ultra-careful if you are using the same one for schooling, hacking and events.

Only a snaffle bridle is allowed for the dressage phase of novice one-day events. The dressage judges will probably be more impressed to see the horse in an ordinary cavesson noseband, so I would prefer to use this for the test and then switch to a grakle (cross-over) noseband for the show jumping and cross-country. I would make an exception when I am riding a horse that has a tendency to open his mouth or cross his jaw. He can do neither if he is wearing a grakle, so I might use it for all three phases. Because leather reins can become so slippery in wet weather, I prefer to have black rubber ones for dressage. They are narrower and more elegant than the wider rubber reins (normally brown) that I use for the cross-country.

All the horses wear a breastplate for canter work-outs, as well as for show jumping and cross-country. It must be correctly fitted; I have seen horses competing with their shoulders severely restricted because their breastplates were far too tight.

Some of our horses may wear a martingale or a martingale attachment, which is fitted to the breastplate for the show jumping and cross-country. We always use stops on the reins; these prevent the rings from sliding too far forward and getting caught up on the bit. I have the martingale loose enough to avoid interfering with the horse's natural head carriage, but tight enough to prevent me from getting my teeth smashed in! Turning is often difficult when the horse gets his head too high, so the martingale also helps to overcome steering problems – but I have to admit that it is there only because he is not sufficiently well schooled to go without it.

SADDLE, NUMNAH, SURCINGLE AND GIRTHS

A general-purpose saddle is perfectly adequate for all three phases of a novice one-day event; indeed, you can use one all the way through to the Olympic Games. If you do have a dressage saddle, it has the advantage of getting your leg into a better position to use for the various test movements. For that reason, I prefer to use one specifically built for dressage in the first phase, whether I am riding a novice or advanced horse. I then use the Barnsby Leng competition saddle, which I helped to design, for show jumping and cross-country. It has a flatter seat than the traditional jumping saddle, allowing me to move my weight further back at drop fences.

I always use foot grips on the stirrup irons and a fairly substantial numnah, about one inch thick, under the saddle. It is made of foam with a cotton cover, so that it can be washed easily. If the saddle has to fit two horses, you can use pads (which are available in a variety of materials) between the numnah and the saddle. Obviously, the same saddle cannot be used on two horses of a completely different size, but pads are a great help when small adjustments need to be made to stop the saddle rubbing. Pads also help to compensate for the changing shape of the horse's back as he becomes increasingly fit.

Though elasticated web girths are widely used in the racing world, I am not keen on them myself. I prefer to use two girths made of webbing, plus an elasticated web surcingle, when I am riding across country or schooling over fences at home. Normally I use leather girths for dressage and show jumping at one-day events. There is no reason, however, why you should not use your webbing pair (assuming that they are not bright pink or purple) for these two phases as well.

BOOTS AND BANDAGES

We always use exercise boots on all four of the horse's legs when schooling at home, out hacking or working in for the dressage test. They are removed at the last possible moment before the test, thus reducing the risk of a horse going suddenly lame through knocking himself.

I have seen this happen to others and experienced it myself on one occasion, when I was riding Night Cap around the outside of the arena at Locko Park prior to his test. As I asked him to strike into canter, he hit himself on an old splint and was as lame as a crow for about one minute. Since boots have to be removed before you begin your laps around the outside of the arena, there was nothing we could have done to prevent this happening. But it does underline the importance of keeping the horse's legs protected for as long as possible.

For show jumping, our horses' forelegs are protected by half-boots, which cover the tendons but leave the cannon-bone exposed. We also use over-reach boots on each horse that has shown a tendency to get a hind toe dangerously close to a foreleg. The back legs remain uncovered for show jumping.

We use tendon protectors and over-reach boots on the forelegs for the cross-country phase and also for canter work-outs, gallops and cross-country schooling. The protectors cover the cannon-bone and incorporate a special strip of plastic at the back which gives extra protection to the tendon. There is no boot in the world that could withstand the enormous impact of a horse striking directly into himself with a hindleg while at full gallop, which has been estimated at ten tons per square inch, so we have to be content with doing the best we can to protect his forelegs. We put elasticated bandages over the front protectors and they are always sewn in place to make them secure. I could never set out with confidence if the bandages were held in place by sticky tape, which some people use.

The hindlegs are protected by woof boots or back tendon protectors for cross-country jumping, whether schooling or in a competition. For canter work-outs and galloping, we would use back brushing boots.

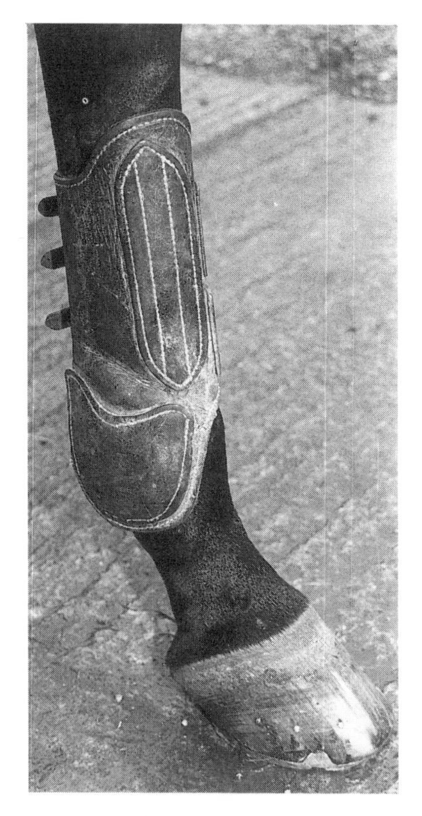

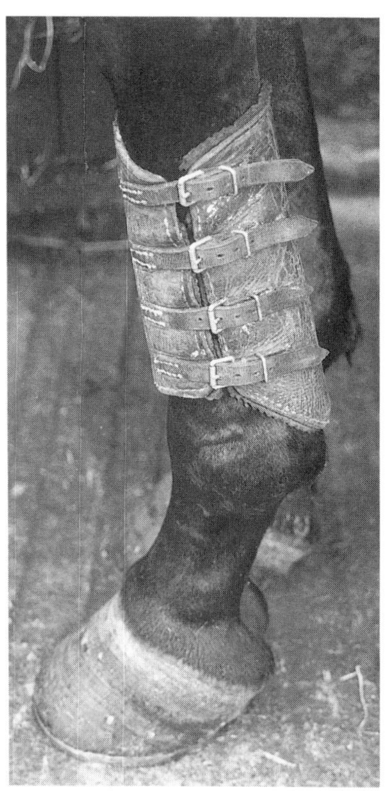

Right: *Front and hindleg Speedicut leather boots with buckles, which are suitable for cross-country at competitions*

Below: *Simple brushing boots for everyday use*

STUDS

Our horses wear studs for all three phases of eventing and any other occasion when jumping on grass. The difference they make in giving a horse a secure foothold is particularly noticeable in show jumping, whatever the ground conditions. Studs are not particularly expensive, and since they give you so much added security I would always try to have a wide variety – ranging from the pointed ones used for hard ground to the large squares or rounds used in soft going. You are then equipped for all conditions.

Our policy is to use smallish square road studs in front for all three phases. If the ground is boggy or bone hard, I might change these front studs for dressage and show jumping, but I am reluctant to do so for cross-country, although it has occasionally been known. The main cause for my reluctance is knowing that the bigger the stud, the more painful it will be if I fall off and the horse treads on me! I also want to allow his front feet to slide very slightly as he lands; if there is no such give when he is jumping at speed across country, it will put excess strain on his forelegs.

The back studs we use depend on the going. Having made our selection, we screw one stud into a hole on the outside of each shoe. In the United States they use two per shoe, presumably because the ground tends to be much harder there, which would make the foot permanently unlevel if the British system were adopted.

RUGS

The novice would need only to be trace or blanket clipped, so he requires less clothing than the fully clipped advanced horse. He would still, however, need a minimum of four rugs at a novice one-day event, a top woollen rug (which can be the one that he wears in his stable), a New Zealand rug (which he can also use when turned out in the field), a thermal rug and a sweat sheet. In the United States and other countries with a less temperate climate, a wider range of clothing is necessary.

A smart travelling rug is nice to have, but not essential. The same applies to a waterproof sheet, which is useful for keeping the saddle dry for short periods when you don't want to bother with a New Zealand rug. A quarter sheet is also useful when riding in before the dressage on a freezing cold day.

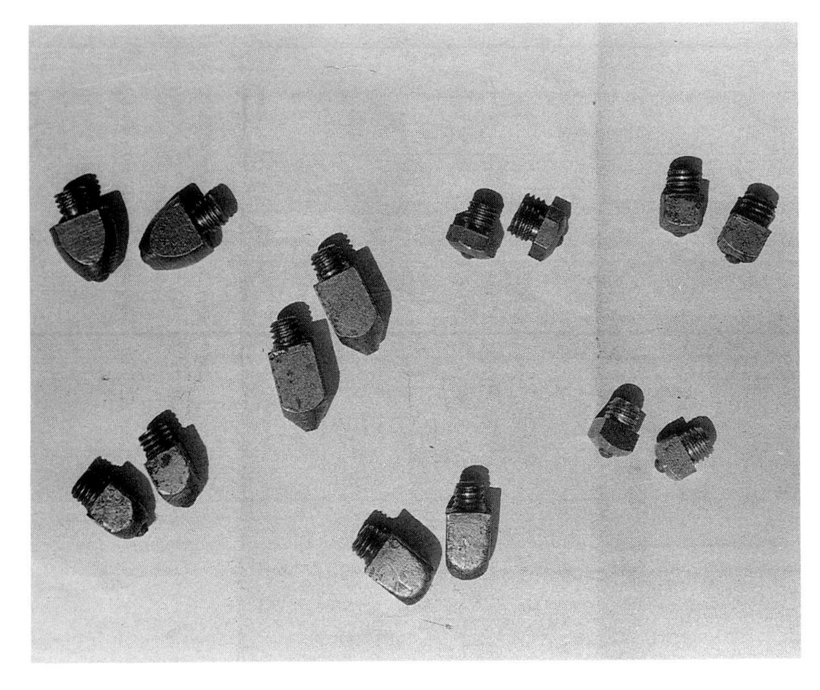

A collection of studs used for all three phases

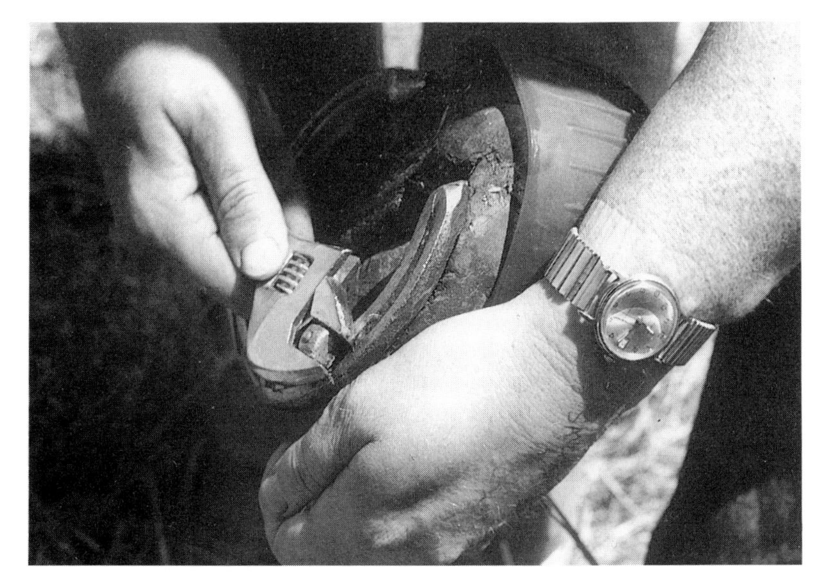

Fitting studs with an adjustable spanner

American horses have two studs fitted to keep the foot level

SUMMARY

Check up-to-date rule-book for correct clothing and tack.
Bitting:
Make sure mouthpiece does not put pressure on any sensitive area of the mouth.
Make sure the bit is the correct size for the horse.
If the horse has a problem with his mount, try to pinpoint the cause.
Use as mild a bit as possible.
If a stronger bit is required for control, aim to go back to something milder at a later stage.
Suitable bits for the young horse: unjointed rubber snaffle, jointed rubber snaffle with cheek bars or D-rings, jointed snaffle in metal, loose-ring snaffle, French bridoon.
Other equipment:
Check all straps and stitching regularly.
Make sure the breastplate is correctly fitted.
If any small adjustment is needed to make the saddle a better fit, use pads between the numnah and saddle.
The martingale (if used) should be loose enough to avoid interfering with the horse's natural head carriage.
Protective boots:
For hacking and schooling – exercise boots.
For show jumping – over-reach and show jumping boots on forelegs.
For cross-country (schooling and competitions) – tendon protectors and over-reach boots on forelegs; woof boots or back tendon protectors on hindlegs.
For canter work-outs and galloping – tendon protectors and over-reach boots on forelegs; back brushing boots on hindlegs.
Studs:
Try to have a wide variety, suitable for all types of going.

9

PREPARING FOR THE FIRST NOVICE EVENT

Most horses seem to mature while they are wintering in the fields. When they come back into work as five-year-olds, they are usually ready to be asked for more quality in their paces. Sometimes, particularly with the older horses, they go brilliantly on the flat when they start working again after their three weeks of walking on the roads. Problems from the previous autumn seem to have disappeared, only to resurface a few weeks later.

At this stage we would be considering whether the five-year-old is likely to be ready for a spring one-day event in late April or early May – which will depend on his physical and mental maturity. He would normally restart his schooling in January, so there is plenty of time to get him fit should we decide to run him.

SCHOOLING FOR THE NOVICE DRESSAGE TEST

We are now looking for improved quality, with better engagement of the hocks, a round outline and smoother transitions. If we can get these things right through constant repetition, it will save a great deal of hassle later on. Though the lessons need to be varied, the trainer should never lose sight of the basic requirements.

The horse does not need to learn any new movements for the first test he will perform at novice level. We can therefore concentrate on keeping him straight, in rhythm and balanced with the weight of the rider while following random parts of the test.

If the test were to include working canter on a 20-metre circle, I would probably begin by doing that shape at walk and trot. When this is accomplished satisfactorily, I move on to canter. I also use half-circle variations at slower, then faster, paces – plus the various shapes the horse was doing as a four-year-old, which are all done at walk.

A nice outline in working canter

I never go through the whole test that will be used in the competition. The horse would be quick to learn it and so start to anticipate each movement. Instead I make up my own tests and practise them in a small arena of 20 by 40 metres, which is the size used for novice tests. Though lettered markers are unnecessary for schooling purposes, it makes sense to practise with some objects of a similar size – such as traffic cones or old paint tins – so that the horse is used to seeing something on the ground beside the arena.

Although some lengthened strides at trot may be required in the novice test, I do not ask for them until the horse has established rhythm and co-ordination within that pace. I would not discourage lengthened strides if he showed a natural ability to produce them, but I would not press him. Unless it comes naturally, the horse will find it difficult to avoid going on to his forehand as he lengthens, whereas we want him back on his hocks. I do, however, include periods of walk, trot and canter on a long rein, which allows him to stretch his head and neck forward and down. It is at this stage that we discover how well he can balance himself.

TRANSITIONS

I attempt only simple transitions at this stage, for example walk to trot or trot to canter. Walk to canter (or vice versa) is obviously more difficult and it will probably not be attempted until later, depending on the horse's aptitude. Before that happens I need to teach him to stay straight and maintain the same outline while making the easier transitions, whether upwards or downwards. These should not be abrupt; we want the horse to flow smoothly and rhythmically from one pace to another, which will probably take a few months to achieve. If you ask the horse to make more difficult downward transitions (for instance from canter to walk) before the simple ones are established, he is likely to lose his balance, lean on the bit and fall into a heap.

ESSENTIALS

The horse must go forward without any hesitancy; he must be obedient, in balance and stay straight. These things will have an effect on everything he does in the future. If he is not straight in his basic flatwork – if he cannot do correct turns and circles – there is little hope of him doing shoulder-in and half-pass correctly when the time comes.

He will be similarly handicapped if he is too much on his forehand, without his hocks being sufficiently engaged. You can increase the impulsion that comes from the hocks, and so lighten the forehand, by moving your weight slightly back and down a little while your hands control the energy created by your legs, which are applied on the girth. In giving and taking with a soft hand, together with correct use of upper body weight and balance, you are containing that energy and preventing it from escaping into a faster speed. There is therefore increased power in the horse's hindlegs, which are the engine that propels him forward. Some horses have natural impulsion; others have to acquire it through training.

Work over trotting poles will also help to engage the hocks and lighten the forehand, this time without any interference from the rider. Hillwork, introduced after five weeks when he is fit enough to cope with it, provides additional help in bringing the hocks further under the horse's body. He will also discover how to balance himself, with the weight of the rider, while going up and down hills.

JUMPING

Once the five-year-olds are fit, I jump them once a week in the school – for no more than twenty minutes and often for much less. Horses that are good jumpers would not be required to go over grids and a course of single fences on the same day. They do one or the other for about ten minutes, which is quite enough.

At this age the horse's jumping muscles can become fully developed with exercise, but they must be built up gradually through his jumping lessons. He will then be capable of tackling more difficult fences. I continue to use grids – which incorporate uprights, spread fences and bounces – to improve the horse's athleticism, always approaching them in trot. The lesson is never the same two weeks running; as already mentioned, I have a better chance of maintaining his interest if he is not aware of what is coming next. I have to play it by ear, taking account of the temperament and ability of the individual horse.

If he tends to have a rather long, flattish jump, I might choose to play around with grids for one lesson. You can help to overcome most problems by building fences up within a grid. If the distances between them are correct, the horse will meet each one at the right spot and gain tremendous confidence as a result.

The actual distances will depend on the height of the fences as well as the horse's conformation, age, experience and length of stride, so you should not rely too much on the tape measure. For fences not exceeding 3 feet 3 inches (1 metre) the average distances for one non-jumping stride are as follows:

Between cross-rails (or from cross-rails to first fence): 18 feet (5.5 metres).
After first fence: 21 feet (6.4 metres).
After second and subsequent fences: 23 feet (7 metres).
For a bounce (as distinct from one non-jumping stride): 10-12 feet (3.05-3.7 metres).

All the above can be lengthened by one foot (0.3 metre), depending on the horse, but it would be rare to shorten them. The shorter the distance, the more effort will be required for him to collect himself in time for the next jump. The diagrams show sample grids of progressive difficulty that could be used for a novice horse.

The week after concentrating on gridwork, I might do one quick grid to a round of fences. You need no more than four jumps – an oxer, a vertical and two together to form a double – to have a short course. The two single fences can be jumped from both directions and possibly the double as well, depending on how it is built.

Small indoor show-jumping contests will be used as part of the horse's training. Before that happens, I will have done some homework by introducing him to the different types of obstacles he might meet when he gets there. You have to use your ingenuity with the materials available to build something resembling a wall. Brightly painted poles and planks will help to prepare him for the vivid colours that you often see in indoor jumping. Doubles and combinations, which green horses tend to find particularly spooky, will also be included in the fences we jump during our preparations at home.

Grids suitable for novice eventers. All distances can be shortened or lengthened by one foot, depending on the horse and height of the jumps. Lower fences required for a green novice

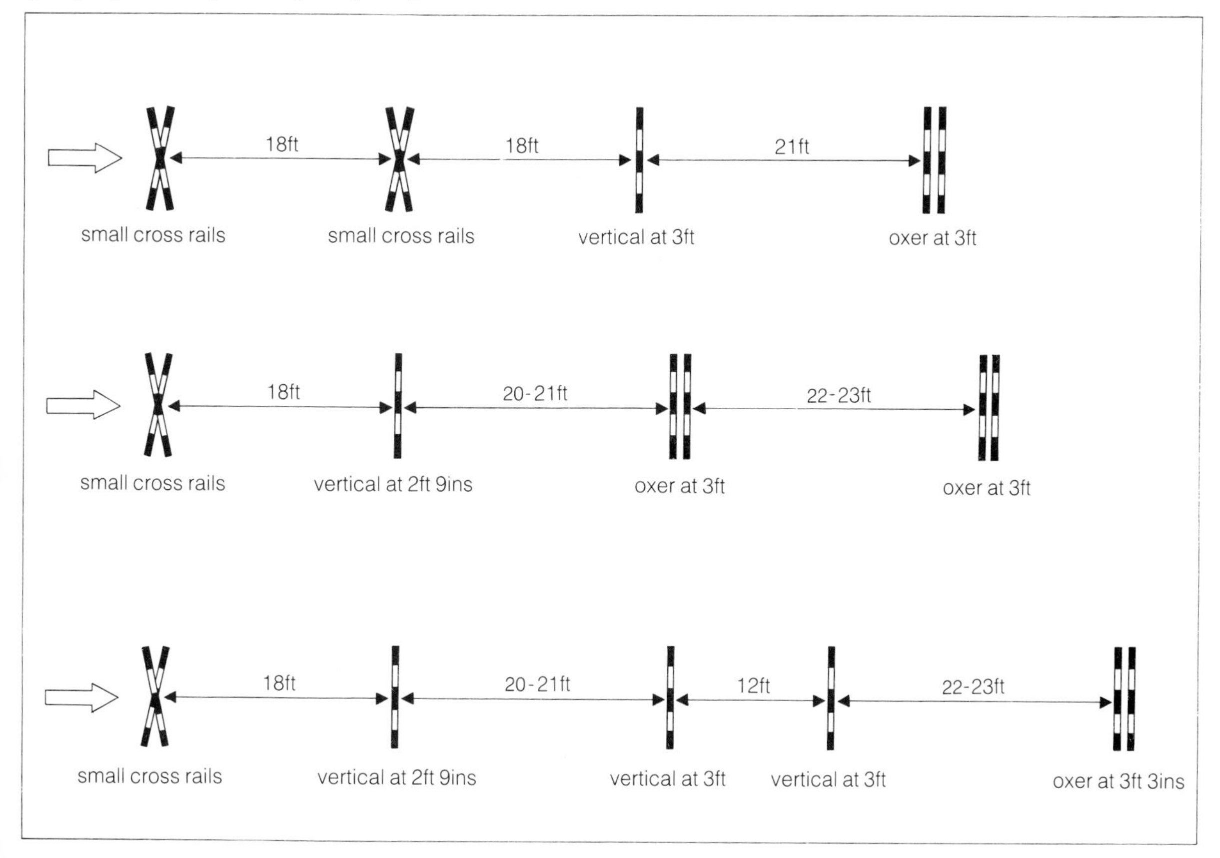

Horse competing in his first small jumping derby, which combined show jumps and cross-country fences. Such contests are an excellent way of building up confidence

Needless to say, I would pick contests with small courses for the horse's introduction to the amazing new experience of going to a show. Travelling in the lorry can be an eye-opener in itself. When he arrives, there might be as many as a hundred horses milling around, which he also finds incredible. I am sure that the baby novices feel totally exhausted when they return from one of these outings. My aim is to build up gradually to the point where the horse can jump safely round a Newcomers' course, with fences up to 3 feet 7 inches (1.1 metres). I take part in these contests with the aim of jumping clear rounds and without getting too carried away when it comes to going fast against the clock.

Flatwork should have taught the horse to stay on a straight line whenever he is moving forward. That is instilled into him at walk, trot and canter, and repeated yet again when jumping. If the horse is cantering round a corner and heading for a jump, he must come off the corner, stay straight and keep going forward. He must not be allowed to give himself extra room by veering to one side in front of the fence, because this can quickly become a habit. He has to learn to shorten or lengthen whenever necessary, so that he can stay on the same line to jump the fence.

The horse should not be reprimanded if he hits a fence. You only need to try knocking down a pole by hitting it with your own legs to know that it hurts. The horse also finds it painful – so he has, as it were, reprimanded himself. He will be anxious to avoid making the same mistake again; if he does repeat the error, it is likely to mean that your training is at fault.

Angles are not introduced until I am happy with the horse's canter and feel completely confident that he will stay on a straight line. I use two small fences for this exercise, placed as shown in the diagram. The horse takes only one fence at an angle for his first two attempts. At this stage you should avoid going too near the fence that is not being jumped, because you might then have to turn the horse away from it. This could lead to him running out at a fence when you do intend to jump it.

If he stays straight both times, the horse is asked to jump the two fences on an angle, maintaining a straight line from his approach to the first until after landing over the second. This exercise prepares him for jumping his first corner, which I would attempt only when I was happy with the way he coped with angles. I would need to have jumped at least half a dozen angles successfully before tackling a corner for the first time,

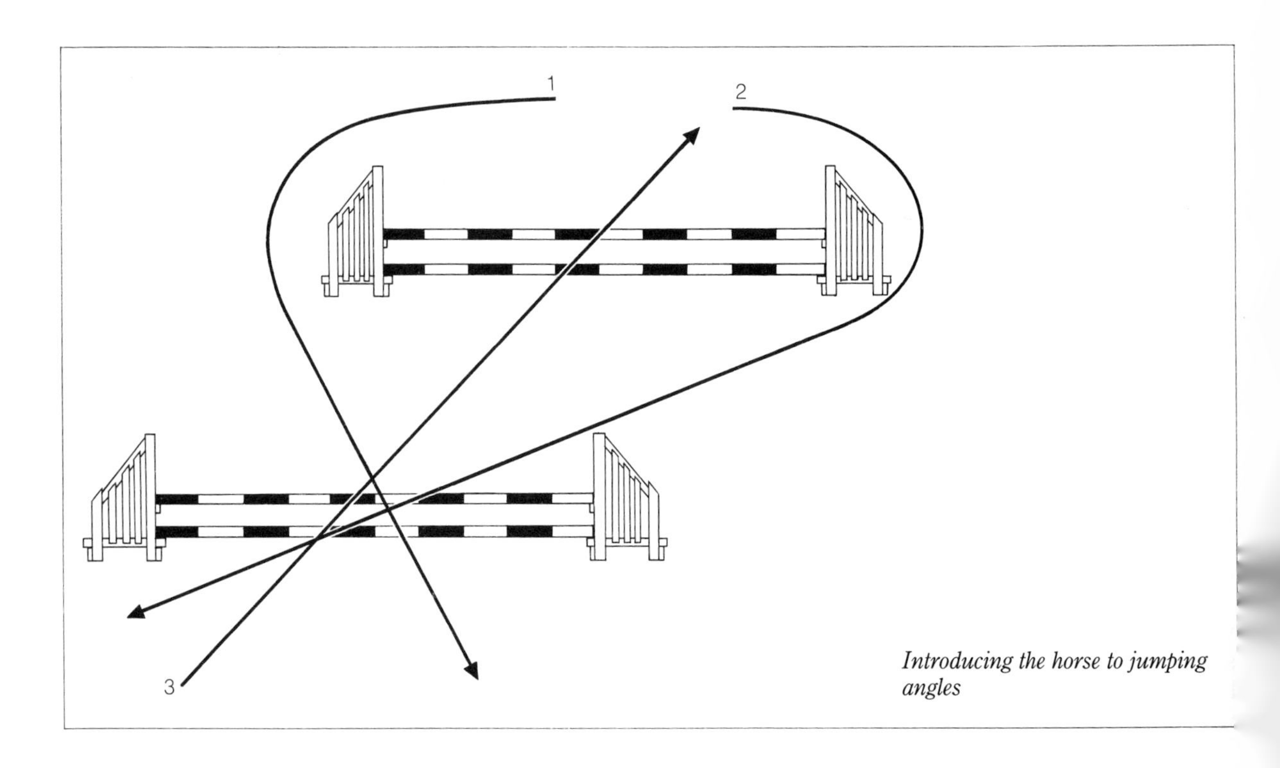

Introducing the horse to jumping angles

which is always done in the school at home using show jumps rather than fixed cross-country obstacles.

I also like the horse to learn flying changes at this stage, so that he can organise himself without any difficulty if he lands over a fence with his left leg leading and has to make a right turn almost immediately – or vice versa. This movement is taught at an early age in the United States, which I believe is a good idea. In Britain we are inclined to think that we would then run into trouble when we had to do counter-canter in the FEI dressage test, but the fact that the horse has learnt flying change as a youngster should not be an excuse; he ought to be trained well enough to stay on the outside leg when the rider tells him to do so.

I therefore encourage him to use flying change whenever he finds himself on the wrong lead for turning after a fence. The horse will normally make the change when I give the aids; if he fails to do so it should not become a big issue. He will learn this movement in his own good time and it will then become easy for him, so that he automatically uses flying change to put himself right for turns.

CROSS-COUNTRY SCHOOLING

I would probably take the horse for his first *proper* cross-country school early in his fifth year, but he would have to be confident while cantering into coloured jumps at home and popping over natural fences. I would also want to be reasonably sure that he was willing to listen to me and that he would stay straight. I very rarely use a stick on these occasions. It is far better to encourage the youngster with legs and voice, so that he can build up his own self-confidence. He is likely to be in a state of nervous tension when he goes for his first proper cross-country school, and use of the stick would serve only to make him more tense.

Unlike the rider, the young horse has not seen it all before; these strange new obstacles can therefore be a perfectly understandable cause of alarm. This obviously has to be taken into account when you select the fences to be jumped. You cannot expect him to perform the way he does in the familiar surroundings of the school at home, so you will probably need to start with obstacles that can be taken from a standstill.

I remember taking one five-year-old to Wokingham, where I attempted to ride him down some fairly steep steps. The horse stopped on top of the first step and looked down in horror. He had no idea how to cope and worked himself into a state of agitation because he didn't know what he was supposed to do. I simply patted the youngster's neck and talked to him. Eventually I could almost feel the tension drain away as he began to realise that the route ahead was not impossible after all; then he jumped down the steps. Had I used a stick, he would have started to think about what I was doing. By sitting quietly, I gave him the chance to concentrate on the steps and work out his own way of getting down. He was far less tense when faced with the same obstacle a second time.

I would never present a horse to a fence unless I believed we had a 99 per cent chance of getting over it. This is not too much of a problem at the kindergarten stage, when the youngster is asked to jump only fences that can be taken from a standstill. It becomes far more difficult when you move on to the next stage and are attempting larger obstacles, which cannot be jumped without turning back should the horse grind to a halt.

'If in doubt, don't' is one of the Ivyleaze mottos, which refers to everything we do with the horse and is particularly relevant to schooling over cross-country fences. If I am concerned that the horse might stop at a particular fence, I don't attempt it. It is far better to spend more time jumping smaller obstacles of different types than risk a setback through the horse refusing. This does not mean holding him back to the point of mollycoddling him. There has to be a balance whereby the horse keeps progressing, albeit cautiously.

This involves a form of continual assessment on the part of the rider. You also have to keep in mind the horse's experience over different types of fences, so that you can gauge the right time to present him with the next challenge. I always aim to make the horse familiar with one new experience at a time. I would therefore want to be sure that he was happy jumping a ditch on its own before attempting to jump one that had some poles above it.

The components of other cross-country fences can be separated in the same way. Before jumping off a bank into water, I would want the horse to have jumped down an ordinary step and to have paddled through water. I would probably ride him through the water at the bottom of the bank before jumping into it. If all went well, I might attempt a log into water when this particular horse went for his next cross-country school. He would be familiar with each of the three components on its own – that is the log, the drop and the water – and he has jumped down a bank into water. Assuming that he still seemed full of confidence, I would expect him to pop over the log and into water without the slightest hesitation.

I apply the same policy to every type of cross-country fence, tackling the individual components first and then putting them together. This does not mean that our horses never encounter problems, but it does help to avoid them. Before entering for a novice one-day event, I would need to be sure that the horse had successfully tackled every type of obstacle he is likely to meet on the cross-country course. But I would not try to rush him through his schooling. If the careful step-by-step process does not prepare him quickly enough for a spring event, he can wait until later in the year – or maybe as long as the following spring.

The rider should already have tried to acquire the ability to see a stride in order to assist the horse, though there will always be times when it is better to sit still and let him get you out of trouble. On one occasion, while cross-country schooling at Wylye many years ago, I plucked up courage to ask the dual world champion Bruce Davidson how he managed to see a stride from so far back. He told me that he always looked at the top part of the fence and did not take his eye off it. I then followed his example and discovered that it worked. Since then, I know that whenever I miss my stride it is because my eye has wandered for a split second.

Whether in show jumping or cross-country, I reckon that you need to have your eye on the fence from at least six strides out – and keep it there. On very rare occasions I have come down a long gallop to a cross-country fence and seen my stride fourteen paces out; because my eye was already on the obstacle, I knew that we were going to meet it right as long as I kept looking at the spot where I aimed to jump.

You should not need to look down to check the length of your reins or see which leg the horse is leading on; you should know these things by feel. Nor should it be necessary to look down in order to switch your stick to the other hand or shorten your reins. Your eye needs to be concentrating entirely on the fence, particularly if you are jumping at an angle or across a corner, otherwise you could be heading for disaster.

When you have seen your stride, you should make any necessary adjustment within the rhythm of the horse's movement. This shortening or lengthening should be so smooth and subtle that it is virtually imperceptible; you are not supposed to look up and keep kicking!

FITNESS WORK

A horse can get fit only through regular work; he needs to be ridden six days a week for two to three months to get ready for a one-day event. This may sound like a long slog, but I know of no other way to produce a horse that is capable of sustained cantering. The slow build-up also reduces the risk of strains, which are far more prevalent in horses that work only spasmodically.

If you cannot find the time to ride six days a week, you will need to enlist the help of someone who may not be able to help with the schooling but can at least walk the horse on the days when you are not available. It is easy to be deceived by a horse that is obviously fresh and seems to have a great deal of energy. This does not prove the animal is fit; usually it means exactly the opposite. He may be eager to charge off across the fields, but he would probably be on his knees ten minutes later. Unless he has been given too much hard feed, the fit horse has a calmer approach and does not expend his energy in one short burst.

All our horses are given three weeks of walking exercise when they come back into work after a holiday, but this does not mean ambling along the roads. The process of learning to go forward, keep straight and stay in rhythm continues while soft muscles become firm again. The daily sessions build up from forty-five minutes to between one and one and a half hours, with one day off each week so that the horse can relax in the field. If you have no field, he will need to be taken out for hand walks instead, preferably where he can pick at some grass.

I would probably introduce about three sessions of flatwork in the school during the fourth week – mainly at walk and trot, with a little cantering. These lessons should last no more than twenty minutes, otherwise the horse will begin to suffer from muscle fatigue. If he misbehaves, he would still be put away after twenty minutes and then brought out later the same day. He is still hacked out every day. For the Ivyleaze horses, this continues to mean walking on the roads in the early part of the year when we need to avoid Badminton Park because preparations for the three-day event are in full swing. We have no tracks for riding except in the park, where all the horses can do some trotting and cantering later in the year. We never trot on the roads except when going uphill otherwise it causes undue jarring. In an ideal world, we would use tracks and fields after the first three weeks of roadwork had helped to harden the horse's legs.

During the fifth week, the horse has one session of hillwork. His other daily sessions will include hacking, plus flatwork or jumping a small grid in the school. A canter work-out is introduced during the sixth week, by which time the horse should be reasonably fit. The last five to six weeks of his training and fitness programme involve a mixture of hacking, canter work-outs, hillwork, schooling on the flat, jumping in the school and park, small competitions and cross-country schooling. He still has a rest once a week and one hard feed is omitted that day.

A typical programme is given at the end of this chapter, but it should be used only as a rough guideline. The actual amount of work – and the way the programme is devised – will depend on the individual horse. We have been known to omit canter work-outs altogether for one particular horse who was naturally very fit and hyper-active. This horse did hillwork, which helps to expand the lungs and improve the heart-rate, as well as building up muscles. He did dressage schooling and went for hacks, which included some cantering – but there was no need for the more strenuous canter work-outs which most of the Ivyleaze horses do when they are preparing for a one-day event. As a general rule, we would expect a five-year-old to need more canter sessions than a mature nine-year-old, who might just need to be kept ticking over.

Our five-year-olds might be given a short uphill spurt, but they are never taken for five-furlong gallops. They will not be going flat out at a one-day event and galloping would make them too fit for the job we want them to do, which is a big mistake. It would probably get them thoroughly wound up as well, undoing all the work that has been aimed at making them calm and co-operative.

CANTER WORK-OUTS

It is essential to find good ground for cantering, otherwise you are likely to do more harm than good. Hard ground is the most dangerous, especially when it is uneven with the grooves you find in ridge and furrow or pits left by horses' hooves.

Whether it is smooth or ridged, hard ground can cause such problems as a sprained fetlock joint, a pulled muscle, damaged tendon or an injured foot. Very deep ground is also likely to cause sprains, and it could leave the horse equally open to risk of a muscle or tendon injury. Though we may have to take our chance in a competition, I am loath to take any risks in training. I would rather replace one of the twice-weekly canter work-outs with hillwork and search for somewhere suitable to canter once a week. It could mean driving some distance in the lorry, but it would be worth the extra effort.

The horse needs to be thoroughly warmed up before cantering, with a minimum of twenty-five minutes' walking and five to ten minutes' trotting. He will also have to walk afterwards until he has completely recovered from his exertions.

I use a simplified form of the method known as Interval Training for canter work-outs. This involves working the horse in short bursts, resting briefly, then restarting before the heart and lungs have returned to normal pulse and respiration. This builds up the capacity of the heart and lungs without having to put undue stress on the horse's limbs.

When he makes a sustained physical effort, the heart and lungs have to work to capacity in order to pump oxygen through the bloodstream to the muscles, which release the energy required. If there is insufficient oxygen, the horse will have tired and aching muscles. By giving him a period of rest between canters, the spent energy can be restored to the muscles by the heart and lungs, which are still working overtime while the horse is walking quietly round.

Interval Training has been used for many years by human athletes and is now the accepted method for training event horses. By building up the capacity of heart and lungs through this method, the horse reaches a level of fitness that enables him to make the sustained effort required in competition. It is, I promise, less complicated than it sounds.

Our usual routine for a novice is to start with a three-minute canter, followed by a two-minute break and another three-minute canter. This builds up over five or six weeks to three five-minute canters, with a break of two minutes in between – or up to three minutes in hot weather if the horse is still blowing hard. This should be sufficient to get him fit for a one-day event – unless the terrain is flat, in which case you may need to work up to three six-minute canters. If the horse looks like being too fit, we would substitute a canter day with hillwork, trotting up the gentle slopes and walking up the steeper stretches.

The first two canters on the novice are at a speed of 400 metres per minute (15 m.p.h.) and subsequent ones at 500 metres per minute (19 m.p.h.). Nowadays I can normally judge how fast I am going, but it is still possible to be caught out. When I rode Murphy Himself, who has an enormous stride, I discovered that I was actually going much faster than I realised. It is easy enough to check, by finding a stretch of grass where you can canter while someone drives a car beside you at the relevant speed. You can do this in a field with a road alongside; there may be a hedge between the horse and the vehicle, but the rider has no need to see anything more than the aerial.

ASSESSING THE HORSE'S FITNESS

I always time the recovery rate of the horse after his canter work-out, so that I know how fit he is. The simplest way to do this is by watching his rib cage to check his respiration, which means counting the number of puffs he takes per minute with the help of a stop-watch.

You need to know his normal rate before he canters. This will probably be somewhere between ten and fifteen breaths per minute, but it could be more with a younger horse. The rate then needs to be counted after cantering, when the horse's breathing will obviously be much faster than normal. Because we do not want him standing still for too long after he has cantered, I count the breaths he takes in fifteen or thirty seconds and multiply by four or two, rather than wait for a full minute. The rate of recovery is the time he takes to return to his normal breathing.

During this recovery time, the horse will be walking quietly home. Having made a note of the time when he finished cantering, I will pull up five minutes later and count for another period of fifteen or thirty seconds. That second count gives me a clue as to when I should stop again. If he is not blowing too much I might take another count two minutes later; if the breaths are still coming quickly I would probably wait for a further five minutes. It is basically a matter of stopping every so often until the horse has returned to normal breathing and noting the time this has taken since he finished cantering. I regard this information as essential because I rely on recovery rate more than anything else to gauge the horse's level of fitness.

Given fairly cool weather, the average recovery rate for a fit horse is between seven and fourteen minutes. In order to assess his progress, you need to keep a record each time he canters. This would have to include weather conditions as well as the recovery rate, because it can make such an enormous difference. I have known a horse recover in seven minutes and one week later, in a sudden spell of humid weather, take as long as seventeen minutes.

FITNESS OF RIDER

Though the horse may appear to be doing most of the work at an event, the rider will need to be fit in order to give clear directions and other assistance from the saddle. If you have only one horse to ride at home, you will need to supplement this with some other activity. This could take the form of jogging twice a week, or you could choose to get fit through other activities like squash, aerobics, swimming, cycling and skipping. If you happen to be riding two or more horses each day, you should be fit enough to compete in a novice one-day event without any additional exercise.

Learning to fall, possibly by taking judo lessons, is another useful preparation. You are far less likely to get hurt in a fall if you have learnt how to roll out of harm's way and are not too tense.

DECIDING ON THE FIRST EVENT

It is up to the horse to tell you whether he is nearly ready to compete in his first BHS one-day event. I would want him to be confident over a variety of cross-country fences, otherwise he will need to be given more time. It is crazy to risk a fall or a stop across country through competing before the horse is ready for the challenge.

I would like his preparation to include other competitions in addition to the indoor jumping. If you can find them, hunter trials and dressage competitions (preferably with jumping) will help in his education. Mini one-day events, such as those run by riding clubs, are even more important. I always try to fit in two of these, but I don't always achieve it; when you have advanced horses to ride as well, it is not easy to arrange contests for the youngsters. I would, however, regard at least two mini events as essential for a novice rider.

I would certainly have taken my novice horse for three or four cross-country schools and would always advise an inexperienced rider to do about twice that amount. Since I do not believe in schooling round a BHS course, the youngster has to be properly prepared before he gets there. I will not expect him to win a prize at his first event, because I will be taking him fairly slowly, but I do expect him to do a reasonable dressage test and to go clear in the show jumping and cross-country.

In the United States, where you can compete in novice and training classes before the preliminary (which is the equivalent of our novice), the preparatory cross-country schooling is not so crucial. With two classes below the level at which most of our horses start, the plan of campaign is therefore very much easier to organise.

Fortunately we do have a growing number of pre-novice events in this country, which give young horses an excellent introduction to the sport. If unable to find one within easy reach, I would look for suitable courses at the ordinary BHS one-day events that are nearer home. I would want the youngster's first three competitions to be over the sort of cross-country fences that will help to build up his confidence, without asking too many difficult questions. You can find out which are the best courses for the first-time novice by telephoning fellow competitors or event organisers.

I would not run a five-year-old more than four times during the spring season, assuming that he was ready to start at all. He is still maturing at that age and it will be quite enough for his brain to cope with. If the ground was good, I might give him six outings during the autumn.

SUMMARY

Flatwork

Continue to work on basic priorities: forward movement, rhythm, balance, contact, straightness.
Now look for better engagement of hocks, round outline and smoother transitions.
Practise movements in test at slower paces (e.g. if 20-metre circle required at canter, start with same shape at walk and trot).
Make up tests to practise in 20 by 40 metre arenas.
Introduce lengthening and shortening when ready in each pace.
Work on simple transitions, aiming to make them more fluent and rhythmic.
Work to improve impulsion and lighten forehand.

Jumping

Work on grids (which incorporate uprights, spreads and bounces) to improve horse's athleticism.
Keep lessons varied – jump different grids and/or small courses.
Take part in small indoor jumping contests.
If horse stays straight, angles are introduced.
If he has jumped at least six angles satisfactorily, corners are introduced.
Horse has first proper cross-country school when confident over show jumps and small natural fences.
Before first novice event, aim to ride in at least two mini one-day events and take horse for four to eight cross-country schools (depending on experience of rider).

Other points

The rider must be fit.
The first few events must be over courses that are suitable for first-time novice, or incorporate plenty of alternatives.

Sample fitness programme

(Horse has one rest day per week.)

Weeks 1-3:	Walking on roads (forty-five minutes building up to between one and one and a half hours).
Week 4:	Hacking – plus three twenty-minute sessions of flatwork in the school, mainly at walk and trot.
Week 5:	Introduce one day of hillwork and jumping a small grid during lessons in the school at home.
Weeks 6-10:	Introduce canter work-outs twice a week. Use a varied programme which also includes hacking, flatwork, lungeing, jumping in the school, cross-country schooling, hillwork and competitions (such as indoor jumping, hunter trials, dressage with jumping and mini one-day events).

Typical programme for the last five weeks

Monday	Rest day.
Tuesday	Light hack and flatwork.
Wednesday	Canter or workout.
Thursday	Longer hack, possibly short schooling session.
Friday	Hack and jumping, otherwise occasional cross-country school or show jumping.
Saturday	Flatwork.
Sunday	Canter or competition.

10

COUNTDOWN

As the event approaches, you would be well advised to go through a check-list in order to make sure that the horse is fully prepared. Here are some of the questions you might ask yourself.

Is the horse the right weight – neither too thin nor too fat?

Has he practised all the skills? In other words, has he done all the movements in the dressage test, gained experience at home and in competitions for the show jumping, and (most important of all) have you done your best to jump a wide variety of cross-country fences?

Have you practised in a small dressage arena of 20 by 40 metres, which is the size used for the novice test?

Does his recovery rate after canter work-outs suggest he is fit enough?

What is the state of the horse's shoes? They should not be brand new, but they should not be worn down either. If one of our horses needs to be shod, we always arrange for it to be done at least a week before the event.

Are you using the right tack? It is unwise to try using anything new at the event, so everything needs to be ridden in well in advance. New girths, boots and numnahs can rub; a last-minute change of bit can confuse and distract the horse.

Have your own clothes been ridden in beforehand? If it rains, new boots that have not yet acquired a slight stickiness can slide on the saddle.

Does the saddle fit correctly? The shape of the horse changes as he gets fitter, so you need to check this on a regular basis. A thick numnah and pads can compensate for the changing contours of the back.

Has the horse been plaited up a few times at home? If he is plaited only before a competition, you are as good as telling him that he is off to a party, so you can't blame him if he gets excited.

Do you know the rulebook inside out? If not, it is frighteningly easy to get eliminated. You only need to ride into the dressage arena with your stick or be seen jumping the practice fence the wrong way round to face instant dismissal. Regulations can change, so we all need to go through the new rulebook with a fine-tooth comb each year.

Has the vehicle been checked? Apart from checking oil, water and tyres and filling up with petrol or diesel, you need to make sure that everything is functioning. Partitions, locks and the ramp mechanism may need oiling, especially at the start of the season, otherwise accidents can occur. The vehicle also needs to be checked after the event, so that there is time for any necessary work to be done before the next competition.

Have you arranged for someone to help you at the event? Even if I were riding only one horse, I would never consider trying to manage single-handed. You do not necessarily need someone with a knowledge of horses; any willing helper who is reasonably level-headed can assist with the practice jump and lend a hand in many other ways. Those who turn up on their own are regarded as an organiser's nightmare.

If you are using spurs, have you practised at home with them? If the horse is unaccustomed to spurs, he might add some unscheduled bucks and kicks to the dressage test.

Have you learnt the dressage test? I always memorise it the day before and I get somebody to hear me go through it while we are travelling to the event.

EQUIPMENT CHECK-LIST

With so many things to take, it is all too easy to forget some vital piece of equipment – though I doubt whether many people would do as my mother once did and forget the horse! If we hadn't been forced to turn back because I had forgotten my hat, we might have reached the showground before discovering that she had failed to hitch the trailer containing the horse on to the back of our ancient Land Rover.

Nowadays the equipment – and the horses – are transported in our smart Citibank Savings horsebox, which is filled with plenty of cupboards to take everything we need. As the following list shows, you have to do some extensive packing even for a one-day event. Spares are important; you might otherwise have to pull out of the competition because of a broken strap.

For the rider

Correct clothes for all three phases
Waterproof jacket and boots
Whip (and spare)
String gloves, with spare pair if riding more than one horse
Sewing equipment (in case a button pops off)
Spare breeches (in case you take a dip in the water jump)
Bib for displaying your number
Spurs (optional at novice level)
Spare pin for hunting tie, if worn
Women also require hair nets, with spares

For the horse

Saddle(s) – if using general-purpose saddle for all phases, take spare girths and stirrup leathers
Bridle(s) – if using the same bridle throughout, take a spare (borrowed if necessary)
Spare bits
Numnah and pads
Breastplate
Surcingle
Martingale attachment, if used
Headcollar and rope
Cavesson and lunge line
Lunge whip
Side-reins
Nylon halter for washing down horse
Rugs for all weather conditions: top woollen, New Zealand, thermal, sweat sheet

Roller
Bandages
Four exercise boots for working in before dressage
Two half-boots for show jumping
Two over-reach boots for show jumping and cross-country
Two front tendon protectors for cross-country
Two back tendon protectors for cross-country
Studs and spanner
Vet box
Spare shoes
Sponges
Scraper
Towels
Grooming kit with extra hoofpick
Hoof oil
Stencil for putting diamond shapes on quarters (optional)

We always sew the strip of Velcro securely on bandages over front tendon protectors and on hind brushing boots. This prevents them from coming undone, which may otherwise happen – particularly in wet weather

Hay and hay-nets
Prepared hard feeds for lunch and tea
Buckets
Water – for drinking and washing down
Fly spray, if weather is hot
Plaiting box
Leather punch
Good-luck mascot!

Other essentials
Vaccination certificates
Omnibus schedule for copy of dressage test
Rulebook
Gas on lorry, if it has a cooker

It might seem as though you need an articulated lorry to contain all this – but believe it or not, it does fit into a fairly compact unit.

Right: *It is important to school over different types of obstacles before tackling your first event*

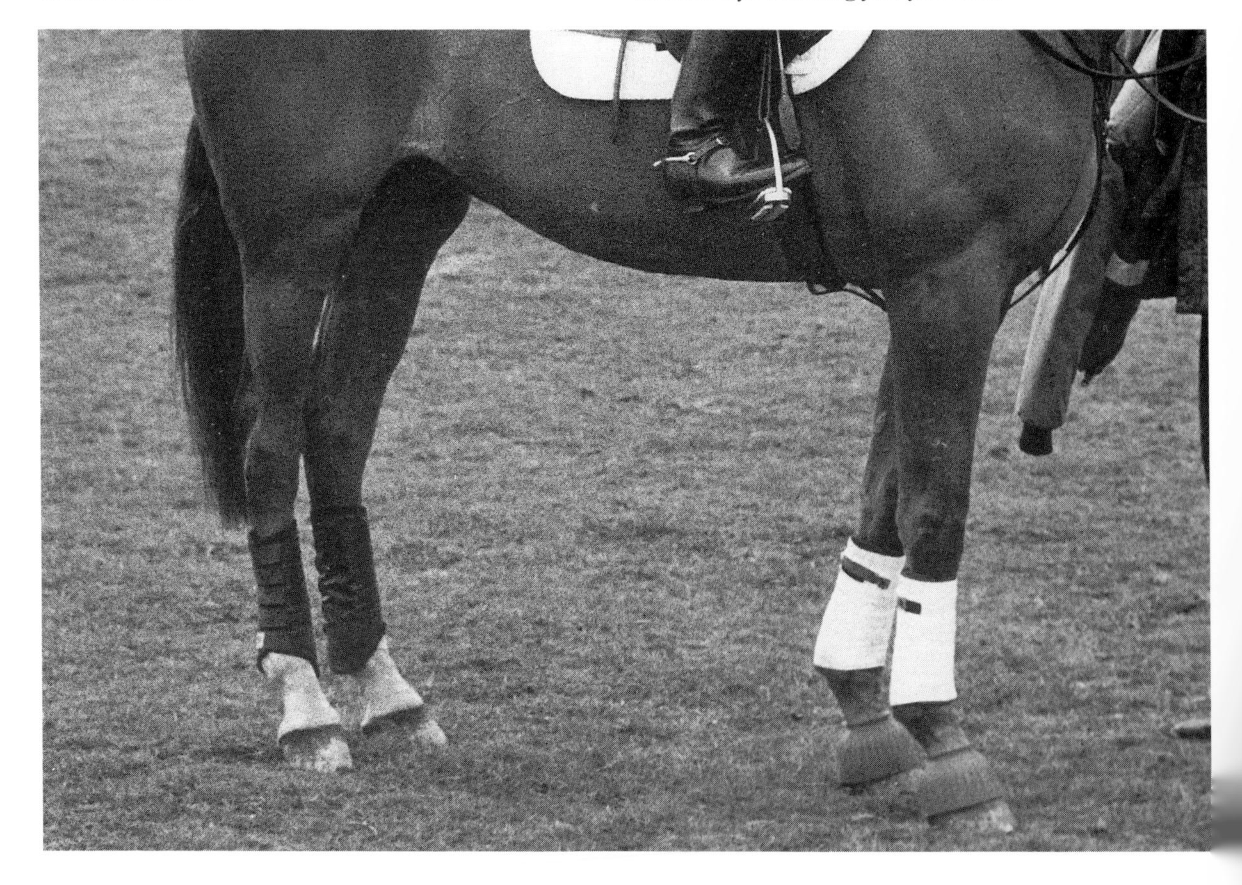

WALKING THE CROSS-COUNTRY COURSE

If the event is local, it makes good sense to walk the cross-country course the day before. You will then have more time to ponder over any fences that might worry you. Most horses and riders have their Achilles heel and by now you will be well aware of any shortcomings in this direction. Your horse may be suspicious of drop fences or ditches or water; he may have a tendency to hang on one rein more than the other. You may also have a psychological hang-up about a particular type of fence, which has to be recognised and accepted along with the horse's vulnerable spot. There will almost certainly be something when you walk the cross-country course that worries you as soon as you set eyes on it. The subconscious is quick to alert us to problems on these occasions.

If there is an alternative slow route at the fence that makes your heart miss a beat, you would be well advised to take it – especially if this is your first event. Bearing in mind that prevention is better than cure, it is much wiser to avoid the problem than to risk a refusal that could completely disrupt the horse's training. When there is no alternative, you may manage to solve the problem by standing in front of the fence and thinking about it in a positive and sensible way.

For instance, if the horse has a tendency to jump to the left and this particular fence seems difficult for that reason, you might decide to switch your stick to the left hand after the previous obstacle. Then you will be ready to slap him on the left shoulder if he feels like swerving – or perhaps, just having the stick on the left side will prevent him from drifting.

If the horse is anxious about jumping into water, you could ask yourself whether you should approach it in trot, so that he has the opportunity to see what lies ahead before you use your legs strongly to get the forward impulsion needed for the jump. Generally speaking, the faster you go at this type of fence the more likely it is that the horse will stop. He puts on the brakes because he wants to look at the fence, and once those brakes are applied it becomes extremely difficult to get one's foot back on the accelerator.

Perhaps the horse may dislike jumping from light into shadow. In this case you must again make sure that he is given every chance to see where he is going; if you are coming off a turn before jumping into a wood, for instance, you can make the turn wider to give him time for a better view. If you cannot figure out the best way to jump a particular fence, you can always ask another rider for advice. Those of us who have been in the sport for a number of years are often asked this type of question and we are happy to try to find the right answer.

You should not feel you are being feeble by trotting into certain cross-country fences; experienced riders do so quite frequently with their novices. It can be the best way to approach a jump into water or such obstacles as a coffin, downhill steps or a sunken road. The reason for using trot can be explained by taking the coffin as an example.

This type of obstacle invariably has a drop on the landing side of the first element, with a ditch to follow. The drop and the ditch are the two ingredients that are most likely to worry the horse, so the object is to get as close as possible to the first fence in order to give him the chance to see where he is going. This is best done from trot with a novice horse; he should come in steadily, with impulsion, and take off close to the first fence from a point where he can see over it. Having landed safely over this first element, you simply need to encourage him to keep going forward over the ditch and the fence that follows.

If you were to approach in canter, you might see a long stride and be tempted to go for it. If the horse is brave, he will probably take off where you ask him, which means that he will be airborne when he suddenly sees the drop and ditch. His automatic reaction is to put his undercarriage down in order to protect himself, which means dropping his hindlegs and hitting them on the fence. He will probably hurt himself in the process and will not be quite so keen to jump the next coffin he meets.

On the other hand, he may not think it such a good idea to take off some way back, at the point you suggest. Instead he might try to put in a short stride, which brings him too close to the first part of the obstacle; he therefore has a refusal. The same principle applies to any drop fence, including those into water. There is a tendency to feel that the faster you go and the further the horse stands back at such fences, the less likely he will be to have a refusal. In fact, the opposite is true. By going fast and asking for a huge leap over drop fences, you will also increase the strain on his forelegs as he lands.

Opposite above: *Unaffiliated shows help the horse to become accustomed to a party atmosphere*

Below: *The horse would have to be completely straight and obedient before attempting a corner, as shown here*

A young horse negotiating his first angle at a novice event. The rider's legs encourage the horse forward on the approach, while hands maintain contact. At take-off, the rider is climbing a little up the horse's neck but achieves a good landing. At this point four eyes are looking towards the next fence. As they move away from the obstacle, the rider's upper body is a little too far forward. The hips could be a little further back at the next take-off, but horse and rider show good style in the air

A typical grid for schooling an advanced event horse: place pole, 9 feet (2.75 metres) to cross-rails, 18 to 19 feet (5.5 to 5.8 metres) to vertical, 10 to 12 feet (3.05 to 3.66 metres) to oxer

A double bounce: cross-rails, 10 to 12 feet (3.05 to 3.66 metres) to cross-rails, the same distance to oxer. Place poles are used to keep the horse straight on approach

A good example of extended trot, marred by the horse's head being behind the vertical

An example of the correct outline

Always keep contact with the horse's mouth, while allowing freedom of head and neck

The horse should be kept in balance between hand and leg

Many falls occur by tripping up steps out of water. The horse needs to be supported by the rider's hand and leg

Jumping a corner with Master Craftsman. This one was low and very wide, demanding extreme accuracy and straightness

Should you decide that the horse might need a reminder at one particular fence, because it represents his Achilles heel, you should make up your mind not to use the stick too early. If you give him a whack six strides out, he will wonder what he has done wrong and will be reacting to you rather than the fence. By the time he switches his attention back to the obstacle, he will be almost on top of it. He is therefore far more disconcerted than if you were to give him one slap in the final stride, when it coincides with his preparation for take-off.

You have to work out a plan that suits both you and your horse. You also have to put your own doubts behind you; the horse can hardly be expected to jump with confidence if the person on top is anticipating trouble. If the doubts persist, and if there is one fence on the course that still fills you with foreboding, you don't have to run the horse. It would be better to withdraw, so that you can have more time to practise over that type of obstacle. You can always do the dressage and show jumping, and leave the cross-country for another day and a different event.

The course builder should be kind to novice horses by using distances that ride well in combinations, but it still makes sense to check them. If nothing else, it will give you some useful practice before reaching the higher levels where distances will become more crucial. The simple yardstick is to take 12 feet (3.66 metres) as the canter stride for an average horse. If you can walk a fairly accurate pace of 3 feet (i.e. 1 yard or 0.9 metre), four of your strides will be about equal to one for the horse.

Whether walking a cross-country or show-jumping combination, I put my back against the landing side of the first fence and take two paces to the spot where the horse is likely to land – which is 6 feet (1.83 metres) away from the base of the obstacle. Another four paces will be equal to one stride for the horse; if that brings me 6 feet (two paces) away from the next fence, it will probably be a fairly simple double with one non-jumping stride between the two elements.

I then have to consider all the other factors, including whether my horse has a shorter or longer than average stride. The distance is also affected by the type of obstacle, the terrain and the going. If you have two spread fences, the correct distance in the double will be shorter than between two verticals. This is because the highest point of the horse's parabola in a double of spreads will be before the rail on the landing side of the first fence and after the one on the take-off side of the next. If it is jumped downhill the horse's strides will be longer; if uphill or on heavy going, the strides will be shorter.

If all the fences seem jumpable, you will need to complete your course walk by going through the finish and beyond it so that you can decide where to pull up. You need to consider where the other horses are likely to be milling around and you have to take the location of the horseboxes into account, because your partner will probably be keen to get back to his mobile home. You will risk damaging the horse by braking too quickly; failure to pull up within the space available can be even more perilous, as I know to my cost.

I once came flying through the finish in a novice event on Beneficial, only to find that the path on the left, where I had planned to pull up, was full of people – and the right-hand alternative was obstructed by vehicles. I was forced to go straight on towards a rope, which Ben failed to see until the last moment. It tripped him on to a gravel road, where he scraped the top off both knees. My mother had to call the vet while I was taken to hospital, where they discovered a hairline fracture of my left wrist.

We knew that the danger to Ben (and yours truly) could have been much worse, but it was still upsetting to see a brave young horse hurt – perhaps psychologically as well as physically. The wounds took a long time to heal and Ben, then a five-year-old, was sidelined for the rest of the year.

TIMETABLE

Having phoned for your starting times, you will need to work out your timetable for the day. We always aim to arrive at least one and a half hours before our first horse does his dressage test, and earlier if he needs to be worked in for more than an hour or if I need to walk the cross-country and/or the show jumping before I start. We allow an extra half-hour for the journey, in case we run into fog or a traffic jam or have a flat tyre.

Unless the horse is gassy and needs more work, I plan to be on him one hour before his dressage test. Assuming that he is doing the other two phases on the same day, I get on half an hour before the show jumping and am at the start of the cross-country, ready to warm up, twenty minutes before we are due to go. The time I get mounted for this final phase will depend on how far I have to ride to get to the start.

If (as occasionally happens) the dressage is the day before, I allow one hour to warm up for the show jumping. If both dressage and jumping are on a different day the horse needs at least one hour to warm up for the cross-country.

We are lucky enough to have the girls to help us, and we always write out a timetable for them so that they know when each horse has to be tacked up and ready for each phase. I no longer need one myself, because events have become so much a part of my routine. If you are competing in your first one-day event, it would certainly help to write a timetable incorporating details from the next chapter.

THE EVENING BEFORE

If you are due to make an early start, the horse can be plaited up the night before. You can also polish your boots and begin loading the lorry. The more you can do at this stage, the less you will have to worry about in the morning. I always have to think ahead in order to avoid any sense of panicking against time, otherwise I find it impossible to be calm and relaxed.

SUMMARY

Read check list at start of chapter.
Make sure you have all the necessary equipment, in good condition.

Walking cross-country course:
Be aware of your own and your horse's Achilles' heel.
Be prepared to take slow alternative routes (if available) at fences that worry you.
Think about each fence positively.
Consider whether some fences should be approached in trot, so that the horse has a chance to see what he is jumping.
If in doubt, ask a more experienced rider for advice.
If the horse is likely to need a reminder from the stick, plan to give it on the final stride.
Check distances in combinations, remembering that the average horse's stride is 12 feet. Allow 6 feet from fence for landing and another 6 feet for take-off.
Take account of your own horse's length of stride and the terrain. He will take longer strides downhill and shorter ones uphill or on heavy going.
Remember that the distance needs to be slightly longer between verticals than between spreads.
Walk through the finish and decide where to pull up, remembering that you will risk injury by braking too quickly.

Timetable:
Aim to arrive at least one and a half hours before the horse does his dressage (or earlier if he needs extra work before his test).
Allow an extra half-hour for hold-ups on the journey.
Aim to be on the horse at least one hour before the dressage.
Assuming you are riding all phases on the same day, allow half an hour to warm up for the show jumping and 20 minutes for the cross-country, plus the time it will take to walk to the start.
If dressage is the previous day, allow one hour to warm up for the show jumping.
If both dressage and jumping are the previous day, allow at least one hour to warm up for the cross-country.

11

THE DAY OF THE EVENT

Our horses often breakfast early on the day of the competition. The actual time obviously depends on the length of the journey and their starting times; but if we are leaving home at seven o'clock, the morning feed would be given at six o'clock. Most horses finish eating within thirty minutes, which leaves the other half-hour to get them ready for the journey and load them into the horsebox. We prefer them to eat in their own stable, but if we have to leave much earlier they are fed in the lorry, where we are lucky enough to have the facilities to feed en route.

The remaining hard feeds have to fit in with the horse's starting times. He is not fed less than one and a half hours before being ridden or less than three and a half to four hours since his last feed. When he has finished the cross-country, he has to wait for an hour before he gets another meal. The hay ration is reduced until after he has completed the cross-country, but if the travelling time is two hours or more, he may be given a small hay-net of 3-4 lb to eat on the way there assuming that he was going fairly late across country.

CLOTHES FOR TRAVELLING

The horse has to be dressed for the journey. Our eventers wear a headcollar (with rope for tying) and long travelling boots, which are shaped to cover part of the knees and the hocks as well as completely encasing the coronets. We also use tail bandages (which should not be wound tightly) or tail guards. Clothing, if required, depends on the individual horse and the prevailing temperature. Horses are like people; some shiver at the first hint of a chill breeze, while others seem to stay warm in Arctic conditions. We use only light rugs in the lorry, but weather conditions will determine whether the horse is to wear a sweat sheet, a porous thermal rug or, perhaps, one made of light towelling.

The horse is ready to travel. He is wearing leg protectors, roller and tail guard

ON ARRIVAL

One member of the Ivyleaze team has to go in search of the secretary's office to collect our numbers. If the office is a fair distance from the lorry park, this could take a quarter of an hour – but I am usually lucky enough to have somebody else to do it for me. I may have to walk the show jumping (and the cross-country if I have not already done so) before I get myself togged out for dressage and the horse is tacked up.

We always put on the bridle before unloading. Once out of the lorry, the travelling boots are removed and replaced by exercise boots. The horse is then saddled and studs suitable for the ground conditions are screwed into his shoes. If he needs extra work to calm him down before the dressage, we will have allowed time for him to be lunged for half an hour before I get on. He is likely to be far more excited than when being lunged at home, but he is still not supposed to kick and buck at the end of the line.

Most horses that behave well at home will continue to do so when lunged at an event. Occasionally, however, one of them lights up and has to be worked with slightly less restraint for maybe five or ten minutes, simply to avoid a battle which would create tension. As an alternative to lungeing a gassy horse, you could put on a cavesson and lead him around – for as long as an hour if you feel like it. He can look at all the sights until he loses interest in them. Boredom is often the best way of getting an over-excited horse to calm down.

DRESSAGE

Though I am on the horse for an hour before his test, only half that time is spent in serious work. The rest is for mooching around on a loose rein, so that the horse can absorb all the sights and sounds. Meanwhile I would need to know who is riding the test before mine (because we do not always go in numerical order), and I would make sure I knew how to get into the arena, which is often roped off. I also keep an eye on the arena to see if there is any dip in the ground and to check how close the judge's car is parked. I might trot past a parked car as part of my schooling if the horse tends to be a bit spooky.

During the half-hour of serious work I will be trying to achieve smooth transitions, correct shape, calmness, suppleness and obedience. I practise some of the movements I will be doing in the test, but not in the same order. Before I go into the arena, the horse's boots are removed and he is quickly smartened up. I get rid of my whip, so that I will not be eliminated for riding into the arena with it still in my hand. There may be a steward responsible for checking the horse's bit at this stage, which I will have found out in advance. I will also have made myself known to the dressage steward.

While I am riding the test, I am always thinking of the next movement – as well as my aids and whether the horse is straight and in rhythm. I have to keep thinking ahead so that I can prepare him for transitions, normally by giving some signal three strides before. The actual signal would depend on the animal. For downward transitions, it might mean sitting a little deeper in the saddle and putting my leg on the horse for the last three strides, but only if I were on one that did not mind me doing this. You have to know your horse and be flexible. If I am going to canter, I might prepare him by leaning a little on the inside seat-bone three strides earlier.

The smallness of the arena often comes as a shock to both horse and rider. Everything seems to happen far more quickly than at home, where you might start with two complete circles of trot before your next movement. Transitions and changes of direction come far more rapidly in the test. You may also have the added problem of an inattentive horse, who is keener on looking at the strange surroundings than listening to you. You nevertheless have to ride him as he is on the day; it is easy (but unhelpful) to flap if he misbehaves or refuses to listen. You are not in the arena for a schooling session, so you have to carry on and do the best you can.

BETWEEN PHASES

We never leave a horse tied to the lorry unless someone will be there to keep an eye on him, otherwise he goes back into the vehicle, which is where he is normally fed. Wherever he is left to relax between phases, he must be tied short enough to prevent accidents. We once saw a horse attempting to get out of a lorry through the groom's door, which might have had catastrophic consequences. If you tie the horse on a long rope outside the vehicle so that he can eat grass, you run the risk of his legs becoming entangled in it. There have even been cases of horses sustaining rope burns on their pasterns as a result. We always tie to a loop of string, rather than directly on to the ring at the side of the horsebox; if the horse runs back, he will then break the string rather than injuring himself or ruining his headcollar.

Injuries can happen in the bat of an eyelid. I remember holding Priceless at an event while someone was getting a bucket out of one of the cupboards on the side of the lorry. Because he was feeling itchy, he suddenly began rubbing himself up and down on the edge of the open cupboard, and before I could stop him, he had cut his lip. Had it been a quarter of an inch higher, the cut would have been in the corner of his mouth and he would have missed the Olympic Games.

Keeping things tidy throughout the day – and getting everything ready in advance – will avoid much unnecessary hassle. Unless the ground is too muddy, anything left out can be put under the ramp or beneath the vehicle to keep dry if it rains. You should always be aware of the weather; if it is cold and blustery, you will need to keep checking that the horse is warm enough and put on an extra rug if necessary. Water buckets – one for drinking and the other for washing off bits and sponging down the horse – should be filled in advance so that they are ready when required. The same applies to hay-nets.

SHOW JUMPING

Most novice horses have a tendency to sidle towards home (which means the exit from the ring and the horsebox while away at an event) so I would bear this in mind when walking the show-jumping course. Otherwise, if I am turning away from home I might find that I am 6 feet off my planned route; the horse therefore has to jump the fence at an angle and is more likely to have it down. If I decided to turn a little earlier, I could probably avoid that problem.

You need to know your horse inside out so that you can consider all his foibles while you are inspecting the fences. Perhaps he is more difficult to turn one way than the other, which is quite usual with novices; if so, that has to be taken into account when deciding where you will make your turns. There will be at least one double on the course, where you can measure the distance by taking 1-yard strides between the two elements – as with cross-country combinations.

Clearing a show jumping fence in good style

Maybe the double consists of a parallel into an upright. If you are on a big forward-thinking horse – or one that tends to balloon over parallels – you should be aware that you may need to make adjustments on the approach to avoid getting too close to the upright. On a green and slightly stuffy horse, you might have to approach it with a little more pace. You also need to consider the length of your horse's stride – and his eagerness to get on with the job – to know the pace at which you will need to approach each jump.

My course walk would also take account of the going, whether the ground was flat or sloping, the location of the practice fences and of the start and finish. Studs are selected to suit the going in the arena rather than the practice area – unless it is like concrete outside and a quagmire inside, which is hardly likely to happen. The large studs we use for soft going would cause severe jarring on hard ground.

If possible, I like to watch a couple of horses jumping before I get mounted. It may reveal some aspect I had missed when walking the course; it also gives me a chance to hear the starting signal – and to know whether it happens to be a bell, hooter, whistle or whatever. I have good reason to be diligent over such things, having once been eliminated for starting before the signal was given.

Assuming I had done the dressage (and working in beforehand) that morning, I would get on the horse when twelve to fourteen riders are left to go before it is my turn. By now the horse will be tacked up in his jumping and cross-country saddle and bridle, numnah, breastplate, surcingle, martingale attachment (if worn), half-boots and over-reach boots. I will be dressed in accordance with the rulebook, carrying a stick no longer than the allowed length, wearing the right hat and (I hope) the right number. If you are riding more than one horse, it is all too easy to forget to change the number because it is out of sight.

I always have someone to help me in the collecting ring. Though you can ask a bystander for help with the practice fence, it puts you under far more pressure if you don't have your own assistant to move the poles to the height you want them. It is also useful to have a rug brought to the collecting ring when it is cold or wet, and a fly spray when the weather is hot – together with a hoofpick in your assistant's pocket.

It is courteous to make yourself known to the collecting ring steward, who will in turn help by letting you know how many horses are in front of you. I have my first jump over the practice fence when there are six horses left to go. It may be tempting to start earlier, but it would be a mistake. You end up either jumping too much or losing the benefit of this brief sharpening exercise. I would normally have four practice jumps at this stage (all taken at canter) – over a vertical at about 3 feet and then at 3 feet 3 inches, followed by an ascending oxer (with the back rail higher than the front) and a square oxer. If all went well, I would then walk round on a loose rein until the second-last rider in front of me had nearly finished the course. Then I would jump the vertical – once only if the horse cleared it and a second time if he made a mistake. A vertical is better than a spread fence for getting the horse sharp and quick on his feet.

Once through the start, you have to keep thinking ahead. As you land from one fence, you will need to be thinking about where to turn in order to get a good approach to the next obstacle. Having landed over the last, you must be sure to go through the finish. That may sound obvious, but I have seen people miss going between the electronic timers – usually because the horse has swerved to one side.

CROSS-COUNTRY

Before the cross-country, I need to put on my back protector and jersey, maybe change my number and spurs, and make sure I have a pair of clean, dry string gloves ready. At this point I always remove my hunting-tie pin for safety reasons; I have heard of people piercing their windpipes with a pin when falling on the cross-country.

The horse then has to be booted and bandaged. I always do my own bandaging; I regard this (and the checking of tack and equipment) as the rider's responsibility. When all this is done, I sit down on my own with the programme for a last mental run through the course. Whether I am riding a novice or advanced horse, I always go through it fence by fence, remembering any markers I have noticed and the lines I plan to take.

Someone will have checked that the cross-country is running to time. It may be cold and wet; if so, the last thing I want is to arrive at the start a quarter of an hour early and end up with a frozen horse. I always ride to the start at walk, which could take two or ten minutes, so I need to know how far away it is in advance. My twenty-minute warm-up consists mainly of cantering, with some trotting and one sharp burst at a faster speed. Then I usually have a jump or two, though this might be omitted if the show jumping has gone well (and was on the same day) or if the practice fence was

Jumping a lane crossing at a novice one-day event. A good position over the first obstacle, showing a closed leg and contact with the horse's mouth as the rider sits up and waits for the landing. The lower leg has slipped back a little on landing but both leg and hand positions are good as the horse takes one non-jumping stride. The lower leg is a little too far back at the second take-off but the hips could be further back

unsuitable – perhaps because it was flimsy or the ground was bad. If I had done both the dressage and show jumping the previous day, it would obviously be necessary to extend the warming-up time. In this case I would have to give the horse four or five jumps to get him stoked up for the cross-country.

After warming up, the girths have to be checked before I walk quietly round, with a rug on the horse's quarters and a jacket on myself if the weather is cold or wet.

I never wear a stopwatch at one-day events, whether riding in novice or advanced classes. I prefer to go by the feel the horse is giving me; I want him to stay within a rhythm and maintain a pace at which he feels comfortable. You quite often see young horses being hurried out of this rhythm – and thereby ruined – at novice one-day events. They do not necessarily achieve the fastest times by going at breakneck speed between fences, because they are more or less obliged to slow down and show jump each obstacle – thus defeating the object of the exercise.

If the horse has been taught to go in a rhythmic way across country, the miles per hour can be increased as he gains experience. We have had horses whose times proved slower than they felt at the first few events. Then, slowly but surely, they began to get faster without my being aware of it. With one particular horse, I knew that I had made slightly shorter turns and had taken less time to set him up for certain fences, because he was becoming more educated. I was nevertheless surprised to learn how much faster we had gone. The rhythm felt exactly the same; it was only the miles per hour that had changed because he had gained enough confidence to maintain a faster pace.

I always aim to set off across country as though I mean business. The first two fences are usually straightforward, and if you can ride them in a positive way and get two good jumps, the horse will be given an automatic boost to his confidence. I aim to keep thinking ahead to the next fence, but there is a tendency for me to harp back to the previous obstacle if I had a bad jump. That usually results in the next one being just as bad. Having acknowledged the first mistake, you have to try to put it behind you and be positive about the next fence.

Forward thinking on the cross-country course will include remembering the lines you planned to take. You will also be concentrating on keeping the horse in rhythm, watching out for stray spectators and for any steward waving an emergency flag to indicate that you must pull up. Spectators are always a worry at one-day events; if I am riding towards a bend in a wooded track, I might shout something (such as 'coming round') because often people fail to realise that a horse is galloping towards them. Events would not be the same without spectators, so we should do our best to avoid mowing them down!

Having completed the course, it is essential to pull up correctly. Your horse will be in serious danger of breaking down if you kick on through the finish, then drop the reins and pat him while he comes back into trot. That is when he starts propping, applying the brakes with his forelegs and causing them undue stress. The same risks occur when you are pulling up downhill or when making a sharp turn. After urging the horse through the finish, you should sit quietly and ease him back on the bridle – on a straight line or a gentle turn.

I dismount as soon as the horse has stopped, loosen the girths, surcingle and noseband, run up the stirrups and lead him back to the horsebox. He should be fit and therefore not blowing too badly, but I will need to time his recovery rate (as we do after canter work-outs) to get a more accurate idea as to how his fitness is going. We do this with every horse at every event.

HORSE CARE

There is still plenty to do when the horse gets back to the lorry after completing the cross-country. The Ivyleaze routine goes something like this:

Remove saddle.
Put on sweat sheet or rug, depending on weather.
Walk horse round until he has stopped blowing.
Remove boots and bandages.
Check legs for injuries – if cut or scratched, wash with warm water and apply cleansing agent.
In hot weather, sponge horse down with cold water.
Wash neck (whatever the weather) if horse has sweated profusely.
In cold weather, wipe over the rest of the horse with damp sponge.
Scrape off excess water.
Dry ears with towel (the type used to dry hands rather than tea-cups).
Wipe eyes and nose with another sponge or cloth.
Pick out feet and remove studs.
Apply ointment or poultice if necessary.
Offer small amount of water.
Prepare horse for journey by putting on travelling boots, etc.
Feed small hay-net.

The horse is given his hard feed one to one and a half hours after he has completed the cross-country. If he is the last of our horses to go and the journey is less than two hours, he would probably wait to be fed at home in the peace and quiet of his own stable. On longer journeys we give all the horses a hay-net of 4-5 lb to eat while travelling and a hard feed to whichever horse is due for one. If we did not have mangers built into the lorry, we would be unlikely to attempt some makeshift arrangement, which usually results in food being tipped on to the floor and wasted; instead, the hard feed would wait until our return home. On journeys lasting three hours, we make one stop to offer the horses some water. We also check their temperature every quarter of an hour, by our usual method of putting a hand under the rug to feel their shoulders and loins to make sure they are neither too hot nor too cold. Once back at home, the horse's legs are checked again for any sign of swelling, bruising or any other type of injury, and appropriate action is taken if necessary.

SUMMARY

General:
Arrive at least one and a half hours before dressage test.
Have correct clothes, tack and equipment for each phase.
Remove hunting tie pin before cross-country.

Feeding:
Hay ration reduced until after cross-country.
No hard feed less than one and a half hours before being ridden.
Allow three and a half to four hours between hard feeds.

Dressage:
Ride for one hour before test, using about half an hour for dressage schooling.
If extra time required to calm horse down, lunge for half an hour or lead in hand for up to one hour.
During half-hour of dressage schooling, work on achieving smooth transitions, correct shape, calmness, suppleness and obedience.
Make yourself known to the dressage steward.
Keep protective boots on until last minute.
Remember to get rid of whip before entering arena.
During test, keep thinking ahead to next movement.
Prepare for transitions three strides in advance.
If horse is difficult, stay calm.

Show jumping:
During course walk, decide on pace required and where to turn into fences.
If possible, watch others jumping before getting mounted.
Start riding about half an hour before your turn (or one hour if dressage test ridden the previous day).
Make yourself known to collecting-ring steward.
Jump practice fence (about four times) when six horses still to go before your turn.
Keep horse moving (walk on loose rein when not jumping).
Final practice jump(s) when the second-last rider before you has almost completed the course.
During round, keep thinking ahead to the next fence.
Be sure to go through finish.

Cross-country:
Have last mental run through course before mounting.
Allow time to ride horse to start (at walk), leaving twenty minutes for warm-up on arrival.
If dressage and show jumping ridden the previous day, allow one hour for warm-up.
Set off to ride all cross-country fences in a positive way.
Keep thinking ahead to next fence.
After completing course, avoid pulling up too quickly.
Dismount when horse has stopped.
Loosen girths, surcingle and noseband, and run stirrup irons up leathers.
Time the recovery rate.

Horse care:
See details opposite.

12

FROM NOVICE TO INTERMEDIATE

The horse will gradually be able to start tackling novice courses at a faster pace, taking the quick routes at fences that offer alternatives. You might let him tackle a corner at his first event, assuming that it is an easy one, but I would be more likely to wait until his third outing – and I would not try a corner then unless I had complete confidence that he would stay straight and be obedient.

If the horse is competing on consecutive weekends, there is no necessity for a mid-week canter work-out. His normal programme for the week might then be as follows:

Monday	rest day
Tuesday	light hack
Wednesday	hack and dressage
Thursday	hillwork
Friday	hack and jumping lesson
Saturday	flatwork and short hack
Sunday	competition

When planning your own programme, you need to bear in mind that one-day events improve the horse's fitness quite dramatically. Once he has begun competing, you probably need do no more than keep him ticking over. Even when the horse is aiming for a three-day event, I would not canter between one-day competitions that are on consecutive weekends. The horse needs to be given four easy days after an event; you should also avoid any strenuous work after his rest day.

If the horse began competing as a five-year-old, I will not consider riding him in an intermediate class until the following year. He will have his winter rest, go through the process of getting fit again and start the spring season with maybe five novice events, including one open novice. The number of outings will depend on his physique and the going; he would certainly do no more than three events on consecutive weekends – and he would do those (at a steady pace) only if the going was very good. The decision as to whether or not to attempt an easy intermediate class at the end of the spring season

Good and bad halts. The incorrect halt (above) *shows the horse overbent and not standing four square*

will depend on how well he is going across country. Once again you have to listen to the horse.

NEW MOVEMENTS

We aim to teach the new movements required in intermediate tests about six months before the horse performs them in competition. This obviously requires a bit of guesswork, but nothing has been lost if we get it wrong and he is not ready to tackle a more difficult cross-country course in six months' time. In some cases, if the horse is straight and doing correct turns and circles, I might start shoulder-in at walk at least a year earlier. It is only because we are training a pentathlete, instead of a specialist in one discipline only, that we often need to bide our time before introducing new movements. In eventing we always have to be careful to avoid asking the horse for too much at once.

If you ask for a lateral bend to the left while in halt, the horse will invariably fall in by moving his shoulders in that direction. That is why you need your weight and your legs to achieve lateral flexion without shoulder movement. Gymnastic exercises, such as shoulder-in and travers, will help to improve the horse only if you are secure and balanced within the aid. If you are crooked, he will become so as well – in the process losing his flexion, tempo and impulsion.

Shoulder-in at walk is the first new lateral movement the horse learns after leg yielding. It requires him to move on three or four tracks, with his inside hindfoot following in (or close to) the track of his outside forefoot. It is useful in helping to establish control over the horse's forehand and quarters, as well as encouraging him to use his hocks and become more supple.

The best way to start shoulder-in is from a circle or corner, because you will already have the correct bend through the horse's body from head to tail. When the horse returns to the side of the arena from a circle, he can be asked to continue on the same curve until his inside shoulder is off the track as if starting another circle. He is then in the correct position to perform shoulder-in along the side of the school, with the whole length of his body flexed around the rider's inside leg and bent away from the direction in which he is moving.

The rider maintains the same flexion by use of the inside rein, while the outside rein stops the horse continuing on the circle and corrects excessive bending. The inside leg is applied on the girth, asking him to move his forehand forward and away from it. The outside leg, used behind the girth, maintains impulsion and prevents the quarters from swinging outwards. Do not ask for more than a few steps at a time when the horse is first learning this movement. It requires him to make quite an effort, since he has to bring his inside hindleg well forward under his body in order to carry his own and his rider's weight during this exercise.

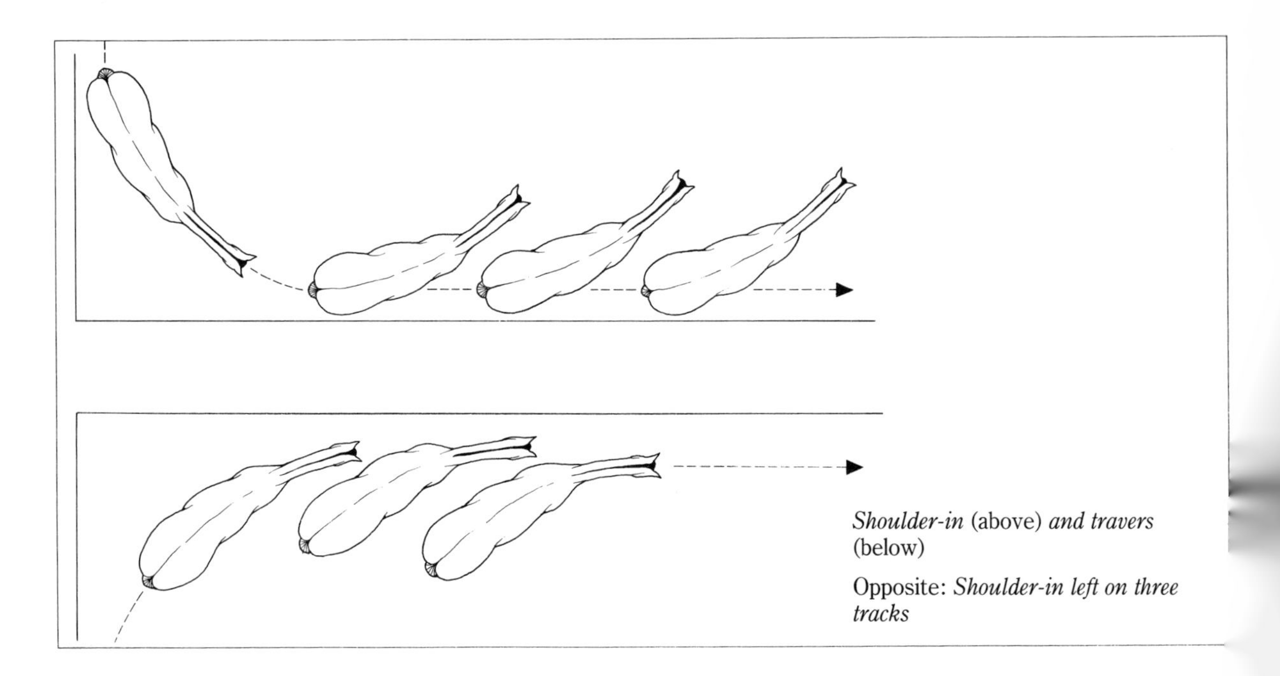

Shoulder-in (above) *and travers* (below)

Opposite: *Shoulder-in left on three tracks*

The horse should maintain his rhythm and impulsion, without leaning to the outside of the arena. Though his forelegs cross over during this exercise, his hindlegs should not do so; when he crosses his hindlegs, he is avoiding making the effort required from his inside hock. This is more likely to occur when the rider asks for too much bend or too wide an angle; the horse's hindlegs should virtually remain on the track they would normally take if he were moving straight ahead. Having completed the shoulder-in steps, the horse's forehand should be brought back to the track so that he moves straight ahead rather than continuing with the same bend on a circle. Otherwise he will learn to anticipate circling at the end of this movement. Once he understands what is required of him, the horse can be asked to perform the same exercise at trot.

Travers (or quarters-in) is a useful exercise although not included in eventing dressage. It helps to increase the horse's attentiveness to the rider's legs and gives improved suppleness in his rib cage. In this movement the horse's quarters are off the track and his hindlegs cross over, whereas his forelegs do not. He is bent slightly around the rider's inside leg, towards the direction in which he is going.

This exercise can also begin with a circle, which will put the horse into the correct position to start the movement. Instead of straightening his quarters as you complete the circle, they should be kept inside the

Left: *Shoulder-in right on four tracks*

Below: *Travers right*

track by the rider's outside leg, applied behind the girth, so that they follow on a different track from the forehand. The inside leg is used on or near the girth to keep the horse moving forward, while maintaining impulsion and bend. The inside rein keeps the horse moving in the correct direction, while the outside rein controls his speed and the amount of bend. The rider should put extra weight on the inside seat-bone.

You should always bring the horse's hindlegs back on to the track before reaching a corner, and you must avoid asking the horse to perform this exercise for too long, otherwise he will become a little too free with the unaccustomed mobility of his hindquarters and may be more difficult to keep straight in his basic work. Travers will probably be easier on the left rein, because of the tendency in most horses to prefer taking contact with the left hand. As with shoulder-in, we ask for this movement in walk first and then trot.

The *half-pass* would normally be attempted at walk only while the horse is still a novice. It requires him to move diagonally, so that he goes forwards and sideways at the same time, with each outside leg crossing over in front of its inside pair to make four separate tracks. You can get the correct position for half-pass in the same way as you would for shoulder-in – that is, by riding a circle and continuing on the same curve after returning to the side of the arena until the horse's inside shoulder is off the track. Thereafter the two movements take the horse in different directions. For half-pass he moves across the school on the diagonal, with his body bent slightly towards the direction in which he is going. Impulsion and forward rhythm must be maintained.

The rider's outside leg drives the horse sideways, while the inside leg asks for forward movement and impulsion. The inside rein asks for the correct bend, and the outside rein controls the forward movement and, if necessary, corrects the amount of bend. You should be looking the way the horse is going, with shoulders turned slightly in that direction and a little more weight on the inside seat-bone, with inside hip forward.

The movement is incorrect if the quarters are leading. In other words, for right half-pass, which is normally the easier direction, the horse's forehand should be slightly further to the right and in advance of his quarters. His head should not be tilted; this fault usually occurs because the horse has not been taught to stay straight and maintain an even contact on both reins in his basic flat work.

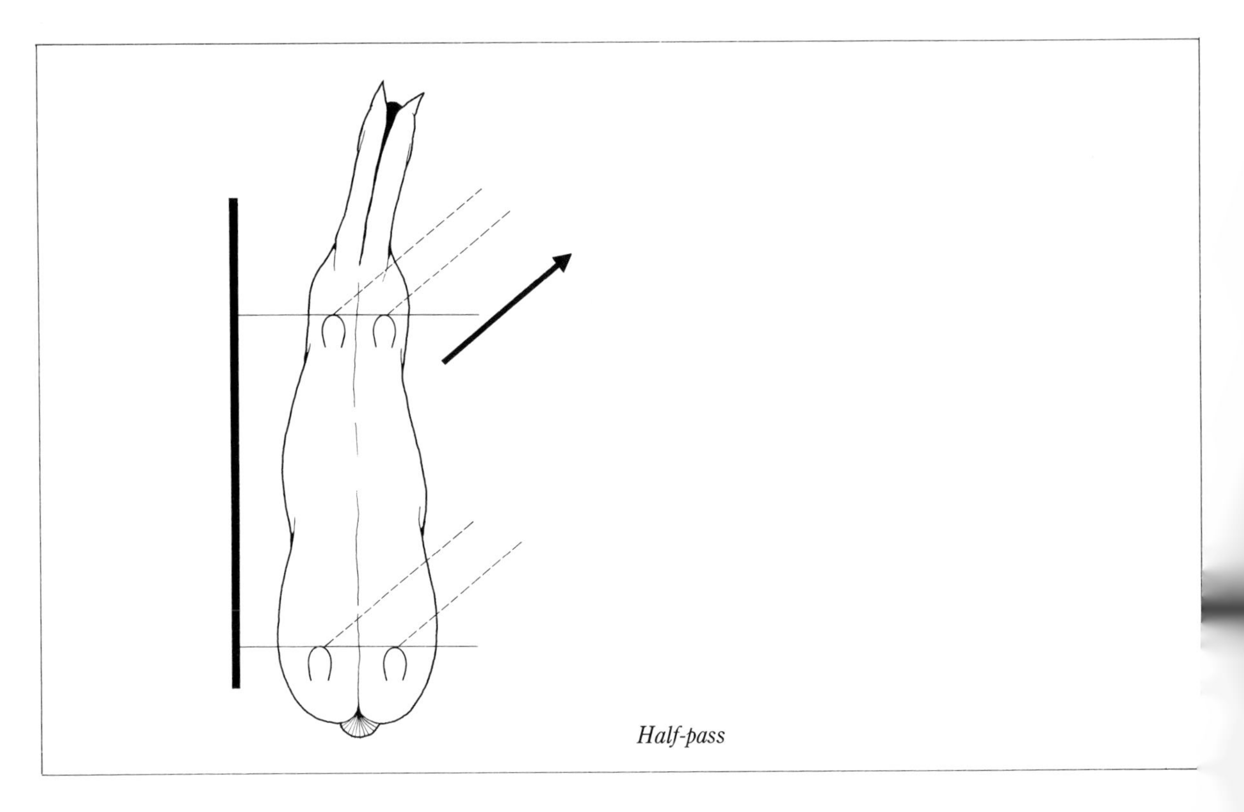

Half-pass

Medium paces will be required in trot and canter. By now the horse will have learnt to shorten and lengthen within both; in the process he will have taken some medium strides. They are longer than those used for working paces, but shorter than those required for the more advanced extended work. We will be looking for a rounder outline and improved impulsion from the hindlegs, which have to be properly engaged before the horse is capable of medium paces. If his hocks are not underneath his body but are trailing behind him, he will lose his balance and rhythm, move on to his forehand and produce quicker (rather than longer) strides. It is the elevation within these strides that gives him the necessary time to lengthen. The rider has to activate the horse's hindlegs to stoke up the engine before asking for medium paces. Once achieved you should allow a little with the hand without losing contact.

Transitions become more demanding at intermediate level. Halt to trot and walk to canter are now included in the upward transitions and trot to halt in the downward ones. They require more energy than the simple transitions used in novice tests, but the principle remains the same. It is important to alert the horse, by using one or two half-halts, so that he knows something different is about to be required of him. He should then be attentive to your aids given through your legs, weight and use of the retarding rein. He is therefore aware of the slight increase in emphasis that tells him, for instance, that you want him to come back from trot to halt rather than trot to walk.

The *simple change* is not so much a new movement as taking less time for something the horse has done already. He has to canter on part of a circle with the inside leg leading, come back through trot to walk for a few strides as he changes the diagonal and then strike off into canter with the opposite leg leading. He has already learnt the aids for striking off on one particular leg, but will obviously need extra impulsion to do this after the transition down to an active walk.

He will also have to learn *five-metre loops* at canter, this time without changing legs. This movement is an easier version of the counter-canter which is required at advanced level and is the first time the horse is asked to canter on a curve with the outside leg leading. We teach this in easy stages, moving only 2.5 metres off the outside track at first so that the curve is gentle. The horse should be bent in the opposite direction to the curve he is following. The position of both horse and rider remains on the leading leg as regards aids and flexion. The horse is encouraged to change direction with the outside rein and inside leg. You have to take care that he stays true to the bend and neither falls in with the shoulder nor swings out with his quarters.

Rein-back is also required for intermediate tests. We are inclined to leave this until the last of the new movements to be learnt, because we always want the horse to be thinking of going forward and it can therefore work against us; he can also use it as a form of resistance. We teach rein-back by having someone on the ground who can help to explain exactly what the rider is asking the horse to do. Without such assistance, he is likely to become confused. It is when he fails to understand what is required of him that he is likely to start rearing.

The rider gives the voice command 'back', and the usual aids. This means moving your weight slightly forward to give the horse's back full freedom, applying both legs on the girth and using your hands, with a give and take, to restrain any forward movement. On no account should the horse be pulled backwards.

The person on the ground can reinforce these aids by pushing with a hand in front of the horse's shoulder or, maybe, giving a light tap with a stick in the same area. He is likely to put his head up like a periscope when first asked to back and it helps to be aware of that. If he takes only one step back, make much of him before asking for another one. It will help him to know exactly what you want when these aids are applied and you will soon be able to dispense with the voice command. As the horse becomes more familiar with the movement the aids will gradually become less obvious.

DECIDING ON THE FIRST INTERMEDIATE CLASS

It is easy to persuade yourself that a horse is ready for his first intermediate after he has done, say, eleven novice events. In theory that is probably right, but it does not necessarily work out in practice. I prefer to take no notice of how many or how few events the horse has contested and concentrate instead on the feel he gives me over cross-country fences. Unless he collects the points that upgrade him too quickly, which can be a big problem, the decision to enter for an intermediate will depend on his confidence and character, not miles on the clock.

k Savings

k Savings

Teaching the horse to rein back using poles on the ground to keep him straight. The handler encourages him to move backwards by tapping him lightly on the shoulder

We like our horses to do their first intermediate before they upgrade; if they find the bigger fences slightly unnerving, you can then restore their confidence by going back to novice events. The decision to tackle an intermediate has to be based on each horse's confidence across country. You can ride the dressage test without doing him any harm and the extra two inches on the show jumps are hardly going to give him the heebie-jeebies, but if he makes a mistake across country you would have good reason to feel depressed. A stop would mean that he now knows how to refuse; a fall could seriously undermine his confidence.

I am careful to choose one of the smaller intermediate courses, with plenty of alternative routes, for the horse's first attempt at this level. I would be unlikely to consider an open intermediate, since it caters for all grades, including advanced, and is usually more difficult.

Canter work-outs

Because the intermediate course is longer, the horse needs to be slightly fitter than he was for novice one-day events. When training on flat ground, he might therefore move up from three five-minute canters to three of six minutes, still with a break after each one. If you are able to use an uphill incline for work-outs, it may be more effective to go from three six-minute canters to two of seven minutes. The actual programme needs to be adapted to the individual horse as well as the terrain; some get fitter on three canters and two breaks, while others are better suited by two canters and one break.

I rely as much on the horse's breathing as I do on my stopwatch when deciding on the length of these breaks, which would never be less than one minute or more than three. The actual time comes somewhere between these two parameters. If the weather is hot, the horse will be blowing harder and longer. He is therefore given a slightly longer rest than he would have on a cold day.

It continues to be important to time the recovery rate and be flexible with one's programme. You still do not want the horse to be three-day-event fit for a one-day test, nor do you want him to be struggling in any way at the end of the cross-country. If the recovery rate is the same for the longer canter work-outs, he is probably at the right level of fitness for an intermediate one-day event. Once into intermediates, he can be kept ticking over with just two seven-minute canters (with the usual break between) on weekends when he is not competing.

The horse responds to the rider's aids, remaining straight as he moves back between the poles

Jumping

The horse will be jumping considerably faster over novice cross-country courses than in the show-jumping arena. This extra speed results in a flatter jump; we therefore need to regain his natural bascule. This is done through gymnastic exercises, which also improve his suppleness and the elasticity of his muscles. The grids described in Chapter 9 – and your own variations on the same theme – will be of particular value to the long-striding horse. The shorter-striding and more athletic type would probably derive greater benefit from canter combinations on varying distances.

A combination suitable for a horse that is still at novice level would have three fences that gradually increased in height. The first might be at 3 feet 3 inches (0.97 metre), with the next two at 3 feet 6 inches (1.07 metre) and 3 feet 9 inches (1.15 metre). A distance of 26 feet (7.9 metres) would require a longish non-jumping canter stride between elements. As the horse gains experience, the heights and distances can be varied so that he learns to shorten and lengthen his non-jumping strides.

Four- to six-bar jumping, over between four and six vertical fences of about 3 feet 6 inches (1.07 metres), is another good exercise. It encourages the horse to be tidy in his jumping and again, by varying the distances, to adjust the length of his stride whenever necessary.

Perhaps you have noticed some imperfections in his style. Maybe he loses power by drifting to one side as he approaches a fence, thus reducing the forward thrust that enables him to clear the obstacle easily. Placing poles, as mentioned in Chapter 5, can be used to correct this tendency – and they can be brought a little closer together so that they give the horse less room for drifting.

It could be that the horse wastes time and effort by jumping too high. Normally this means that he is failing to fold up his legs sufficiently to clear the fence from a lower parabola. His body has to go higher to compensate for his trailing limbs. He can be trained to fold up his legs, but you have to be careful that he does not lose his natural bascule and begin jumping with a hollow back. Gridwork is the best way of curing the first problem without creating a different one.

The horse that consistently jumps too high will invariably land further away from his fences than a horse

with a lower parabola. If you gradually shorten your distances within the grid, he will realise that he has to find some way to land closer to the fence he has jumped. He therefore has to jump lower. When he first experiments with this solution, he may hit the fence. The only alternative left to him is to keep the lower parabola and fold his legs away. Similar results can be achieved by putting a place pole on the ground 9-10 feet (2.7 – 3.05 metres) after the fence, so that he has to land a little shorter to get inside the pole and avoid treading on it. In order to do this, his body has to be lower; if he is to clear the fence, his legs have to be folded out of the way.

The horse has to work hard in grids and all your efforts will be undone if you ask him to carry on for too long. If you do encounter a major jumping problem, it will not be solved by continually working on it. Quite often it is much better to go hacking for a few days and then come back to jumping when the horse is mentally and physically refreshed.

His technique and confidence should be improving, through schoolwork and competitions, while he is competing at novice level. Before riding him in an intermediate event, I would aim to take him schooling over some intermediate fences. I might also ride him in some slightly more ambitious show-jumping contests, but only if he happened to be fairly exuberant by nature. You can overdo show jumping to an extent where the more you do, the less tuned up the horse becomes. Often he will be much better when fresh to this discipline, so I normally aim to have two or three show-jumping shows over a period of six weeks before eventing starts and leave it at that for the rest of the season.

Cross-country course walk

The intermediate fences will certainly be bigger than anything the horse has tackled at novice level, but they will probably contain the same ingredients as those he has jumped before. If you think carefully enough, you can normally associate each fence with something similar which the horse has already tackled in a lower or easier form. Once the association is made, you can consider how you rode the easier version – and how the horse jumped it.

Make sure your horse has sufficient confidence to take the big step up from novice to intermediate

Horses are creatures of habit and I am sure they remember jumping the mini versions of the more difficult fences they meet as they go up through the grades. If you can also relate to some past experience, you will feel more confident about jumping the fence and have a better idea as to how you should ride it. There could be a more difficult approach or landing which needs to be taken into account, so you may have to tackle it slightly differently to the one in the past which had basic similarities.

If the horse has successfully negotiated the quick routes at novice level, there is no reason why he should not do the same in his first intermediate. You should not, however, attempt to play the hero. If the course designer has a bright idea which seems horrific to me, I will certainly opt for a slower alternative on my green intermediate horse, if one is available.

THE COMPETITION

The first intermediate contest is likely to tell you whether or not your basic training has been a success. Flaws in that department tend to become more transparent as you move up through the grades, because the horse cannot perform the more difficult movements unless the foundation is right. You will also be given further clues about his courage and confidence. It is not unusual for him to be slightly unnerved by his first experience over an intermediate cross-country course, which is why I always like to have the option of going back to novice afterwards. I normally aim for an easy intermediate at the end of the spring season before the horse has upgraded, and unless he appears to be in his element over the bigger obstacles, I would take him in one or two novice events at the start of the autumn. After that I would hope that he is able to tackle more intermediates with renewed confidence.

This horse is gaining all-important confidence over one of the bigger novice courses

SUMMARY

New movements required for intermediate tests are taught six months (or more) before horse competes at this level.
Decide whether the horse is ready by the way he goes across country.
Try to compete in intermediate before horse has upgraded.
Choose small intermediate course for first attempt.
Increase time for canter work-outs to three at six minutes, with breaks between.
Continue to time the recovery rate.
Use gymnastic jumping exercises to improve suppleness and elasticity.
Work to improve jumping style.
School over intermediate cross-country fences before competing in intermediate event.
Go back to one or two novice events (unless upgraded) if horse needs to regain confidence.

Aids for new movements
Shoulder-in:
Start from turn or circle when horse's forehand has moved off the track.
Weight on inside seat-bone.
Maintain bend with inside rein.
Ask for lateral movement with outside rein.
Inside leg on girth asks horse to move away from it and go forward.
Outside leg behind girth maintains impulsion and controls quarters.
Finish by straightening and then move straight ahead.

Travers (schooling exercise):
Start from turn or circle before horse's quarters are back on the track.
Weight on inside seat-bone.
Inside leg on or near girth maintains impulsion.
Outside leg behind the girth keeps quarters off the track.
Inside rein asks for direction and flexion.
Outside rein controls speed and bend.

Half-pass (at walk only until more experienced):
Start from shoulder-in position, then move across school on the diagonal.
Slightly more weight on inside seat-bone with inside hip forward.
Inside leg on the girth for forward movement and impulsion.
Outside leg behind girth drives the horse sideways.
Inside rein asks for correct bend.
Outside rein controls speed and bend.

Medium paces (trot and canter):
Create impulsion with legs.
Prevent this energy escaping into a faster speed with give and take of hands, without loss of contact.

Five-metre loops at canter:
Horse remains bent on the side of his leading leg.
He is asked to follow a curve in opposite direction with use of rider's inside leg and outside rein.

Rein-back:
Start with assistant on ground to reinforce the aids.
Weight slightly forward.
Legs applied behind the girth.
Give and take with hands to restrain forward movement.

13

CLOTHING AND EQUIPMENT – INTERMEDIATE AND ADVANCED

By now the rider's wardrobe will include a red or black jacket (for men) and a black or blue coat (for women) in order to compete in intermediate events. There is no obligation to add anything more to your own outfit than light-coloured gloves and a pair of spurs (both of which you may already have) in order to compete at the higher levels.

You do, however, have the option of wearing a top hat and tail coat in the dressage phase of advanced one-day events and intermediate or higher levels of three-day eventing. Men have an additional expense if they exercise this option, because they have to get white breeches and black boots with mahogany tops to complete the outfit.

The top hat and tails – whether borrowed, hired or bought second-hand – are obviously not essential, but they do help the overall picture and make you feel the part. The tails may flap a little, so you might like to practise at home in your tail coat. If, however, you go hacking in a long Australian mac during wet weather, the horse will already be accustomed to feel the movement of your clothing on his flanks.

Correctly dressed for the show jumping on Griffin (above) *and for the cross-country on Master Craftsman*

BITTING

I have already mentioned the bits I would try on a novice horse if an unjointed rubber snaffle did not prove satisfactory. These are a jointed snaffle in rubber, then metal, followed by a loose-ring snaffle and a French bridoon. I would hope that one of these was a success and that I could stay with it forever, but it will not always be the case.

If I wanted more control I would probably try a Magenis or Copper D Roller, followed by a Dr Bristol. These three stronger varieties of the snaffle can be effective on a horse that pulls with his head tucked in – usually as the result of being ridden in draw reins at some stage.

The horse that pulls with his head in the air is easier to correct. In this case I would use a rubber Pelham, with a loose curb chain or one with rubber backing. The curb chain encourages the horse to bend from the poll, which automatically lowers his head. If something stronger is needed to control the horse and bring his head down, a gag could be the answer – although it does occasionally have the reverse effect and encourage the head to go higher.

Though the stronger bits may make the rider feel more secure, they can have a detrimental effect on the horse's performance across country. You want him to be forward-thinking, so that he can go out and attack the fences with confidence. If he finds the mouthpiece uncomfortable or too strong, he will be concentrating too much of his attention back to the rider's hands and too little on the fences ahead. You do not want him to be too light on the rein; he needs to be happy on a reasonable contact so that he can take you into the fences.

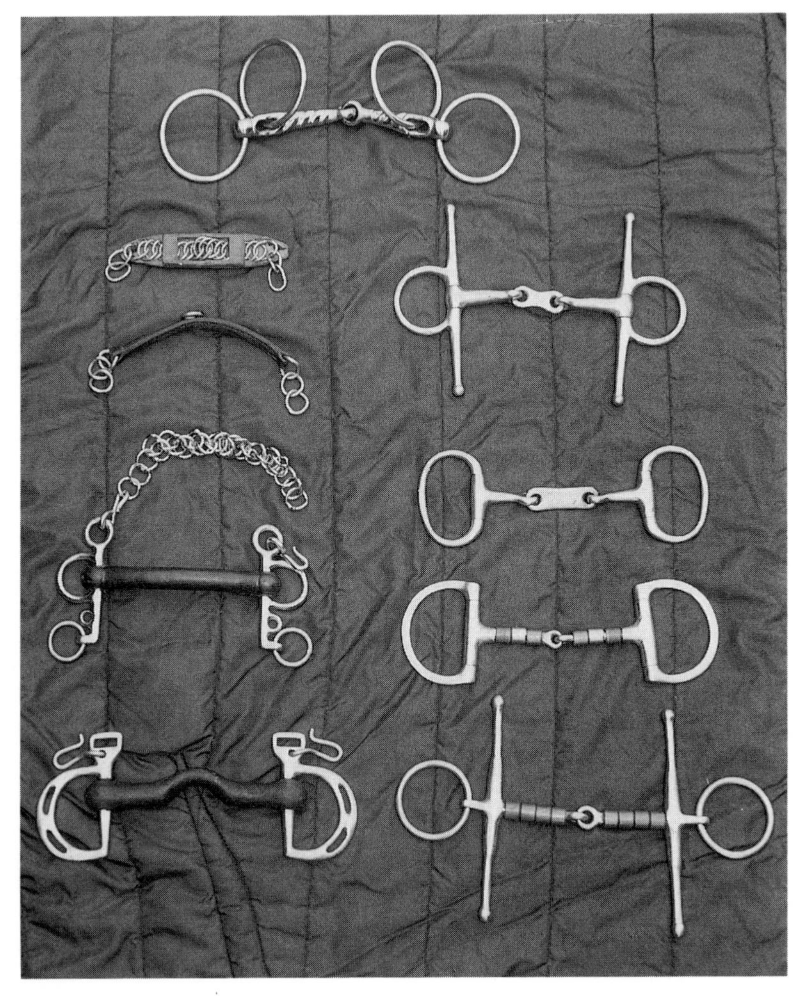

Bits and curb chains that can be effective on a horse that pulls. Top: *Scorrier.* Left (top to bottom): *rubber-backed curb chain, leather curb strap, vulcanite Pelham, vulcanite Kimblewick.* Right (top to bottom): *French bridoon, Dr Bristol, Copper D Roller, Fulmer copper roller snaffle*

Barnsby-Leng general purpose and cross-country saddles

STOPWATCH

You will need a stopwatch when you compete in three-day events, but do make sure that you can read it when you are on a horse and moving at speed. The figures may seem as clear as daylight while you are standing still, but they can turn into a useless blur when you attempt to read them during the speed, endurance and cross-country phase. A spare stopwatch will also be necessary; it will be too late to buy or borrow a replacement if the first one goes wrong on the day.

I always wear my stopwatch over my jersey and an inch or two above my wrist, where there is nothing that can accidentally stop it. Pressing the wrong button by mistake can be very costly, so practise at home to make sure that you can avoid this mishap, and that you can read the figures.

OTHER EQUIPMENT

It makes sense to double up on every essential item of equipment before competing in a three-day event. You will have spent months preparing for the contest and would find it galling, to say the least, if you had to retire because part of your tack had broken and you had no replacement.

You have the option of using a double bridle for the dressage phase of advanced events, but I have very rarely done so. It obviously gives you more control and collection, but I suspect that it might give too much. We are not looking for anything like the degree of collection required for Grand Prix dressage and, rightly or wrongly, I feel things can go wrong if I get more collection than I need by using a double bridle. The horse needs to accept the bit with confidence and be forward-

Ready to start the dressage on Night Cap. Note the rider's tail-coat and the straight front flap of the dressage saddle

thinking for his jumping.

I would prefer to carry on using a simple snaffle bridle with an ordinary cavesson noseband, but it has to be said that I have rarely achieved this beyond novice level. More often than not, I use a Flash noseband instead of the cavesson, which gives me a little more control and avoids the possibility of the horse crossing his jaw or opening his mouth.

I like to use white girths for the major three-day events. Since they are purely cosmetic and a pain to keep clean, you may prefer to spare yourself the effort and expense! A sheepskin noseband can be of practical benefit in the show jumping, as I found with Master Craftsman. He was bred to race over steeplechase fences, which could mean taking off 15 feet in front of the fences and landing another 15 feet away from them. Because of his breeding, Crafty has a tendency to bascule too late, so that the highest point of the parabola comes after the fence instead of immediately above it. The sheepskin encourages him to lower his head in order to see where he is going. This in turn encourages an earlier bascule and shorter landing.

A minimum weight is required in certain contests (including the cross-country phase of advanced one-day events) so you will need a weight cloth and lead unless you are heavy enough to make the minimum of 11 stone 11 lb (75 kilos) without them. I normally practise with the extra weight for a couple of canter work-outs, but not necessarily for the jumping – unless I have a lightweight horse and want to see how he will carry the extra pounds over fences. Usually dead weight is easier for him; unlike the live weight of the rider, it is less inclined to move and throw him off balance.

We occasionally grease the horse's legs at one-day events, depending on his style of jumping and the ground. It could do no harm – and it might do a lot of good – if you were to use it every time your horse jumps a cross-country round. Should he hit a fence, the grease will help his legs to slide over it.

Our horses always have their legs greased for three-day events. Normally we use Vaseline – but you do get the odd one who is allergic to it, so we need to check beforehand. If there is an allergy, wool fat is used instead, and applied sparingly during the ten-minute halt. There is no reason why your horse should look as though he is ready for the frying-pan! The grease (applied mainly to the stifles, inside the back legs, the knees and forearms) is a bore to remove. You will need a blunt knife and washing-up liquid to help get rid of it.

My three-day event equipment also includes an Easyboot (or similar product) which can be used on the second section of roads and tracks if the horse happens to lose a shoe on the steeplechase. It has a metal clip, rather like a ski boot, and can be fitted far more quickly than a normal shoe. A measuring wheel is another useful item of equipment to take to three-day events, but you should be able to spare yourself the expense by finding a friend who has one.

We also need a bootlace on cross-country day, but it goes nowhere near the horse's feet. It is used to secure the bridle by tying it to the top plait on his neck. If you do have a fall, it is all too easy to pull the bridle off the horse; with the bootlace to secure it, you will at least have the option of remounting and carrying on!

Leaving the start box for an advanced cross country, with the necessary weight-cloth and weights

SUMMARY

Consider whether it is necessary to practise at home in tail coat (if worn).
Keep bit as mild as possible.
If more control needed for horse that pulls with his head tucked in, one of the following bits may prove effective: Magenis, Copper D Roller or Dr Bristol.
If horse pulls with his head up, a pelham and loose curb chain or one with rubber backing may be the answer.
A gag, which is more severe, normally lowers the horse's head – but it can sometimes have the reverse effect.
Make sure you can read your stop-watch when galloping.
Double up on all essential equipment.
Practise with lead weights, if required, to achieve minimum weight.
Decide on type of grease to be used on horse's legs and make sure he is not allergic to it.

A measuring wheel is necessary for walking the steeplechase and cross-country courses. If you don't possess one, try to find a friend who does. Dot and I found such a friend in Clarissa Bleekman, who is walking the course with us

14

PREPARING FOR THE FIRST THREE-DAY EVENT

If I had only one horse to ride and everything went according to plan (which rarely happens), my youngster would do a maximum of ten novice events as a five-year-old, with more novices and two or three intermediates the following year leading up to a novice three-day event. I would want him to enjoy his first three-day test and he has a better chance of having fun over a novice course, which is the shortest and least demanding of all. It is only because the advanced horses have to take priority at certain times of the year that our novices usually gain their first three-day experience at intermediate events. We are careful to choose one with a lenient cross-country course.

This might well be a one-star CCI (the international equivalent of an intermediate three-day event), so we apply for the horse's passport almost as soon as I start thinking in that direction. It may take some time to come through from the International Equestrian Federation (FEI) and I don't want any worry about whether it will arrive on time.

A nice frame in trot, showing harmony between horse and rider

FLATWORK

I have ridden a variety of different tests at intermediate three-day events (or one-star CCIs); on one occasion, we had to do the FEI test that was used at Badminton. You will therefore need to find out – as early as possible – which test is to be used, so that you have plenty of time to practise all the movements.

The requirements of the test should not, however, be the only factor which determines when the horse is taught new movements. Generally speaking, he learns them when he is ready – which might be more than a year before he has to perform them in eventing. Meanwhile he can use them at dressage shows, which is an excellent way of giving him arena experience and monitoring his progress. Though I have left some of the more difficult movements and transitions to Chapter 16, you may decide to begin tackling them well before the horse does his first novice or intermediate three-day event. Indeed, some of them may be included in the test, in which case the horse will need to learn them at this stage.

Whatever level he has reached in his training, the horse has to be given time to get his hocks engaged before he is asked to perform any lateral movements or other demanding exercises. This might take only five minutes with a short-based horse and anything up to forty minutes for one with a long base. Master Craftsman comes into the latter category and he always requires plenty of basic work to get his hocks engaged. When working him in before a dressage test, I would probably wait until the last ten minutes before practising any lateral movements.

FITNESS PROGRAMME

All our horses follow the same initial programme, whether or not they are being aimed for a three-day event. It is only when they reach one-day fitness that they go their separate ways, with a completely different routine for those who need to become three-day fit. You can reach this goal in a number of different ways, using the facilities that you have available.

Hills are a tremendous asset; they improve the horse's lungs, heart and muscle tone, enabling you to get him fit with fewer canter work-outs. The ideal place for faster work would be all-weather or grass gallops with a gradual uphill incline. You may not have this facility, in which case you will have to improvise. Two different charts, which were both used to produce horses that were fit enough for advanced or championship three-day events, are included in Chapter 17. You may be able to adapt one – or possibly use parts of both – to suit your own needs.

If the horse has never run in a three-day event before, it may be more difficult to get him fit. His programme could therefore be the same as that used for the advanced horse who is getting ready to cover a longer distance. Thoroughbreds, however, are rarely a problem in this direction, so their work would probably be slightly less. Suggested reductions for novice and intermediate three-day events are given with the two charts in Chapter 17.

STEEPLECHASE

If you have been getting close to (or within) the optimum time at one-day events, you will already have been doing some galloping. Another gear, however, will be necessary to get inside the time required for the steeplechase section of a three-day event. You need to practise at the faster speed and learn to gauge the miles per hour, metres per minute or seconds per furlong, according to which method of measurement suits you best. Using miles per hour means that you can check your speed alongside a car, as mentioned in Chapter 9. For metres per minute you will have to use markers and a stopwatch – but you can dispense with conversion tables, since this is the way in which speeds are given in the rule book. Seconds per furlong are of use if you are able to work the horse on gallops with furlong markers; alternatively, you can easily measure the distance and put up your own markers at home.

The faster speed should not be a problem if you are riding a Thoroughbred; he already knows how to gallop because it was bred into him, so he needs to concentrate on the endurance aspect. The half-bred, who is capable of cantering for long distances at a slower speed, needs to work on maintaining a faster pace.

You will probably have had the chance to jump some steeplechase fences while riding across country at one-day events; if so, the horse has no great need to go schooling over a steeplechase course. It would, however, be of benefit to you as a rider, especially if you are fairly inexperienced.

Above: *At this stage your canter/gallop work-outs change in order to build up to three-day event fitness*

Left: *The roads and tracks make the three-day event a test of endurance as well as speed. Make sure that your horse has done sufficient fitness training to be able to cope with the endurance aspect*

Right: *If you have never ridden over a steeplechase course, it will help to have some practice at the speed required*

FITNESS OF RIDER

The cross-country, speed and endurance phase of a three-day event is much more tiring than a one-day contest. You will therefore need to do exercises or other activities to build up your own strength, no matter how many horses you are riding each day at home. You can choose anything that takes your fancy – skipping, running, swimming, sessions on a standstill bike, or whatever – and you need to do it at least three times a week.

FEEDING

We have always been loath to give feeding charts, because quantities vary so much for each individual horse. As already mentioned, you have to judge the amount of hard feed he needs by his appearance and the feel he gives you when ridden. If a horse has already competed in a three-day event from Ivyleaze, we would have the advantage of being able to refer back to the quantity of hard feed he was receiving at that time – and the notes which tell us how well he fared on it.

Using the rough guidelines of 1 lb of hard feed per day for every mile the horse is covering, you would obviously expect the quantity to be increased for the longer distance in a three-day event and the preparatory canter work-outs. This is usually (although not automatically) true. Since we are always aiming to get the horse fit enough for the job in hand on the minimum quantity of hard feed, we would very rarely give more than 12 lb a day and it could be much less.

The quantity of bulk food depends on his waistline rather than his level of fitness. If he has a weight problem, he will obviously not be allowed to gorge himself on grass all day. His hay and 'slops' will also be given more sparingly, since we do not want him to be burdened with surplus weight during the competition.

PRACTICE FOR THE HORSE INSPECTION

It is all too easy to make a sound horse look lame. He only needs to turn his head – normally towards the person who is trotting him up – in order to start moving on an angle, which invariably makes him appear unlevel. I always practise the trot up at home, leading the horse from the wrong side so that it will not be a habit for him to turn his head to the left. Apart from wanting his head and neck to be absolutely straight, I need to assess his best speed. The Ground Jury insists on working (rather than medium) trot, but the miles per hour can vary within that pace. Some horses shorten if they go too fast; others look very sluggish (and rather strange) if they trot up slowly. Quite a few do not trot well in hand, maybe because they are gawping at everything in sight or because they need the extra impulsion from the rider's legs and look like slobs without it. Practice at home may not make perfect, but it will certainly help.

TIMETABLE

A three-day test, especially one with two days of dressage, might seem quite leisurely in theory when compared with your one-day outings. In practice, it can become fairly hectic. The horse will need exercise, possibly twice a day if he is feeling scatty; you need to walk the cross-country at least three times, the steeplechase at least twice and take a second look at the roads and tracks, after having been driven round them once in a crowded vehicle. These all have to be fitted in with the briefing, horse inspection, one short final gallop and the competition itself. The horse will also require a certain amount of attention in his stable!

Once aware of the overall pattern, we do at least have time to plan the finer details of our programme one day at a time. This will be done each evening, when making detailed plans for the following day.

A tack box helps to keep the equipment tidy when travelling

CHECK-LIST

In addition to all the items required for a one-day event (see pages 91-92) I would take the following:

For the rider:
Top hat and tails (optional)
Stopwatch (and spare)

For the horse:
Spares of all essential equipment
Easyboot
Weight cloth and weights
Vaseline or wool fat
Rubber gloves
Blunt knife and washing-up liquid
Sheepskin and Flash nosebands
White girths for dressage (optional)
Bootlace (for tying bridle to top plait)
Animalintex (see page 148)

Other essentials:
Horse's passport (if required)
If hot, check that ice will be available

SUMMARY

Apply for horse's passport, if required.
Check on dressage test to be used and learn any new movements required.
When schooling on the flat, make sure the horse's hocks are engaged before he is asked to perform lateral movements or other demanding exercises.
Work out fitness programme (see charts in Chapter 17).
Practise at speed required for steeplechase.
Remember that Thoroughbreds need to work on endurance aspect.
Half-breds need to work at maintaining a faster speed.
Rider requires some fitness training.
Horse will probably need an increased ration of hard feed because of extra distance being covered.
Practise for horse inspection.
Check that all necessary equipment is available and in good condition.

Practise canter work-outs with your weight-cloth (if required) to make sure that it is comfortable for both horse and rider – and that the weight is evenly distributed

15

FIRST THREE-DAY EVENT

We normally arrive at novice or intermediate three-day events the day before the briefing, unless we are travelling abroad. For an overseas fixture I like to allow an extra two days so that the horse has time to recover from his journey. The arrival time is planned to allow for a quiet hack when we reach the venue, so that the horse has a chance to unwind before he settles into his temporary stabling. The following day's routine, which has to be planned round the times fixed for the briefing and first horse inspection, will include going out for one or two hacks on the horse and walking the cross-country course on foot.

Unless the horse has a thoroughly laid-back attitude, I will probably decide to take him out twice and give him time to soak up the atmosphere. It will be far more stimulating than anything he has seen or heard at one-day events, with many more marquees, vehicles, flowers, flags and people. The sight of all this extra activity may not worry him nearly as much as the additional noise. The horse has much sharper hearing than a human; he can catch the slightest click of your tongue or a quietly spoken word while training. Strange sounds seem to alarm him far more than unfamiliar sights, so he needs the chance to hear as well as see what is going on around him. I rarely attempt any flatwork with my horse on the first day. He needs time to settle into his surroundings, and if I tried to school him before he was relaxed, it would probably end up in a battle – which is the last thing I want at this stage.

THE DAY OF THE BRIEFING

Having a less than perfect memory, I believe in taking a notebook with me to the briefing so that I can jot down any relevant items of information. These may involve fences with interlocking penalty zones and will certainly include details as to where we can exercise and gallop. We may also be told about tack checks and dope tests.

Immediately after the briefing competitors pile into a convoy of vehicles to be driven round the roads and tracks. With so many people and vehicles, it is not always possible to see where you are going or the position of the kilometre markers. Often you can do no more than get a feel of the terrain. You have to go back before cross-country day to get a clear view of the route and the markers. I like to drive around, but some people prefer to ride. Whichever means of conveyance you use, you need to be sure of the route; it is easy to by-pass a turning flag, which would be a very disappointing way to get eliminated.

Between the two sections of roads and tracks, we all pour out of the vehicles and walk the steeplechase course. This usually turns into a time for gossiping rather than a careful study of the fences, so I will go back once or twice to give it more concentrated attention. By then I hope to have learnt the locations of the quarter, half and three-quarter points of the course, through the help of a reliable friend who has walked round with a measuring wheel. These places will be pinpointed in my mind when I walk round again. The information that I wear on my arm when setting out on the speed, endurance and cross-country phase will tell me the number of seconds I should take to reach each of these three points and the finish.

I give the horse his final practice trot up on the day of the briefing and first inspection. He is then plaited up; oil is put on his hooves and diamond shapes on his quarters so that he will look smart when he appears in front of the Ground Jury. We allow half an hour for him to be walked round beforehand to loosen and relax his muscles. He wears a rug or light sheet depending on the weather and, because he is so fit and may be feeling skittish, he wears exercise boots on all four legs. These are whisked off at the last moment; at the same time any mud or gravel is removed from his feet. Even tiny stones can cause mild discomfort and make the horse slightly unlevel, which could be enough for him to be failed by the Ground Jury.

CROSS-COUNTRY COURSE WALK

The time fixed for the horse inspection will dictate whether I walk the cross-country beforehand or afterwards. I try to avoid going in a group when I am looking at it for the first time. I prefer to sit down and eat a sandwich, while most of the other riders set out together on foot to inspect the fences. My concentration is ruined if I go with the others; I begin to absorb their ideas instead of focusing on my own first impressions. I therefore walk either on my own or with Dot and, maybe, a friend. I do not, if I can help it, discuss the course with them.

The first impression is enormously important because you are seeing the course as the horse will see it on cross-country day. By then it will be much more familiar to you, because you will have walked it two or three more times, whereas the horse will not have his first view of the course until the actual competition. I try to concentrate on my horse's likely reaction to each fence and the way I intend to ride him. Combination fences sometimes cause me a few moments of panic before I have worked out my line, but there is usually one route I prefer on the first walk round – and, oddly enough, I very rarely change my mind. When I walk round the second time, I am more than happy to listen to everyone else's views; at that stage I want to glean as much information about the course as possible. Having absorbed the opinions of other riders, I need to adapt them to my own horse and his individual needs – taking his length of stride into consideration as well as any doubts or fears he may have.

I normally walk the complete course three times. As well as knowing every detail of the route that I plan to take, I want to be familiar with the alternatives. Sometimes a particular route proves hazardous on the day; maybe it has become dangerously slippery because of overnight rain, or perhaps the distance is not working out quite as well as I had anticipated. It is no good being warned of an unexpected hazard unless you know the way to go in order to avoid it. Knowing the alternatives will also be important if you have a refusal, because it might then be prudent to take an easier route.

It helps to have some idea of the number of people likely to turn up on cross-country day. If there are likely to be big crowds, you need to use your imagination to see how they will affect your view of the fences. If the course is roped off, you should also consider whether the position of the ropes will have any effect on the way you approach each obstacle. At certain fences, there may be crowds leaning against the ropes and they could completely block your view of the obstacle as you approach it. Remember, too, when you are walking round that everything will happen much quicker when you are on a galloping horse.

FINAL PREPARATIONS FOR DRESSAGE DAY

On the evening before the dressage, I make a point of checking all the equipment and making sure that my own clothes are ready and that my boots are clean.

A few other details also need to be considered in advance, such as where to put one's number. It is no good attaching it to the bridle shortly before you are due into the arena, only to find that the horse shakes his head because he doesn't like it being there. Nowadays I sew mine on to the numnah, where it is completely out of the way. It will need to be taken off and sewn on to a clean one after the cross-country.

If your horse is wearing a Union Jack, that too has to be tried on beforehand with the saddle on top. There is otherwise the risk of a last-minute panic when you tack up for the dressage and discover that it doesn't fit. You may then be ten minutes late getting mounted, which means that you are upset and the horse senses it, so he is unsettled as well. Although some people function better when they are feeling slightly hassled, most of us need to be thoroughly organised to stay ahead of ourselves and avoid flapping.

The contents of our dressage bucket are likely to include:

Fly spray and applicator (having checked that this is a permitted spray)
Omnibus schedule with dressage test
Hoof pick
Sponge and cloth
Stencil and brush for putting diamond shapes on quarters
Spare bits
Spare bridle
Drink (for rider)
Spare spurs
Boot polish and duster
Saddlesoap and sponge
Leather punch

In cold or wet weather, there would also be a rug available for the horse and a jacket or mac for the rider.

DRESSAGE

The plans I make for the morning of the dressage test depend on the individual horse. He might be better for an early-morning hack or a gallop; he may need only to be walked out in hand at that stage. It is important to know what best suits your own partner and to be aware that his behaviour may be adversely affected by the atmosphere.

The serious work begins about one hour and ten minutes before the time for my test, when I start riding in. Though some horses take longer, I regard this as the ideal time. It allows for forty minutes of proper dressage schooling interspersed with periods of walking on a loose rein, which would occupy another twenty minutes. The remaining ten minutes are for smartening up. At that stage I change my coat while the horse is given a quick wipe over to remove foam from his neck and bit; his hooves are oiled, feet picked out, diamond shapes put on his quarters and, finally, his tail bandage and exercise boots are removed.

There can be an uncomfortably thin dividing line between a nice active test and boiling over. The horse is fitter than he was for one-day events and there will be more people to create an exciting atmosphere; he might light up when he hears applause as the competitor before you leaves the arena. As always, you have to ride him as he is on the day, concentrating your attention on the next movement and making sure that you prepare him for it three strides in advance. Whatever the horse does, you must keep your cool. He is far more likely to become overwrought if he senses that you are uptight.

FINAL PREPARATIONS FOR CROSS-COUNTRY DAY

Every item of tack and equipment needs to be checked on the evening before the cross-country. We get out everything that will be needed and go through it, checking that all the items (including the spares) are in good order, deciding on studs to be used, making sure that we have the right bridle, bit, saddle, numnah, weight cloth and lead, girths, surcingle, etc. We also check the bandages, tendon protectors and over-reach boots, as well as the items that will be taken to the start of the steeplechase course and to the ten-minute-halt box before the start of the cross-country.

The steeplechase bucket will probably contain:

Spare shoes with studs already fitted
Easyboot
Spare bridle and reins
Spare surcingle and girths
Small first-aid box
Bandages
Scissors, needle and cotton
Leather punch
Spare stopwatch
Spare stick

All these items will be taken to the ten-minute-halt box once I have departed on phase C, the second section of roads and tracks. The following will be taken directly to 'the box':

Drink for the rider (non-alcoholic and non-fizzy!)
Glucose drink for the horse

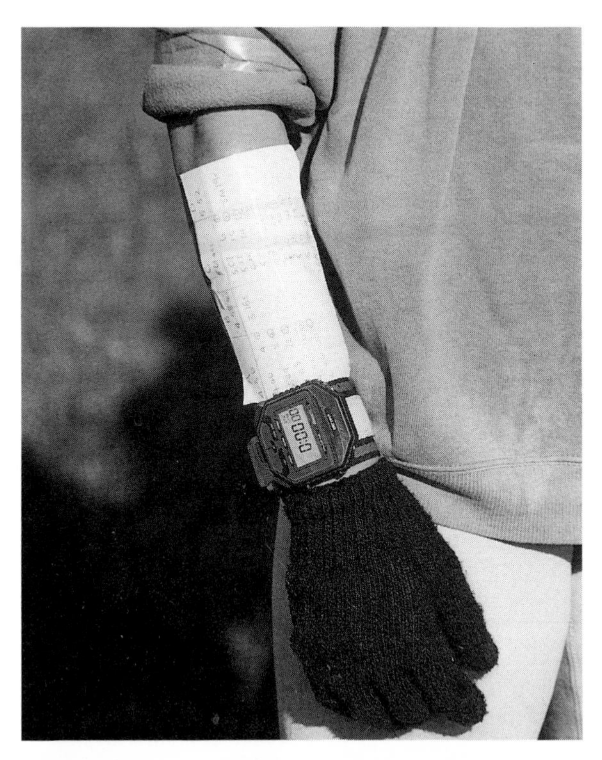

Charts and stopwatch will help this competitor to finish within the time on the roads and tracks, and the steeplechase

Leave yourself time after weighing out, so that the equipment can be checked before you set out on phase A

Spare saddle and stirrup leathers
Spare numnah
Stud spanner and studs
Towels
Sponges
Vaseline or wool fat
Rubber gloves
Headcollar and lead-rope
Spare over-reach boots
Towels
Sponges
Scraper
Buckets (2)
Programme with course plan
Coat for rider
Sweat sheet
Rugs
Ice packs (if hot and/or humid)

As well as checking the equipment, we have to work out our programme for the day. It includes the time the horse will be plaited and when his bandages will be put on, with sufficient time allowed for them to be sewn in place without anyone's thumbs being pricked because it has to be done in haste. We also need to know when the horse will be fed and how much he will be given, which will depend on his starting time for phase A. His hay ration is always cut right back before the cross-country, though he would have had an extra 2 lb or so to eat between six and nine o'clock the previous evening. He has just 2-3 lb during the night and, whatever time he is due to start, only a handful of hay before his hard feed the following morning.

My own times for the actual event, which will be sellotaped to my arm the following day, have to be written out after consultation with the programme. Three-day events can vary, with some roads and tracks more lenient than others, so you always need to check the times. My charts are worked out with the intention of reaching the steeplechase one or two minutes early and of having an extra two minutes in hand before the cross-country. I believe in keeping the details as simple as possible. For phases A and C, the two sections of roads and tracks, I need to know how long it should take to reach each kilometre marker. At most events I aim to take four minutes per kilometre, which is slightly faster than the time allowed.

Writing out the charts does not exactly tax my brain. I simply put: 0-1 (kilometres) = 4 (minutes), 1-2 = 8, 2-3 = 12 and so on. Since I will be trotting all the way on phase A, there will be no variations between the markers unless the time happens to be generous, allowing for a five-minute kilometre which would include some walking. On phase C, where I always do some cantering and walking, I include five-minute and three-minute kilometres, which I mark with an asterisk. For the steeplechase, I simply write down the time it should take to reach four points on the course – a quarter, half and three-quarters of the way round, plus the finish.

These details will be sellotaped to my arm just below my elbow, before I set out on phase A. Whether or not there is a cloud in the sky, I always keep winding the tape around my arm until the charts are completely covered. If it should happen to rain, they cannot be ruined.

I work out my cross-country times by deciding where I should be at the end of each two-minute period, again with the help of some reliable friend who has been round the course with a measuring wheel. I have this information before I walk the course the second and third times. I can therefore look for some particular feature – it might be a large tree or a barn – which will help me to fix these places in my mind. Some riders work out where they should be at the end of each minute on the cross-country. I find that too complicated because it means remembering so many places on the course, so I feel happier with my slightly simplified version. The cross-country times are not attached to my arm. They are written on a card so that I can have a last look at them during the ten-minute halt and make sure that they are indelibly printed on my mind.

SPEED AND ENDURANCE

I like to be on the horse ten minutes before we set out on phase A. If we are going early, he will have been led in hand for at least twenty minutes beforehand. If our starting time leaves us in the second half of the competition, he will have been hand-walked for half an hour in the early morning, before most of the spectators arrived. The horse will know that this is the big day and he may produce some fireworks in his excitement. With this in mind, we always have a lungeing cavesson and lunge line on the horse while he is hand-walked. If he does play up, the person leading has plenty of room to play with on the lunge line. With an ordinary halter, one sharp flick of the horse's head can be enough for the handler to lose the lead-rope – and the horse.

Unless he feels as though he needs a short period of canter to help him settle down, I normally trot the whole way for the first section of roads and tracks.

Some riders give their horses a pipe-opening gallop at the end of phase A before reaching the steeplechase, but I am not among them. I keep trotting, trying to think of nothing else except reaching each kilometre marker at the right time (but rarely succeeding). The steeplechase is looming ever nearer and I find myself wishing that it was over; it may be exhilarating to gallop over steeplechase fences, but I find it nerve-racking at a three-day event because too much depends on getting it right. By arriving one or two minutes early, I have time to check my girths and make sure everything is in order before setting out on phase B.

Mistakes most often occur on the steeplechase through the rider seeing a long stride and going for it. The horse may think he is too far away and put in another stride where there is no room for one, falling as a result. If I see a good stride I would ride for it, but the long ones are best ignored. It is better to sit quietly and try to let the horse decide; I would rather he fiddled and popped over the fence than ride for a long stand-off.

If you are required to make two circuits of the steeplechase course, you may find that your newcomer to three-day eventing applies the brakes when he reaches half-way. This is a normal tendency in novice horses; the steeplechase is new to them and, having been slightly surprised at being asked to jump at a faster speed than ever before, they expect to finish when they get back to the start box. They fail to understand that they are required to do a second circuit.

This is also potentially dangerous and you have to be cautious. You should ask yourself whether the horse is genuinely tired or simply confused at being asked to go round a second time. Usually the inexperienced three-day eventer slows down because he thinks he has reached the finish, in which case you need to shake him up a little so that he regains his momentum. If you suspect he is jaded, he can be eased for ten to fifteen strides – but he must be stoked up again before he reaches the next fence.

Unless the going has become deep on the inside track, I find the best place to be on the steeplechase course is close to the rails. It means you can go a little slower because you are taking the shortest route. If there are any sharp bends, I would still hug the rails but slow down a little into the turns and pick up again coming out of them. This gives the horse a little relief during his long gallop. During this phase I will again be thinking of my timing; I want the horse to finish just

Ready for the start of phase A, with weight-cloth in place and number attached to breastplate

PECK
150

within the time, without being over-stressed. I also remind myself to be very careful about the last fence and to ride it positively; it is easy to think that the steeplechase is over before you have jumped the last. Once it really is finished, you need to avoid pulling up too quickly or dropping the reins. The horse should come gently and quietly back to a balanced canter and then to walk. If it is hot, I will be handed a bag of ice which can be rubbed along the horse's neck to help him cool off.

There is a special area at the end of the steeplechase where someone can check that his bandages and shoes are still securely in place and that there are no cuts or grazes. We use white bandages, so that blood is instantly visible if the horse cuts a leg. If this is the case, we obviously need to check that the injury is not severe and that the horse is sound before I carry on. I am fully aware that time lost at this stage will have to be made up on the second section of roads and tracks, so I am anxious to avoid any delay.

If the horse loses a shoe, there should be a farrier in the vicinity who can quickly nail on one of the spares from the steeplechase bucket – but I normally prefer to use the Easyboot for phase C and make sure that somebody arranges for a farrier to be at the ten-minute halt when I arrive there. By arriving two minutes early, I know that I will have twelve minutes in the 'box', so will not have to make up for lost time if the horse is reshod there before the cross-country.

On the second section of roads and tracks, I tend to canter and walk as well as trotting. These paces are less jarring and I think the horse appreciates a change. As on phase A, I try to think of nothing else but reaching the kilometre markers at the appointed time (again without great success – the cross-country course has a habit of intruding on my thoughts!).

TEN-MINUTE HALT

A vet will check that the horse is sound and not unduly distressed as we complete phase C and come into the ten-minute-halt box; unless there is some obvious problem, he will not need to be looked at again before we set out across country. Once I have been given permission from the vet to dismount, the rest of the Ivy-

leaze team goes into action to deal with the horse, while I get information as to how the course is riding – from the *chef d'équipe* (if I am in a team), otherwise from Dot or one of my fellow competitors who has already jumped the fences. I then have a short time to sit down and relax, have my drink, take a last look at the course plan and my times for reaching those pre-appointed places on the course.

It is much harder work for those looking after the horse. Unless I had fearful problems on the steeplechase and have decided that the bit needs to be changed, the bridle is left on (with the noseband undone) and a lead rope is attached to the bit. The girths and surcingle are loosened sufficiently to allow the saddle to be raised so that the horse's back (as well as his ears and head) can be towel dried. Some people remove the saddle, but I prefer to leave everything on the horse.

If the weather is hot, he is offered a couple of sips of water – as a mouthwash rather than a thirst quencher. Perhaps because they know that the job is not yet completed, they rarely want to take a deep draught of water at this stage. The horse is also given a small drink of glucose. In hot weather, ice is applied to the pressure points – on top of his head, down his neck and between his back legs.

Unless it is particularly hot and humid, the horse will be washed down only under his throat, along his neck and between his back legs. Care has to be taken to avoid getting anything else wet, especially the bandages which are likely to be soaked if the washing down is over-enthusiastic. The horse's legs are again checked for cuts or bruises and his feet are picked out; the bandages, studs, shoes and tack are also checked and grease is applied to his legs. The back studs will probably be changed if the going is soft. We avoid doing the roads and tracks on large studs, because it causes excess jarring and the horse's feet would be at an angle for the entire journey, both of which would affect his

Left: *Time to sit down and have a drink during the ten-minute halt*

Below: *A sip of glucose for the horse during the ten-minute break*

fetlocks and hindlegs. This is all done quickly because we do not want the horse to be standing still for long. Once finished, he is walked around quietly – with a rug on if the weather is cold.

I like to be alerted to the time five minutes before I am due to start the cross-country, so that I can get mounted four minutes beforehand, which gives me the chance to check my stirrups and have a trot. Some riders like to wait until one minute before they start, but I prefer to have longer, so that I can have my trot to remind the horse that the day's work is not yet over. It will also help to get his adrenalin flowing, his blood circulating and his muscles warm, so that he is ready to tackle the cross-country. You cannot expect him to go straight from a standstill into a gallop and start jumping fences.

CROSS-COUNTRY

Though I set out to get within the time on both steeplechase and cross-country, I will abandon the attempt if it means pushing a tired horse. It is always important to bear in mind that the first quarter of the course will ride slower because you have had to get started, and that the time can always be made up. You should know your horse well enough to realise when he is genuinely tired; you should also recognise the occasions when his

Horses have to be brave to jump into water. Welton Houdini shows that he can do it with his eyes closed!

apparent lethargy is simply a reluctance to go away from home. If you push a genuinely tired horse you are asking for trouble; it is far better to incur a few time faults than to risk taking him beyond his limit. At some stage on the course he will need a breather, which could mean taking it easy for 15 or 30 strides. It is up to you to assess just how long to give him.

As at a one-day event, both rider and horse should be forward-thinking. That means giving positive consideration to the fence ahead, not harping back to any hiccups you have left behind you. You need to know exactly how you are going to ride each obstacle before you come to it, so your mind needs to be focused in the right direction. If you are lucky enough to find yourself in harmony with a brave, careful and generous horse, it will be a thrilling ride.

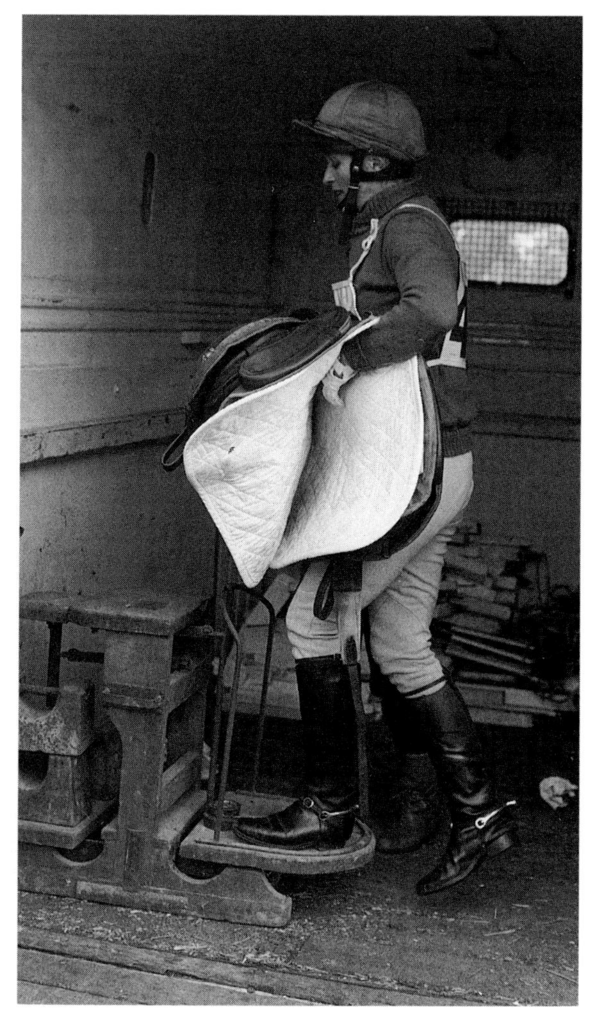

HORSE CARE

When the horse has finished the cross-country, we like to get his tack, boots and bandages removed swiftly so that he can be sponged down as quickly as possible. This is usually done from a bucket rather than a hose-pipe, because we can then adjust the temperature of the water if necessary. We would gladly use a hose if we were sure that the water would be really cold in hot and humid weather, and tepid to warm on a freezing day.

The first priority in hot weather is to get the horse's temperature back to normal, by applying ice packs on the pressure points and by sponging him down. If using a bucket, it will contain some ice to keep the water really cold. The sponging process continues until the water is cool as it comes off the horse; we then stop immediately.

At the 1973 European Championships at Frauenfeld in Switzerland, where it was exceptionally hot and humid, it took all of half an hour to get Night Cap's temperature back to normal. He was also blowing for longer than usual, which you have to expect under such conditions. On a cooler day, we anticipate that the horse will have stopped blowing about fifteen minutes after the cross-country and that his temperature will have come down before that.

Once he is cool enough, excess water is removed with a scraper and the horse is rubbed dry with towels. He is offered a small drink of water when he has stopped blowing heavily, but is not allowed to gulp it down; if he is very thirsty, he will be given a sip every five to ten minutes. Occasionally the horse does not want to drink at all, so we use a wet sponge to give him a mouthwash. We then put on a sweat sheet and walk him round quietly until his breathing is completely back to normal.

Once he has recovered from his exertions, he is allowed to pick some grass before being taken back to his stable. We will obviously have kept an eye open for any sign of injury, but his legs needs to be checked more thoroughly before the grease is removed using a blunt knife and then warm water and washing-up liquid. More grease will tend to come through to the surface later that day and again the following morning,

All competitors are required to weigh out and in before and after the speed, endurance and cross-country phase

requiring repeated washing.

We do not give the horse any pain-killing 'bute', but we normally apply cold poultices to his legs after the cross-country phase of a three-day event. Animalintex, which is the easiest type of poultice to use, will help to ease any bruising and make him feel more comfortable. You do occasionally come across a horse who is allergic to it, so you need to check beforehand. Otherwise, he can end up with a leg that is puffy from Animalintex rather than injury.

The horse can now be left in peace to have some hay and a full drink of water before his hard feed, which is given one and a half hours after completing the cross-country course. We will, however, bring him out for a ten-minute hand-walk every hour for the next three hours. He will also be trotted up in the evening to check that he is still sound.

Injuries, such as cuts or punctures, obviously need to be treated. Having checked that there are no remaining thorns (or anything else potentially dangerous), the wound has to be cleaned with warm water before an antiseptic ointment is applied. We will call the vet if the wound is deep, or if it contains any dirt or gravel. Should we find any sign of heat or swelling, we use gamgee that has been soaked in ice-cold water. This is wrapped around the affected area and kept in place by bandaging over it. The process is repeated every hour, with further applications of ice-cold gamgee.

On one occasion Priceless hit his stifle at Burghley and we spent hours on end applying hot towels and ice-packs alternatively before he was left in peace for the night. We do not believe in staying up all night to get the horse through the final vetting. He might have a hot poultice applied at midnight to relieve bruising for something like a big knee, but we would leave him until five o'clock at the earliest before giving any further treatment. The horse needs his rest, and if he cannot be left alone for five hours, it would be better not to present him to the Ground Jury.

We are always up early the following morning. It is all too easy to be lulled into a false sense of security because the horse seemed perfectly sound in the evening, only to find that he has stiffened up overnight. Most horses are fairly stiff after the cross-country, and some are much worse than others. We had one experience of the effect this can have during a three-day event at Breda in the Netherlands. The horse had been fine at his evening trot up after the cross-country, but was distinctly unlevel at 5.45 the following morning. Everybody shook their heads and said he would never pass the inspection. We nevertheless walked him in hand for a while and he seemed a little better; he had some massage at breakfast time, and when I rode him later that morning he seemed to be back to normal. He was duly passed by the Ground Jury and was never known to be unlevel again.

You need to know how supple the horse is under normal conditions. Some youngsters can be as stiff as the seasoned compaigners, so age is not necessarily an accurate guide. If you know how the horse usually moves, you will obviously have a better chance of knowing when he has returned to normal. Nothing can be done if there is something radically wrong with him, but stiffness can generally be dealt with through massage and exercise if you have left yourself sufficient time.

The bigger horses tend to stiffen up more than the smaller ones. The size of the box may be a contributory factor: if your horse is big and the box is small, he will virtually be doing a pirouette the whole time. He therefore gets stiff because he is unable to walk forward and use his shoulders properly. Lungeing can sometimes help, but not always. Working on a smallish circle can actually make the horse worse, whereas he might have improved had he been ridden on a larger area. If he is still very stiff after light exercise, including some trot and canter, he will be withdrawn. The horse could easily damage himself if asked to show jump in this condition.

If there are no problems, our normal routine would be to give the horse a trot up first thing in the morning. Afterwards I might go for a forty-five-minute hack – with maybe half an hour of walking, followed by a little trotting and, perhaps, a short canter all on a nice long contact. Having loosened up, the horse will feel more comfortable when he is put away to have his breakfast before being prepared for the final horse inspection. Needless to say, we would not present him for the vetting unless he was fit and sound.

SHOW JUMPING

The programme rarely gives information as to when the course will be open for inspection, so you have to find that out for yourself and be ready at the appropriate time. You also need to be suitably attired, since you normally have to be in riding kit when you walk the course.

Inspecting the show-jumping fences is much the same at the first three-day event as it was when the horse took part in his first competition – except that

The horse will be tired on the third day of a three-day event and may feel quite different over his fences. It is a great bonus to have a reliable show jumper, such as Priceless who had only one rail down during his five years in major championships

you will now know his reactions rather better. If he has a tendency to drift towards the collecting ring, it should by now have become easier to work out, with the help of past experience, where you should turn into an obstacle to compensate for any drifting. You should also be more familiar with the way he jumps combination fences, and with the length of his stride. These factors determine the pace required for each obstacle, which is particularly crucial at combinations. By approaching the first element at the correct pace, you arrive at the right take-off point for those that follow.

The extra experience does not, however, make you foolproof against mistakes. I would expect to walk the course three times and, if possible, watch some other competitors jump before I get on the horse. Apart from seeing how the course is riding, I am still anxious to listen to the starting signal so that I know what it sounds like.

If the horse has already gone for a hack that day, I work him in for one hour before he jumps, mostly at walk. His muscles need to be kept loose, but you don't want him on his knees with fatigue after the previous day's exertions. I normally start my practice jumping when there are eight horses to go before my turn; I then have my last couple of jumps when there are two left in front of me. The number of fences I jump depends on the horse. He may need to unloosen by having a couple more jumps than was necessary at a one-day event, but I have to be careful to avoid wearing him out. In that respect, the fewer I do the better. When he is not jumping, he must be kept walking; there should be no standing still in the collecting ring.

Do not forget to salute the judges when you ride into the arena. Once you have your signal to start, you will (as always) be concentrating your attention on where to turn and how you will approach the fence ahead.

SUMMARY

On arrival, take horse for a quiet hack to give him a chance to unwind.
On day of the briefing, horse would probably benefit from two hacks rather than one – to get used to strange sights and sounds.
First horse inspection: allow half an hour for horse to be led in hand beforehand.
Keep exercise boots on his legs until the last moment.
Remove any mud or gravel from his feet just before trot-up.
Walking the course: Remember your first impression of the fences, this will be how the horse will see them on cross-country day. All observations and information on the fences must be adapted to the individual horse.
Be familiar with the alternatives, whether or not you plan to take them.
Consider whether your view is likely to be restricted by the crowds.
Make sure you know the route on the roads and tracks.
Decide places on steeplechase and cross-country courses which you should reach at specific times.
Dressage: Check all your equipment the evening before. Decide whether the horse would benefit from an early morning hack or short gallop before his test.
Allow about one hour and ten minutes for riding-in (40 minutes for proper schooling, 20 minutes for walking on a loose rein and 10 minutes for smartening up).
Speed, endurance and cross-country: Check all equipment the evening before.
Write out charts for time to be taken on roads and tracks, plus steeplechase.
On morning of the cross-country, horse is led out in hand for 20-30 minutes.
Rider mounts about ten minutes before the start of phase A.
Steeplechase: Do not attempt to ask the horse to stand back at the fences.
If allowing horse to reduce speed for a short distance, make sure he regains his momentum before the next fence.
Ride the last steeplechase fence positively.
Avoid pulling up too quickly or dropping the reins.
Ten-minute halt: Ask the vet for permission to dismount.
The horse's noseband is undone and a lead-rope attached to the bit.
Girths and surcingle are loosened.
The horse is towel dried.
He is given a few sips of water and a small drink of glucose.
In hot weather, ice is applied to the top of his head, down his neck and between his back legs.
He is washed down under the throat, along his neck and between his back legs.
His legs are checked for cuts and bruises and his feet picked out.
Bandages, studs, shoes and tack are checked – and studs changed if necessary.
Grease is applied to his legs.
Horse is walked round quietly.
Rider mounts four minutes before start of cross-country, which allows time for a quick trot.
Cross-country: Remember the horse will need a short 'breather' at some stage.
If he is genuinely tired, make sure you do not push him beyond his limit.
Concentrate on correct speed, balance and accuracy on approach to each fence.
Horse care: Remove tack, boots and bandages.
Sponge horse down.
In hot weather, first priority is to get horse's temperature back to normal.
When he is cool enough, excess water is removed with a scraper and the horse is towel dried.
He is offered a small drink of water.
Horse is walked round (in a sweat sheet) until breathing has returned to normal.
His legs are checked thoroughly and, if any injury is found, appropriate treatment is given.
Grease is removed from his legs.
Cold poultices are applied to his legs.
Horse is given water and hay in his stable, followed by hard feed.
He is brought out for a ten-minute hand walk every hour for the next three hours.
He is trotted-up in the evening.
A poultice may be applied at night to relieve bruising.
The horse is trotted-up early the following morning.
He is given light exercise to loosen up – and massage, if necessary to relieve stiffness.
Show jumping: Horse is worked in for one hour, mostly at walk.
Jump practice fence when eight horses are still to go before your turn.
Have last two jumps when two horses are still to go.
Keep horse moving at walk when not jumping.
Remember to salute the judges.
During round, concentrate on the fence ahead.

16

INTERMEDIATE TO ADVANCED

The move from intermediate to advanced one-day events is a major step for the horse. We do, however, have the advantage in Britain of being able to use open intermediate classes as a stepping-stone between the two – and even when upgraded, we can always come back to them if the horse needs to regain his confidence.

He will have to be established at intermediate level, which will probably mean contesting about four competitions without any problems, before I consider the next step. I will then find some open intermediates, which are usually slightly larger than intermediate tracks but not as big as advanced. The horse will probably have contested about ten competitions at either intermediate or OI level before I consider an advanced class. As his rider, you have to go by your own gut reaction to decide whether he is ready for the bigger challenge.

Master Craftsman did his first advanced in the spring of 1987 when he was still only a seven-year-old, which is considerably younger than most. Obviously we can all make mistakes, but I could see no reason why he should not go ahead. He had the right character and confidence to attempt it – and he proved that he was able to cope. Another horse of the same age might not have been ready to move out of novice classes until that autumn. You have to give each individual as much time as he needs. He must not be hurried out of his rhythm across country and he must be secure at each level before he moves up the next rung of the ladder.

I always choose a suitably encouraging advanced track, with plenty of alternatives, for the horse's introduction to the highest level of one-day eventing. In that way he can learn to tackle higher fences without necessarily having to answer bigger questions.

WORK ON THE FLAT

By now I would hope that the quality of the horse's paces is more secure, that he has more weight on his hindlegs and a greater degree of lateral bend from head to tail. We are looking for a horse with a shorter base, plus more impulsion and elevation within his work. At this stage I can probably start to be a little greedier in the accuracy at markers, so that transitions are made at a precise spot rather than close to it. In theory, the more the horse is on his hocks, thereby lightening his forehand, the lighter the contact will become on the reins – though there always must be a contact, whether he is doing Grand Prix dressage or a novice one-day event. In practice, the weight on the reins will depend as much on the preference of the individual horse as on his stage of training.

As always, we start teaching new movements at least six months before the horse is likely to perform them in a dressage arena. He will normally start half-pass at trot when he is just into intermediates; having learnt the movement at walk as a novice, he is already familiar with the aids. He also has to learn counter-canter on a serpentine, which is an extension of the five-metre loops included at intermediate level.

Counter-canter requires the horse to follow the curve of a circle, while bent in the opposite direction and leading on the outside leg. The first loop of the serpentine will be in true canter, with the inside leg leading, and the rider should stay in the same position to maintain the correct bend for the counter-canter loop. If, for instance, the horse was required to lead on his near (left) leg, the rider's weight would be on the left seat-bone, while the left hand and right leg ask for a bend to the left throughout the whole serpentine. On the counter-canter loop the rider asks for a change of direction (but not of bend) by use of the right rein supported by the left leg.

Both *extended walk and canter* have to be learnt for advanced tests. They follow on from the medium paces, which the horse has already learnt, and require greater length of stride and energetic impulsion from the hindquarters. The horse's head and neck have to be stretched further forward because, as already mentioned, his feet cannot strike the ground further ahead than his nose. In extended walk, his hindfeet touch down well ahead of the hoof prints left by the forefeet.

The rider's legs have to create the extra energy for extended paces, with the hands then allowing for the necessary stretching of head and neck but still maintaining contact. You will be asking for greater extension than you have requested for medium paces, which is possible only because of the extra impulsion and elevation the horse acquires through training. You should not ask for too much at once, or ask for it in the same place each time. You want the horse to be constantly listening for the messages you give through the aids rather than anticipating them.

The horse's repertoire will also have to include the more difficult transitions of *canter-halt* and *halt-canter*. Moving into halt without any trot or walk strides will require the usual half-halts to prepare the horse for the transition. His training will have taught him to be alert and listening; he will learn to distinguish your aids for downward transitions by the amount of emphasis you put on them; they will obviously be lighter for canter to trot and stronger for canter to halt. He will also recognise the aids for striking off into canter and, as long as you have created enough energy through the use of your legs, he should respond to this message whether it is given in halt, walk or trot.

At this stage we would probably teach *half-pirouette*, in which the horse's forehand moves round his hindlegs in a half-circle. This is a useful gymnastic exercise, which helps to give control over the horse's quarters and forehand. You can work towards this movement by schooling in small circles which gradually decrease in size, without the horse losing either impulsion or rhythm. The movement will require him to flex in the direction he is moving, with about the same amount of bend as for shoulder-in.

To begin a half-pirouette from walk, the rider would restrain the forward movement with the outside rein. The inside rein then leads the forehand round, while the outside one controls the speed and, if necessary, corrects the flexion. The inside leg is applied on the girth to create impulsion and stop the horse falling in. The outside leg, used behind the girth, prevents him from swinging his quarters out – thereby avoiding the effort required to perform the movement correctly. His inside hindleg, which acts as the pivot, should move in such a small half-circle that it appears to remain on the same spot.

CANTER WORK-OUTS

The additional distance for the cross-country section of an advanced one-day event means that another increase is needed in the time for canter work-outs. These will build up to three seven-minute canters (on flat ground) or two of eight minutes (on an uphill incline) with breaks in between. The horse can then be kept ticking over at one-day-event fitness with two seven-minute canters on weekends when there are no competitions. As always, his recovery rate will be timed when he has finished cantering. I would not run him more than twice on consecutive weekends at intermediate or advanced level, except on the rare occasions when I feel my horse needs more experience and I know the going is good. If I do run him three weeks in a row in order to further his education, I will not go fast across country.

More often than not, the horse will be aiming for a second three-day event – which will probably be

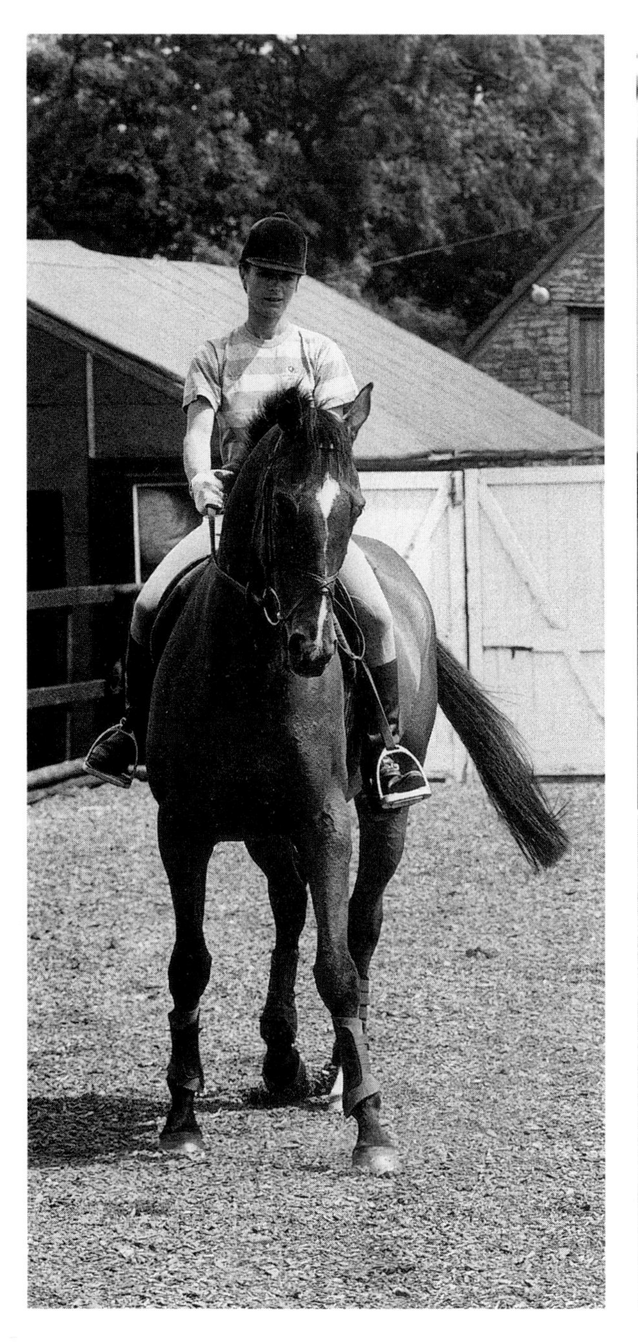

Left half-pass in walk

Right half-pass in walk

another (or first) intermediate or the international equivalent, a one-star CCI. Once he has contested his first one-day event of the season, he will therefore move on to one of the programmes (or a combination of them) described in Chapter 17.

JUMPING

Grids continue to be a means of developing the horse's technique and overcoming any problems in his jumping. Since he has reached a higher level, the difficulties incorporated in each line of fences should seem no more demanding than those he encountered when he was an inexperienced youngster going through grids that were suitable for a novice.

Different distances, used judiciously, will teach him to shorten or lengthen his non-jumping strides. Place poles can also be used (see diagram on page 156), to encourage him to stay straight and jump with a rounded bascule. He may need correction because he has discovered a method of clearing fences without making the effort to use himself by rounding his back.

Priceless was a past master at this; he had learnt that he could take off two feet before the correct place and so clear the fence with a flat back. This meant that he also landed two feet further away, which might not have mattered too much over single fences but could have been a problem in combinations. Priceless, however, had that sorted out as well. Having landed too close to the next fence, he used to take a minute non-jumping stride so that he could put in another flattish jump. When we put a place pole 9 feet (2.74 metres) in front of the second element in a double, he was obliged to take a proper non-jumping stride and so use himself correctly.

Four- to six-bar jumping is a useful exercise for teaching the horse to be tidy and mentally active. At advanced level I would have each fence at about 3 feet 11 inches (1.2 metres). The distances between them can be varied in order to teach the horse to adapt the length of his non-jumping stride. Combinations can also be employed for this purpose, using different heights and distances.

At the early novice stage, the fences in a combination will have to get progressively higher. Once the horse has a few events behind him, this rule no longer applies; you can begin to play around with different heights and distances. At advanced level I might, for example, have a combination with the first fence at 3 feet 6 inches (1.07 metres), the second at 3 feet 10 inches (1.17 metres) and the third at 3 feet 9 inches (1.15 metres). A distance of 23 feet (7 metres) between the first two elements would require one short non-jumping stride; 38 feet (11.6 metres) between the

Opposite and above: *Advanced grid work using poles to keep the horse straight, as well as place poles on the ground, which encourage athleticism and a better style of jump. Some horses require full freedom of head and neck in grid work, without interference or support from the rider*

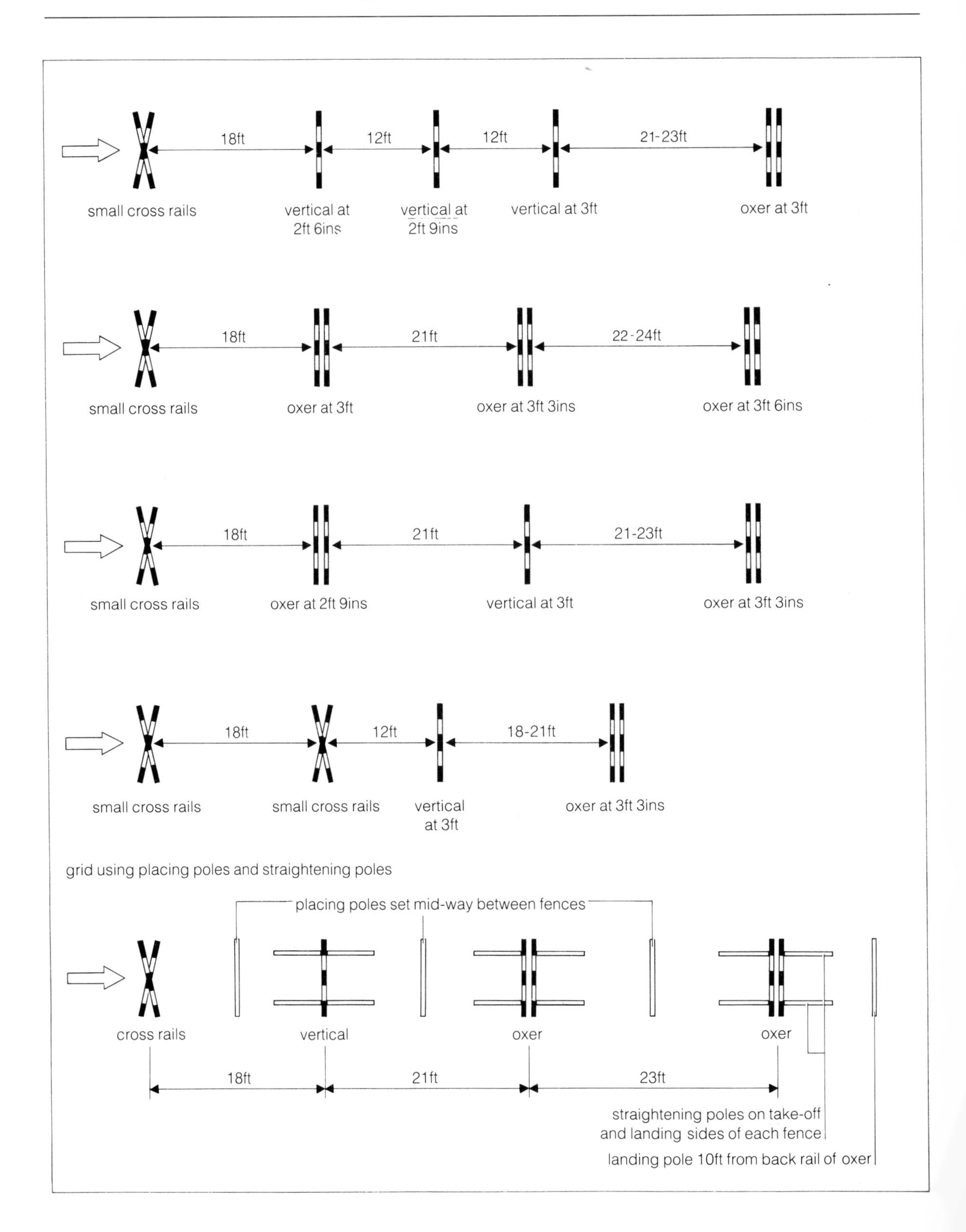
18ft
12ft
12ft
21-23ft
small cross rails
vertical at 2ft 6ins
vertical at 2ft 9ins
vertical at 3ft
oxer at 3ft
18ft
21ft
22-24ft
small cross rails
oxer at 3ft
oxer at 3ft 3ins
oxer at 3ft 6ins
18ft
21ft
21-23ft
small cross rails
oxer at 2ft 9ins
vertical at 3ft
oxer at 3ft 3ins
18ft
12ft
18-21ft
small cross rails
small cross rails
vertical at 3ft
oxer at 3ft 3ins
grid using placing poles and straightening poles
placing poles set mid-way between fences
cross rails
vertical
oxer
oxer
18ft
21ft
23ft
straightening poles on take-off and landing sides of each fence
landing pole 10ft from back rail of oxer

second and third elements would allow for two long strides. The following week I might change these distances, making the first one 26 feet (7.9 metres) for one long stride and the second 34 feet (10.4 metres) for two short ones. I would also change the heights, always remembering that the higher the fences the greater the distance required between them – and vice versa.

The combination can be used as part of a small course which also includes single fences. The horse has to learn how to cope with the different shapes of these single obstacles by getting in close to spread fences and backing off at verticals. If he goes beyond the correct take-off point at a vertical, the highest part of his bascule is likely to occur beyond the fence instead of above it. If he is jumping a true parallel, the highest point should be reached above and between the two top rails, hence the need to take off a little closer.

Some horses habitually get too close to verticals, mainly because they are not solid in appearance. A spread fence commands far greater respect and encourages him to back off simply because it looks so much more imposing. At verticals, the horse often goes a foot (0.3 metre) beyond the point where you have asked him to take off, making it harder for him to clear the fence because it requires a greater degree of athleticism. He can be taught to back off by putting a small fence about 6 inches (0.15 metre) in front of the vertical. Alternatively, you can put a pole on the ground about 2 feet (0.6 metre) in front of the fence so that he takes off in front of it and discovers, in the process, that it is easier to clear this type of obstacle by standing back the extra foot or so. Do not be tempted to put the pole any further forward, because he might mistake it for a place pole and therefore think that he has to go across it before taking off. If he continues to leave his bascule too late, you can try putting another pole 10 feet (3.05 metres) away from the landing side of the fence. This will encourage an earlier parabola because the horse will need to land in front of the pole.

Now that he has reached a higher level, I take a different sequence of practice fences during my show-jumping warm-up. I start the advanced horse with a vertical followed by a smallish square oxer – or, in other words, a spread fence with both front and back rails at the same height. This requires him to make the highest part of his parabola above and between the two rails. I then jump an ascending oxer, which has the back rail 1-1½ feet (0.3-0.45 metre) higher than the front. He now has to leave his parabola until later, otherwise he may trail his hindlegs and hit the back rail. If he does touch it, he will learn by his mistake and remember that he needs to arc correctly, keeping his hindlegs well in the air. The oxer will be squared up again for my next jump to remind him that he also has to jump the first rail cleanly. Finally, as the rider before me enters the arena, I will have my last jump over a vertical.

Grids for intermediate and advanced horses. Straightening poles can be used on the ground or propped with one end on the fence. All distances can be shortened or lengthened by one foot depending on the horse and height of the jumps

Because the time allowed is tighter, you have to be prepared to take shorter turns in the show-jumping phase of advanced competitions. Assuming that the basic schooling is correct, this should not be a problem. The horse will have become more athletic through training, so he no longer needs the same amount of space for his approach.

CROSS-COUNTRY

It is easy to be intimidated by an advanced course unless you start by considering each problem separately. I always like to look at the fence on its own before I consider the outer problems, although these will obviously have to be taken into the reckoning when I decide how to ride it.

I will certainly not want the horse to throw a huge leap at any fence with a drop. If he is landing on to a slope, the further out he jumps the bigger the drop before he touches down and the greater the strain on his forelegs. As already mentioned, I am also anxious to let him see where he is going before he takes off. At advanced level, any fence that incorporates a drop is therefore approached in a short active canter, with plenty of impulsion. In this way you can adjust the horse's stride easily to get close to the fence, so he can see over the top and is comfortable about jumping it. You are looking for a nice rounded jump that does not take you too far away on the landing side.

Typical drop fences include coffins, sunken roads and jumps into water. Problems at these obstacles invariably arise because the rider has approached at the wrong speed, usually by going too fast. That encourages the horse either to take off too early or to put on his brakes, because he wants to see where he is going before launching himself into the unknown. The correct pace depends on the distances in combination fences (such as a coffin) and whether you are approaching a spread or an upright. If the course builder has been kind to the horses, you should not have to jump a spread fence with a drop. It requires a faster pace than an upright and puts greater strain on the horse, because the extra speed produces a greater impact on landing.

The same guidelines apply when you are jumping a drop fence into water, whether it be a vertical or a spread. If the horse stands off a long way back at the fence, he will enter the water at a faster speed and therefore experience more of a dragging sensation, which could cause a fall. I always aim to get fairly close to the obstacle, but I do not want the horse to land too steeply; he should learn to put his forelegs out in front of him as he makes his descent, which will help to reduce the dragging effect. Once he has landed, experience will have taught him to expect a flat surface under his feet.

When jumping a drop, you need to straighten your arms to allow for the extra length of rein the horse requires as he stretches his head and neck. Usually that is enough, but there are rare occasions when you have to slip your reins a little to avoid interfering. You also have to open the angle of your body so that it is more or less perpendicular as the horse makes his descent, ideally with your shoulders no further back than your hips. The lower legs should be forward, with heels well down to take the impact on landing.

It can be a costly mistake to lean forward again too early. If the horse is going to stumble, he normally does so within the first stride; by regaining your normal position as soon as he lands, you therefore run the risk of shooting straight over his head. It is safer to wait for a couple of strides, then move forward and gather up the reins. Sometimes you have to jump a fence at the bottom of the drop, but the same thing applies. There is no time to gather up the knitting; it is better to stay with the contact and position you have, so that the horse can keep flowing forward and remain in balance.

If you decide to tackle a corner, you will have to make careful preparations when you inspect the fence on foot. This means working out an imaginary line (as shown on the diagram) which will take you over both

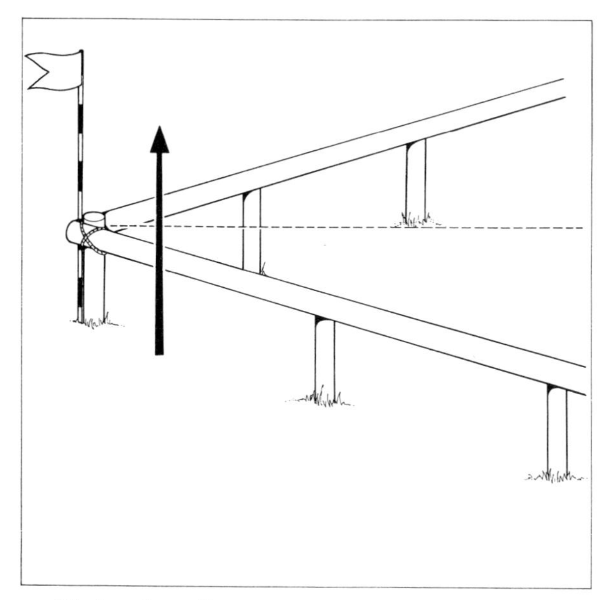

The imaginary line over a corner

arms of the fence at the same distance from the corner flag. Having assessed the angle, you have to find some means of recognising the correct line when you come to ride it. This means finding something to look for on the fence itself, then lining it up with an object (usually a tree or a telegraph pole) in the distance.

The same applies to jumping angles. Having assessed where the distance between angled fences is best suited to your horse, you will need to find some way of lining it up so that you will be able to recognise your chosen route from the saddle. It is too late to try looking for the angle when you are on the horse; you need to be on the correct line from at least 12 strides away and your eye should stay on the fence, so that you can make any necessary adjustments of stride in plenty of time. If you leave it until the last minute, the horse will be thrown off balance and may well run out.

It is the rider's responsibility to come into every fence at the correct pace. The horse will then take his cue from the person on his back and learn to know the type of fence to expect. More pace is needed for wide spreads or combinations requiring a long stride, but you should not encourage him to stand off. If you are jumping a wide ditch or water with no fence above it, you want to take off close to the edge and so keep the amount of effort required to the minimum.

At uphill steps, you should aim to get fairly close. If the horse stands back, he will land rather flatly and want to apply his brakes. By taking off close to the first step you will get a more rounded jump and land in balance, giving yourself room either to push or to check for the next jump up.

At steps going downhill, the horse should be allowed to come in quietly and drop down without any hassle. The rider's position should be the same over this type of obstacle as all other drop fences, with arms straightened to allow the horse sufficient freedom of his head and neck without losing contact. Having come down the steps at a steady pace, you may have another fence to jump at the bottom – and it could be on a long distance. You will therefore have to use your legs as soon as you land from the final step in order to create the necessary impulsion for the fence ahead.

Left: *Master Craftsman wears a weight-cloth for an advanced one-day event. Make sure you don't make the big step up from intermediate to advanced too early*

We all feel elated after a good round across country!

SUMMARY

Use open intermediate classes as a stepping-stone to advanced.

Decide whether the horse is ready for advanced by the way he goes across country.

Choose an encouraging track, with plenty of alternatives, for his first advanced.

Work to increase impulsion and elevation in flatwork.

Increase time for canter work-outs to 3 at 7 minutes or 2 at 8 minutes, with breaks between.

Continue to time recovery rate.

If aiming for a three-day event, revise fitness programme after his first one-day contest.

Use grids to develop horse's jumping technique.

Use different distances to teach horse to shorten and lengthen his non-jumping strides.

Place poles encourage straightness and a rounded bascule.

Four- to six-bar jumping helps the horse to be tidy and alert.

Horse is taught to get close to spread fences and back-off at verticals.

During warm-up for show jumping, take practice fences as follows: vertical, square oxer, ascending oxer, square oxer, vertical.

On the cross-country course, aim to get close to drop fences and approach them in a short active canter.

Choose the correct pace for the approach, depending on the width of the fence and distances in combinations.

Work out your exact line if planning to jump a corner or angle.

Aids for new movements

Counter-canter with near (left) leg leading:
Weight on left seat bone.
Left hand and right leg ask for bend to the left.
Right hand supported by left leg asks horse to move in required direction.

Extended walk and canter:
Rider's legs create energy.
Hands then allow for horse's head and neck to stretch, while maintaining contact, so that he can achieve extension.

Canter-halt:
Prepare horse with half-halts.
Give aids for downwards transition, with slightly more emphasis than for canter walk.

Halt-canter:
Use legs to create energy.
Give usual aids for strike-off into canter.

Half-pirouette (gymnastic exercise) from walk:
Restrain forward movement by use of inside rein.
Inside rein leads forehand round.
Outside rein controls speed and corrects bend.
Inside leg, applied on the girth, creates impulsion and prevents falling in.
Outside leg, behind the girth, prevents quarters swinging out.

Master Craftsman clears a footbridge, which required the same accuracy and straightness as a corner

Beneficial makes a big splash. You should never think that you have successfully negotiated this type of obstacle until you are out of the water

Griffin displays perfect technique. It would be wonderful if all my horses jumped the same way over every fence!

Master Craftsman shows that he is forward-thinking, which is always a great asset

17

PREPARING FOR A CHAMPIONSHIP THREE-DAY EVENT

By now I would hope that my equine pentathlete has developed all his skills and that he shares something of my own joy in setting out across country. He has been moving up the rungs of the ladder slowly and, in the process, he should have acquired increasing confidence in his own ability and learnt to appreciate that I am anxious to look after him (and myself!). There will have to be mutual trust between us before we attempt the ultimate goal and contest a championship three-day event.

The number-one priority is that he should be confident and happy across country; my own gut reaction will tell me whether or not this is the case. If I believe he is ready, I still have to keep an open mind during the time of preparation. This will include some advanced one-day events, which are the best means of telling whether he has the necessary self-assurance to tackle a championship cross-country course. If he lacks confidence, he is likely to have a refusal or a fall which will undermine him even more.

It is rare for a horse to benefit from a cross-country mistake, but it does occasionally happen when you have a brave and confident jumper who never so much as considers refusing. Master Craftsman answers that description and he certainly learnt to be more careful after he clobbered the fourth cross-country fence at the Seoul Olympics and we nearly parted company. He had never hit anything that hard before and he has obviously remembered the experience, without being too upset by it.

NEW MOVEMENTS

The horse may already have learnt the full repertoire of movements required in eventing dressage. These include *extended trot*, which is a continuation of medium trot and requires even more impulsion and elevation. As always, this is produced by the use of the rider's legs, which have to stoke up the engine before the movement begins. Impulsion is needed to drive the horse forward; elevation is necessary in order to give him sufficient time in the air for his extended strides. It is therefore important to have these two essentials before allowing with the hand (while maintaining contact) and so releasing the power as you ask him to move into extended trot.

The horse may begin to run when you are first teaching this movement, in which case you have to slow down and repeat the process. He will normally get the message quite quickly; having reached the stage in his training where his hocks are well engaged, he realises that this is something he can do naturally. The strides – and the outline of the horse – should become longer, without loss of rhythm or change of tempo.

Counter canter has already been learnt on a serpentine, when the horse was required to lead on the outside leg during the second loop. He now has to be able to strike off into canter, with the leg nearer the outside of the arena leading, while moving on a straight line. This means applying the normal aids to canter on a particular leg (see Chapter 4 summary) and maintaining the same position when you reach the corner of the school. If turning right there, the horse's left (near) leg will be leading and he should be bent a little to the left – in the opposite direction from the way he is going. As on the counter canter loop of his serpentine, the rider's right rein and left leg ask for a change of direction but not of bend.

Right: *You need to slip the reins and adopt an upright position to jump safely into water. It has to be said that in this example I have rather overdone it!*

Below: *Every rider needs instruction for all three phases*

REFUSALS

Though I do everything in my power to avoid refusals, they do sometimes happen – and they are not necessarily the end of the world! If you examine the reasons that led to the horse stopping and try to be constructive, it is even possible that good might come out of it and you learn by the mistake. You may realise that the horse was not ready for that type of fence, or that it was beyond his scope. It may therefore give you a new attitude as to how you need to ride him in the three-day event, which could mean taking some slower routes in order to jump a clear round.

There is no point in making excuses. If you come to the conclusion the mistake was due to rider's error, you have to own up to it no matter how demoralised you might feel. If you can find no fault in your own riding – or your judgement – you have to try to discover what other reason may have caused the horse to stop.

I had to go through this mental process after Griffin refused at a one-day event coffin during the build-up to his first Badminton. This meant remembering my own gut reaction to the fast route through the coffin when I walked the course; I had felt that the distances between the elements were wrong and that it would be silly to attempt it. I nevertheless went against my instincts. The horse was due to run at Badminton, so I felt he should go the quick way just to prove he was ready for the big challenge. In retrospect, I had no right to attempt it, so I had to own up to bad judgement and put the blame fairly and squarely on my own shoulders. The experience was to help at Badminton, however, because it gave me an insight into how I should ride the horse. I was less foolhardy this time and, having taken some of the slower routes, Griffin jumped a clear round to finish tenth.

BITTING

As a general rule, horses pull less at a three-day event than they do at one-day horse trials; they are more settled simply because they have done the steeplechase and the roads and tracks. I therefore prefer to use a slightly milder bit when it comes to the full-scale test. Unfortunately, however, the theory is not foolproof. Griffin had been strong at one-day events before I took him to his first Badminton, where I expected him to pull less. In fact the reverse was true and I felt distinctly under-bitted on the Badminton cross-country. If the horse is pulling you are bound to be slower because it takes longer to set him up for each fence, and I probably had an extra ten time penalties as a result.

The one place where I prefer to be almost overbitted is on the gallops where I do my work-outs. Once the horse learns to pull to an extent where he is going faster than you wish, the extra speed can become a habit. If I found he was too strong on the gallops, I would certainly use a bit that gave me more control when I did my next work-out; I might then go back to the previous bit as long as I felt that I would be back in control.

In common with most other riders at three-day events, I believe in having a stronger bit on hand in the ten-minute halt box. I do not often use it, but it is there as a precaution in case the horse had been out of control on the steeplechase. Apart from anything else, it helps my peace of mind to know that I can change the bit if I wish to do so.

IN CASE OF INJURY

If your training programme is interrupted through injury, you may be forced to withdraw from the competition – though much will depend on the timing. It would matter less if the horse had already contested two one-day events, because they will have brought him on in fitness and he will not lose too much by walking for a week on a reduced ration of hard feed. But if he were to miss two runs as the result of lameness – or be off work for more than a week – it would be better to accept the disappointment and withdraw from the three-day event. Both his fitness programme and his preparation will have been badly disrupted, and in my opinion it would be asking for trouble to attempt a three-day event (particularly a championship) under such circumstances.

If a minor injury occurs at a less crucial time, I would want the horse to be sound for at least a week before running in a one-day event. He would also need to do a canter work-out beforehand, and be perfectly level when trotted up the following day.

THE TRAINING PROGRAMME

The horse will be tested in five departments – dres sage, steeplechase, endurance, cross-country an show jumping – so his preparation has to take them a into account. The actual time allowed for each on depends, to a large extent, on the individual horse. He may, for instance, need a good half-hour of loosening-up work before his hocks are sufficiently engaged to attempt any lateral work or other movements which also require a considerable amount of impulsion.

You need to be flexible. If he is not going well in the arena, it is sometimes better to put him away and bring him out again later, so that you avoid a battle. It is remarkable how the horse can improve after a couple of hours in his stable. I would not, however, make a habit of putting him away every time he misbehaves, because he might begin to see this as a way of avoiding work. For the same reason, I would also try to get him to co-operate in some easy movement before returning him to his box.

If he is to be fit enough for a three-day event, the horse will need at least two canter work-outs per week, unless he has competed in one-day trials. We allow four fairly easy days after events to give him time to recover and the best part of a week before doing any more fast work. We also avoid any strenuous work after his day off.

Ideally, I would aim to ride the horse in three advanced one-day events. He might have contested one open intermediate beforehand, but that will depend on the horse. If he tends to be over-cautious, the OI will give his confidence a boost. If he is ultra-brave, it would probably be wiser to do only the advanced contests which have substantial fences that he will have to respect.

As already mentioned, Thoroughbreds are easier to train for speed. They need to work on the endurance aspect at cross-country pace, whereas the half-bred has to learn to maintain a faster pace. These factors have to be borne in mind when you work out your fitness programme for the horse. You will also need some extra form of exercise to get fit yourself, since riding on its own is insufficient preparation for the rigours of a three-day event.

The horse's schedule has to take account of the facilities available, which is why I am giving two separate charts. You can adapt either one to your own needs or use a combination of both. The key points to remember are that you must have sufficient canter/gallop work-outs and competitions – and that you need to allow time to practise all disciplines. The horse still requires at least twenty-five minutes' walking and five to ten minutes' trotting before he begins a canter work-out. You will continue to time his recovery rate, as explained in Chapter 9, in order to monitor his fitness.

Chart One was used for training a horse in the United States on flat and firm ground. Because of the distances involved in travelling to events, the horse spent some time away from home, but the facilities were similar

It is all too easy to overtrain the horse at this stage. A spell in the field works wonders for his mental attitude, helping to keep him relaxed and happy

throughout his build-up to a two-star three-day event.

This system follows the Interval Training method, beginning with two five-minute canters and gradually building up to three of nine minutes. A break of three minutes is allowed between each canter or gallop. Because it involves slower and longer canters than Chart Two, it is helpful for building up stamina in Thoroughbreds and avoids the problem of them becoming over-excited from too much galloping. There were no hills to help get the horse fit and, in order to give him sufficient work, travel days had to count as his rest days towards the end of the programme.

Chart Two made use of an uphill incline for canter work-outs while I was preparing a horse for Badminton. By using faster speeds as well as the uphill slope, it was possible to use shorter distances than those in Chart One because the horse was working harder during his work-outs. This puts less strain on his limbs, but the system would not be recommended for hard ground. The faster speeds would make it too punishing on the horse's feet. If the going is satisfactory (or if you can use all-weather gallops), Chart Two will help the non-Thoroughbred to gallop at the speeds required in a three-day event.

Breaks of two to three minutes are given between each canter or gallop. This is mainly because it takes about two minutes to walk down the hill on the all-weather gallops that I am lucky enough to use. I would never canter downhill because it puts too much strain on the horse, so the breaks are there whether I want

Master Craftsman at an advanced one day event, which forms an essential part of the build-up to a three-day championship

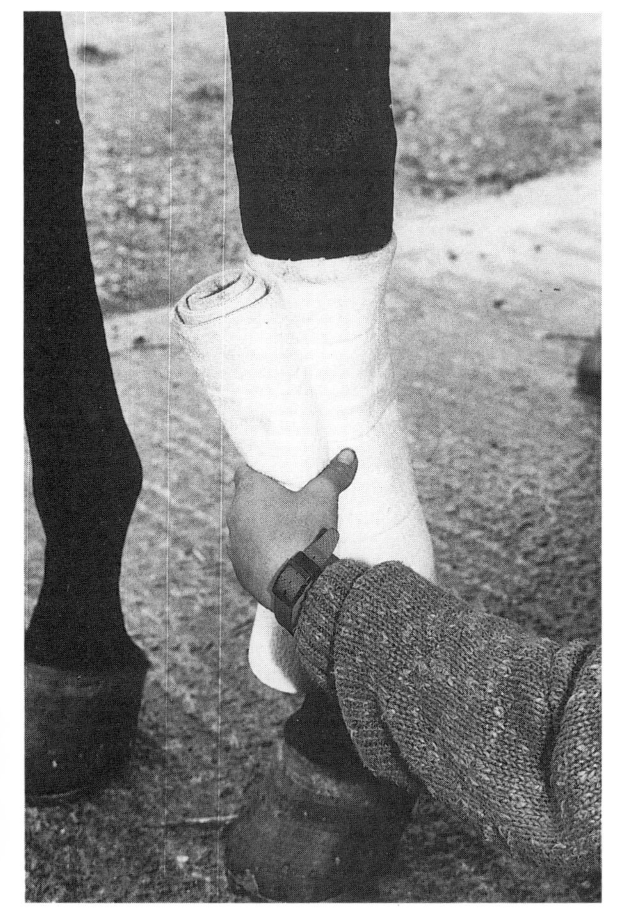

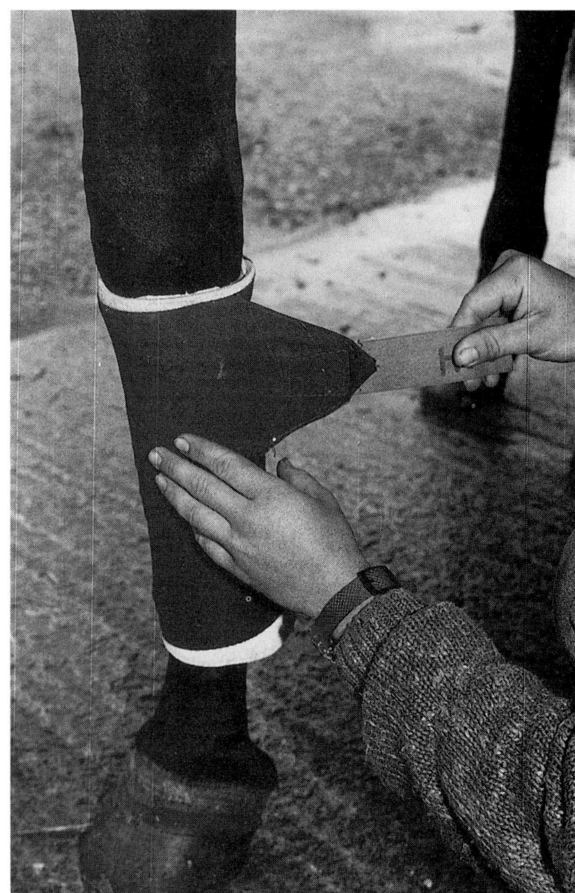

Accidents can happen, so we always bandage our horses during the run-up to a three-day event

them or not. I would probably make some changes if I were to use a full circular gallop, but, never having done so, I have no idea what they would be.

Chart Two is a combination of Interval Training and the method used for getting racehorses fit. It allows time to practise all the disciplines, which would not be possible if I followed more closely the racehorse programme and cantered three times a week. I tried that once and, though it was certainly effective in getting the horse fit, I found it left insufficient time for dressage and jumping. So I now have two canter sessions a week, plus one session of hillwork which has the same desired effect on the horse's lungs and heart. He can do some jumping or dressage on the same day as the hillwork, which would not be possible if he were doing a more strenuous canter work-out.

Sometimes, if the horse does not have much Thoroughbred blood – or if he needs to be more keyed up mentally – you may feel that a third canter day is essential. If using Chart One, however, you need to be careful about increasing the work. The horse is already cantering for a longer time and over a greater distance than those following Chart Two, so his condition could deteriorate. If I were adding a third work-out to Chart Two, I would give the horse two one-mile canters at 545 metres per minute, which would be approximately three minutes each.

The last five work-outs in Chart Two have four different speeds over a distance of one mile, with the third as the fastest. The reason for not leaving this until last is that I want to avoid galloping the horse at speed on tired legs.

You can normally reduce the work when preparing for a *novice* or *intermediate* event. You should, however, bear in mind that it can be difficult to get a horse three-day-event fit for the first time and he may therefore need as much work as those preparing for a championship. If this were not a problem, I would make the following changes.

Chart One: reduce each canter/gallop by one minute. Chart Two: use exactly the same system until four sets of fast work are introduced in week 15. Change all work-outs from this stage to: one mile at 545 metres per minute, one mile at 690 m.p.m. and one mile at 545 m.p.m. (approx 3, 2¼ and 3 minutes).

Chart One

Weeks 1-3: Walking (¾–1½ hrs)

Week 4: Hacking and dressage

Week 5: Jumping (small grids) with dressage and hacking

Week 6

Monday	30 mins dressage, 45 mins walking
Tuesday*	2 × 5 mins at 400 m.p.m.
Wednesday	30 mins dressage, 45 mins walking
Thursday	Jumping (grids and single fences at canter)
Friday	30 mins dressage, 45 mins walking
Saturday*	2 × 6 mins at 400 m.p.m.
Sunday	Rest day (turned out)

Week 7:

Monday	35 mins dressage, 50 mins walking
Tuesday*	6 and 7 mins at 400 m.p.m.
Wednesday	40 mins dressage, 45 mins walking
Thursday	Jumping (single fences and grids at canter)
Friday	45 mins dressage, 45 mins walking
Saturday*	2 × 7 mins at 400 m.p.m.
Sunday	Rest day (turned out)

Week 8:

Monday	45 mins dressage, 45 mins walking
Tuesday	Travel: 24-hr journey to clinic and event
Wednesday	Rest day (with hand walking)
Thursday	1½-hr hack
Friday	45 mins dressage, 45 mins hacking
Saturday*	2 × 5 mins at 400 m.p.m. and 1 × 5 mins at 500 m.p.m.
Sunday	1¼-hr hack

Week 9:

Monday	30 mins dressage, 1 hr walking
Tuesday	Dressage clinic (lessons)
Wednesday	Dressage clinic (lessons)
Thursday	Jumping (single fences at trot and canter)
Friday	45 mins dressage, 30 mins hacking
Saturday	1½-hr hack
Sunday	Rest day

Week 10:

Monday	45 mins dressage, 45 mins walking
Tuesday*	3 × 6 mins at 400 m.p.m. (last 5 mins at 500 m.p.m.)
Wednesday	45 mins dressage, 45 mins walking
Thursday	Jumping (over courses)
Friday	1 hr dressage, 30 mins walking
Saturday	**First event**: dressage and show jumping
Sunday*	**First event**: cross-country

Week 11:

Monday	Rest day: 1 hr hand walking
Tuesday	1½-hr hack
Wednesday	1 hr dressage, 30 mins walking
Thursday	1 hr dressage, 30 mins walking
Friday	1 hr dressage, 30 mins walking
Saturday*	3 × 6 mins at 400 m.p.m. (last 5 mins at 500 m.p.m.)
Sunday	1½-hr hack

Week 12:

Monday	1 hr dressage, 30 mins walking
Tuesday*	2 × 6 and 1 × 7 mins at 400 m.p.m. (last 5 mins at 500 m.p.m.)
Wednesday	1 hr dressage, 30 mins hacking
Thursday	Jumping (grids)
Friday	45 mins dressage, 45 mins hacking
Saturday*	3 × 7 mins at 400 m.p.m. (last 5 mins at 500 m.p.m.)
Sunday	45 mins dressage, 45 mins hacking

Week 13:

Monday	1 hr dressage, 30 mins hacking
Tuesday*	3 × 7 mins at 400 m.p.m. (last 5 mins at 500 m.p.m.)
Wednesday	Travel: 8-hr journey to next event
Thursday	Jumping (grids and single fences)
Friday	1 hr dressage, 30 mins walking
Saturday	**Event**: dressage and show jumping
Sunday*	**Event**: cross-country

Week 14:

Monday	Rest day: 1 hr hand walking
Tuesday	1½-hr hack
Wednesday	45 mins dressage, 45 mins hacking
Thursday	1 hr dressage, 30 mins hacking
Friday	45 mins dressage, 45 mins hacking
Saturday*	3 × 8 mins at 400 m.p.m. (last 5 mins at 500 m.p.m.)
Sunday	1 hr dressage, 30 mins hacking

Week 15:

Monday	45 mins dressage, 45 mins hacking
Tuesday*	3 × 8 mins at 400 m.p.m. (last 5 mins at 500 m.p.m.)
Wednesday	Light jumping (single fences)
Thursday	Travel: 6-hr journey to next event
Friday	1 hr dressage, 30 mins walking
Saturday	**Event**: dressage and show jumping
Sunday*	**Event**: cross-country

Week 16:

Monday	Travel: 11-hr journey home
Tuesday	1½-hr hack
Wednesday	1 hr dressage, 30 mins hacking
Thursday	1 hr dressage, 30 mins hacking
Friday	Light jumping (single fences)
Saturday	45 mins dressage, 45 mins hacking
Sunday*	9 mins at 400 m.p.m., 9 mins at 450–570 m.p.m., 9 mins at 500–690 m.p.m.

Week 17:

Monday	45 mins dressage, 45 mins hacking
Tuesday	1 hr dressage, 30 mins hacking
Wednesday*	9 mins at 400 m.p.m., 9 mins at 450–570 m.p.m., 9 mins at 500–690 m.p.m.
Thursday	1 hr dressage, 1 hr hacking
Friday	Jumping (single fences and course)
Saturday	1 hr dressage, 1 hr walking
Sunday*	9 mins at 400 m.p.m., 9 mins at 450–570 m.p.m., 9 mins at 500–690 m.p.m.

Week 18:

Monday	Travel: 4-hr journey to three-day event
Tuesday	45 mins dressage, 1¼ hrs walking
Wednesday	1 hr dressage, 1 hr walking
Thursday	1 hr hacking (morning), 30 mins hacking plus dressage (afternoon)
Friday	**Dressage test**
Saturday	**Speed, endurance and cross-country**
Sunday	**Show jumping**

* fitness programme (allow 3-minute break between each canter/gallop)
m.p.m. metres per minute

Chart Two

Weeks 1-3: Walking (¾–1½ hrs)

Week 4: Dressage and hacking

Week 5: Light hillwork, jumping (small grids) with dressage and hacking

Week 6:

Monday	Rest day
Tuesday	1½-hr hack, dressage or lungeing
Wednesday*	2 × 1 mile at 500 m.p.m. (= 2 × 3 mins)
Thursday	1½-hr hack, lungeing
Friday*	Hills and dressage
Saturday	1½-hr hack, flatwork, jumping (grids)
Sunday*	2 × 1 mile at 500 m.p.m. (= 2 × 3 mins)

Week 7:

Monday	Rest day
Tuesday	1½-hr hack, dressage or lungeing
Wednesday*	1 and 1½ miles at 500 m.p.m. (= 3 and 4½ mins)
Thursday	1½-hr hack, lunge over poles
Friday*	Hills and dressage
Saturday	1½-hr hack, flatwork and jumping
Sunday*	1½ and 1¼ miles at 500 m.p.m. (= 4½ and 4 mins)

Week 8:

Monday	Rest day
Tuesday	1½-hr hack, dressage or lungeing
Wednesday*	2 × 1½ miles at 525 m.p.m. (= 2 × 4½ mins)
Thursday	1½-hr hack, lunge over poles
Friday*	Hills and dressage
Saturday	1-hr hack, dressage, jumping (grids)
Sunday*	2 × 1½ miles at 525 m.p.m. (= 2 × 4½ mins)

Week 9:

Monday	Rest day
Tuesday	1½-hr hack, dressage
Wednesday*	1½ miles at 525 m.p.m. and 1½ miles at 545 m.p.m. (= 4½ and 4 mins)
Thursday	1½-hr hack, lunge over poles
Friday*	Hills and dressage
Saturday	Show-jumping show
Sunday*	2 miles and 1½ miles at 525 m.p.m. (= 6 and 4½ mins)

Week 10:

Monday	Rest day
Tuesday	1½-hr hack, lungeing
Wednesday*	2 × 2 miles at 545 m.p.m. (= 2 × 6 mins)
Thursday	1½-hr hack, lunge over poles
Friday*	Hills and dressage
Saturday	Show-jumping show
Sunday*	2 × 2 miles at 545 m.p.m. (= 2 × 6 mins)

Week 11:

Monday	Rest day
Tuesday	1½-hr hack, dressage or lungeing
Wednesday*	2½ miles and 2 miles at 545 m.p.m. (= 7 and 6 mins)
Thursday	1½-hr hack, light lungeing
Friday*	Hills, plus dressage or jumping
Saturday	1½-hr hack, dressage
Sunday*	**First one-day event** (Open Intermediate)

Week 12:

Monday	Rest day
Tuesday	1½-hr hack
Wednesday	1½-hr hack, lunge over poles
Thursday	1½-hr hack, dressage, light hillwork
Friday	1½-hr hack, jumping
Saturday	1½-hr hack, dressage
Sunday*	1 mile at 545 m.p.m., 1 mile at 570 m.p.m., 1 mile at 650 m.p.m. (= 3, 2¾ and 2½ mins)

Week 13:

Monday	Rest day
Tuesday	1½-hr hack, lunge over poles
Wednesday*	1 mile at 545 m.p.m., 1 mile at 570 m.p.m., 1 mile at 680 m.p.m. (= 3, 2¾ and 2½ mins)
Thursday	1½-hr hack
Friday*	Hills, plus dressage or jumping (grids)
Saturday	Light hack, dressage
Sunday	**One-day event** (Advanced)

Week 14:

Monday	Rest day
Tuesday	1½-hr hack
Wednesday	1½-hr hack, lunge over poles
Thursday	1 hr hack, light hills, dressage
Friday	1½-hr hack, jumping
Saturday	Hack, dressage (lesson)
Sunday*	**One-day event** (Advanced)

Week 15

Monday	Rest day
Tuesday	1½-hr hack
Wednesday	1½-hr hack, dressage (lesson)
Thursday*	Hillwork and hack
Friday	Hack, short dressage, gridwork
Saturday*	½ mile at 400 m.p.m., 1 mile at 545 m.p.m., 1 mile at 690 m.p.m., 1 mile at 545 m.p.m. (= 2,3,2¼ and 3 mins)
Sunday	1½-hr hack

Week 16

Monday	Rest day
Tuesday	1½-hr hack, lunge over poles
Wednesday*	½ mile at 400 m.p.m., 1 mile at 545 m.p.m., 1 mile at 690 m.p.m., 1 mile at 545 m.p.m. (= 2,3,2¼ and 3 mins)
Thursday	1½-hr hack
Friday	1½-hr hack, light gridwork
Saturday	1-hr hack, dressage
Sunday*	**One-day event** (Advanced)

Week 17

Monday	Rest day
Tuesday	1½-hr hack
Wednesday	1½-hr hack, lunge over poles
Thursday*	Hillwork, hack
Friday	Hack and dressage (lesson)
Saturday	1½-hr hack, jumping
Sunday*	1 mile at 500 m.p.m., 1 mile at 545 m.p.m., 1 mile at 690 m.p.m., 1 mile at 545 m.p.m. (= 3¼, 3, 2¼ and 3 mins)

Week 18

Monday	Rest day
Tuesday	Travel to three-day event, 2 × 1-hr hacks
Wednesday	Hack in morning, dressage in afternoon
Thursday	Hack in morning, dressage in afternoon
Friday	**Dressage test**. Pipe-opener – ½ mile at 680 m.p.m.
Saturday	**Speed, endurance and cross-country**
Sunday	**Show jumping**

* fitness programme (allow 2–3-minute break between each canter/gallop, depending on weather conditions and how hard the horse is blowing)
m.p.m. metres per minute (times given are approximate)

SUMMARY

Decide if the horse is ready by the way he jumps across country. If he has a refusal, try to assess the reason constructively.
If the horse is too strong during canter and gallop work-outs, use a stronger bit but aim to change back to a milder one.
Remember that horses usually pull less on the cross-country at a three-day event, though there are exceptions.
Plan fitness programme according to facilities when working out your chart.
Aim to ride in three advanced one-day events during your build-up to a championship, possibly preceded by an open intermediate.
Continue to time recovery rate after canter/gallop work-outs.
Chart One:
This method is used on flat and hard ground.
It is helpful for building up stamina in Thoroughbred horses.
For a novice or intermediate three-day event, reduce each canter/gallop by one minute (unless the horse needs the extra work).
Allow 3-minute break between each canter/gallop.
Chart Two:
This method makes use of an uphill incline for canter work-outs and steeper slopes for a weekly session of hillwork.
It is helpful for encouraging non-Thoroughbred horses to gallop at the speeds required.
It is not recommended for hard ground, because the faster speeds would make it too punishing on the horse's feet.
For a novice or intermediate three-day event, reduce final work-outs as shown on page 168 (unless the horse needs the extra work).
Allow 2-3 minute break between each canter/gallop.

Aids for new movements
Extended trot:
Rider's legs create energy for increased impulsion and elevation.
Hands then allow (while maintaining contact) so that the power created is released into the extended movement.

You may be part of a team at a championship. If so, remember that you are now riding as a team member and not as an individual

18

THE ULTIMATE TEST

I prefer to arrive at championship three-day events two days before the briefing, if time permits and the organisers are prepared for horses and people turning up early. Since most championships run from Thursday to Sunday, with the briefing on Wednesday, this normally means getting there on a Monday afternoon. I like to have time to unpack on arrival and to take the horse out for a quiet hack so that he can survey his surroundings.

Horses are not easily fooled; they know that a big competition is coming up as soon as they arrive at the venue. As a result, they are often quite loopy the following day. I rarely do any schooling the day before the briefing; instead, I go out for one or two really nice hacks. If the horse is behaving like a complete nutcase, I might even take him out three times – for no more than forty-five minutes each time, most of which would be walking.

I know that I feel the benefit, just as much as the horse, from having that complete day to unwind and get my bearings. The day of the briefing is always fairly hectic. I therefore try to plan my programme with a view to avoiding any situation in which I might find myself panicking against time, thereby causing hassle for everyone else. I would prefer to ride half an hour earlier than necessary, rather than have any worry about getting to the briefing on time. If I find myself with half an hour to spare, so much the better; I can sit down, relax and read a newspaper. I also make sure that I have enough time to eat; if you find yourself in too much of a hurry to stop for food, you invariably end up feeling hungry and bad-tempered.

Equally important, we make sure that whoever is looking after the horse knows exactly what is required. It is unfair to expect anyone to read your mind, so I write down the exact times I will be riding, what tack the horse should wear and when he has to be plaited up and ready for the horse inspection. The Ivyleaze team believes in communication as well as careful planning!

The details given in Chapter 15 are common to all three-day events, so there is no point in repeating them. This chapter is therefore concerned with the extra problems that may be encountered in championships, mainly because of the enormous crowds who turn up to watch them.

DRESSAGE

There is a tendency for all of us to feel we must try to improve our horses' dressage while we are at a championship and waiting to ride the test. This is mistaken; we cannot improve at such a late stage and would be better employed trying to keep the horse relaxed and happy, so that we can hold on to what we already have.

Much of his preparation for the arena can be done without any dressage. This includes mooching about, hacking and anything else that helps him to unwind mentally while he soaks up the atmosphere. You know that he is capable of performing all the movements in the test, but he will not do them satisfactorily if he is tense.

The biggest battles I have had at championships were the result of spending too long in the practice arena. Nowadays I try to do no more than half an hour all told while I am at the event. The horse tends to become claustrophobic if you spend too long there, as I know to my cost. His movements then become restricted as tension seeps in. I may not have had a disagreement with him all year but his tension, combined with my own anxiety to do a good test on such an important occasion, can result in a serious battle of wills.

Most horses are more settled if they go out for an early-morning hack on the day of the dressage test. I was up with the sparrows to ride Master Craftsman before he did his test at the Seoul Olympics and I know he was much better for the outing. It is always a good idea to plait up the horse before going out for the first time. He will be well aware that plaiting means something important is about to happen, and if it is done well in advance there is time for the message to sink in.

All horses are affected, to a greater or lesser extent, by the electrifying atmosphere of a championship three-day event. Some positively enjoy it; others become introverted and nervous. Night Cap came into the latter category and we racked our brains trying to find a way of getting him to relax before the dressage. It was noise which upset him most, particularly the sound of clapping from the stands.

Having tried various ploys with little success (four days at the Bath and West Show, visits to local football matches, a clapping rent-a-crowd at home and brass-

Master Craftsman at Badminton. You need to be aware that big crowds create an electric atmosphere

Always make sure that your horse's head is straight when you trot him up for the horse inspection

band music in his stable), we discovered a more effective solution. The horse was given a couple of hours to get used to the atmosphere, during which time he was led around the collecting-ring and allowed to pick some grass. In the past we had used similar tactics, but for a shorter length of time, and Night Cap had then been taken back to the stables. This tended to undo all the good work because he became overwrought again when he returned to the collecting-ring area. By keeping him there, he was able to continue listening to the clapping until he became bored by it. He was much less tense as a result and therefore able to show the judges that he was capable of performing a really good test.

SPEED, ENDURANCE AND CROSS-COUNTRY

There are normally two fences that alarm me when I walk the cross-country at a championship three-day event for the first time. This is especially true of Badminton, where the declared aim of Colonel Frank Weldon (who was course designer and director there for many years) was 'to frighten the living daylights out of the riders without hurting any of the horses'.

As at every stage of eventing, I have to look at each fence in a positive way – first on its own and then considering the problems surrounding it. Normally you can relate each obstacle to something similar the horse has jumped in the past (even if it is only a diminutive version), which will give you encouragement and help you to know how to ride it. Invariably the problem fences begin to look less alarming once you have done some positive thinking. The only time I can recall being completely flummoxed was at the Seoul Olympics where, for the life of me, I could not figure out a route across the Wondang Walls. It was an immense relief when the first element of this combination was removed, because it then seemed jumpable.

I walk championship courses four times – on the day of the briefing, on both days of dressage and on the morning of the cross-country. By the time I ride round it, I should be familiar with each fence and know exactly how and where I plan to jump it. I should also know all the alternatives available, in case there is some good reason for abandoning my chosen route.

On most cross-country courses there is an area in front of each fence which allows you to straighten up and get your line right. Badminton is an exception; the ropes there are positioned cunningly and there is far less room to sort yourself out. Quite often the fences are out of sight until you are almost on top of them. You can find yourself galloping towards an obstacle and all that is visible is a bulge in the ropes, with people leaning over it to obliterate your view. I am convinced that there would be at least 25 per cent more clear rounds at Badminton if there were no ropes and no spectators.

You have to use your imagination when you walk the course and consider how your view will be altered when the crowds pour into the Duke of Beaufort's park. A fence might be visible from a long way back when you are looking at it without many people around, but they could be standing ten deep on either side of the ropes on cross-country day. If you have failed to consider this aspect – and the much faster pace you will be going when you are on your horse – you could find it alarmingly disconcerting on the day.

The other imperative is (as ever) to be aware of your own and your horse's capabilities and limitations, so that they can be taken into account when you plan how to ride the course. In an ideal world I would always want to tackle big cross-country fences on a horse that answered the same description as Priceless: fairly small, with enormous scope, and incredibly athletic. Unfortunately, such horses are not often to be found, so we all have to accept certain limitations and learn to adapt to them.

From the time he starts to event, the long-striding horse will more than likely find galloping and jumping spread fences easy, but he might lack the athleticism that would help him through a short coffin. The short-strider will probably find spreads more of a problem, but he should be able to make snappy turns – so it might be advisable for him to follow a twisty alternative route where one is available.

Unless you are going early, it should be possible to watch the early runners on closed-circuit television. This can be useful, but it can also be dangerously misleading if you are watching a horse that is totally different from your own. You could, for instance, watch a short-striding horse through a water complex and make the mistake of relating it to your own long-striding partner who happens to be much bolder, which is something I have done myself.

Information about how the course is riding can be potentially dangerous for the same reason. If you do get advice, it must come from somebody who knows your horse well and can be relied upon to give you all the relevant details. It is no good being told how a horse jumped through a combination if your informant fails to mention that the animal was short-striding and that he was looking both chicken-hearted and knackered!

Sometimes you do find a double or combination rides

differently from the way you had expected. The two sets of white rails into the lake at Badminton in 1989 came into this category. In theory the distance between them should have been exactly right for one normal stride, but both the white rails and the water encouraged horses to back off and it rode as a long stride. It was therefore important to approach it with rather more pace than had at first seemed necessary, otherwise the horse was liable to try putting in a second stride. A number of them did this, usually with painful results.

Once you are on your way, riding along the first section of roads and tracks, the steeplechase fences will be the first obstacles to worry about. By now the horse will have done enough three-day events to realise that he quite often has to do two circuits on a steeplechase track. If he slows down at the end of the first circuit, it is therefore more likely to be through tiredness or unsoundness than misunderstanding. You need to make a rapid assessment as to whether he is lame or, perhaps, feeling the effects of the weather or the ground. At the same time, you must not underestimate the fence ahead. If you let him ease up for a short distance, he must get back into gear so that he will have enough power and momentum to clear the obstacle.

There will not be a vast number of spectators watching the steeplechase, but the cross-country, when you reach it, is likely to be awash with people. Riders inevitably become infected by the atmosphere, which increases one's nerves before the cross-country. Once started, however, you are concentrating too much on the fences to be conscious of the crowds. You know they are watching only to the extent that you want to be seen to give a good performance – which, if anything, helps you to ride better. If I happen to hear clapping after I have jumped a difficult fence, it gives me a tremendous lift.

With Ian Stark, Lorna Clarke and Anne-Marie Taylor (now Evans). We all have to check and re-check our line of approach to each fence, as well as making sure that we know all the alternative routes

Above: *Master Craftsman as a comparatively young and inexperienced horse competing in the Olympic Games at Seoul*

Right: *The show-jumping phase is extremely nerve-racking and can mean the difference between success and failure*

The horse, on the other hand, can be seriously distracted by the people. He can start looking at the crowds instead of concentrating on the obstacles and then start spooking. There is a danger that he will back off both the crowds and the fences, which can be a real problem and there is nothing you can do about it. These horses tend to become introverted and anxious when there are large crowds. I had expected Griffin to react this way when I first rode him at Badminton and was delighted when he turned extrovert instead, which meant that the crowds helped rather than hindered him.

As always, it is the rider's responsibility to prepare the horse for each fence. He needs your help to establish the correct speed, balance and accuracy on the approach; it is too late to put him right once you reach the obstacle. The correct pace going into a bounce will depend on the distance between the elements, but you should always aim to take off close to the first part so that you do not land too near the one that follows. Ideally, the horse should touch down exactly half way between the elements; he will then be well placed to bounce over the next part of the fence.

Sometimes the horse begins to lean on the bit towards the end of the course when he is tiring, which makes him feel very heavy in front. The rider must be able to use legs and hands to get him back on his hocks and off his forehand, otherwise he will be in serious danger of hitting one of the remaining fences. This is one of the reasons why the rider needs to have done some fitness training!

SHOW JUMPING

I normally walk the show-jumping course four times at a championship, which helps me to overcome the terror of losing my way! We are all well aware at this final stage of the competition that the advantages of a superlative dressage and heroic cross-country performance can easily be laid to waste in the show-jumping arena. Every fence has to be studied carefully, often in the knowledge that each rail on the floor is likely to move you down in the placings. There is normally far more nervous tension before the show jumping than the cross-country, which is bound to communicate itself to the horse.

The parade of competitors before the show jumping can also prove rather too stimulating for your partner, despite his efforts of the previous day. Having become over-excited while in the arena with all the other horses, it can be a real problem to get him to concentrate when you go back in there to jump the fences. It is worth remembering that the calmer you can keep him during the parade, the better he is likely to be in the final phase of the competition. Some horses thrive on the charged atmosphere of the big occasion and jump better for it; others can lose their concentration and become very spooky. Riders are obviously aware of the crowded stands before they jump but, once in the arena, there is fortunately so much to think about that most of us manage to forget the people.

Only one person can win, but plenty of others are likely to feel equally ecstatic when the competition is over. If your horse has responded gamely to the ultimate challenge, you have to be thrilled with him. And when all the clapping is over and you come down to earth, you will still be left with a deep-rooted sense of satisfaction.

SUMMARY

(To be used in conjunction with summary for chapter 15 on page 150)
Dressage:
Do not spend too long in practice arena.
If horse is tense, give him plenty of time in the collecting-ring before his test so that he can get used to the atmosphere.
An early morning hack before the test can also help him to settle.
Speed, endurance and cross-country:
When walking cross-country course, look at each fence on its own before considering the problems surrounding it.
Consider the position of the ropes. They may allow limited space for straightening up before the fence; if people are leaning over them, your view may be obliterated.
Take your horse's limitations into account when planning how to ride each fence.
If watching on closed-circuit television, be sure to relate what you see to your own horse's length of stride. The same applies to any information received.
If the horse begins to lean on the bit through tiredness, use legs and hands to get him back on his hocks.
Show jumping:
If there is a parade of competitors beforehand, try to keep the horse as calm as possible during it.

The trot-up on the last day of a three-day event is almost another phase in itself. Griffin has just passed this hurdle at the World Championships in Stockholm, where his gleaming coat and bright eye prove that he was at peak fitness

19

OVERSEAS TRAVEL

Horses have a far easier journey when they travel by air rather than road and sea. Because they are physically incapable of vomiting, a rough sea crossing can be particularly distressing and you could have a very sick horse with a high temperature at the end of it. Enforced inactivity at a time when the horse is three-day-event fit can also cause serious problems unless his hard feed is reduced and you make sure he gets some exercise.

FOOD FOR THE TRAVELLER

The horse has his normal amount of hay and other bulk food when making a long overseas journey, but we always reduce the hard feed two to five days beforehand, and we allow the same length of time to bring him back to his normal ration. The actual quantity – and the time span during which the hard feed is reduced – will depend on the length of the journey, how well the horse travels and the facilities for exercise en route.

Our plans have to be adjusted if there is an unexpected hold-up due to bad weather. We would not want to sail with the horses if a gale is blowing – and it is unlikely that we would be allowed to do so. Waiting for the next sailing (and calmer seas) can add half a day or more to the time between departure and arrival.

We had one such hold-up when we took Griffin to the Boekelo three-day event in the Netherlands. He was due to be on the night ferry from Harwich, which would have meant arriving at Boekelo at about four o'clock the following afternoon. We planned to give him half his normal ration of hard feed; he had been ridden that morning and it should have been possible to give him an hour's ridden exercise when he arrived. The plan had to be amended when the lorry was not allowed on to the night ferry; Griffin would have to spend the night in it parked at the docks, before sailing the following morning. His hard feed was therefore cut right back to two pounds, because we would arrive after dark and therefore too late for the horse to be lunged or ridden; his only exercise would be walking in hand.

If the horse is leaving home before daylight and arriving late the same evening, he will probably have his normal rest-day ration of hard feed on the day he is travelling. It is on longer journeys that we need to cut back further, especially if walking in hand is likely to be his only exercise. However, we have to bear in mind that he will probably lose condition if he has scarcely any hard feed for three days in a row.

This will not be a problem if there are at least five days to bring him back to his peak before the competition. The flight to Australia for the 1986 World Championships lasted thirty-three hours and Priceless's hard feed could be cut right back during the journey, since he would have four weeks before the competition began. Master Craftsman had twelve days in Seoul before competing in the Olympics, so he also had time for his hard feed to be reduced gradually before the flight and slowly increased to normal rations after he arrived.

Time is not usually on our side when we are competing on the Continent of Europe. If our journey by road and sea to one of these events is to take two days or more, we will try to find somewhere to lunge the horse each day he is travelling. He can then have a small amount of hard feed without running too much risk of encountering the problems associated with a build-up of protein. The most obvious of these is azoturia (also known as 'tying up'), which causes severe cramps in the horse's back and quarters when he returns to work.

When they are at home, all our horses have a minimum of ten minutes' grazing a day. It is rarely possible to lead them out to pick grass when they are travelling to an overseas event, so we make up this deficiency by feeding carrots when we are abroad. The carrots are a special treat for away trips; we do not feed them at Ivyleaze, but they help to compensate for taking the horse away from the comfort and security of his own stable. They also add some interest to meals that may otherwise consist mainly (or totally) of 'slops'. It is still important for him to have his four feeds each day because he will be looking forward to meal-times.

If the weather is likely to be hot and humid, we automatically give electrolytes, which replace essential elements lost through sweating. We would start giving them before leaving home and continue until a few days after our return, following the instructions on the packet.

We like to take all the horse's food with us on overseas trips, though this is not always allowed. It is therefore important to check the regulations of any countries you will be entering, as well as making sure that you have all the necessary documentation. If we cannot take the food with us, we do our level best to obtain

Travelling by air. Horses have to be carefully monitored throughout the long journey to avoid such potential dangers as dehydration and sickness

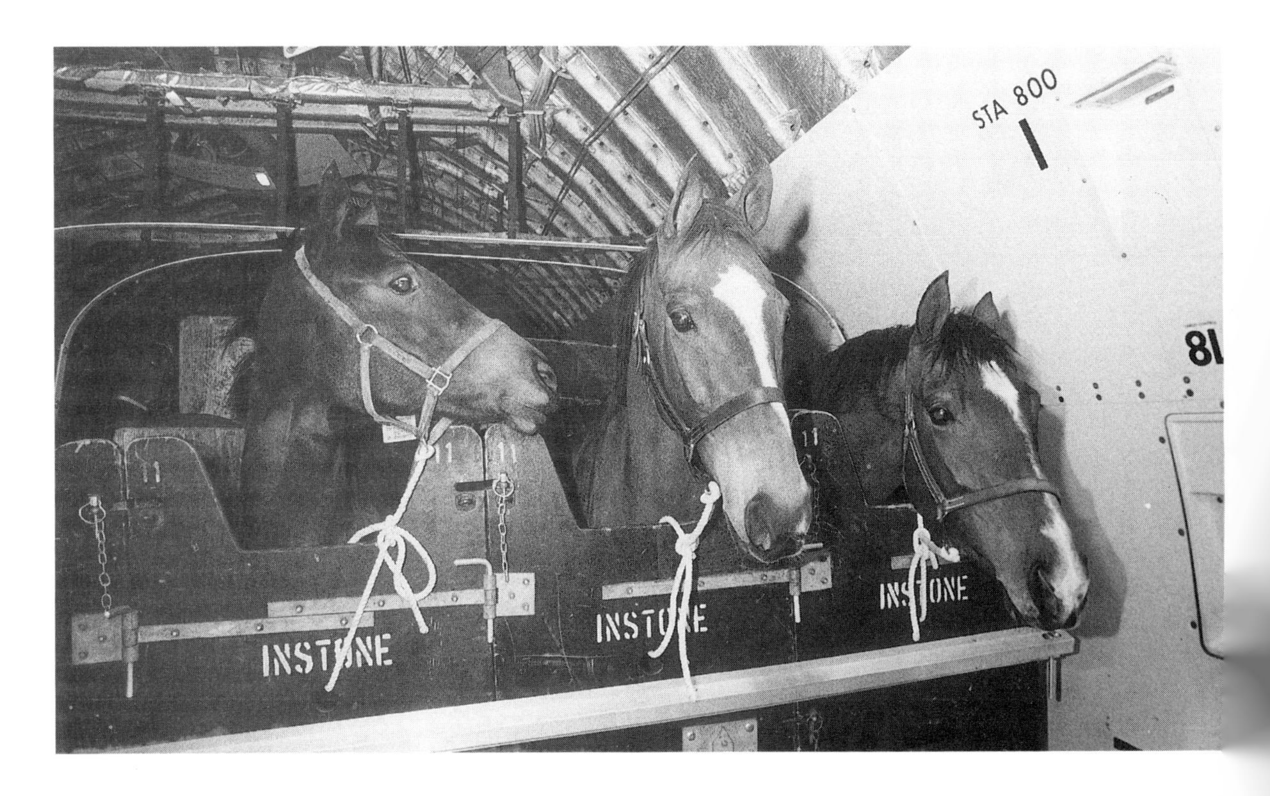

some in advance from the country of destination, so that it can be introduced gradually into his diet before he leaves home, thus avoiding any sudden change. We are not allowed to bring hay back into this country, but HorseHage is permitted because it is packed into sealed bags. We therefore make sure that we have some HorseHage with us, otherwise the poor horse will have nothing to eat when he arrives at the port.

Normally we give the same amount of hard feed on the outward and return journeys. The horse will be tired by the time he makes the trip home, so it is not the time to start deviating from a system. Tiredness makes him less resistant to germs, so there may be more potential danger at this stage than on the outward journey.

READY TO TRAVEL

For extra protection and to prevent the legs swelling, we always put gamgee and bandages under the horse's travelling boots when going abroad. As with most shorter journeys, he wears a light rug (when necessary) that is suitable for the weather and his own individual needs. Temperatures can change more frequently when you are travelling a long distance, so regular checks are made to see whether the horse needs warmer or cooler clothing.

He will cope with the stresses of travel much better if there is good ventilation. A stuffy atmosphere is the main cause of stress, especially in aircraft which are invariably too hot. Low temperatures are not a problem; you can always put on extra rugs to keep the horse warm. It is when he is hot that you are likely to run into trouble, because heat has a debilitating effect which causes stress.

We do not use a tail bandage on these trips, for fear of it tightening or slipping to an area where it can cause pressure and prevent the proper circulation of blood. Horses have been known to lose their tails through having bandages put on too tightly, so we prefer not to take the risk. The tail is a delicate area, with no tough skin to protect it – and very little hair, because we remove most of it for the sake of appearance. Instead of a bandage, we use a tail guard which is attached to a light roller to keep it in place.

Unloading at the end of a successful trip

EXERCISE

We always exercise the horse before loading him into the lorry for a long journey, even if it means getting up at three o'clock in the morning and leading him round for half an hour by torchlight. If off-loading is allowed at the sea port, he has another half-hour of hand-walking before embarkation, and he may be taken out again at the end of the crossing.

There could still be two or three days of road travel ahead, depending on the destination. If this is the case, the horse is exercised each morning before setting out and again in the evening, when arriving at the overnight stop or the venue itself. We also stop at lunch-time and off-load to give the horse another short session, which means that he has a total of 1–1½ hours' exercise each day. We would hope to find somewhere suitable for lungeing in the morning or at lunch-time; he can then be exercised more vigorously than hand-walks allow. If he is left standing in the lorry all day, he can hardly be expected to arrive at the destination in peak condition.

REST-TIME

The horse will need a holiday when he returns home. This does not mean putting him straight out in the field; after competing in a three-day event, whether at home or abroad, he needs about a week to be let down slowly before he is turned out. At the start of the week he is hacked out for about an hour and a half. This gradually decreases to half an hour, and at the same time, his hard feed is slowly reduced. This avoids an abrupt change, which would be likely to upset his system.

The break will help to refresh him both physically and mentally. It may be beneficial for the rider as well! Though the horse still needs loving care and attention while he is turned away, it is obviously far less time-consuming than when he is in training. The spring season is always particularly welcome, because horses and humans have had a good rest.

Holiday time. The rider may fancy Barbados, but all the horse asks for is a nice big field

SUMMARY

Feeding:
Reduce hard feed 2-5 days before a long overseas journey and allow the same time to come back to normal ration.
On shorter journeys (24 hours or less) give normal rest-day ration while travelling.
If permitted, take food and hay for the whole trip.
Clothing for the journey:
Use gamgee and bandages under the travelling boots.
Choose a light rug (if needed) suitable for temperature and individual horse.
Check horse regularly to make sure he is neither too hot nor too cold.
Use a tail guard in preference to a bandage.
Exercise:
The horse should be exercised for at least 20-30 minutes three times a day – before setting out, again at lunch-time and on arrival at over-night stop or destination. Try to find somewhere to ride or lunge the horse for one of the three sessions (the other two can be hand-walking).
Rest-time:
Before turning the horse out for his holiday, gradually reduce hard feed and exercise over a period of about one week.

CONCLUSION

The training methods described in this book have been used successfully on a number of different horses, but they are not written in concrete. We are always learning and therefore must be ready to adapt. As Egon von Neindorff wrote, 'The horse already knows how to be a horse, the problems of horsemanship are entirely those of the rider.'

GLOSSARY
(including terms used in America)

Backing-off: refers to the horse shortening his stride to reach the correct point of take-off, instead of going beyond it. He does this with his hocks underneath him, maintaining impulsion.

Bandage: known as a leg wrap in America.

Bascule: derived from the French word for see-saw, this describes the movement of the horse as he clears a fence.

Bounce: jumping two fences without taking a non-jumping stride between them.

Box (loose): equivalent of the American stall.

Brushing: when the horse knocks a fore or hindleg with the one opposite.

Dressage arena: see diagrams.

Drop fence: any fence where the ground on the landing side is lower than the take-off.

Easy boot: a boot made of plastic that can be used, as a temporary measure, to protect the sole of the foot if the horse loses a shoe.

Falling in: describes a horse when he shifts his balance to the inside shoulder on a turn or circle, while moving his head to the outside. His feet normally follow his shoulder by drifting inwards.

Falling out: describes a horse when he moves his shoulder outwards on a turn or circle. When falling in or out, the horse will not be straight (q.v.).

Gamgee: cotton wool with a covering of gauze, which is used under bandages.

Hack: the equivalent of an American trail ride.

Half-halt: an almost imperceptible moment in which the rider's legs create energy in the horse, while the hands restrain him. It produces momentary collection and is a useful preparation for transitions.

Hard feed: energy-giving food that has a fairly high protein content (e.g. oats, event nuts and some mixes).

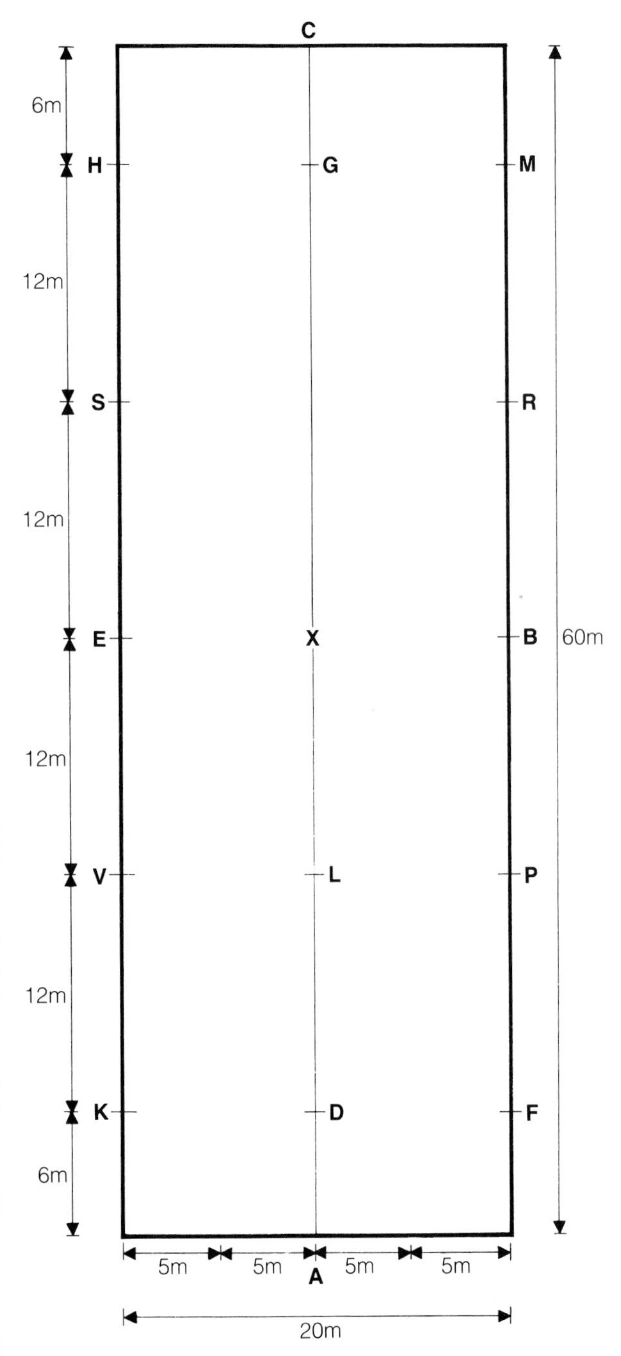

Large Dressage Arena 20m x 60m

Headcollar: known as a halter in America.

Horsebox: a van used for transporting horses.

Inside (or outside): refers to the inside (or outside) of the horse's bend when making a turn or circle – or when performing lateral movements.

Let down: the gradual decrease of hard feed and exercise which is followed by a holiday for the horse when he is turned out (or 'turned away') in a field.

New Zealand rug: a waterproof rug used on a horse when he is turned out in cold weather.

Numnah: a thick pad, which is fitted underneath the saddle to prevent rubbing. Unlike the smaller pads which can be used to improve the fitting of the saddle, the numnah covers the whole saddle area.

Over-reach: when the toe of a hindfoot strikes into the heel of a forefoot.

Over-reach boots: a bell-shaped rubber boot, used to protect the front heels from an over-reach injury. It is called a bell-boot in America.

Plaiting: known as braiding in America.

Spooking: refers to a horse that views unfamiliar fences with great suspicion, normally shortening his stride on the approach. He may then make a quick dart to get over the obstacle, losing his rhythm and often going crooked in mid-air.

Straight: the horse is said to be straight when his hindfeet follow in the same track as his forefeet, whether on a straight line or a circle.

Surcingle: also known as an over-girth, this passes over the saddle and round the horse's body to be secured by a buckle. Usually made of webbing or elasticated web, it gives extra security to the saddle.

Ten-minute halt box: also known as 'the box', this is an enclosed area where riders weigh out at the start of the speed, endurance and cross-country phase and weigh in at the finish. It is also used for the compulsory ten-minute halt before the cross-country.

Yard: the equivalent of an American barn, it refers to the entire stable area.

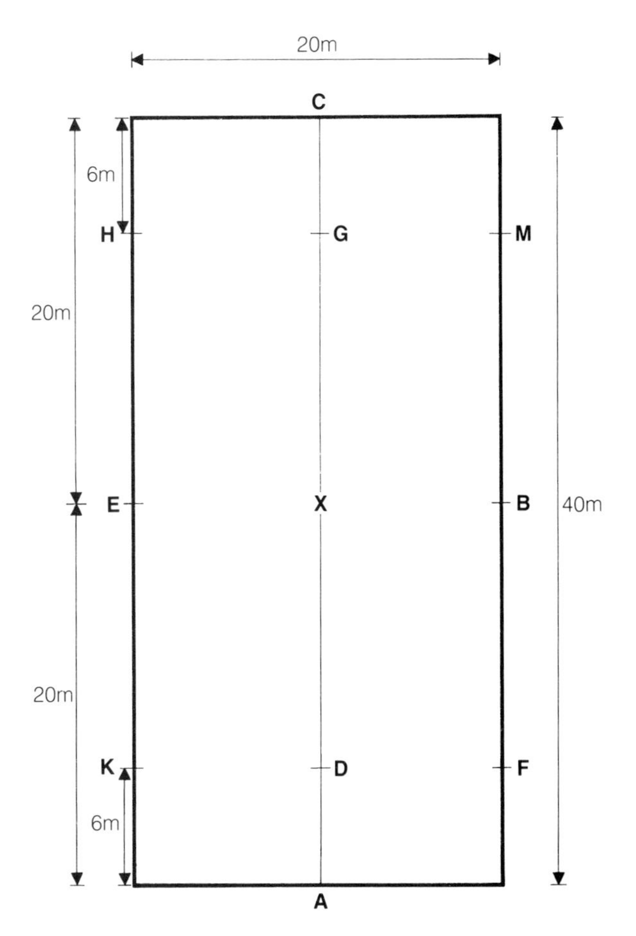

Small Dressage Arena 20m x 40m

INDEX